THE ROUTLEDGE HANDBOOK OF HUMAN-ANIMAL RELATIONS IN THE BYZANTINE WORLD

Animals have recently become recognized as significant agents of history as part of the 'animal turn' in historical studies. Animals in Byzantium were human companions, a source of entertainment and food – it is small wonder that they made their way into literature and the visual arts. Moreover, humans defined themselves and their activities by referring to non-human animals, either by anthropomorphizing animals (as in the case of the Cat-Mice War) or by animalizing humans and their (un)wanted behaviours.

The Routledge Handbook of Human-Animal Relations in the Byzantine World offers an in-depth survey of the relationships between humans and non-human animals in the Byzantine Empire. The contributions included in the volume address both material (zooarchaeology, animals as food, visual representations of animals) and immaterial (semiotics, philosophy) aspects of human-animal coexistence in chapters written by leading experts in their field.

This book will appeal to students and scholars alike researching Byzantine social and cultural history, as well as those interested in the history of animals. This book marks an important step in the development of animal studies in Byzantium, filling a gap in the wider research on the history of human-animal relations in the Middle Ages.

Przemysław Marciniak is a Research Professor and the Director of the Center for Byzantine Studies at the University of Silesia in Katowice, Poland. His research focuses on Byzantine performative culture, the reception of Byzantium, and recently on animals and nature in the Byzantine world. His publications include articles on Byzantine entomology and a co-edited volume on the reception of Byzantium in the popular imagination.

Tristan Schmidt is Assistant Professor at the University of Silesia in Katowice, Poland. His research is focused on human-nature relations in the Byzantine world as well as on the aristocracy and military leadership between the 11th and 13th centuries in Byzantium. His animal-related publications include a monograph on animal imagery in Byzantine political discourse, and studies on concepts of animal agency in Byzantine texts and on ecological awareness in Byzantine society.

THE ROUTLEDGE HISTORY HANDBOOKS

THE ROUTLEDGE HANDBOOK OF MEDIEVAL RURAL LIFE
Edited by Miriam Müller

THE ROUTLEDGE HANDBOOK OF EAST CENTRAL AND EASTERN EUROPE
IN THE MIDDLE AGES, 500-1300
Edited by Florin Curta

THE ROUTLEDGE HANDBOOK ON IDENTITY IN BYZANTIUM
Edited by Michael Edward Stewart, David Alan Parnell and Conor Whately

THE ROUTLEDGE HANDBOOK OF MEMORY ACTIVISM
Edited by Yifat Gutman and Jenny Wüstenberg

THE ROUTLEDGE HANDBOOK OF TAXATION IN MEDIEVAL EUROPE
Edited by Denis Menjot, Mathieu Caesar, Florent Garnier and Pere Verdés Pijuan

THE ROUTLEDGE HANDBOOK OF THE BYZANTINE CITY
From Justinian to Mehmet II (ca. 500 - ca.1500)
Edited by Nikolas Bakirtzis and Luca Zavagno

THE ROUTLEDGE HANDBOOK OF BYZANTINE VISUAL CULTURE
IN THE DANUBE REGIONS, 1300–1600
Edited by Maria Alessia Rossi and Alice Isabella Sullivan

THE ROUTLEDGE HANDBOOK OF GENDER AND SEXUALITY
IN BYZANTIUM
Edited by Mati Meyer and Charis Messis

THE ROUTLEDGE HANDBOOK OF HUMAN-ANIMAL RELATIONS
IN THE BYZANTINE WORLD
Edited by Przemysław Marciniak and Tristan Schmidt

For more information about this series, please visit: www.routledge.com/Routledge-History-Handbooks/book-series/RHISTHAND

THE ROUTLEDGE HANDBOOK OF HUMAN-ANIMAL RELATIONS IN THE BYZANTINE WORLD

Edited by
Przemysław Marciniak and Tristan Schmidt

Routledge
Taylor & Francis Group

LONDON AND NEW YORK

Designed cover image: Cod. Marc. Gr. Z. 479 (=881), fol. 6r. (su concessione del Ministero della Cultura – Biblioteca Nazionale Marciana. Divieto di riproduzione).

First published 2025
by Routledge
4 Park Square, Milton Park, Abingdon, Oxon OX14 4RN

and by Routledge
605 Third Avenue, New York, NY 10158

Routledge is an imprint of the Taylor & Francis Group, an informa business

British Library Cataloguing-in-Publication Data
A catalogue record for this book is available from the British Library

Library of Congress Cataloging-in-Publication Data
Names: Marciniak, Przemysław, 1976– editor. | Schmidt, Tristan, 1988– editor.
Title: The Routledge handbook of human-animal relations in the Byzantine world / edited by Przemysław Marciniak and Tristan Schmidt.
Description: Abingdon, Oxon ; New York, NY : Routledge, 2025. |
Series: Routledge history handbooks | Includes bibliographical references and index.
Identifiers: LCCN 2024027715 (print) | LCCN 2024027716 (ebook) |
ISBN 9780367519643 (hardback) | ISBN 9781032881898 (paperback) |
ISBN 9781032881898 (ebook)
Subjects: LCSH: Human-animal relationships–Byzantine Empire. |
Human-animal relationships–Religious aspects. | Human-animal relationships in art. |
Human-animal relationships in literature.
Classification: LCC QL85 .R678 2025 (print) | LCC QL85 (ebook) |
DDC 590–dc23/eng/20240621
LC record available at https://lccn.loc.gov/2024027715
LC ebook record available at https://lccn.loc.gov/2024027716

ISBN: 978-0-367-51964-3 (hbk)
ISBN: 978-1-032-88189-8 (pbk)
ISBN: 978-1-003-05587-7 (ebk)

DOI: 10.4324/9781003055877

Typeset in Sabon
by Newgen Publishing UK

The Open Access version of chapter 9 was funded by Uppsala University, Faculty of Languages.

CONTENTS

Contents

FIGURES

CONTRIBUTORS

Henriette Baron is a zooarchaeologist. She is the head of the exhibitions department at the Leibniz-Zentrum für Archäologie Mainz (LEIZA) since 2020. She has also conducted archaeozoological research on human-animal relationships, with a focus on the early Middle Ages and the Byzantine Empire.

Nicolas Drocourt is an Associate Professor of Medieval History at the University of Nantes, France, and member of the lab Centre de recherches en histoire internationales et atlantiques (CRHIA) (UR 1163). His research investigates the history of the Byzantine Empire and its diplomacy between the seventh and the beginning of the thirteenth century. He has recently co-edited *A Companion to Byzantium and the West, 900–1204* (Leiden-Boston: Brill, 2022), and has published *L'autre Empire du milieu. La diplomatie byzantine (VIIe-XIIe s.)* (Rennes: Presses universitaires de Rennes, 2023).

Stephanos Efthymiadis is Professor at the Open University of Cyprus, Programme in Hellenic Culture. He holds a PhD from the University of Oxford (UK) and BA degrees in Law (University of Athens) and in Greek Literature (University of Crete). He has published numerous studies on Byzantine hagiography, historiography and prosopography. He co-edited the volume *Niketas Choniates: a Historian and a Writer* (with Alicia Simpson, La Pomme d'or, Geneva 2009). A volume of collected articles on Byzantine hagiography appeared in the Variorum Collected Studies series in 2011. He is the editor of the two-volume *Ashgate Research Companion to Byzantine Hagiography* (2011 and 2014). His most recent monograph is *The Hagiography of Byzantine Cyprus (4th–13th c.)* published in Greek by the Cyprus Research Centre, Nicosia; its English translation, to be published with Cambridge University Press, is in press. He is currently preparing a monograph on the political and social history of Hagia Sophia of Constantinople (532–1453).

Johannes Koder is a member of the Austrian Academy of Sciences (Institute for Medieval Studies/Division Byzantine Studies). His current research interests include the Hagiographical

Kontakia of Ps.-Romanos Melodos and the pronunciation of the Greek language in Late Antiquiy and the Byzantine period.

Maciej Kokoszko, University of Łódź, Poland is interested in the history of food and medicine (especially in dietetics) in Antiquity and early Byzantium.

Stavros Lazaris is a researcher at the Centre National de la Recherche Scientifique (CNRS) and Professor of Byzantine History at the Catholic University of Paris. His research and teaching concern the history of Byzantine science, including the study of original textual and visual documents related to the field. Since his doctoral thesis, he has worked on medieval illustrations and their place in the transmission of medical and scientific knowledge to Byzantium. Horses and their role in the army and the economy of Late Antiquity and Byzantium is another major theme of his research, including equine science technology and the history of veterinary medicine. He has also written on Visual Studies and, since his habilitation, he is interested in the Christianization of pagan scientific literature.

Henry Maguire is Emeritus Professor of the History of Art at Johns Hopkins University. Throughout his career he has been interested in the relationships between art and literature in Byzantium, although he has also written on other topics, including ivories, mosaics, Byzantine secular art, and attitudes toward nature in Byzantium.

Przemysław Marciniak is Research Professor of Byzantine Literature and the Director of the Centre for Byzantine Studies at the University of Silesia in Katowice. His research focuses on performance, satire, the reception of Byzantium, and animal studies.

Charis Messis holds a PhD in Byzantine Studies from the École des Hautes Études en Sciences Sociales in Paris and a habilitation from the Sorbonne University. He is now teaching Byzantine literature at the National and Kapodistrian University of Athens, Greece. His research interests concern Byzantine history and literature, especially the history of gender, along with other social and anthropological aspects of the Byzantine world.

Ingela Nilsson is Professor of Greek and Byzantine Studies at Uppsala University. Her research interests concern processes of rewriting, storytelling, and narratology.

Christodoulos Papavarnavas is a postdoctoral researcher at the Austrian Academy of Sciences, Department of Byzantine Research. He was awarded a doctorate with honours of the Federal President of Austria (*Promotio sub auspiciis praesidentis*) from the University of Vienna in 2019. His research focuses on late antique and Byzantine narratives, especially hagiography, audience response, performance, space, gender, emotions, and the senses. He has taught and published on Byzantine literature. His monograph *Gefängnis als Schwellenraum in der byzantinischen Hagiographie* was published by De Gruyter (2021).

Katarzyna Piotrowska is Assistant Professor at the University of Silesia in Katowice. She specializes in twelfth-century Byzantine literature. Her recent publications include books and papers on Theodoros Prodromos' *Katomyomachia* and the translation of satirical dialogues.

Zofia Rzeźnicka, University of Łódź, Poland is interested in ancient and early Byzantine medicine (especially dietetics), food history, and cosmetology.

Tristan Schmidt is Assistant Professor at the University of Silesia in Katowice. His research is focused on the Byzantine military aristocracy and military leadership in 11th to 13th century Byzantium and on human-faunal relations in the Byzantine world. His animal-related research includes a monograph on political animal imagery in the Byzantine political discourse, 11th–13th c. (Wiesbaden: Harrassowitz, 2020), studies on concepts of animal agency in Byzantine texts, on animal speech and on ecological awareness in Byzantine society.

Nancy Patterson Ševčenko is an independent scholar. Her interests have focused on painted narratives of the lives of saints, on issues of art and liturgy, and on manuscript illumination.

Bligh Somma is Assistant Professor of Philosophy at Fordham University in New York. His research interest lies in psychology, ethics, and animals in philosophy of the Islamic world.

Kirsty Stewart is an independent researcher based in Edinburgh. Her research focuses on nature and beauty in the later Byzantine period, addressing humour, religion, and physical appearance.

Melpomeni Vogiatzi is a researcher at Ludwig-Maximilians-Universität München. Her research interest lies in classical to Byzantine philosophy, especially ethics, logic, and psychology.

Arnaud Zucker is Professor of Greek Literature at University Côte d'Azur (Nice). His key research topics are ancient zoology, ancient astronomy, mythography, and folk etymology. He is heading an international research network on ancient zoological knowledge (*Zoomathia*) and is director of the journal *Rursuspicae*.

ACKNOWLEDGMENTS

This volume is part of the UMO-2019/35/B/HS2/02779 project funded by the National Science Center Poland. We would like to thank Michael Greenwood, our editor at Routledge, for his patience and assistance. Additionally, we extend our gratitude to the University of Silesia in Katowice (Excellence Initiative, IDUB) for providing financial support for the proofreading of this volume.

ABBREVIATIONS

AASS	Acta Sanctorum
AB	Analecta Bollandiana
BF	Byzantinische Forschungen
BHG	Bibliotheca Hagiographica Graeca
BMGS	Byzantine and Modern Greek Studies
BSl	Byzantinoslavica
Byz	Byzantion
BZ	Byzantinische Zeitschrift
CFHB	Corpus Fontium Historiae Byzantinae
CMG	Corpus Medicorum Graecorum
CML	Corpus Medicorum Latinorum
CPG	Clavis Patrum Graecorum
CSHB	Corpus Scriptorum Historiae Byzantinae
DOML	Dumbarton Oaks Medieval Library
DOP	Dumbarton Oaks Papers
GRBS	Greek, Roman and Byzantine Studies
JÖB	Jahrbuch der Österreichischen Byzantinistik
Jsav	Journal des Savants
LBG	Lexikon zur Byzantinischen Gräzität
ODB	The Oxford Dictionary of Byzantium
ODL	The Oxford Dictionary of Literary Terms
PG	Patrologia Graeca
PLP	Prosopographisches Lexikon der Palaiologenzeit
REB	Revue des Études Byzantines
RSBN	Rivista di Studi Bizantini e Neoellenici
SJBMGS	Scandinavian Journal of Byzantine and Modern Greek Studies
ZPE	Zeitschrift für Papyrologie und Epigraphik

BYZANTINE ANIMAL STUDIES
History Beyond Humans

Przemysław Marciniak and Tristan Schmidt

Ἡροδότου Ἁλικαρνησσέος ἱστορίης ἀπόδειξις ἥδε, ὡς μήτε τὰ γενόμενα ἐξ ἀνθρώπων τῷ χρόνῳ ἐξίτηλα γένηται, μήτε ἔργα μεγάλα τε καὶ θωμαστά, τὰ μὲν Ἕλλησι τὰ δὲ βαρβάροισι ἀποδεχθέντα, ἀκλεᾶ γένηται, [...].

This is the publication of Herodotus of Halicarnassus' systematic observations [historia], so that the events do not fade from human [memory] in the course of time, nor that the great and marvelous deeds, both of the Hellenes and of the barbarians, lose their fame [...].[1]

When Herodotus introduced his monumental *Histories* that would become the reference for the tradition of historiography that followed in the Western world, the *pater historiae* described very anthropocentric aims: to preserve, first and foremost, the memory of the deeds of humans. This focus on human and, to a certain degree, divine agency does not imply that his work does not include references to nonhuman animals as well.[2] On the contrary, they feature in considerable numbers in zoological and ethnographical excursuses; they are described as parts of everyday human life and interpreted as portents for human fate.[3] History as such, however, is largely shaped by gods and humans.

This minor role that animals play, not just in Herodotus but in Euro-Mediterranean (and global) historiography in general, correlates with the fact that "history" itself is considered a deeply anthropocentric concept, alien to every other living being except humans. Authors interested in zoology in and beyond the Greek tradition admired animals' abilities to plan ahead for their own good and pointed to the memory of certain animals.[4] But even though zoological research has hitherto expanded a great deal, we still do not possess any cues indicating that nonhuman species have developed a concept comparable to what we call "history."[5] Despite being limited to the human perspective, however, it is evident that even an anthropocentric concept of history is by no means an exclusively human affair; in many regards, the social, economic, political, and cultural events and developments recorded in our sources have to be acknowledged as the results of aggregated action, conscious and unconscious interventions, and mutual dependencies within human-nonhuman networks.[6]

DOI: 10.4324/9781003055877-1

Assessing the roles of nonhuman animals within these networks and interrelations is the aim of historically oriented human-animal studies. There are many ways to approach and understand the significance of animals in history: from an archaeological exploration of the material remains of animals, to the investigation of human concepts of nonhuman beings, to the search for a concrete intervention of animal action in historically relevant processes such as human-animal coevolution, or warfare and transport.

In Byzantine studies, the animal world and human-animal interaction has rarely been the focus of scholarly attention, even though the interactions between the Byzantines and the animals surrounding them could hardly be ignored when it comes to the agrarian economy;[7] rural and urban everyday life; or specific practices such as hunting, animal parades, or the horse races and animal fights in the empire's hippodromes.[8] An explicit focus on the Byzantine fauna (and flora), however, has been a topic in only few, albeit influential studies; the extensive works by Henry Maguire and Eunice Dauterman Maguire on Byzantine secular and religious art should be mentioned here, as well as the important, although by now partly outdated work of Phaidon Koukoules.[9]

A first in-depth and interdisciplinary approach toward the animal world in Byzantium was undertaken in 2008, resulting in a conference on "Animals and Environment in Byzantium" (Ζώα και περιβάλλον στο Βυζάντιο [7ος-12ος αι.]) at the Institute for Byzantine Research at the National Hellenic Research Foundation in Athens. The proceedings (2011) are still the main reference for anyone interested in the animal world.[10] The topics addressed in the volume provide an overview of core questions of animal-related research in Byzantine studies. A clear focus lies on everyday practices in agriculture and animal husbandry, pastoralism, hunting and fishing, as well as the use of animal products. Other chapters explore the semiotic dimension of animals in the political debates of Byzantium, as metaphors in rhetorical culture, as well as in religious doctrine and interpretation of the *kosmos*.[11]

Since the publication of this groundbreaking volume, the methodological and theoretical development of historical animal studies has progressed, primarily in neighboring disciplines of cultural history and in literary studies. This can be attested in numerous handbooks and compendia that appeared during the last decade.[12] In Byzantine studies, animal-related research has profited first and foremost from ongoing studies of source material (texts, figurative representations, and archaeological material). The emergence of perspectives on the role of nonhuman animals in historical processes, even though this largely happened outside the discipline, affected the small community of animal-focused researchers of Byzantium as well, opening new windows for scientific exploration.

Taking into account these developments in- and outside Byzantine studies, the present collective volume is meant as a continuation of the work done so far in Byzantine animal studies. Contributing to *The Routledge History Handbook Series*, our aim is to present current approaches and research interests in the field and give an overview of the available source material, including its potential for research in the upcoming years.

When in 2011 Taxiarchis Kolias opened "Ζώα και περιβάλλον" with an assessment of the state of animal-related studies, he pointed to the limited amount of research literature in comparison with studies in the Central and Western European Middle Ages.[13] This statement has not lost its validity. One likely reason is the limited size of our discipline in terms of funding and personnel compared with the overall number of potential research topics. Another aspect is the fact that pioneering theoretical and methodological works from animal-related studies of medieval history and literature such as Jeffrey Cohen's

"Medieval Identity Machines" (2003) and Udo Friedrich's "Menschentier und Tiermensch" (2009), were received in Byzantine studies only after some delay and only sporadically. On the other hand, scholars from neighboring disciplines have not shown much interest in Byzantium. A certain separation of Byzantine animal studies from neighboring disciplines becomes manifest also in the fact that the comprehensive "Cultural History of Animals in Antiquity" and the following volume on the Middle Ages (2007) hardly contains any reference to the Eastern Mediterranean world.[14]

This, however, does not mean that Byzantine animal studies did not progress during the last decade. Significant developments can be identified with regard to the study of relevant source material. As for the texts, new studies on zoological knowledge traditions, the Greek *Physiologus* and the *Hexaemera* literature, as well as the rich corpus of animal-related satirical texts have deepened our understanding of perceptions, conceptualizations, and representations of animals in Byzantium.[15] Important new editions, e.g., of the *Zoological Collection* commissioned by Emperor Constantine VII (forthcoming), the Nomos Georgikos (2020), and the Eugenian recension of Stephanites and Ichnelates (2022), have further broadened the source base available for study.[16] A special place in Byzantine studies has always been held by horses and horse-related veterinary studies, and the extensive research on the *hippiatrica* has considerably broadened our understanding of human-horse relations.[17]

Significant progress has also been made with regard to non-textual sources, above all, in the field of zooarchaeology. When H. Baron published her overview on the sites within the Byzantine Empire in 2010, she described the situation as "heterogenous, both chronologically and geographically."[18] In fact, access to well-documented and studied zooarchaeological material has been limited ever since; as the preservation of animal bones (especially small species that require sieving) and plant remains became a concern only in recent decades, standardized practices of collecting and documenting such finds are still a desideratum.[19] Baron nevertheless detected a positive development both in methods of collection and in the quantity of the material.[20] Her own study greatly facilitated the systematic inclusion of zooarchaeological data in animal-related research. Subsequent findings, in particular the important excavations at Yenikapı (Harbor of Theodosius), Istanbul (2004–2013), where great care was taken with the archaeozoological remains, further increased the amount of data on the use and living conditions of different species in Byzantium and, in this specific case, in its capital.[21]

The developments in Byzantine animal studies between the "first wave" (culminating in the 2011 collective volume) and the "second wave" (on which the present handbook is based) manifest themselves not only in the expansion and further study of available source material but also in a widening of the research perspectives and topics in which animals have become considered relevant. While previous research was characterized by a strong (albeit not exclusive!) focus on animals in everyday life and agriculture as well as in animal semiotics,[22] these traditional topics are now complemented by new, previously neglected areas of research. The thematic expansion concerns new areas of the natural environment (such as the hitherto sparsely researched aquatic/maritime space), previously neglected species (such as insects), an ongoing interest in different aspects of equids in the Byzantine world, and a strengthened importance of animals in the study of court rhetoric and techniques of prognostication in medieval societies.[23]

The current handbook takes into account both the new research foci that have developed during the last decade and the evolution in sectors of study where animals are traditionally considered relevant. The semiotics of animals clearly belongs to the latter category. Already a long time ago, André Grabar and Otto Treitinger have acknowledged the important messages of political and social power embedded in the imagery of animals in representations of imperial victory or in aristocratic hunting practices.[24] Fittingly characterized as an "Age du Symbole" (M. Chenu), the Euro-Mediterranean Middle Ages were in many ways focused on the meanings inherent in animal signs, including in political speech, public performance, moral discourse, and religious doctrine.[25] Taking into account the constant interaction of semiotic concepts and material human-animal contact, *Tristan Schmidt* provides a systematic view over the manifold ways in which animals were considered significant in the Byzantine world and provided meaning even beyond their own actions and appearances.

One of the bases of any symbolic interpretation of animals in the Byzantine world was zoological knowledge, the sources of which can be found in everyday human-animal interaction, in the ancient traditions of zoological, and in the Christian concepts of the *kosmos* that dominated the discourse from Late Antiquity onwards. The question of what counts as "zoological literature" and how to separate the corpus of extant texts from Byzantium is still open for debate and has been answered differently in scholarship.[26] A striking characteristic is the Byzantines' openness to combine and harmonize seemingly different approaches – see, for instance, Michael Glykas' employment of the Aristotelian concept of *physis* to support the idea of a God-created *kosmos*.[27] As *Arnaud Zucker* demonstrates, however, the collection and compilation of ancient zoological texts continued in Byzantium even without the intervention of the Christian discourse. Discussing the works of Timotheus of Gaza (6th century) and the zoological collection commissioned by Emperor Constantine VII (10th century), he delineates the status of the ancient zoological tradition in Byzantium and contextualizes two important source texts pertaining to the small group of "non-physiological and non-catechetical" zoological literature.[28]

The Christian zoological perspective, by contrast, can justifiably be called the leading paradigm of conceptualizing the fauna during the Byzantine era. Research in this regard has always revolved around the famous *Physiologus* and its morality-based interpretations of the zoological knowledge at the time. *Stavros Lazaris* has pushed forward the research on this central text tradition in recent years. In the present handbook, he delivers the latest insights on this influential text tradition, delineating the core characteristics of Christian "natural science."[29] Far less prominent than the *Physiologus* is the rich literature in Byzantine *Hexaemera*, i.e., extended comments on the Book of Genesis that often go far beyond the biblical text and include ample zoological knowledge and interpretations. As one of the first researchers of Byzantium, Henry Maguire understood the crucial role of these texts within the Christian discourse on nature.[30] Lazaris provides a long-required overview of the most important texts, not just from the hitherto better researched formative phase in Late Antiquity but deep into the Middle Byzantine era.

With the chapter by *Bligh Somma* and *Melpomeni Vogiatzi* on animal rationality in Byzantine and Islamic philosophy, we include a topic that has received ample attention in recent years in Ancient and Western medieval studies, but less so in our field.[31] The *ratio* or *logos*-based agency has been a core element of human-animal separation since Antiquity, even though the inferior status of animals resulting therefrom was never unanimously accepted. Recent studies on 13th-century Western Europe emphasize the great diversity of opinion and lively scholarly debate on the topic.[32] Somma and Vogiatzi take a look at no less vibrant discussions on the intellectual faculties of animals in Byzantium and the Islamic world. Comparing strategies of adopting, developing, and emancipating from the ancient Aristotelian and Neoplatonic foundations, they connect two key areas of intellectual development in the Middle Ages to the wider discourse on human-animal categorizations within global animal studies.

The focus on semiotics and zoological knowledge in the first section of the handbook provides up-to-date overviews of the most basic themes of animal-related studies in a specifically Byzantine context. The following section zooms further in, focusing on the discursive diversity of approaches to the animal world visible in the extant sources. This perspective points to the fact that in all societies, human perceptions and representations of animals are characterized by a multitude of collective and individual concepts. In each concrete testimony, representations of animals therefore encompass only a fraction of all conceptual elements attributed to an animal or a species in a given milieu or society. For the present handbook we consider it crucial to lay out the different discourse- and media-dependent ways of perceiving, "thinking with," and presenting animals in the socioecological context of Byzantium. A necessary prerequisite for such a multi-perspectival approach is the scientific progress made in many areas of Byzantine cultural studies in recent years. Thanks to this, it is now possible to diversify animal-related research and dig deeper into specific discourses on animals and investigate their respective audiences.

One such specific, and at the same time very influential, discourse is treated in *Christodoulos Papavarnavas'* chapter on the functionalization of animal characters in hagiographic martyr narratives. Working on a subgenre of hagiographical narratives, he discusses the central role given to animals in these texts, namely that of "main and secondary heroes or as antagonists" that significantly contribute to the martyrs' spiritual progression. In this way, Papavarnavas shows how the extension of holy status even to nonhuman animals became thinkable in the context of Eastern Roman hagiography.

Stephanos Efthymiadis' assessment of the presence of animals in Byzantine historiography deals with another literary approach toward animals that influenced both the mindsets of the Byzantine elites and the modern researcher's picture of Byzantium. Whereas previous research was mostly limited to general stocktaking of animal descriptions in specific works, Efthymiadis presents a much broader comparative view on historiographic texts.[33] He emphasizes that the metaphorical use of animals widely overshadowed their real-life descriptions. In this, Byzantine historiographers followed the standards set by the "Greco-Roman tradition of the anthropocentric collection of facts." Discussing the adoption of a more personal, biographical model of history writing in the 11th and 12th centuries, Efthymiadis shows that developments within the anthropocentric mode of writing history had noticeable consequences for the literary manifestations of animals.[34]

Another, no less anthropocentric, way of looking at animals and letting them "speak" is through satirical texts. *Kirsty Stewart* responds to increased interest in late Byzantine sociocritical animal poetry. Important studies have attempted to categorize texts such as the *Entertaining tale of the Quadrupeds*, the *Poulologos* or the *Tale of the Honorable Donkey* as reflections of late Byzantine politics and society, as literary reflections of quarrels about social rank and status at court, and as representations of a nascent genre that blended animal stories with rhetorical exercises, satire, and sociopolitical comment.[35] Recent studies on Theodore Prodromos' mock epic Katomyomachia and on Byzantine continuation of the Lucianic tradition of paradoxographical encomia show that in the literary circles of the

11th and 12th centuries, it was already fashionable to connect humorous narrative agendas with catchy animal stories.[36] Based on these developments in research, Stewart discusses the general role of animals within Byzantine humorous literature with its manifold functions, from social criticism to didactic purposes and entertainment.

Ingela Nilsson and *Charis Messis* delve into the Byzantine world of hunting. This topic has enjoyed limited popularity in previous scholarship.[37] Nilsson and Messis, using various sources, differentiate between two types of hunting: the ordinary hunt, whose primary motivation was economic (hunting for food), and the heroic/imperial hunt, which served more representational purposes. Hunting was also a topic of several *ekphraseis* and played its part in late Byzantine literature. Hunting is a multilayered event that includes performative aspects and that arises from the most basic human need: food.

Texts clearly dominate the corpus of sources relevant for animal studies in the Byzantine world, and yet, Eastern Roman society was also an intrinsically visual culture. Two chapters by *Henry Maguire* and by *Nancy Ševčenko* address these issues. However, although they both engage with the visual material, they are fundamentally different. Maguire's chapter discusses animals depicted in mosaics in different roles as metonymy, metaphors, and talismans. And while (a correct) interpretation might require some additional knowledge, these representations are displayed in places available to the wider public (e.g., in churches). Conversely, Ševčenko's chapter deals with the depictions of animals accessible to a relatively small group of people – those who were in possession of costly illustrated manuscripts. Ševčenko not only discusses the various types of animals represented in the manuscripts but shows possible links between these representations and historical events, as was the case of a zebra in a 14th-century manuscript, which could echo an earlier story of a shipload of exotic animals that was detained by the Byzantine emperor.

The third section of the handbook engages with the materiality of animals in the Byzantine world. While literary and artistic sources often idealize nature or focus on its symbolic meaning, most Ancient and Medieval humans were in close contact with countless animals on an everyday basis, which influenced their lives to a high degree. The material presence of animals has been the object of studies on everyday life in Byzantium, beginning with the seminal work of Ph. Koukoules in the 1950s.[38] For historic research in general, we can detect an increasing awareness of the impact of animals on human life and their significance for economic, cultural, and political developments all over the globe. This is visible, for instance, in the emergence of new concepts such as "deep history," where coevolutionary processes between humans and animals are investigated that greatly influenced the development of both human and nonhuman species,[39] or in the emphasis of historians of equids that without their functions in terms of baggage carrying and draft, "many societies in history could never have prospered."[40]

In the present volume, *Maciej Kokoszko* and *Zofia Rzeźnicka* approach the materiality of animals by investigating their role as part of the Byzantine diet. This thorough chapter written by the eminent specialists in the field details many instances of consuming meat, dairy products, and eggs. Even more importantly, it strives to reconstruct the meat-eating patterns of the Byzantines. Quite unsurprisingly for premodern societies, meat was consumed mostly by the better-offs and only sporadically by the lower strata of society. This chapter is also noteworthy because animals here are what Carol Adams calls "absent referents." As she explained in *The Sexual Politics of Meat: A Feminist Vegeterian Critical Theory*, "Behind every meal of meat is an absence: the death of the animal whose place the meat takes."[41]

Kokoszko's and Rzeźnicka's contribution discusses an undeniably material aspect of the animal presence in Byzantium. However, a significant portion of the source material comes – understandably – from texts. This is partly, though not entirely, due to the fact that Byzantine zooarchaeology is still an underdeveloped field. *Henriette Baron*, the first scholar to provide a systematic overview of animal remains in Byzantium, titled her chapter, in a sort of dialogue with Kokoszko's and Rzeźnicka's work, "More Than Food." Baron surveys the remains of various species, not only those consumed by the Byzantines but also beasts of burden as well as cats and dogs. Her chapter gives a physical dimension to many other chapters in this volume by adding the "material animal" to their literary/semiotic versions.

Nicolas Drocourt approaches the materiality of animals from the perspective of gift giving and exchange across and beyond the Byzantine world. In his chapter devoted to "animals on the move," Drocourt examines the diplomatic exchange of animals in the Mediterranean area, illustrating how animals served diplomatic purposes. Certain animals were more common, such as horses, while others were given quite rarely, such as elephants. The choice of a particular animal carried political significance, as Drocourt points out. Intriguingly, some of the skeletal remains found in Theodosius harbor might have belonged to animals received by the Byzantine emperors as gifts.

Drocourt mentions in his paper the public display of some of the animals that came to Constantinople. However, most animals mentioned in his chapter were inaccessible to the wider population of Constantinople. And yet, Constantinople (and other Byzantine cities and settlements), like other premodern spaces, must have teemed with animals – both invited and uninvited.[42] The chapters by *Katarzyna Piotrowska* and *Przemysław Marciniak* address two particular groups of animals: rodents (mice) and pets, respectively.[43]

Piotrowska offers a comprehensive overview of the presence of mice in Byzantine culture and literature. Interestingly, mice are featured as protagonists in many literary works, particularly from the Middle Byzantine period. However, as her chapter reveals, much of the knowledge of mice can be inferred from paroemiographical material. The main focus of Marciniak's chapter is on the emotional bonds between animals and humans in Byzantium. This is of particular importance, as research on pet keeping has hitherto been conducted mostly outside Byzantine studies. In his chapter, Marciniak discusses the appropriateness of using the terms "pets" or "domestic animals" within Byzantine society. Perhaps more importantly, he illustrates the peculiar approach of the Byzantines to pet keeping.[44]

The chapters collected in this handbook present a field of animal-related Byzantine studies that has expanded during recent decades, partly by continuing and deepening existing research aims, partly by taking new directions that widen and diversify the area of interest. This development must be regarded in the context of an increased interest in human-environmental relations and the wider discourse on current environmental change. Another factor to be considered is the wish for a more balanced perspective on the relations and interdependencies between human and nonhuman actors that came with posthumanism. Both aspects have contributed to a general reorientation of "history," away from seemingly autonomous actions of humans and toward processes and developments that took place within heterogenous networks including the hitherto neglected animal world.

The increase in research topics can be attributed less to a coordinated research program and more to the initiative of individual researchers and research teams. In comparison with thematic expansion, however, one has to admit that the adaptation of methods and perspectives developed in critical literary and cultural human-animal studies outside our field is still in its initial phase. Even though ecocriticism, a pluriform initiative emerging from literary studies that focuses on human-environmental relationships in texts, has gradually made its way into research on premodern literatures, Byzantine texts have only sparsely been analyzed in this way.[45] The pioneering work by Adam Goldwyn on Medieval Greek Romance opened the path for the exploration of human-animal relations in Byzantine literary texts. His discussions on the animalization of Digenis Akritas as a hunter/predator and the "sympoietic" entanglement of humans, plants, and animals in literary descriptions of garden spaces present new, ecocritically informed readings of the boundaries and entanglements between humans, animals, plants, and objects in Byzantine texts.[46] Thomas Arentzen, Virginia Burrus, and Glenn Peers recently applied a similarly environmentally oriented perspective to "Byzantine Tree Life," discussing, among other things, the important concept of human-plant hybridization in the example of Byzantine dendrites.[47]

These approaches are not exclusively animal-related, but they further shift the perspective away from the concept of "history" as the result of the activity of autonomous humans toward an acknowledgment of nonhuman, animal agency. A recent colloquium at the Metochi Study Centre on the island of Lesbos (" 'The sound of your leaves implores the Creator': Relating to the Non-Human in Byzantium" 2023) progressed in the topic of environmental-oriented readings of Byzantine "wilderness writing," ecocritical perspectives on the *Physiologus*, and methodological questioning of emic and etic views on animal agency in Byzantium.[48] These endeavors are flanked by other initiatives, e.g., by Maroula Perisanidi who combines animal studies with disability studies, again taking a direction toward a post-humanist view that differentiates between a clear human-animal divide and alternative readings that counteract these boundaries.[49]

Even though this is a still nascent research trend, these initiatives show that Byzantine animal studies is a developing field. The coming years will show whether this development will turn into a "third wave" of Byzantine animal studies, in the course of which the research becomes (even) more methodologically aware, theory based, and more strongly embedded in tendencies of environmental studies and cultural human-animal studies outside the subject of Byzantine studies. The research of the last decade, which has led to the results presented in this handbook, has created a solid basis for this endeavor.

Notes

1 Herodotus, *Histories*, 1.1.
2 On the role of the gods in Herodotus, see M. H. Munn, *The Mother of the Gods, Athens, and the Tyranny of Asia: A Study of Sovereignty in Ancient Religion* (Berkeley et al., 2006), 296–316.
3 See the zoological description of the crocodile, including the ancient "peoples" that held them sacred or, by contrast, hunted them, in Herodotus, *Histories*, 2, 68–70, or the role of snakes and horses as portents in ibid., 1.78.
4 See, for instance, Aristotle, *History of Animals* 589a (memory) and *Nikomachean Ethics* 1141a (forethought); Aelian, *On the Nature of Animals* 7.7 (on birds foreseeing changes in the weather); Michael Glykas, *Annales*, ed. I. Bekker, *Michaelis Glycae Annales* (Bonn, 1836), 93 (on the outstanding memory of the elephant).
5 See on this topic A. Steinbrecher and G. Krüger, "Tiere," in Europäische Geschichte Online (EGO), ed. Leibniz Institut für Europäische Geschichte (IEG) (Mainz, 2015), 10. URL: www.ieg-ego.eu/steinbrechera-kruegerg-2015-de. On the problem of writing history from an animal perspective, see E. Fudge, "A Left-Handed Blow: Writing the History of Animals," in *Representing Animals*, ed. N. Rothfels (Bloomington, IN, 2002), 3–18, esp. 6: "Animals have, as far as we know (and this is the only perspective available to us) no sense of periodization."

6 On the historical significance of animals, see the "synergic work of human and nonhuman animals" in agricultural production, described by H. Lang, "Economic History," in *Handbook of Historical Animal Studies*, ed. M. Roscher, A. Krebber and B. Mizelle (Berlin and Boston, 2021), 181–95, citation at 183 and, regarding the influence of material and symbolic animals on political history from the 20th century on, A. Skabelund, "Political History," ibid., 165–179, esp. 166. See, with a focus on concepts of historical agency, P. Howell, "Animals, Agency, and History," in *The Routledge Companion to Animal-Human History*, ed. H. Kean and P. Howell (London and New York, 2019), 197–221, at 200.

7 See, for instance, the regular mention of animals in A. E. Laiou (ed.), *The Economic History of Byzantium: From the Seventh to the Fifteenth Century*, 3 vols. (Washington, D.C., 2002), clearly with a focus on the human social organization of economy and economic practices.

8 For the horses used in the hippodrome races, see Ph. Koukoules, *Βυζαντινών βίος και πολιτισμός*, 6 vols (Athens, 1948–1955), 3:37–49. See also the early work by J. Théodoridès, "Les animaux des jeux de l'hippodrome et des menageries impériales a Constantinople," *BSl* 19 (1958): 73–84; for hunting, wild animals, pastoralism, and agriculture, see Ph. Koukoules, *Βυζαντινών βίος και πολιτισμός*. vol. 5. For more recent histories of everyday life, see, for instance, J. Koder, *Die Byzantiner: Kultur und Alltag im Mittelalter* (Vienna et al., 2016), esp. ch. 4. For political representation in Byzantium where the role of animal (symbols) is considered, see A. Grabar, *L'Empereur dans l'art Byzantin* (Paris, 1936), esp. ch. II d., "L'Empereur a cheval," or the discussion on hunting imagery in E. Patlagean, "De la Chasse et du Souverain," *DOP* 46 (1992): 257–63.

9 H. Maguire, *Earth and Ocean. The Terrestrial World in Early Byzantine Art*. Monographs on the Fine Arts 43 (University Park and London 1987); E. Dauterman Maguire and H. Maguire, *Other Icons. Art and Power in Byzantine Secular Culture* (Princeton, NJ and Woodstock, Oxfordshire, 2007); Koukoules, *Βυζαντινών βίος και πολιτισμός*.

10 I. Anagnostakis, T. G. Kolias, and E. Papadopoulou (eds), *Ζώα και περιβάλλον στο Βυζάντιο (7ος–12ος αι.). / Animals and Environment in Byzantium (7th–12th century)*. National Hellenic Research Foundation, Institute for Byzantine Research, International Symposium 21 (Athens, 2011).

11 For the latter, see, for instance, the chapters by Leontisi on the use of bird imagery, Koutrakou on animal imagery in political discourse of the Middle Byzantine era, or Lozanova-Stancheva on the basilisk.

12 See the increasing number of compendia and handbooks on the topic (*Handbook of Historical Animal Studies*, ed. M. Roscher, A. Krebber and B. Mizelle; *The Routledge Companion to Animal-Human History*, ed. H. Kean and P. Howell (London and New York, 2018); the 6-volume *Cultural History of Animals*, ed. L. Kalof and B. Resl (London, 2007).

13 T. G. Kolias, "Ο άνθρωπος και τα ζώα στο Βυζάντιο," in *Ζώα και περιβάλλον στο Βυζάντιο*, ed. I. Anagnostakis, T. G. Kolias and E. Papadopoulou, 15–22, at 15.

14 J. J. Cohen, *Medieval Identity Machines*. Medieval Cultures 35 (Minneapolis and London, 2003); U. Friedrich, *Menschentier und Tiermensch. Diskurse der Grenzziehung und Grenzüberschreitung im Mittelalter*. Historische Semantik 5 (Göttingen, 2009); L. Kalof (ed.), *A Cultural History of Animals in Antiquity* and B. Resl (ed.), *A Cultural History of Animals in the Medieval Age*. A Cultural History of Animals, 1–2 (Oxford and New York, 2007). For an overview of the "animal turn" in medieval studies, see recently P. Byrne, "Chronologies of the Animal Turn," *postmedieval* 13.1–2 (2022): 141–61, esp. at 143–44.

15 On the zoological studies, see A. Zucker, "Zoology," in *A Companion to Byzantine Science*, ed. S. Lazaris (Leiden and Boston, 2020), 261–301; on the *Physiologus*, see S. Lazaris, *Le Physiologus grec*, vol. 1, *La réécriture de l'histoire naturelle antique*; vol. 2, *Donner à voir la nature*. Micrologus Library 77.1–2 (Florence, 2016 and 2021); a third volume is planned. On the Hexaemera, see the bibliography in the chapter by S. Lazaris in the present handbook. On animals and satire and mock epic in the Lucianic tradition see below, note 35.

16 For the *Zoological Collection*, see the chapter by A. Zucker in the present volume. For *Stephanites kai Ichnelates*, see A. Noble with A. Alexakis and R. P. H. Greenfield (ed. and trans.), *Animal Fables of the Courtly Mediterranean: The Eugenian Recension of Stephanites and Ichnelates*. DOML 73 (Cambridge and London, 2022).

17 See A. McCabe, A *Byzantine Encyclopaedia of Horse Medicine: The Sources, Compilation, and Transmission of the Hippiatrica*. (Oxford, 2007); S. Lazaris, "Learning and Memorising

Hippiatric Knowledge in the Late Antiquity and in Byzantium," in A. McCabe (ed.), *Le cheval dans les sociétés antiques et médiévales* (Florence, 2015), 269–94.

18 "in zeitlicher als auch in räumlicher Hinsicht als heterogen." H. Kroll (nun Baron), *Tiere im Byzantinischen Reich: Archäozoologische Forschungen im Überblick.* Monographien des Römisch-Germanischen Zentralmuseums 87 (Mainz, 2010), 227.

19 H. Kroll (nun Baron), *Tiere im Byzantinischen Reich*, 5.

20 Ibid. See also her chapter in the present volume.

21 So far, 65,535 identified specimens (NISP) have been analyzed. V. Onar, "Animals in Food Consumption During the Byzantine Period in Light of the Yenikapı Metro and Marmaray Excavations, Istanbul," in *Multidisciplinary Approaches to Food and Foodways in the Medieval Mediterranean*, ed. S. Y. Waksman (Lyon, 2021), 331–42. See also V. Onar, H. Alpak, G. Pazvant, A. Armutak, N. Gezer İnce and Z. Kızıltan, "Animal Skeletal Remains of the Theodosius Harbor: General Overview," *Turkish Journal of Veterinary and Animal Sciences* 37 (2013): 81–85. For cattle, see V. Onar, K. Oya Kahvecioğlu, D. Kostov, A. Armutak, G. Pazvant, A. Chrószcz and N. Gezer İnce, "Osteological Evidences of Byzantine Draught Cattle from Theodosius Harbour at Yenikapı, Istanbul," *Mediterranean Archaeology and Archaeometry* 15.2 (2015): 71–80.

22 See, for instance, the multitude of papers in *Ζώα και περιβάλλον στο Βυζάντιο*, ed. I. Anagnostakis, T. G. Kolias, E. Papadopoulou and P. Byrne, "Chronology," 143, which fittingly remarks that "agricultural historians can fairly claim to have been toiling away with animals long before the coining of the 'animal turn,' having addressed questions of breeding, the place of animals within complex systems of economic exchange, and how the basic rhythms of agricultural life shaped human and animal experience."

23 For the expansion into the marine space, see the conference held in Katowice in 2021 (Aquatic Animals in the Global Middle Ages) and the almost parallel conference "Transmission des savoirs sur les poissons et les animaux aquatiques, textes et images (Antiquité, Moyen Âge, XVIe siècle) – Colloque Zoomathia – Ichtya" in Caen. See for instance the forthcoming publication on the octopus by P. Marciniak (*BZ*, forthcoming 2024), and T. Schmidt "Constantinople and the Sea: Narratives of a Human-Nonhuman Ecosystem?" *SJBMGS* 8 (2022): 67–103. For research on equids, see M. Perisanidi, "Byzantine Parades of Infamy through an Animal Lens," *History Workshop Journal* 90 (2020): 1–24; the Byzantium-related papers in the collective volumes by A. Ropa and T. Dawson (eds.), *Echoing Hooves: Studies on Horses and Their Effects on Medieval Societies.* Explorations in Medieval Culture 22 (Leiden and Boston, 2022), and also in S. Lazaris (ed.), *Le cheval dans les sociétés antiques et médiévales* (Florence, 2015), both showing the close relation between horse-related studies in Byzantine studies and neigboring disciplines (Ancient, Medieval Studies). For research on insects, see P. Marciniak, "Byzantine Cultural Entomology (Fourth to Fifteenth Centuries). A Microhistory of Byzantine Insects," *DOP* 77 (2023): 177–193; P. Marciniak, "I nomi degli insetti a Bisanzio tra sviluppi eruditi e tradizioni popolari," *RursuSpicae* 5 (2023), https://journals.openedition.org/rursuspicae/2749. For animals in rhetoric, see T. Schmidt, *Politische Tierbildlichkeit in Byzanz. Spätes 11. bis frühes 13. Jahrhundert.* Mainzer Veröffentlichungen zur Byzantinistik 16 (Wiesbaden, 2020) and F. Leonte, " '... For I have brought to you the fugitive animals of the desert': Animals and Representations of the Constantinopolitan Imperial Authority in Two Poems by Manuel Philes," in *Animaltown: Beasts in Medieval Urban Space*, ed. A. M. Choyke and G. Jaritz (Oxford, 2017), 179–87. For animals and techniques of prognostication, see S. Costanza, "Nitriti come segni profetici. Cavalli fatidizi a Bisanzio (XI-XIV sec.)," *BZ* 102 (2009): 1–24 and M. Grünbart, "Mantic Arts: Traditions and Practices in the Medieval Eastern Christian World," in *Prognostication in the Medieval World: A Handbook*, vol. 1, ed. M. Heiduk, K. Herbers and H.-Ch. Lehner (Berlin and Boston, 2021), 446–52, at 449–51.

24 See A. Grabar, *L'empereur*, 57–61; O. Treitinger, *Die oströmische Kaiser- und Reichsidee nach ihrer Gestaltung im höfischen Zeremoniell. Vom oströmischen Staats- und Reichsgedanken* (Darmstadt ²1956), 167, 184. See also the reference to the works by Maguire and Dauterman Maguire in n. 9.

25 M.-D. Chenu, *La Théologie au douzième siècle* (Paris, 1957), 161, referring to the views held by elites ("écolâtres et mystiques, exégètes et naturalistes, profanes et religieux, écrivains et artistes") in the twelfth-century Latin West.

26 See H. Hunger, *Die hochsprachliche profane Literatur der Byzantiner*, vol. 2. Byzantinisches Handbuch im Rahmen des Handbuchs der Altertumswissenschaften, 5.2 (Munich, 1978), 265–70 who distinguishes into 1) writings based on ancient anecdotes and paradoxographical stories (Timotheos fragments, Sylloge of Constantine VII, Manuel Philes, etc.), 2) comments on Aristotle, 3) practical writings (veterinary medicine), and 4) theology (here mostly Hexaemera); see also Kolias, "Ο άνθρωπος και τα ζώα," 15–22 who separates the "theoretical" literature on animals (including the collection of ancient zoological knowledge and the Christian interpretations in the *Hexaemera* and the *Physiologus*) from texts on the "practical" contact, including *hippiatrika*, *ierakosophia*, etc.

27 Michael Glykas, *Annales*, ed. Bekker, 97:22–98:6, arguing that, according to Aristotle, God (in Aristotle: *physis*!) gave every being the appropriate functions of the body. Discussed in Schmidt, *Tierbildlichkeit*, 55–56. On Glykas, see also the chapter by Stavros Lazaris in the present volume.

28 Note that A. Zucker is currently preparing a new edition of the *Zoological Collection* of Constantine VII.

29 See Lazaris, *Le Physiologus grec*, and for older literature, for instance, *Physiologus*, ed. F. Sbordone (Milan, 1936); N. Henkel, *Studien zum Physiologus im Mittelalter. Hermaea* 38 (Tübingen, 1976).

30 See Maguire, *Earth and Ocean*, especially on Late Antique and Early Byzantine *Hexaemera*.

31 See, for ancient cultural studies, S. T. Newmyer, *The Animal and the Human in Ancient and Modern Thought. The 'Man Alone of Animals' Concept* (New York and London, 2017); R. Sorabji, *Animal Minds and Human Morals: the Origins of the Western Debate* (Ithaca, NY, 1995); for medieval studies, see, e.g., I. P. Wei, *Thinking about Animals in Thirteenth-Century Paris: Theologians on the Boundary Between Humans and Animals* (Cambridge, 2020); A. Oelze, *Animal Rationality: Later Medieval Theories 1250–1350* (Leiden, 2018).

32 See the review by Byrne, "Chronologies."

33 See A. Littlewood, "Vegetal and Animal Imagery in the History of Niketas Choniates," in *Theatron. Rhetorische Kultur in Spätantike und Mittelalter*, ed. M. Grünbart; Milennium-Studien zu Kultur und Geschichte des ersten Jahrtausends n. Chr. 13 (Berlin and New York, 2007), 223–58; A. Littlewood, "Imagery in the Chronographia of Michael Psellos," in *Reading Michael Psellos*, ed. C. Barber and D. Jenkins; The Medieval Mediterranean 61 (Leiden and Boston, 2006), 13–56. See also the collection and interpretation of animal imagery in middle Byzantine historiography and chronicles (among others) by N. Koutrakou, "Animal Farm in Byzantium? The Terminology of Animal Imagery in Middle Byzantine Politics and the Eight 'Deadly Sins,'" in *Ζώα και περιβάλλον στο Βυζάντιο*, ed. I. Anagnostakis, T. G. Kolias and E. Papadopoulou, 319–77.

34 S. Efthymiadis in the present volume.

35 See G. Prinzing, "Zur byzantinischen Rangstreitliteratur in Prosa und Dichtung," *Römische Historische Mitteilungen* 45 (2003): 241–86, at 260-86; U. Mönnig, "Literary Genres and Mixture of Generic Features in Late Byzantine Fictional Writing," in *Medieval Greek Storytelling. Fictionality and Narrative in Byzantium*, ed. P. Roilos. Mainzer Veröffentlichungen zur Byzantinistik 12 (Wiesbaden, 2014), 163–82, at 175–79. For the sociopolitical interpretations, see also N. Gaul, "The Partridge's Purple Stockings: Observations on the Historical, Literary and Manuscript Context of Pseudo-Kodinos' Handbook on Court Ceremonial," in *Theatron*, ed. M. Grünbart, 69–103, esp. 91–94, regarding the *Poulologos* in the context of court propaganda at the time of John Kantakouzenos and drawing connections to the *Ps.-Kodinos* treatise.

36 P. T. Marciniak and K. Warcaba (now Piotrowska), "Theodore Prodromos' Katomyomachia as a Byzantine Version of Mock-Epic," in *Middle and Late Byzantine Poetry. Texts and Contexts*, ed. A. Rhoby and N. Zagklas (Turnhout, 2018), 97–110. P. T. Marciniak, "The Paradoxical Encomium and the Byzantine Reception of Lucian's Praise of the Fly," *Medioevo Greco* 19 (2019): 141–150; Idem, "A Pious Mouse and a Deadly Cat: The Schede tou Myos, attributed to Theodore Prodromos," in *GRBS* 57 (2017): 507–27.

37 The most recent article presents a rich collection of sources but almost nothing in the way of a deepened analysis of these sources, see A. Kotłowska and K. Ilski, "Institutional and Functional Aspects of Hunting in Byzantium," *Al-Masaq: Islam and the Medieval Mediterranean* 36 (2024): 42–67. See, furthermore, E. Patlagean, "De la Chasse et du Souverain"; G. T. Dennis, "Some Notes on Hunting in Byzantium," in *ΑΝΑΘΗΜΑΤΑ ΕΟΡΤΙΚΑ. Studies in Honor of Thomas F. Mathews*, ed. J. D. Alchermes, H. C. Evans and T. K. Thomas (Mainz, 2009), 131–34; A. K.

Sinakos, "Το κυνήγι κατά τη μέση βυζαντινή εποχή (7ος–12ος αι.)," in *Ζώα και περιβάλλον στο Βυζάντιο*, ed. I. Anagnostakis, T. G. Kolias and E. Papadopoulou, 71–86.

38 Ph. Koukoules, *Βυζαντινών βίος και πολιτισμός*; Koder, *Die Byzantiner. Kultur und Alltag im Mittelalter* (Vienna, Cologne, and Weimar, 2016), and several contributions in *Ζώα και περιβάλλον*.

39 See, for instance, M. C. Stiner and G. Feeley-Harnik, "Energy and Ecosystems," in *Deep History: The Architecture of Past and Present*, ed. A. Shryock and D. L. Smail (Berkeley et al., 2011), 78–102 at 84–86, referring to the development of human behaviors "not typical of primate relatives" in their competition with other carnivore animal species (cats, hyenas, members of the dog family, etc.) and, in the same volume, F. Fernández-Armesto with D. L. Smail, "Food," in ibid., 131–59, at 147 on the development of interdependencies between humans and the animals consumed. See also Howell, "Animals," 200.

40 T. Dawson, "Baggage Animals –The Neglected Equines. An Introductory Survey of Their Varieties, Uses, and Equipping," in *The Horse in Premodern European Culture*, ed. A. Ropa and T. Dawson (Kalamazoo, 2020), 45–54, at 45 and similarly M. Röben, "The Horse behind the Text: Animal Agency in Early Medieval Historiography," *Cheiron* 1.1 (2021): 68–85, at 68.

41 Carol J. Adams, *The Sexual Politics of Meat: A Feminist Vegeterian Critical Theory* (New York, 1990).

42 Generally on domestic animals/pets, see K. Kynast, "Geschichte der Haustiere," in *Tiere. Kulturwissenschaftliches Handbuch*, ed. R. Borgards (Stuttgart, 2016), 130–38; A. Zelinger, "History of Pets," in *Handbook of Historical Animal Studies*, ed. M. Roscher, A. Krebber and B. Mizelle, 425–38.

43 For some general remarks on mice, see C. Carpinato, "Topi nella letteratura greca medievale," in *Animali tra zoologia, mito e letteratura nella cultura classica e orientale*, ed. E. Cingano, A. Ghersetti and L. Milano (Padua, 2005), 175–92. A general survey on dogs in Byzantium was recently offered by A. Rhoby, "Hunde in Byzanz," in *Lebenswelten zwischen Archäologie und Geschichte. Festschrift für Falko Daim zu seinem 65. Geburtstag*, ed. J. Drauschke, E. Kislinger, K. Kühtreiber, et al. (Mainz, 2018), 807–20; on cats, see E. Kislinger, "Byzantine Cats," in *Ζώα και περιβάλλον στο Βυζάντιο*, ed. I. Anagnostakis, T. G. Kolias and E. Papadopoulou, 165–78.

44 See, for instance, K. Kynast, "Geschichte der Haustiere," in *Tiere. Kulturwissenschaftliches Handbuch*, ed. R. Borgards, 130–38; A. Zelinger, "History of Pets," in *Handbook of Historical Animal Studies*, ed. M. Roscher, A. Krebber and B. Mizelle, 425–38.

45 For medieval studies, see the multi-book review by V. Nardizzi, "Medieval ecocriticism," *postmedieval* 4 (2013): 112–23, already from 2013 discussing volumes from the late 2000s and early 2010s. Note also Nardizzi's remarks on ecocritical theory in studies of Renaissance literature and the particular focus on Shakespeare. For ancient studies and ecocriticism, see C. Schliephake (ed.), *Ecocriticism, Ecology, and the Cultures of Antiquity* (Lanham, MD, 2017).

46 A. J. Goldwyn, *Byzantine Ecocriticism: Women, Nature, and Power in the Medieval Greek Romance* (Cham, 2018), esp. 56–57; 210–14.

47 T. Arentzen, V. Burrus and G. Peers, *Byzantine Tree Life: Christianity and the Arboreal Imagination* (Cham, 2021), esp. 107–62. See also V. Della Dora, *Landscape, Nature, and the Sacred in Byzantium* (Cambridge, 2016) on environmental perceptions in Byzantium, in particular on the imagining of nature and the "theology of nature of the Orthodox Church, in which Byzantine world views were grounded" (p. xiv).

48 Organized by Laura Borghetti and Thomas Arentzen, soon to be published in an edited volume: *Ecologizing Late Ancient and Byzantine Worlds*. We refer to the papers by Laura Feldt, Thomas Arentzen, and Tristan Schmidt.

49 As part of her ongoing project "A Cultural History of Disability in Byzantium" (as a Wellcome Trust Research Fellow 2022–2025).

PART I

What do "They" Mean to "Us"?

Theoretical Perspectives on Byzantine Animals

1

MORE THAN MEETS THE EYE

The Semiotics of Animals in Byzantium

Tristan Schmidt

Animals as Signs and Concepts in Human Thought and Speech: The Case of Byzantium

Conducting research on animals in Byzantine sources, one hardly misses their strong presence as bearers of anthropocentric meaning and common usage in figurative and symbolic communication.[1] In this capacity, we find animals omnipresent in texts, in the visual arts, and as living creatures in performative contexts. For their human observers, animals were undeniably more than "just" signs and symbols, but parts of a material and social reality that shaped their everyday lives. Yet the prominence of animals' semiotic dimension in our source material clearly indicates that their role as carriers of meaning, sources of guidance, and instruments of communication was crucial to the ways Byzantine humans approached the animal world around them, both in thought and action.[2]

The prominent position of animals as signs, symbols, and metaphors in Byzantine sources is reflected in the various scholarship dedicated to Byzantine religious and secular art[3] and literature.[4] Numerous individual studies have considered the meanings attributed to specific animal "species", among them the single- or double-headed eagles,[5] lions,[6] cheetahs,[7] horses,[8] dogs,[9] boars/pigs[10] and dragons.[11] Unlike research on animals in the medieval West, however, no comprehensive study on the cultural history of animals in the Byzantine world exists, nor has a lexicon specifically dedicated to animal signs and symbols been compiled.[12]

What and how did animals "mean" in Byzantium? It would be the task of a whole field of interrelated disciplines to define Byzantine specificities within the development of Euro-Mediterranean animal semiotics. In fact, a brief analysis of the extant sources shows that even *within* the society commonly referred to as "Byzantine", the system of animal signs was by no means homogenous. The diversity of "animal readings" proposed by clerics, aristocrats, ethnographers, historiographers, or farmers, among others, reflects the diversity of Byzantine society itself. Its multitude of discourses and overlapping symbolic systems were limited only by a broader sociocultural framework that established certain communal standards of meaning. This combination of diversity and common ground must be taken into account when exploring "Byzantine" way(s) of "reading" animals.

From a methodological point of view, a focus on animals as signs always risks treating them as mere objects of human definition, "pressed into symbolic service as metaphors, or

DOI: 10.4324/9781003055877-3

"

as figures in fable or allegory".[13] Academic initiatives such as "Zoopoetics" and "Cultural Literary Animals Studies" have increasingly critiqued a widespread perception of animals in texts and artworks as passive, anthropocentric signs. They justifiably point out a contrasting need to properly consider "material" animals in their own right and acknowledge their impact *on* human thought and expression.[14] The present study follows a harmonizing approach, considering animals as both material *and* semiotic nodes (D. Harraway), or hybrids (R. Borgards).[15] It directs the attention to the (attributed) significance of animals, while keeping in mind their physicality as well as their ability to shape human perception and expression. The study of animals' social and cultural meanings consists, therefore, of both an analysis of human-animal relations in the material world as well as one concerning developments in (human) ideas and concepts.[16]

The structure of the present chapter takes this duality into account, focusing first on the impact of material relations on the semiosis of animals, before concentrating more specifically on human-derived animal concepts, their developments and adaptations. The study can only offer a beginning attempt to systematize the diversity of sources and perspectives on the Byzantine material. Following the concept of material-semiotic nodes, this assessment investigates the complex relations between animals as literary constructs and their physical counterparts: how did semiotic meaning and zoological knowledge interact with each other? What role did physical animals play in their semiosis, and in which ways were they impacted by human concepts and interpretations? Considering the diversity of the Byzantine source material, its creators and audiences, we furthermore have to ask, "whose" symbolic systems are actually represented in the extant texts and material remains; can we even speak of "a" Byzantine way of approaching animals, or do we have to differentiate further? Addressing these questions will enable us to assess better the requirements of a comprehensive approach to Byzantine animal semiotics, and delineate how Byzantine studies contribute to the wider field of historical human-animal studies.

From Material to Symbolic: Human-Animal Contact, Collaboration, Consumption

Joint work in fields, travel, or hunting, in addition to the use of animal products, provided major sources and points of reference for the derivation of animal signs, symbols, and metaphors in human expression. Meaningful forms of human contact with animals could include real, everyday experience; they could be imagined and transmitted in texts and images; and they could be deliberately performed, as parts of human self-display.

Semiotic interpretations deriving from human-animal contacts in the material world often established a conceptual association between the humans and animals involved, resulting in two different possible outcomes. On the one hand, individual animals or species often appear as positive reflections of the morals, values, and social status of their human owners and companions. On the other hand, human contact with and closeness to animals could be interpreted pejoratively, in terms of a shared animality that is incompatible with social norms and values.

Beastly Barbarians

Pejorative notions connected to human proximity to animals frequently appear in sources with an ethnographic focus. These texts were generally produced by intellectual and, in many cases, urban elites. Well aware of the centuries-old discourse on the "barbarian", they integrated references to real or imagined human-animal contact as categorical markers evoked to assimilate humans (individuals or groups) within the animal world, and provide a basis for their social, cultural, and ontological separation. One text produced around the 5th or 6th century, a "Christian monastic tale" attributed to the learned monk Neilos of Ancyra, provides one example of such "social othering". Comparing semi-nomadic Bedouins on the Sinai Peninsula to local monks, Neilos calls attention to their savage lifestyle and lack of agriculture; to them "hunting desert animals and devouring their flesh"; to the consumption of their "camels [...] for food" almost uncooked. For Neilos, the Bedouins, "like dogs", lived with, from, and similar to animals, thus evidencing their own "bestial and bloodthirsty way of life".[17]

Similar motifs can be found in the descriptions of Pechenegs, Cumans, and Mongols by the Constantinopolitan intellectuals Michael Psellos (11th c.), Niketas Choniates (12th/13th c.), and George Pachymeres (13th/14th c.). All three stress the close contact to animals shared by these groups, in a clearly biased way. According to the Greek writers, these groups only subsisted on hunting, clothing themselves in the furs of animals; assaulting their enemies in a predatory manner and hiding in the woods, they themselves behaved like wild beasts. Living in close quarters with their horses, they transgressed the accepted rules of human-animal contact by eating their flesh, drinking their blood, and, in Choniates' polemic words, engaging in sexual intercourse with them.[18]

Written from a settled Christian perspective, it is no surprise that such negative assessments of human-animal closeness surface most prominently in literary analyses of (semi-)nomadic groups. These represented the opposite of the Greek authors' ideals through their mobile lifestyles, their religious beliefs, and their occupation of isolated spaces such as forests, deserts, or pastures far from human settlements. However, such ideals and their oppositions were fluid and relative. While Choniates highlights the Cumans' closeness to their mounts and their use of fur clothing, conceivably in contrast to the splendor of urban, courtly Constantinople, Neilos pursues a different path, setting up a contrast between "predatory" Bedouins and ascetic and vegetarian eremites, whose lifestyle was arguably no less alien to wealthy Constantinopolitans than the lifestyles of Cumans, Pechenegs, and Mongols.[19]

The authors established contrasts in lifestyle and culture along traditional boundaries between (idealized) humans and animals, or animal-like humans. The close association between humans belonging to the latter category and those animals that were considered integral parts of their lives forms an important factor for our semiotic analysis. The meanings attributed to individual species overlap with the notion of a general animality inherent in animals and humans alike. The impact of classical ethnography and its core assumptions on animality and "barbarism/nomadism" have been extensively studied. Substantial evidence of parallel developments in other medieval societies that drew from the same tradition, remain well-studied and extant.[20]

Beyond the notion of a general "barbaric" animality, the individual categories or species of animals were often attributed further, particular semiotic meanings that resonated in their literary descriptions. Regarding the previous examples, we can see how the texts move from a general concept of instinct-driven predatory animals that informed the pejorative image of nature-dwelling, hunting nomads, to more specific cultural habits and beliefs. One of these is the (quasi) Byzantine taboo on eating horsemeat and animal blood, which likely informed the conceit of horse eating and blood drinking "barbarians" in

Psellos' description of the Pechenegs.[21] The common association of horses with uncontrolled (sexual) instinct is also notable and mentioned in a wide variety of Byzantine texts, and we can reasonably connect it to Choniates' accusation of human-animal intercourse in the case of the Cumans.[22]

We can conclude that the narrative function and significance attributed to the animals described in these passages results from interactions between social realities, on the one hand, and established semiotic traditions, on the other hand. As companions, mounts, and sources of food, animals provided visible markers of human lifestyles, here exemplified by the mobile Pechenegs and their cultural reliance on horses, or the Bedouins' companionship with their camels.[23] At the same time, these animals already belonged to existing symbolic systems that had shaped Byzantine (elite) culture for centuries. These systems helped to frame and process the social realities encountered by our authors, providing literary tools to describe (and evaluate!) these realities in ways that accommodated their and their audiences' needs and tastes.

Aristocrats and Their Steeds

The needs and tastes that guided the interpretations of human-animal contact varied significantly according to the segment of Byzantine society for and by whom a text was composed. This is best exemplified when comparing the negative semiotic meanings associated with horses in the previously discussed ethnographic descriptions to the display of horses in Byzantine elite representation. As a "complex surface for the aristocracy to inscribe their cultural values and norms", horses have been extensively studied with respect to western Medieval elites.[24] Byzantine society never developed a comparable concept of knighthood as in western and central Europe, where horse and rider were considered a military-technical as well as a sociocultural unit.[25] Horses nevertheless functioned as integral parts in the everyday life of Byzantine secular elites and, just as their Western counterparts, as symbolic proxies for their human owners' and companions' status and lifestyle.

Aristocratic self-representation, as figured in textual and visual source material from the 11th century on, confronts us with overwhelming references to horses. A representative example of this interest is found in Theodore Prodromos' (12th c.) encomium to the newly born aristocrat Alexios Komnenos. In a key passage, the orator mentions "Arabian and Thessalian [horses], servants of war" that should be trained together with the child. He points out an obvious closeness between the human and his animal companions, both prepared for their joint future within the aristocratic society.[26]

Prodromos' "Thessalians" and "Arabians" are embedded within a broader cosmos of literary horses that framed his interpretation: Bukephalos, the horse of the popular hero Alexander the Great, who was thought to be of Thessalian breed;[27] or the famous Thessalian cavalry mentioned by Herodotus, a precursor of the cavalry in Prodromos' own time;[28] the latter are also referenced by the 10th-century zoological collection ascribed to the scholar-emperor Constantine VII, including a praise on the Thessalian horse's zeal for battle and excellence.[29] The "literary" Thessalian horses not only reflected the orator's knowledge of authoritative traditions, they also symbolically connected their human "owner" (Alexios Komnenos) to models of a heroic past. At the same time, Thessaly was very likely still a breeding ground for Byzantine war horses and a recruitment area for cavalry at the time of Prodomos' poem. It is therefore conceivable that the newborn aristocrat eventually would have possessed horses that were actually bred in that region.[30]

The literary character of the "Arabian horses" in the poem likewise cannot be denied. They appear in Byzantine texts as prime exemplars of speed and grace, a particularly popular notion from the late 11th century, indicating a literary trend of the period. On the other hand, horses that were considered "Arabian", whether as a specific race or, more generic, as imports from "Arab" lands, had long been objects of diplomatic gift exchange between Byzantium and neighboring courts in Damascus and Ayyubid Egypt.[31] Anna Komnene reports of ambassadors to Palestine who bought "noble-born horses from Damascus, Edessa, and even from Arabia" in the 12th century. Niketas Choniates mentions the gift of two Arabian stallions by the Ayyubid Sultan of Egypt to Emperor Alexios III (r. 1195–1203). Both testify the physical presence of these animals in Byzantium and the great value that Eastern Mediterranean elites attributed to them.[32] A literary echo of their popularity is found in the anonymous dialogue *Timarion*. Describing a procession by the *doux* of Thessaloniki and his entourage, the text explicitly mentions the splendid "Arabian horses" on which the noble participants were mounted.[33]

We cannot determine for certain whether Prodromos' Thessalian and Arabian horses were literary fictions or if the author had specific animals in mind. That both possibilities are equally valid demonstrates the close connections between literary references to these animals and their physical presence in the aristocratic world of Byzantium. Associations between humans and animals were drawn on both levels, be it in a literary poem, or by the display of living specimen. This exemplifies the animals' double roles as material-semiotic nodes "in which diverse bodies and meanings coshape one another",[34] being entangled in a joint semiosis that ultimately affected both sides.

The Semiotics of Threat

Animals played a significant role as companions to humans, as status symbols, and they largely defined the everyday life of those who depended on them for food, means of travel, and workforce. At the same time, many animals were perceived as menacing threats to life and health, and again, semiotic meanings are inseparably connected to material human-animal contact. The variety of perceived threats from the animal world that enriched human metaphorical language and literary imagery is immense, and a systematic exploration beyond the scope of this chapter.

The pejorative image of the "rabid dog", widespread across many Euro-Mediterranean cultures, proves a key example of the rich semiosis resulting from actual contact with such threats. Its frequent association with "mad heretics" or "uncontrolled barbarians" in Byzantine texts shows how the phenomenon of rabies (and behaviors interpreted in that way) influenced human concepts and descriptions of endangered social order, impurity, and insanity.[35] A comparable area of threat-based human-animal contact that resulted in popular, socio-politically connoted imagery, are the hunting activities of the Byzantine elite. Based on real life practice, imagery of violent encounters with the fauna of the forest and the mountains left a considerable imprint in political language. Historiography and the rich encomiastic material of the 11th and 12th centuries permits manifold insight in a discourse where success in overcoming the threats of the animal world reflected and foreshadowed military and governmental abilities by rulers and aristocrats.[36]

The epic of "Digenis Akritas" provides conceivably the most famous example of the ideal "warrior-hunter". The figure of "Digenis" connects and continues millennia-old ideals of human male domination over wild animals and humans alike.[37] Its ongoing reception and popularity in Byzantium was not limited to literary descriptions, such as Eustathios of Thessalonikes' praise of Emperor Manuēl I's son Alexios as a young, Digenis-like hunter-warrior.[38] It remained connected to actual practices by the aristocratic elite. Similar to other pre-modern Eurasian elites, Byzantine aristocrats had their status and self-image closely tied to the semiotics of human-animal confrontation. Actual big game hunting was a marker of their status, providing them the opportunity to "re-enact" the feats of their literary heroes.[39] This way, human-animal encounters during real-live activities and those within a semiotic, literary-artistic discourse influenced and consolidated each other.

The use of animal-shaped *apotropaia* in Byzantium figured yet another area where human contact with a dangerous and threatening fauna produced a complex semiosis and shaped the (re-)production of symbolically charged imagery. The use of animal-shaped *apotropaia* was already known in Antiquity, while similar practices in the Islamic world attest their popularity across the medieval Eastern Mediterranean.[40] Premised on the idea that "things and images" stood in secret, cosmic relationships to "natural forces and entities, including [...] plants, animals, [...] demons, [...] and efficacious words", depictions on amulets or architecture were believed to bear protective capabilities against whatever threat they represented.[41] Animals played a large role to this end. Niketas Choniates (12th–early 13th century) describes an ancient statue of an eagle killing a snake in the Hippodrome of Constantinople. Allegedly enchanted by Apollonius of Tyana (1st century AD), it was believed that the sight of the snake in the eagle's claws "by its example, scared away also the other snakes in Byzantium and persuades them to coil themselves up in [their] holes and hide".[42]

The idea of a counteractive effect of animal-imagery is also visible in the *Patria* of Constantinople, a 10th-c.-collection of (older) anecdotes about the city's statues and buildings. Here, the civic display of bronze sculptures of mosquitos, fleas, and bugs in the city is connected to their claimed power to keep such pests away from the city.[43] Recorded in elite literature, these stories conceivably transmitted popular knowledge that appealed to wider parts of the population. Reflecting everyday contact with often harmful fauna, they once more demonstrate the impact of human-animal contact on Byzantine interpretations of animal signs and symbols, not only evidenced in written sources but also in the built environment of the Constantinopolitan cityscape.[44]

Consuming Animal Matter

Not only did the contact with living animals provide the basis for semiosis and figurative use; the consumption and use of products obtained in the collaboration, exploitation or killing of animals played a significant role as well, proving that animal-semiosis transcended the physicality of the living creatures.[45] As with the cases discussed above, relevant sources predominantly relate the perspective of urban elites, individuals who profited from the animal-driven economy rather than directly participating in it as farmers or herdsmen.

This elitist perspective is visible in the 14th-century *Entertaining Tale of Quadrupeds*, a collection of poetic animal invectives set in the lion king's court. By mutually insulting each other, animals argued for their superiority over other species based not on their abilities and behavior but on the value of their (dead) bodies for human consumers. Some exemplary speeches include those by the hare, the deer, and the boar, who define their own meaning and value by asking whose meat is more prized and whose fur is worn and displayed by members of the Byzantine elite, reflecting the use of animals as status symbols.[46]

That the death of an animal did not necessarily end but rather transformed its semiotic value is a conceit that surfaces also in the historiographic work by Leon Diakonos (10th c.). Describing a scene at court during the 960s, he recounts how Emperor Nikephoros II (r. 963–969) insulted the Bulgarian Tsar as a "leather-gnawing" primitive who was "dressed in animal skins".[47] Not using animal imagery in a strict sense, the author establishes an implicit contrast between a culture that supposedly dressed in barely treated animal furs, and the sophisticated costumes of the Byzantine court. His reference to chewing animal skin can be read as a metaphor for poverty, but it may have been an actual practice in small scale leather production of the time. Fictional or not, animal products clearly retained a semiotic role in Diakonos' narrative, functioning as classifiers of morally coded social behavior.[48]

Similar takes can be found in "reports" by 11th- and 12th-century Byzantine officials sent from the capital to the provinces south of the Danube. Following a narrative of "exile" or "distinction" that required members of the Constantinopolitan elite to set themselves and by proxy their audiences apart from the "uncultivated" province, they heavily relied on imagery of animals and animal-based economies.[49] Theophylaktos (11th c.), an orator in Constantinople and later archbishop of Ochrid, admitted his "fear" that he would soon become a "Bulgarian who stank like sheepskin".[50] His 12th-century colleague Gregorios Antiochos defined life in his new social environment by highlighting its absence of singing birds, a common literary reference to art and rhetoric. These were replaced only by the sounds of sheep, goats, and cattle, living metaphors of the author's "misery".[51]

Studies on Western medieval culture have directed ample attention to the permeable distinction between living and dead animals, especially as regards their semiotic functions. During elite dinners, the display and consumption of certain animals, such as cranes, evidently served to highlight the hosts' elevated social status.[52] Similar phenomena characterized the culinary representation of the Byzantine elite, where the choice of food animals reflected the status of the hosts and his/her guests.[53]

Medieval Western romance literature also emphasizes a moralizing interpretation of the consumption of animals. When Wolfram von Eschenbach, for instance, depicts Parzival eating a partridge during his famous encounter with the noblewoman Jeschute, the scene clearly points beyond aristocratic eating culture.[54] As S. Obermaier points out, the partridge, commonly connected to a notion of "boundless lecherousness", is clearly meant to comment on the aspiring knight's own "uncourtly lack of moderation" and his difficulties to integrate into an idealized world of chivalric culture.[55] Byzantine texts have not yet been researched from this perspective in a systematic manner.[56] The examples discussed above, however, show that also in Medieval Greek sources, dead animals and side products were linked to concepts describing the social affiliation, status, morals, and cultural capabilities of their human consumers. The semiotic functions of "Byzantine" animals did not end with their deaths, but continued to provide meaning and characterize those humans who used, consumed, or displayed them.

From Symbolic to Material: Animals as Signs and the Construction of Social and Moral Concepts

While the previous section focused on encounters between humans and animals (including their products/bodies) as they informed the development of human-centered sociocultural systems, the following section takes the symbolic traditions that buttressed Byzantine interpretations of animals as its starting point of analysis. This change in perspective does not disregard the material foundations of animal semiosis; the focus of interest, however, shifts toward conceptions and attitudes that the Byzantines inherited and adapted from preceding and neighboring cultures. Informed by longstanding beliefs, their interpretations of animals often transcended the physicality of everyday human-animal contact; occasionally we even find reference to "semiotic animals" that existed apart from, or bearing very little relation to, their material counterparts.

Organizing Human Society

The representation of human social organization and ethics figures as a key thematic area where the detachment between material and semiotic animals is particularly evident. Byzantine texts and images concerned with these topics abound with references to morally charged animals and highlight parallels, contrasts, and connections between fauna and human society. The representation of herd and shepherd stood as the most influential model of social organization deriving from contacts between humans and animals. References to this model were usually accompanied by a third element – the threat of predatory animals. The resulting conceptual triangle (shepherd–herd–predators) expresses basic elements of leadership, dependence, community, threat, and protection. These elements not only characterized human-animal relations, but were easily applicable within human Byzantine society; a society that was defined by patron-client relations from the village community up to the highest authorities, one where social, political, and spiritual leadership was grounded in an ability to protect the community, and whose social structure depended on the concept of fixed roles within a divine hierarchy.

The semiotic relevance of the shepherd-herd concept found its roots in the quotidian human experience of herding and animal breeding. The Byzantines, however, inherited it from a longstanding Eastern Mediterranean tradition, developed over millennia into a powerful, symbolically charged model for social organization and leadership.[57] In Byzantium, the Christian appropriation of the shepherd as an emblem of God/Christ and His spiritual and secular representatives on Earth became the most influential variant of the image.[58] Religious authorities used it to express their hierarchical and leadership roles. Describing an abbot's tasks, the famous theologian Theodore Stoudites (8/9th c.) explicitly refers to Jesus as the good shepherd who leads, protects, and provides for his flock. For Theodore, the crosier (ποιμαντικὴ ῥάβδος) is a visible reminder of this role. This perspective was also reflected in the ceremonial handing over of the "shepherd's staff" by emperors to newly appointed patriarchs.[59]

Emperors frequently compared themselves to the Good Shepherd. Leon III (8th c.) explicitly referred to him in the introduction to his law book "Ekloga", defining his role as one granted by God to all emperors to "tend the sheep", just as St. Peter was considered the apostles' shepherd.[60] The "bad shepherd", on the other hand, was a common image invoked to discuss poor leadership; examples feature in the vita of Cyril Phileotes (beg. 12th c.), which describes critiques against the Byzantine governor of Cyprus, or in the polemical descriptions of the ill-renowned Emperor Andronikos I (l. 12th c.).[61]

The inheritance and adaptation of older Mediterranean imagery characterized other animal-based models of social organization as well. When the 12th-century scholar John Tzetzes compared the eagle and lion to "kings and great men",[62] he referred to a traditional understanding of internal hierarchy organizing fauna, widespread in the literary canon of the time.[63] Numerous comparisons of Byzantine emperors to omnipresent, vigilant eagles, fierce, protective lions, or lion hunters in encomia and historiography unite the idea of an "animal kingdom" with specific allusions to the lion and eagle as traditional symbols of rulership in the whole Near East.[64]

The specifically "Byzantine" variant of this imagery surfaces in concrete reference to key symbols of Greco-Roman Christian imperial rule: the eagle resurrecting in God/Christ, a common reference to a new emperor's rule in court oration and poetry;[65] the "lion of Judah" (Gen 49), connecting Byzantine rule to biblical kingship;[66] the lions surrounding Salomon's throne, reproduced at the imperial throne in the Magnaura palace of the 9th/ 10th centuries;[67] or the ruler as a hunter and tamer of beasts, a motif derived from imperial exemplars such as Alexander the Great, King David, or the epic hero Digenis Akritas, who proved their value as warriors and rulers over humans and animals.[68]

Contact with physical animals was no absolute requirement for the development and popularization of such motifs. For example, after the Byzantines' loss of the Levant in the 7th century, lions were no longer frequently found in Byzantine territories, even though some evidence suggests rare sightings in Anatolia up to the 19th century. Their physical rarity, however, did not prevent the diffusion of rich lion imagery, especially in Byzantine literature, throughout the Middle Ages.[69] Even stronger is the argument in regard to the flourishing fantastical imagery, such as the use of dragon-monsters as symbols of threat in political and religious/moral discourse, which obviously developed independent of physical

Figure 1.1 Byzantine Ivory Casket (10th/11th c.) depicting an imperial lion hunt, Troyes Cathedral. Copyright: photo by Fab5669 – Own work, CC BY-SA 4.0, https://commons.wikimedia. org/w/index.php?curid=51458177.

contact.[70] The same relative independence between access to physical animals and the development of animal imagery can be seen in Medieval Western and Northern Europe. Inhabitants from these regions had only very rare possibilities to actually see, for instance, some of the North African species presented to them in the Physiologus tradition, which was nevertheless very popular among them.[71] In Byzantium, the cosmopolitan character of Constantinople and the network of royal gift exchange provided the court and the city's inhabitants at least with some access to a variety of non-native animals. These exchanges introduced occasional possibilities to connect "semiotic animals" to their otherwise little-known physical counterparts.[72] We know from 12th-century travelers' accounts that lions, leopards, and cheetahs were kept on the imperial palace premises.[73] The latter were reportedly used for the hunt. 14th-century Pseudo Kodinos reports of them taking part even in imperial ceremony, a practice known also from other Eastern Mediterranean courts.[74]

Performances of Power

Especially in Constantinople, animals played significant roles in ceremonial, visual, and performative contexts, including parades, ritualized beast fights, and animal shows.[75] Horses and mules were instrumental in representing the status of their riders (sovereigns, legates, ambassadors); allowing or denying imperial guests to appear mounted was a well-established tool in the language of court protocol.[76] On the topic of Byzantine humiliation parades, M. Perisanidi recently highlighted the symbolic and choreographic importance of the selected animals: when the deposed Emperor Andronikos I was paraded through the streets of Constantinople, he was deliberately placed on a camel rather than a horse, as was common for triumphs of the time. This deviation from the norm was an obvious symbolic gesture directed at the spectators, one that further humiliated the deposed emperor.

The intellectual and cleric Michael Choniates went even further in his interpretation, remarking on the fittingness of Andronikos being carried by a camel, given that the deposed emperor was himself "more vengeful than camels".[77] The semiotic tradition of the vengeful camel known from early Church fathers guided Choniates' moralistic reading of the ex-emperor, who was known for his ruthless punishments of opponents.[78] We cannot tell if such a reading was sponsored by the organizers of the parade and understood by spectators. For viewers, the use of a camel was a choreographic element that turned the "pseudo-triumph" into a public humiliation.[79] Choniates, however, applies the full semiotic potential of the "literary" camel for a learned courtly audience who welcomed any attempt to slander the deposed ruler's reputation.

Another case of a multi-layered semiotic interpretation of exotic animals is found in the detailed report by the 11th-century courtier Michael Psellos, who describes the display of an elephant and a giraffe in 1053, diplomatic gifts from the Fatimid Caliph of Egypt to Emperor Constantine IX.[80] Psellos states that, until then, he had only known of elephants from incredible tales (μύθοι). Seeing one in the hippodrome, it seemed impossible to him to conceive of the animal as something "a-symbolic and devoid of further meaning" (ἀξύμβολον, ἀσημείωτον). When the elephant bowed before the emperor's seat "as if by rational intention", it was as though this "giant of the earth" bridged the western and eastern regions of the cosmos; his *proskynēsis* symbolized the emperor's rightful rule (ἀρχὴ δεξία) over territories stretching in both directions.[81]

Referring to myths from India, Psellos may have been informed by a wide range of contemporary symbolic interpretations of the elephant. Drawing from the rich literature on dream interpretation, the so-called *oneirokritikon* of Achmet foresees rich and powerful captives or subjects to an emperor who dreams of leading an elephant. These interpretations contributed to the wider symbolic context connected to the elephant, and it is likely that they prompted Psellos' reading.[82] However, just as in the case of Choniates, it is worth noting that Psellos' reading appeared in a written piece of poetry, to be presented in front of a courtly audience acquainted with literary traditions and political imagery. The way in which the elephant was interpreted by the masses in the Hippodrome remains unknown. The general purpose of the show was to entertain and impress the crowd with displays of exoticism and wonder.

These performances were nevertheless unique occasions for learned and common audiences alike to (re-)connect textual, visual, and orally transmitted interpretations of animals that were not part of their everyday experience to living specimens. Providing contact zones between the literary and the material, the ceremonial animal displays enabled the confrontation of cultural concepts and material realities. They prompted orators such as Psellos to reproduce traditional readings, but they also facilitated real life observations and thus the updating of zoological knowledge and, possibly, reconsiderations of semiotic meaning.

Markers of Virtues and Vices

The belief that individual animal species were characterized by specific attitudes and behavioral patterns also defined their prominent role as interpretative keys for individual (human) moral values and social behaviors. Here, the extant texts tend to isolate certain "characteristic" behaviors and physical traits of individual species, rather than provide comprehensive descriptions. A typical passage to this end is found in a homily on the creation of (land-) animals (Gen 1.24–31) by the church father Basil of Caesarea (4th c.): "the ass [= all asses; T.S.] is idle, [...], the wolf is untamable, [...], the deer is fearful, [...] the dog [is] grateful and does not forget friendship".[83] Similar characterizations appear in the *Physiologus*, ascribing to the wolf a "heart full of treachery and the urge for robbery", or "royal characteristics" to the lion.[84]

A text attributed to John Chrysostomos maps such characteristics directly onto human social behavior, referring to the "rule" (κανών) "that the dogs are the shameless ones [= humans; T.S.], the wolves the rapacious ones, that the snakes the treacherous [...], horses the ones who are mad after women [...]", etc.[85] Such moralistic readings were not limited to religious discourse. Fables, the most popular of which were attributed to Aesop, in addition to new material such as "Stephanites kai Ichnelates", the late Byzantine animal satires "Poulologos", and the "Entertaining Tale of the Quadrupeds" (both 14th c.) all presented animals as speaking and acting exempla of (human) (im-)moral behavior.[86] A rare, preserved visual depiction of moralistic animal imagery in an architectural context can be found in an 11th-century rock-carved chamber in Eski Gümüş, Cappadocia. Originally identified as part of a monastic complex, recent scholarship has advanced a possible secular function of the chamber as a "private salon or reception room of a provincial magistrate".[87]

The images and accompanying texts reproduce Aesop's well-known fables: the farmer and the ungrateful snake; the rapacious wolf and the heron; the tale of the lion's share. Representing the wicked (wolf), the ungrateful (snake), and the power hungry (lion), the animal characters represent typical aspects of human social behavior. To this end, they conveyed moral insight and basic worldly wisdom to their spectators, whether monks, aristocrats, or officials and their guests and clients.[88]

Figure 1.2 Scenes from Aesop's fables in an 11th-c. rock-carved building complex in Eski Gümüş, Cappadocia. Copyright: Reproduction from M. McGough, "The Monastery of Eski Gümüş—Second Preliminary Report," *Anatolian Studies* 15 (1965): 163; Reproduced with permission of the Licensor through PLSclear. The scenes are accompanied by epigrams from the 9th c. metrical paraphrase of Babrius' fables attributed to Ignatios Diakonos.

The personal "literary bestiaries" of "moralistic" animals of only a few Byzantine authors, such as Niketas Choniates and (in part) Michael Psellos, have been examined thus far.[89] Scholarship has focused on individual species and their moralistic interpretations in specific discursive contexts, such as political controversies,[90] advice literature,[91] and Christian discourse.[92] Developments in the semiotic significance, function, and popularity of individual animal signs can be observed over longer periods; the Christian reinterpretation of traditional Ancient animal imagery emblematized the most far-reaching manifestation of such a change.[93]

In this context, the status of the dragon in Byzantine hagiography stands as an illustrative case in the development of an individual animal-sign. As part of a broader set of Indo-European and Near Eastern dragon narratives, hagiographic texts interpreted "Byzantine" dragons as indicators of moral behavior, manifestations of pagan beliefs that dominated town and village communities, or as powerful rivals against whom a saint could prove his or her steadfastness. Its imagery reflects contemporary issues in late Antique Christian discourse. The remarkable increase in importance of dragon narratives in saints' lives toward the middle Byzantine period indicates a further development and a rising popularity of the imagery in the following centuries.[94]

Despite the impression given by Basil of Caesarea and Ps.-John Chrysostomos of clear, dyadic correspondences connecting animal-signs and their meaning, extant source materials reveal significant variations in the meanings attributed to certain species, even within the same texts and by the same author. For example, while the snake generally referred to a negative connotation, the same animal also could be read as a source of moral guidance.[95] Dogs were frequently praised for their loyalty, which influenced their metaphoric use. Yet according to the analyzed narratives, they could represent threats, uncleanliness, demonic

26

forces, and pagan belief. The beginning of this chapter outlined the different discursive treatments of horses. These readings had to align with conceptual elements that Byzantine society attributed to these animals.[96] The limits of this conceptual and semiotic fluidity were defined within a wider cultural epistemology where groups with similar intellectual backgrounds shared a basic mutual understanding of concepts and symbols. Within this framework, the individual application of oral, textual, or visual animal-signs allowed for and required a considerable amount of interpretative flexibility.

Christian Zoology

Such interpretative flexibility in moral- and value-based interpretations characterized another specificity of the Christian world view that dominated the Byzantine perspective on animals: their assumed role as allegoric signs and symbols in the context of salvation history. Although, or perhaps because, animals were widely considered incapable of reason and articulate communication (*alogoi, aloga*) in established Christian zoology, they were viewed as legible elements within the cosmos, capable of revealing divine, metaphysical truths to those who knew how to observe and interpret them.[97]

Such an approach was by no means uncontested, though. While the 3rd-century church father Origen set an influential example for the allegorical interpretation of flora and fauna, Gregory of Nyssa and Gregory of Nazianzus (4th c.) rejected the notion that the cosmos could be considered as more than a general manifestation of God's glory.[98] Their opposition to decidedly moral and metaphysical interpretations of natural phenomena and fauna never became mainstream. A critical stance waged against animal-signs as representations of Christ (e.g. the lamb) or the Holy Ghost (often represented as a dove) during the iconoclastic controversies (7th/8th c.) also did not have lasting effects.[99]

Christian animal exegesis was based on the vast proliferation of biblical animal imagery, but this did not mean that ancient zoology, empirical knowledge, and popular beliefs did not shape its development. The doctrinal discussion on the skinning of snakes (*ecdysis*) illustrates the transformation of zoological knowledge into a Christian perspective. Already described by Aristotle, Christian theologians such as Kosmas of Jerusalem (8th c.) reinterpreted the *ecdysis* as an example of the faithful who set apart their sinful earthly body.[100] Intervention and, in a certain sense, innovation occurred when a semiotic need surpassed empirically observable animal behavior. When the *Physiologus* stated that lion cubs are born dead to be revived by their father after three days, just as Jesus resurrected after the same time, it is unlikely that traditional knowledge was merely reinterpreted. Rather, symbolic interpretation likely guided the adaption of zoological knowledge.[101]

While, from a merely empirical perspective, the validity of such conclusions remains questionable, they are nevertheless grounded in an epistemological assumption that insight gained by interpretation should not be separated from observed or assumed zoological facts. Christian thinkers were aware of the tension between these different fields of knowledge, exemplified by Augustine's discussion on a pelican killing its chicks before resurrecting them. Although he seriously doubted that pelicans acted in such a manner, the narrative's coherence with Christian beliefs and its resulting moral implications made this story a bearer of its own truth, even though this truth was derived from a literary-semiotic animal.[102]

The exegesis of nature as practiced in clerical and monastic circles depended on the entanglement and mutual legitimation of symbolic-interpretative and zoological knowledge. However, the tendency to consider animals as transmitters of metaphysical truth was more widely relevant and influenced a diversity of texts. When the 6th-century historiographer Prokopios of Caesarea skillfully combined the news of the Empress Theodora's death with the story of a whale stranded in the vicinity of Constantinople, he conceivably referred to apocalyptic sea monsters and immoral behavior in the *Physiologus* as a subversive comment on the deceased Empress.[103] In a much more open manner, the iconophilic opponents of iconoclastic emperors often compared the latter to apocalyptical dragons that devoured the faithful.[104] Finally, funeral speeches (*epitaphioi logoi*), regularly referred to the deceased as lions who slept with open eyes, symbols for the resurrected Christ.[105] In this regard, animals fulfilled a role as semiotic links connecting events and people to (assumed) historical events and familiar patterns of salvation history, thereby opening possibilities for their moral evaluation.

Animals' Prognostic Capabilities

For millennia, Eastern Mediterranean cultures shared an understanding of animals as transmitters of supernatural knowledge and bearers of prognostic capabilities, reading them as moral and metaphysical clues within a divinely created cosmos. One aspect of this understanding that surfaced prominently in the extant Byzantine source material, is the role of animals in dreams and visions. Extant Byzantine sources on dream interpretation from the 9th century on provide particularly rich evidence of the revelatory and prognostic function of animals both within and beyond Christian dogma. Merging Hellenistic authorities such as the famous Artemidoros of Daldis (2nd c. AD) with Near Eastern traditions and folk beliefs, these texts are characterized by vast reference to animal signs used to comment on a dreamer's current situation or foreshadow their future.[106]

The interpretative keys provided are generally brief and simple. For example, the dreambook (*oneirokritikon*) attributed to Patriarch Nikephoros of Constantinople (9th c.) explains his vision of crows as indicating the presence of demons, while a "hawk flying from one's hand" supposedly announced general, impending harm.[107] Interpretations depended both on the animal that appeared in the dream and, occasionally, on the social status of the dreamer. The so-called *Dreambook of Achmet* (9th–11th c.) informs the reader that an eagle seen by an emperor means that "he will gain another empire", yet for "someone from the common people, [that] he/she will be emperor".[108] Written by an anonymous (Christian) author, the text includes a considerable amount of Arabic material. Although Ps-Achmet and other texts partly "Christianized" their material to make it more appropriate for Byzantine audiences, the explicit mention of interpretations from India, Egypt, and Persia demonstrate the appeal of labels used to categorize knowledge across preceding and neighboring non-Christian cultures.[109]

The religious elite showed ambivalent attitudes to dreams. Despite the occasional distrust expressed by members of clerical and monastic circles, the possibility that dreams and visions could transmit divine revelations was widely acknowledged.[110] In Byzantine hagiographical literature, dreams and visions of animals are common *topoi*. Saint Makarios' (4th c.) dream of demons attacking his face as though they were ravens seems to echo a similar interpretation in Ps.-Nikephoros (see above). Similarly, a nightly vision by a companion of Symeon the Stylite (6th c.) of the saint and a lion driving out a boar from their cloister's garden might directly refer to the biblical imagery of the boar in God's vineyard.[111]

Also at court, the sources frequently document animals appearing in dreams. Emperors reportedly collected dreambooks in their libraries, possibly as aids for decision making.[112] Animals in dreams were a way to make sense of unexpected events. When the emperor John II's son Alexios died prematurely in 1142, Niketas Choniates reported a rumor that circulated at court about a dream the emperor had shortly before: "his newly crowned son [...], who was mounted on a lion and held it by the ears, because he had nothing else at hand to lead the beast", meaning that he was unable to control it.[113] The episode seems directly connected to Achmet's dreambook, which was likely available at John's court: "if someone rides a lion that is subjected to him, [...] he will become emperor" – a fate that was obviously denied to Alexios.[114]

The murder of Emperor Leon V in Hagia Sophia on December 25, 820 was not without foreshadowing (animal) signs either. According to Theophanes Continuatus, the emperor discovered in his library a prophecy announcing doom for the empire on the day of Christ's birth. The text was accompanied by a depiction of a stabbed lion (λέων) under the letter *chi*. Whether the emperor really consulted experts about his discovery, and if the imperial questor's alleged interpretation of Leon's (Λέων) death preceded or followed the event cannot be determined. But again, an animal-related omen appears as an instrument to explain often unpredictable twists of fate.[115]

This usage of animal-signs also applies to popular stories about eagles appearing out of the blue to shadow people who would eventually rule the empire (Markianos, 5th c.; Philippikos, 8th c.; Basil I, 9th c.).[116] These encounters were commonly interpreted as signs predicting one's destiny to eventually ascend to the throne, or rather, a convenient instrument to explain and legitimize seizures of power by men who lacked proper dynastic claims.

It is difficult to draw a distinction between literary and physical animals here. Many of the reported omens were likely devised *ex post facto*. However, a story by Niketas Choniates may reflect actual practices of avimancy: According to the historiographer, in 1183 a loose hawk was seen entering the imperial palace from the east, and it took several attempts to capture it. Many observers interpreted this occurrence as an omen signaling the fate of the usurper Andronikos Komnenos, who was approaching the capital from Asia Minor at exactly this time.[117] Whether the specific interpretations reported by Niketas, ranging from the imminent failure of the usurper to his success and later downfall, were fact or fiction, is impossible to tell. His usage of the story, however, indicates that it was still common enough for his audience to consider unusual appearances of animals as sources of guidance and orientation when situations were ambiguous and insecure. The literary tradition of animal-omens and the actual observation and interpretation likely rose in parallel, mutually enhancing their attributed validity.

Conclusion and Perspectives

Within Byzantine studies, animal semiotics is predominantly used in interpretations of textual and visual sources on human stories and history. Referenced animals are primarily relevant in their capacity to act as transmitters of information and objects of projection. However, despite the strong dependence of animal semiotics on anthropocentric interpretation and communication, human-derived definitions did not exclusively define what and

Figure 1.3 Future emperor Basil I as a child, protected from the sun by an eagle. Cod. Vitr. 26–2 (Skylitzes Matritensis), fol. 82v. Copyright: Biblioteca Nacional de España.

how animals "meant" in Byzantine society. The examples discussed in this chapter indicate that a constant contact between humans and (other) animals (as material-semiotic hybrids) both shaped everyday life as well as the symbolic systems that facilitated human expression.

As livestock, aids to labor, companions in war and on travel, animals defined many aspects of human existence. Actual practices involving humans and animals (hunting, war, collaborative work, transport) influenced the reproduction and popularity of literary models (epic hunts; horse loving warrior-heroes; stereotypes about nomads' closeness to animals). On the other hand, visual art and literature consolidated perceptions on and interpretations of animals, which had concrete consequences for their physical existence. This is evidenced by the acquisition of certain animal species or their products for purposes of diplomatic exchange and the display of status, as well as in the ceremonial parading of animals, as reproductions of literary symbols of power, morals, and status.

While zoological knowledge often informed animal related semiosis, the reverse like-wise occurred. As demonstrated with respect to Christian moral discourse, the literary invocation of animals could influence changes to zoological knowledge traditions, as with the belief of lion cubs being resurrected after three days. Here, the production of zoo-logical knowledge oscillated between an acknowledgment of assumed animal habits in the physical world and the urge to harmonize messages encoded in nature within established systems of Christian belief. Such tendencies exemplify the thesis that symbolic systems result from "rivaling demands for meaning", rather than standing as internally consistent constructions.[118]

Although the interpretations of animals in Byzantine sources are generally characterized by variety and fluidity rather than unambiguity and homogeneity, the selection of animal-signs and the range of accepted symbolic interpretations were limited by a framework of epistemological traditions and discursive rules. Authors, orators, and artists had to respect these rules if they wished to be understood by their audiences. While it seems that the set of accepted interpretations of animal-signs remained relatively stable throughout the Byzantine period, this did not prevent adaptions in use and even innovations of new signs and readings in that time.

One has to remain aware that interpretations and signs documented in the extant sources are not necessarily representative of the entirety of perspectives within Byzantine society. The diversity of animal-readings reflects a diversity of society itself, including differences in speakers' intellectual backgrounds, narrative aims, and authoritative models. At the same time, the sources must be considered as deriving from discursive power relations within society. It is evident that extant texts largely represent settled, learned Christian elites. The concentration of social, economic, and political power in Constantinople, par-ticularly from the Middle Byzantine period onward, led to a further overrepresentation of voices from its urban elites. This occurred to the disadvantage of often non-literate rural populations, as well as those who followed lifestyles considered inferior by the majority, such as the semi-nomadic groups in Asia Minor and the Balkans. They, too, contributed to the social landscapes of Byzantium but they left only scarce traces in the source material available to us.

In this sense, the corpus of sources limits our ability to speak of a singular Byzantine view of animals. Rather, it indicates a diversity of perspectives within a wider frame-work of knowledge, ideas, and descriptive traditions. Analyzing these diverse views offers insights into the development and preservation of Eastern Mediterranean traditions in animal semiotics, as well as on the migration of ideas between the Near East and Western

Europe. Within the medieval period, Byzantium assumed the role of an important hub for the diffusion of knowledge, but also rose as a key place where knowledge production occurred. Further investigation on the "Byzantines'" ways of perceiving and interpreting animals, and how these interpretations influenced human-animal contact, will contribute to a better understanding of the complex and multilayered relationships between humans and other animals, not only in the Byzantine world, but across the Euro-Mediterranean Middle Ages.

Funding

This chapter was written as part of the project UMO-2019/35/B/HS2/02779 funded by the National Science Center Poland.

Notes

1 Definitions of crucial terms, as used in this chapter:

Symbol describes "anything that stands for or represents something else beyond it"; the meaning is generally a conventional one. Crucial to the literary usage of the term is that "its application is left open as an unstated suggestion", allowing for different interpretations. See C. Baldick, "Symbol," in *The Oxford Dictionary of Literary Terms* (ODL) (Oxford, 2015), www.oxfo rdreference.com/view/10.1093/acref/9780198715443.001.0001/acref-9780198715443-e-1109 As with metaphors, I want to stress the importance of the discursive, cultural context of an interpretation.

Symbolic system, on the other hand, describes a set of meaningful signs and their interpretations that are commonly accepted in a society (or parts of it, or in a specific discourse), at a given time in history. See U. Friedrich, *Menschentier und Tiermensch. Diskurse der Grenzziehung und Grenzüberschreitung im Mittelalter.* Historische Semantik 5 (Göttingen, 2009), 25.

Semiotics is understood as "the systematic study of signs, or, more precisely, of the production of meanings" from signs, "linguistic or non-linguistic" (C. Baldick, "Semiotics (Semiology)," in *ODL* www.oxfordreference.com/view/10.1093/acref/9780198715443.001.0001/acref-978019 8715443-e-1036").

Semiosis is used according to the definition by C. S. Peirce, "Pragmatism," in *The Essential Peirce: Selected Philosophical Writings*, vol. 2 *(1893–1913)*, ed. N. Houser, A. De Tienne, J. R. Eller, A. C. Lewis, C. L. Clark and D. Bront Davis (Bloomington and Indianapolis, 1998), 411 as a process of generating meaning via signs, "an action, or influence, which is, or involves, a cooperation of *three* subjects, such as sign, its object and its Interpretant".

2 For the prominence of the symbolic perspective towards the material reality in the Middle Ages, see M.-D. Chenu, *La Théologie au douzième siècle* (Paris, 1957), 161, on the 12th-century Latin West. He specifically refers to views held by elites ("écolâtres et mystiques, exégètes et naturalistes, profanes et religieux, écrivains et artistes").

3 See the important study on animals as parts of the symbolic system of Late Antique and Early/Middle Byzantine Christian discourse by H. Maguire (*Earth and Ocean: The Terrestrial World in Early Byzantine Art* (University Park and London, 1987)), as well as in secular culture by E. Dauterman Maguire and H. Maguire (*Other Icons: Art and Power in Byzantine Secular Culture* (Princeton, 2007)) and S. Lazaris' study on the illustrations in the Physiologus (*Le Physiologus grec. Vol. 2, Donner à voir la nature.* Micrologus Library 107 (Florence, 2021)). For representations of animals on jewelry and court costume see B. Popović, "Imperial Usage of Zoomorphic Motifs on Textiles The two-headed Eagle and the Lion in Circles and Between Crosses in the Late Byzantine Period," *IKON* 2 (2009): 127–36 and A. Bosselmann-Ruickbee, *Byzantinischer Schmuck des*

9. *bis frühen 13. Jahrhunderts: Untersuchungen zum metallenen dekorativen Körperschmuck der mittelbyzantinischen Zeit anhand datierter Funde.* Spätantike, frühes Christentum, Byzanz. Reihe B, Studien und Perspektiven, 28 (Wiesbaden, 2011), 136–37.

4 On animal imagery in Byzantine textual and visual sources see the important collection of articles in I. Anagnostakis, T. Kolias, and E. Papadopoulou (eds.), *Ζώα και περιβάλλον στο Βυζάντιο (7ος– 12ος αι.)*. National Hellenic Research Foundation. Institute for Byzantine Research. International Symposium, 21 (Athens, 2011); specifically on the socio-political discourse T. Schmidt, *Politische Tierbildlichkeit in Byzanz: Spätes 11. bis frühes 13. Jahrhundert.* Mainzer Veröffentlichungen zur Byzantinistik 16 (Wiesbaden, 2020), and the chapter by K. Stewart in the present volume.

5 See the old but still rather up-to-date literature review of A. Fourlas, "Adler und Doppeladler. Materialien zum »Adler in Byzanz« – Mit einem bibliographischen Anhang zur Adlerforschung," in *Philoxenia. Prof. Dr. Bernhard Kötting gewidmet von seinen griechischen Schülern*, ed. A. Kallis (Münster, 1980), 97–120.

6 See Schmidt, *Tierbildlichkeit*, 75–122.

7 See T. Buquet, "Le guépard médiéval, ou comment reconnaître un animal sans nom," *Reinardus* 23 (2010–2011): 12–47; N. Nicholas, "A Conundrum of Cats: Pards and their Relatives in Byzantium," *GRBS* 40 (1990): 253–98.

8 See T. Kolias, "The Horse in the Byzantine world," in *Le Cheval dans les Sociétés Antiques et Médiévales*, ed. S. Lazaris (Turnhout, 2012), 87–98; S. Lazaris, "Rôle et place du cheval dans l'antiquité tardive: questions d'ordre économique et militaire," in *Ζώα και περιβάλλον στο Βυζάντιο*, ed. Anagnostakis, Kolias, and Papadopoulou, 245–72.

9 See A. Rhoby, "Hunde in Byzanz," in *Lebenswelten zwischen Archäologie und Geschichte: Festschrift für Falko Daim zu seinem 65. Geburtstag*, vol. 2, ed. J. Drauschke et. al. Monographien des Römisch-Germanischen Zentralmuseums 150, 2 (Mainz, 2018), 807–20; T. Schmidt, "Noble Hounds for Aristocrats, Stray Dogs for Heretics," in *Impious Dogs, Haughty Foxes and Exquisite Fish. Evaluative Perception and Interpretation of Animals in Ancient and Medieval Mediterranean Thought*, ed. T. Schmidt and J. Pahlitzsch (Boston and Berlin, 2019), 103–32.

10 See B. N. Blysidou, "Ο χοίρος ως σύμβολο ευδαιμονίας του βυζαντινού ανθρώπου," in *Ζώα και περιβάλλον στο Βυζάντιο*, ed. Anagnostakis, Kolias, and Papadopoulou, 39–50.

11 See K. Tsaka, "Παρατηρήσεις στις απεικονίσεις του δράκοντος σε παραστάσεις της μεσοβυζαντινής τέχνης," in Ανταπόδοση. Μελέτες προς τιμήν της Ελένης Δεληγιάννη-Δωρή (Athens, 2010), 473–94; D. Ogden, "The Function of Dragon Episodes in Early Hagiography," in *Animal Kingdom of Heaven. Anthropozoological Aspects in the Late Antique World*, ed. I. Schaaf (Berlin and Boston, 2019), 35–58; M. White, "The Rise of the Dragon in Middle Byzantine Hagiography," in *Byzantine and Modern Greek Studies* 32.2 (2008): 146–79.

12 Even the *Cultural History of Animals in the Medieval Age*, ed. B. Resl (Oxford and New York, 2007) does not consider Byzantium. For the medieval west see the studies by U. Friedrich, *Menschentier*, J. E. Salisbury, *The Beast within. Animals in the Middle Ages* (New York and London, 1994) and J. Voisenet, *Bestiaire Chrétien: L'imagerie animale des auteurs du Haut Moyen Âge (Ve-XIe siècles)* (Toulouse, 1994).

13 M. Norris, *Beasts of the Modern Imagination: Darwin, Nietzsche, Kafka, Ernst, and Lawrence* (Baltimore, MD, 2019), 17.

14 See F. Middelhoff and S. Schönbeck, "Coming to Terms: The Poetics of More-than-human Worlds," in *Texts, Animals, Environments: Zoopoetics and Ecopoetics*, ed. F. Middelhoff, S. Schönbeck, R. Borgards, and C. Gersdorf (Freiburg i. Br., Berlin, and Vienna, 2019), 11–40, at 15; K. Driscoll and E. Hofmann, "Introduction: What is Zoopoetics?" in *What is Zoopoetics?: Texts, Bodies Entanglement*, ed. K. Driscoll and E. Hoffmann (Cham, 2018), 1–14, at 4; S. Schönbeck, "Return to the Fable: Rethinking a Genre Neglected in Animal Studies and Ecocriticism," in *Texts, Animals, Environments: Zoopoetics and Ecopoetics*, ed. Middelhoff, Schönbeck, Borgards, and Gersdorf, 111–26, at 111–12; A. J. Goldwyn, *Byzantine Ecocriticism: Women, Nature and Power un the Medieval Greek Romance* (Cham, 2018), 142–43.

15 See R. Borgards, *Tiere: Kulturwissenschaftliches Handbuch* (Stuttgart, 2016), 237; D. Harraway, *When Species Meet* (Minneapolis and London, 2008), 4.

16 For images and symbols as keys to underlying concepts (*l'univers mental*) see J. Le Goff, *L' imaginaire médiéval: essais* (Paris, 1985), viii.

17 "οὐ γεωργίαν ἐπιτηδεῦόν [...] ἢ γάρ ἀγρεύοντες τὰ τῆς ἐρήμου ζῷα διαζῶσι σαρκοφαγοῦντες [...] τότε τοῖς ὑποζυγίοις – κάμηλοι δέ εἰσι δρομάδες – πρὸς ἐδωδὴν κατακέχρηνται, θηριώδη καὶ αἱμοβόρον ζῶντες βίον, [...] καὶ θέρμη πυρὸς ὀλίγη σαρκῶν χαυνοῦντες τὸ εὔτονον, ὡς μόνον εἴκειν μὴ πρὸς πολλὴν βίαν ἕλκουσι τοῖς ὀδοῦσιν, ὡς εἰπεῖν, τρέφονται κυνικῶς [...]." Ps.-Neilos of Ankyra, 3.1, ed. M. Link, *Die Erzählung des Pseudo-Neilos. Ein spätantiker Märtyrerroman* (Leipzig, 2005), 35:24–36:4; for an English trans., dating, and discussion, see D. F. Caner, *History and Hagiography from the Late Antique Sinai* (Liverpool, 2010), 73–83 and 94–95. The author may have lived in the Negev desert. The text likely draws on empirical knowledge, but also on literary tradition. See, for instance, Ammianus Marcellinus' description of the Saracens' nomadism (Ammianus Marcellinus, *Res Gestae*, XIV.4.1–7, ed. É. Galletier and J. Fontaine, *Ammian Marcellin, Histoire*, vol. 1 (Paris, 1968), 68–69). On this passage see also A. Kaldellis, *Ethnography after Antiquity: Foreign Lands and Peoples in Byzantine Literature* (Philadelphia, PA, 2013), 63.

18 See Michael Psellos, *Chronography*, 7.68, ed. D. R. Reinsch, *Michaelis Pselli Chronographia*, 2 vols. Millennium-Studien 51 (Berlin and Boston, 2014), vol. 1, 240:13–241:31; Niketas Choniates, *Or. et Ep.*, no. 2, ed. J. L. Van Dieten, *Nicetae Choniatae Orationes et Epistulae*. CFHB 3 (Berlin and New York, 1972), 8:20–26; Niketas Choniates, *History*, ed. J. L. Van Dieten, *Nicetae Choniatae Historia*, 2 vols. CFHB 11.1–2 (Berlin and New York, 1975), 94:85–86; George Pachymeres, *History*, V.4, ed. A. Failler and trans. V. Laurent, *Georges Pachymérès: Relations Historiques*, 5 vols (Paris, 1984–2000), vol. 2, 447.

19 *Ps.-Neilos of Ankyra* 3.4–7, ed. Link 37:1–38:17. For this contrast see Kaldellis, *Ethnography*, 63.

20 On the origins of the concept of the "barbarian" in Greek culture in 5th c. BCE, see E. Hall, *Inventing the Barbarian: Greek Self-Definition through Tragedy* (Oxford, 1989), 1–3. For the concept in Byzantine times, see G. Page, *Being Byzantine: Greek Identity before the Ottomans* (Cambridge, 2008), 43–46 and, with particular regard to animality, Schmidt, *Tierbildlichkeit*, 63–71. For the development of the concept in the western Middle Ages, again with particular emphasis on the aspect of animality, see Friedrich, *Menschentier*, 83–100.

21 Although archaeological evidence has proven the occasional consumption of horse meat in Byzantium (H. Kroll, *Tiere im Byzantinischen Reich. Archäozoologische Forschungen im Überblick*. Monographien des Römisch-Germanischen Zentralmuseums 87 (Mainz, 2010), 169–70; her chapter in the present volume (Baron, "More than Food") and Kolias, "Horses," 93), elite sources such as texts by the 12th-century courtier Theodoros Prodromos (*Historische Gedichte*, no. 19:26 ed. W. Hörandner. Wiener Byzantinistische Studien 9 (Vienna, 1974), 311) accept this "abominable" practice only in situations of utmost peril. On the rejection of blood, which, naturally, was not strictly observed but frequently condemned, see B. Caseau, *Nourritures terrestres, nourritures célestes: La culture alimentaire à Byzance*. Centre de Recherche d'Histoire et Civilisation de Byzance, Monographies 46 (Paris, 2015), 59–74.

22 On the horse, see *Ps-(?) Ioannes Chrysostomos, ad orationem in illud. Sufficit tibi gratia mea, etc.*, ed. J. P. Migne, *PG* 59 (Paris, 1862), 507–16, at 513; *Basilius von Caesarea: Homilien zum Hexaemeron*, no. 9, ch. 3, ed. E. A. de Mendita and S. Y. Rudberg. Die Griechischen Christlichen Schriftsteller der ersten Jahrhunderte, N.F. 2 (Berlin, 1997), 149:15, or Niketas Choniates, *History*, ed. Van Dieten, 141 (Emperor Andronikos I as a "ἵππος θηλυμανὴς"). For the mostly negative Christian interpretation of the camel see M. P. Ciccarese, *Animali simbolici alle origini del bestiario Cristiano*, 2 vols. (Bologna, 2002–2007), vol. 1, 219–38.

23 On attitudes toward horses in Cuman and Pecheneg cultures see V. Spinei, *The Great Migrations in the East and South East of Europe: From the Ninth to the Thirteenth Century*, 2 vols. (Amsterdam, 2006), vol. 1, 145–46; vol. 2, 335–38.

24 "komplexe Einschreibefläche für das ‚kulturelle' Selbstverständnis des Adels", Friedrich, U., "Der Ritter und sein Pferd: Semantisierungsstrategien einer Mensch-Tier-Verbindung im Mittelalter," in *Text und Kultur: Mittelalterliche Literatur, 1150–1450*, ed. U. Peters (Stuttgart, 2001), 245–67, at 266–67.

25 See J. J. Cohen, *Medieval Identity Machines*. Medieval cultures 35 (Minneapolis, MN, 2003), 63 and M. Keen, *Chivalry* (New Haven, CT, 1984), 9; 23–24.

26 "ἵπποι πωλοδαμνείσθωσαν ἰσήλικες τῷ βρέβει, / Ἀρραβικοὶ καὶ Θετταλοί, τῆς μάχης ὑπηρέται [...]." Theodoros Prodromos, *Historische Gedichte*, no. 44:69–70 ed. Hörandner, 408. See

further mentions of the Arabian horses of the emperor John II in war (poem no. 15:48, ed. Hörandner, 273).

27 On Bukephalos' sale by the horse trader Philon of Thessaly see *Plutarchos, Vitae parallelae*, 6.1, ed. C. Lindskog, K. Ziegler, and H. Gärtner (Stuttgart, 1994), 157; on Bukephalos' Thessalian background see the Byzantine etymological lexicon, the 12th-c. *Etymologicum magnum genuinum: Symeonis etymologicum una cum magna grammatica*, no. 208, ed. F. Lasserre and N. Livadaras, vol. 2 (Athens, 1992), 478 and the 13th-c. Ps.-Ioannes Zonaras (*Ps.-Iohannis Zonarae Lexicon ex tribus codicibus manuscriptis*, ed. J. A. H. Tittmann, vol. 1. (Leipzig, 1808), 398b).

28 *Herodotos, Historiae*, 7.196, ed. H. B. Rosén (Berlin and New York, 2008), 279.

29 *Aristophanis Historiae Animalium Epitome...*, II.598, ed. S. Lampros. Supplementum Aristotelicum 1,1 (Berlin, 1885), 147:1–4. For this text, see the chapter by A. Zucker in the present volume. As Zucker notes, the provenance of this passage is unknown.

30 Nikephoros Bryennios, *History*, IV.6, ed. P. Gautier, *Nicéphore Bryennios: Histoire*. CFHB 9 (Brussels, 1975), 269 mentions the Thessalian cavalry in the army of the late 11th-c. rebel Nikephoros Bryennios. It might be that this term includes the territory under the *doux* of Thessaloniki as well. See H. J. Kühn, *Die Byzantinische Armee im 10. und 11. Jahrhundert: Studien zur Organisation der Tagmata*. Byzantinische Geschichtsschreiber, Ergänzungsband 2 (Vienna, 1991), 257. P. Magdalino, "Τα χαρτουλαράτα της Βόρειας Ελλάδας το 1204," in Πρακτικά Διεθνούς Συμποσίου για το Δεσπότατο της Ηπείρου ('Αρτα, 27–31 Μαΐου 1990), ed. E. Chrysos (Arta, 1992), 31–35 points to the *chartoularaton* Ezeros (mod. Xyniai) in Thessaly, an administrative unit connected to the provision of horses and livestock for the army.

31 See Michael Italikos, "Speech to John Komnenos," ed. P. Gautier, *Michel Italikos: Lettres et discours*. Archives de l'Orient Chrétien 14 (Paris, 1972), no. 43, 264:14 and Niketas Choniates, *History*, ed. Van Dieten, 30:91 on horses as a tribute to Emperor John II by the city of Shezer in 1138. For horses as diplomatic gifts see the chapter by N. Drocourt in the present volume and idem, "Les animaux comme cadeaux d'ambassade entre Byzance et ses voisins (VIIe–XIIe siècle)," in *Byzance et ses périphéries: Hommage à Alain Ducellier*, ed. B. Doumerc and C. Picard (Toulouse, 2004), 67–93, at 68; 74–79. For Arabia as an exporter of horses see A. Hyland, *The Medieval Warhorse from Byzantium to the Crusades* (Stroud, 1994), 43. Among current breeders of Arabian horses, the origin legend of the prophet Muhammad's five favorite mares becoming the "matriarchs" of current strains of Arabian horses is rather important. Early Islamic and Medieval sources, however, "point towards a male oriented equine tradition", whereas today's well-known Arabic horse breeds can be traced back no earlier than the late Middle Ages. See H. Hettema, "Al-Khamsa: The Prophet's Mares – Or Were They Stallions?," *Cheiron* 1.1 (2021): 170–79, at 177.

32 "ἵππους τῶν εὐγενῶν ἀπό τε Δαμασκοῦ καὶ Ἐδέσης καὶ αὐτῆς Ἀραβίας." Anna Komnena, *Alexias*, XIV.2.14, ed. D. R. Reinsch and A. Kambylis. CFHB 40.1 (Berlin and New York, 2001), 434:22–23; Niketas Choniates, *History*, ed. Van Dieten, 493. See also ibid., 458 on Alexios' III mounting an "Arabian stallion" during a procession.

33 *Timarion*, ed. R. Romano, Byzantine Neo-Hellenica Neapoletana 2 (Naples, 1974), 56:193, describing their graciousness; for Arabian horses as examples of speed and grace, see also Gregorios Antiochos, "Laudatio funebris Manuelis imperatoris," in *Fontes rerum byzantinarum: Rhetorum saeculi XII orationes politicae*, no. 12, ed .V. E. Regel and N. I. Novosadskij (Leipzig, 1982), 191–228, at 198; *Ioannes Skylitzes continuatus*, ed. E. Th. Tsolakis, Ἡ Συνέχεια τῆς Χρονογραφίας τοῦ Ἰωάννου Σκυλίτση. Εταιρεία Μακεδόνικων Σπουδῶν. Ἵδρυμα Μελετῶν τοῦ Αἵμου 105 (Thessaloniki, 1968), 131.

34 Harraway, *Species*, 4.

35 See in detail Schmidt, "Hounds," 115–22.

36 For the elite hunt in Byzantium, including its representation in court poetry and artworks (e.g. precious tableware), see, for instance, A. K. Sinakos, "Το κυνήγι κατά τη μέση βυζαντινή εποχή (7ος–12ος αι.)," in Ζώα και περιβάλλον στο Βυζάντιο, ed. Anagnostakis, Kolias, and Papadopoulou, 71–86; E. Patlagean, "De la Chasse et du Souverain," in *DOP* 46 (1992): 257–63; M. Grünbart, *Inszenierung und Repräsentation der byzantinischen Aristokratie vom*

10. bis zum 13. Jahrhundert. Münstersche Mittelalter-Schriften 82 (Paderborn, 2015), 198–205; Schmidt, *Tierbildichkeit*, 196–238. See also G. Moravcsik, "Sagen und Legenden über Kaiser Basileios I," *DOP* 15 (1961), 59–126 on historiographic use of hunting episodes around Emperor Basil I (9th c.).

37 For Digenis Akritas' hunting episodes, see *Digenis Akritis: The Grottaferrata and Escorial Versions*, G, 4.73–90, ed. and trans. E. Jeffreys. Cambridge Medieval Classics 7 (Cambridge, 1998), 70–73) and G. 4, 148–212, ed. Jeffreys, 74–78.

38 See *Eustathii Thessalonicensis opera minora. Magnam partem inedita*, ed. P.Wirth. CFHB 32 (Berlin and New York, 2000), 117, likely inspired by *Digenis Akritis* G, 4.73–90. See also Theodoros Prodromos, *Historische Gedichte*, no. 44:82–83, ed Hörandner, 408, comparing a young aristocrat with Achill being fed by hinds and lions.

39 For a global (Eurasian) perspective on hunting as a marker of aristocracy and royalty, see T. T. Allsen, *The Royal Hunt in Eurasian History* (Philadelphia, 2006), esp. 120–31.

40 See C. Faraone, *Talismans and Trojan Horses: Guardian Statues in Ancient Greek Myth and Ritual* (New York, 1992), 3–53; F. B. Flood, "Image against Nature: Spolia as Apotropaia in Byzantium and the dār al-Islām," *The Medieval History Journal* 9.1 (2006): 143–66.

41 A. Griebeler, "The Serpent Column and the talismanic ecologies of Byzantine Constantinople," *BMGS* 44.1 (2020): 86–105, at 89. Other non-apotropaic animal-shaped talismans also had prognostic powers.

42 "ὃν εἶπέ τις ἂν ἰδὼν ἐπιλελησμένον τῶν ἑλίξεω ἑλίξεων καὶ τοῦ δάκνειν ἐς ὄλεθρον καὶ τοὺς λοιποὺς Βυζαντίους ὄφεις τῷ καθ᾽ αὑτὸν διασοβεῖν ὑποδείγματι καὶ πείθειν ταῖς χειαῖς συσπειρᾶσθαι καὶ παραβύεσθαι." Niketas Choniates, *History*, ed. Van Dieten, 651:50–53. See also the famous, still standing, serpent column in the Hippodrome, reported as a cure against snake bites by travelers to Constantinople in the 14th century. Griebeler, "Serpent Column," 98.

43 See *Patria Konstantinoupoleōs*, 3.24 and 200, ed. Th. Preger, *Scriptores Originum Constantinopolitanarum*, vol. 2 (Leipzig, 1907), 221 and 278.

44 For underlying ideas such as the principle of *sympatheia* and *antipatheia* see Griebeler, "Serpent Column," 92.

45 For the use of animal products, see in this volume the chapter by H. Baron, and for animals as food, see the chapter by Kokoszko and Rzecznica.

46 See *An Entertaining Tale of Quadrupeds*, trans. N. Nicholas and G. Baloglou (New York and Chichester, West Sussex, 2003), 176–8. See also the dog threatening to skin the fox and bring its hide to the tanner, while defining its value due to being "raised in the midst of men: in royal courts" and as a hunter by its human masters, ibid., 172, 250. For the *Tale*, see also the chapter by K. Stewart in this volume.

47 "[…] σκυτοτρώκτῃ καὶ διφθερίᾳ […]" Leon Diakonos, *History*, IV.5B, ed. C. B. Hase, *Leonis Diaconi Caloënsis Historiae libri decem et liber de velitatione bellica Nicephori Augusti*. CSHB 11 (Bonn, 1828), 62:5. See also Liutprand of Cremona who reported that the same Nikephoros described the Saxons as poor wearers of fur. *Liudprandi Cremonensis Antapodosis. Homelia Paschalis. Historia Ottonis. Relatio de legatione Constantinopolitana*, ch. 53, ed. P. Chiesa. Corpus Christianorum. Continuatio mediaevalis 156 (Turnhout, 1998), 865–66.

48 For leather-gnawing as a metaphor of poverty and leather production see Michael Choniates, *Sozomena*, ed. S. Lampros, *Μιχαήλ Ακομινάτου του Χωνιάτου. Τα σωζόμενα*, 2 vols (Athens, 1879–80), vol. 1, 53. For chewing as a process in leather production in prehistoric and ancient Europe as well as practiced by modern-day Eskimos see R. J. Forbes, *Studies in Ancient Technology*, vol. 5 (Leiden, 1966), 6; 20.

49 M. Mullett, "Originality in the Byzantine Letter: The Case of Exile," in *Originality in Byzantine Literature, Art and Music. A Collection of Essays*, ed. A. R. Littlewood (Oxford, 1995), 39–58. See also I. Stouraitis, "Roman identity in Byzantium: a critical approach," *BZ* 107.1 (2014): 175–220, at 199–200. The "Bulgarian provinces" were re-conquered by Emperor Basil II in the first quarter of the 11th century.

50 Theophylact of Ochrid, Letters, no. 4:55–62, ed. and trans. P. Gautier, *Théophylacte d'Achrida: Lettres*, CFHB 16.2 (Thessaloniki, 1986), 141.

51 See Gregorios Antiochos, "Letter," in J. Darrouzès, "Deux lettres de Grégoire Antiochos écrites de Bulgarie vers 1173," *BSI* 23 (1962): 276–84, here 279:25–28. On the singing bird, see Niketas Choniates, *Or. et Ep.*, no. 2, ed. Van Dieten, 18, *Eustathii Thessalonicensis opera minora*,

ed. Wirth, 142 and M. Leontisi, "Οικόσιτα, ωδικά και εξωτικά πτηνά. Αισθητική πρόσληψη και χρηστικές όψεις (7ος–11ος αι.)," in *Ζώα και περιβάλλον στο Βυζάντιο*, ed. Anagnostakis, Kolias, and Papadopoulou, 285–317, at 295; 298–99.

52 See U. Albarella, and R. Thomas, "They Dined on Crane: Bird Consumption, Wild Fowling and Status in Medieval England," *Acta zoologica cracoviensia* 45 (2002), 23–38: at 36.

53 For moral connotations and status display in consumption habits see the chapter by M. Kokoszko and Z. Rzeźnicka in this volume, as well as M. Grünbart, "Spartans and Sybarites at the Golden Horn: Food as Necessity and/or Luxury," in *Material Culture and Well-Being in Byzantium (400–1453)*, ed. M. Grünbart, E. Kislinger, A. Muthesius, and D. Stathakopoulos. Denkschriften. Österreichische Akademie der Wissenschaften, Philosophisch-Historische Klasse, 356; Veröffentlichungen zur Byzanzforschung 11 (Vienna, 2007), 135–39. Especially for birds and poultry as a valuable food at aristocratic and imperial tables see Leontisi, "Οικόσιτα," 213–15.

54 Wolfram von Eschenbach, *Parzival,* ed. K. Lachmann and B. Schirok, trans. P. Knecht (Berlin and New York, 2003), 131:24–132:3.

55 See S. Obermaier, "You Are the Animal That You Eat," in *Impious Dogs, Haughty Foxes and Exquisite Fish. Evaluative Perception and Interpretation of Animals in Ancient and Medieval Mediterranean Thought*, ed. T. Schmidt and J. Pahlitzsch (Boston and Berlin, 2019), 133–64, at 139 and B. Nitsche, "Die literarische Signifikanz des Essens und Trinkens im Parzival Wolframs von Eschenbach. Historisch-anthropologische Zugänge zur mittelalterlichen Literatur," *Euphorion* 94 (2000): 245–70, at 255–57.

56 For the large field of study on food regulations in Byzantium see, with a focus on the religious discourse, Caseau, *Nourritures*. On the morale of food consumption in Byzantium, including its "animalistic" connotations, see T. Labuk, *Gluttons, Drunkards and Lechers. The Discourses of Food in 12th-Century Byzantine Literature: Ancient Themes and Byzantine Innovations*. PhD. Dissertation (Katowice, 2019), 31, 35–36.

57 See in general D. Peil, *Untersuchungen zur Staats- und Herrschaftsmetaphorik in literarischen Zeugnissen von der Antike bis zur Gegenwart* (Munich, 1983), 29–35. On Byzantium see overviews in F. Dvornik, *Early Christian and Byzantine Political Philosophy. Origins and Background*, 2 vols. (Washington, D.C., 1966), vol. 1, 266–68 and Schmidt, *Tierbildlichkeit*, 166–69.

58 On God and Jesus as shepherds see Ps. 73.1LXX; Jer. 13.17; Jo.10.1–21.

59 Theodoros Studites, *Letters*, no. 61, ed. G. Fatouros, *Theodori Studitae epistulae*, 2 vols. CFHB 31.1–2 (Berlin and New York, 1992), vol. 1, 172–73; Nikephoros Gregoras, *History*, ed. L. Schopen, *Nicephori Gregorae Byzantina Historia*, 3 vols. CSHB 4; 19, 48 (Bonn 1829–30; 1855), vol. 1, 163 on the appointment of Gregory of Cyprus as Patriarch of Constantinople in 1283. On the association of the bishop and shepherd, see Apostles, 20.28. On the crozier, see also A. Papas, "Liturgische Gewänder," in *Reallexikon zur Byzantinischen Kunst*, 5, 769–72. The staff was a common symbol of authority at court, with the shepherd's staff only one of several models, see R. Macrides, J. A. Munititz, and D. Angelov, *Pseudo-Kodinos and the Constantinopolitan Court: Offices and Ceremonies* (Farnham and Burlington, VT, 2013), 336 and n. 114

60 *Ekloga: Das Gesetzbuch Leons III. und Konstantinos' V.*, ed. L. Burgmann (Frankfurt am Main, 1983), 160:21–23. For further examples of the shepherd-emperor in official documents, see H. Hunger, *Prooimion: Elemente der Byzantinischen Kaiseridee in den Arengen der Urkunden* (Vienna, 1964).100–2.

61 *Life of Cyril Phileotes*, ed. E. Sargologos, *La vie de Saint Cyrille le Philéote moine byzantin (+1110)* (Brussels, 1964), 375; Gregorios Antiochos, "Oratio ad Isaacium...," in *Fontes rerum byzantinarum: Rhetorum saeculi XII orationes politicae*, no. 18, ed. V. E. Regel and N. I. Novosadskij (Leipzig, 1982), 300–4, at 300–1.

62 "Ἐμοὶ δὲ τοῦτο γίνεσθαι δοκεῖ καὶ ἀμφοτέροις / Ὡς βασιλεῦσιν οὖσί τε καὶ ἀλαζόσιν ἅμα, / Ὀρνέων μὲν τοῦ ἀετοῦ, τοῦ λέοντος θηρῶν δέ." John Tzetzes, *Chiliades*, 5:430–32, ed. P. A. M. Leone, *Ioannis Tzetzae Historiae*. Pubblicazioni dell' Istituto di Filologia Classica. Università degli Studi di Napoli 11 (Naples, 1968), 184.

63 See, for instance, *Aischylus: Agamemnon*, ed. L. West (Stuttgart, 1991), 13–15; *Claudius Aelianus, De natura animalium*, 3.1 and 9.2, ed. M. García Valdés, L. A. Llera Fueyo, and L. Rodríguez-Noriega Guillén (Berlin and New York, 2009), 54 and 208; Oppian, *Kynegetika*, 3:7–62, ed.

M. Papathomopoulos, *Oppianus Apameensis, Cynegetica, Eutecnicus Sophistes, Paraphrasis metro soluta* (Munich and Leipzig, 2003), 56–59.

64 On the two animals' role as traditional symbols in political representation and their adoption in the Byzantine world see Schmidt, *Tierbildlichkeit*, 75–76; 123–24.

65 See Ps. 102 LXX and *Physiologus*, ed. F. Sbordone (Milan, 1936; repr. Hildesheim and New York, 1976), 3rd red., 283; See, for instance, the 12th-c. Michael Italikos, "Speech to Manuel Komnenos," ed. P. Gautier, *Michel Italikos: Lettres et discours*. Archives de l' Orient Chrétien 14 (Paris, 1972), no. 44, 276:18–20 or the 14th-c. Manuel Philes, *De proprietate animalium*, ed. F. S. Lehrs and F. Dübner, "Manuelis Philae versus iambici de proprietate animalium," in *Poetae bucolici et didactici [...]*, vol. 3 (Paris, 1862) 1–56, at 4:83–84.

66 See Gen. 49.9 LXX; Rev 5.2–5. See also T. Schmidt, "Father and son like eagle and eaglet: concepts of animal species and human families in Byzantine court oration (11th/12th c.)," *BZ* 112.3 (2019): 959–90, at 977–79.

67 See A. Iafrate, *The Wandering Throne of Solomon: Objects and Tales of Kingship in the Medieval Mediterranean*. Mediterranean art histories. Studies in visual cultures and artistic transfers from late antiquity to the modern period, 2 (Leiden and Boston, 2015), 66–105, pointing to references to Alexander the Great, Solomon, but highlighting that the underlying symbolism was part of an international language of power with rather broad meanings (101). According to her (78), not all of the animals presented in the palace (cranes, snakes, birds) had specific symbolic meaning, but might have been parts of traditional ensembles.

68 For stories on the future emperor Romanos (I) proving his value when killing a lion, or Basil I outclassing his predecessor Michael III by defeating an attacking wolf and taming the emperor's bolting horse, see Moravcsik, "Sagen," 59–126 (note that both Romanos and Basil would later usurp the throne and were in particular need of legitimation); G. Prinzing, "Historiography, Epic and the Textual Transmission of Imperial Values: Liudprand's Antapodosis and Digenis Akrites," in *Reading in the Byzantine Empire and Beyond*, ed. T. Shawcross and I. Toth (Cambridge, 2018), 336–50 at 348–49; Z. Kádár, "Die Umformung der Löwenjagd Alexanders des Großen in der koptischen und byzantinischen Machtkunst," in *JÖB* 32.5 (1982): 387–92; For the taming of beasts (and barbarians), see Schmidt, *Tierbildlichkeit*, 293–306.

69 On the occasional presence of lions in Anatolia in the Byzantine period, see A. E. Schnitzler, "Past and present Distribution of the North African-Asian Lion Subgroup: a Review," in *Mammal Review* (2011): 1–24, at 16–17. For the lion imagery in Byzantine literature, see Schmidt, *Tierbildlichkeit*, 75–122.

70 See ibid., 239–75 and the literature in footnote 11 above.

71 On the Physiologus tradition in the West, see Henkel, *Physiologus*, 21–138; A. Zucker, S. Draelantis, and S. Lazaris, "Introduction," in *RursuSpicae* 2 (2019): https://journals.openedit ion.org/rursuspicae/1047. See also the chapter by S. Lazaris in the present volume. On menageries in the west, see, for instance, on Frederick II of Hohenstaufen, M. Giese, "Die Tierhaltung am Hof Kaiser Friedrichs II. zwischen Tradition und Innovation," in *Herrschaftsräume, Herrschaftspraxis und Kommunikation zur Zeit Kaiser Friedrichs II*, ed. K. Görich, J. Keupp and T. Broekmann (Munich, 2008), 121–172 and K. Görich, "Il leone dell'imperatore Enrico VII. Domande sul contesto del dono di un animale," in *Enrico VII, Dante e Pisa. A 700 anni dalla morte dell'imperatore e dalla monarchia (1313-2013)*, ed. G. Petralia and M. Santagata (Ravenna, 2016), 45–56.

72 On animals in Byzantine gift exchange, including their symbolic implications (e.g. by their colors), see the chapter by N. Drocourt in the present volume; on menageries, see J. Theodoridès, "Les animaux des jeux de l'hippodrome et des menageries impériales a Constantinople," in *BSl* 19 (1958): 773–84; N. P. Ševčenko, "Wild Animals in the Byzantine Park," in *Byzantine Garden Culture*, ed. A. Littlewood, H. Maguire, and J. Wolschke-Bulmahn (Washington, D.C., 2002), 69–86.

73 See, for instance, Ordericus Vitalis, *Ecclesiastical History* 10.20, ed. and trans. M. Chibnall, 6 vols. (Oxford, 1969–84), vol. 5, 330:17–22. See also Niketas Choniates, *History*, ed. Van Dieten, 349:93–95.

74 See *Ps.-Kodinos*, ed. and trans. R. Macrides, J. A. Munititz, and D. Angelov, *Ps-Kodinos and the Constantinopolitan Court: Offices and Ceremonies*, (Farnham and Burlington, VT, 2013), 268:5. On animals taking part in court ceremonies at the 10th-century Caliph of Baghdad, where

Byzantine ambassadors could witness lions, elephants, and giraffes, see Ševčenko, "Wild Animals," 76. For Abbasid influence on animal-based Byzantine court ceremony, see the possible influence on the trees, mechanical singing birds and roaring lions that decorated the imperial palace in Middle Byzantine time, and apparently played a role in 10th-c. court ceremony, discussed in G. Brett, "The Automata in the Byzantine 'Throne of Salomon,'" in *Speculum* 29.3 (1954): 477–87, at 483. Iafrate, *Throne*, 86–88, however, points out that the influence could well be reversed, a "dialectic interchange".

75 See canon 51 of the synod in Trullo (691/2 AD) forbidding attendance at these spectacles, *Concilivm Constantinopolitanvm a. 691/2 in Trvllo habitvm: (Concilium Quinisextum)*, ed. H. Ohme, in collab. with R. Flogaus and C. R. Kraus (Berlin, 2013), 45. See also the 12th-c. travel account by Benjamin de Tudela, ed. and trans. M. N. Adler, *The itinerary of Benjamin de Tudela. Critical Hebrew text, English translation and commentary* (New York, 1909), 12–13.

76 See N. Drocourt, "Le cheval, animal diplomatique entre Byzance et l'Occident (IXe-XIIIe s.)," in *Byzance et l'Occident VI: Vestigia philologica*, ed. E. Egedi-Kovács (Budapest, 2021), 113–30.

77 "'Ἔδει γὰρ τὸν ὑπὲρ τὰς καμήλους μνησίκακον, τοιοῖς δε ζῴοις ἐποχηθέντα, […]." Michael Choniates, *Sozomena*, ed. Lampros, vol. 1, 239:20–21. The text is an encomium for Andronikos' successor Isaakios II. Angelos, who he successfully usurped in 1185.

78 On this interpretation see M. Perisanidi, "Byzantine Parades of Infamy through an Animal Lens," *History Workshop Journal* 90 (2020): 1–25, at 5. On the vengeful camel see *Basilius von Caesarea: Homilien*, no. 8, ed. de Mendita and Rudberg, 128. See biblical references to the camel that was not able to "go through the eye of a needle" in Math. 19:24, Mark 10:25, and Luke 18:25.

79 For the use of animals other than horses (mules, donkeys) in humiliation parades see Perisanidi, "Byzantine Parades," 1–25, at 9–11, with an interesting take on the animals' own visual, olfactory, or auditory experience during the parades.

80 The exhibition of the giraffe might even have had an impact of the depiction of giraffes in Byzantine manuscript illumination. See the chapter by N. P. Ševčenko in this volume.

81 Michael Psellos, *Orations*, no. 1 and 4, ed. G. Dennis, *Michaelis Pselli Orationes panegyricae* (Stuttgart and Leipzig, 1994), 13–14 and 61–62. See also Michael Attaleiates' description of the elephant and giraffe, discussed by by N. Drocourt in the present volume.

82 See Achmet, *Dreambook*, ed. F. Drexl, *Achmetis Oneirocriticon* (Leipzig, 1925), 220–21. The book was translated into Latin at the court of Manuel I.

83 "νωθὴς δὲ ὁ ὄνος, […], ἀτιθάσσευτος ὁ λύκος, […], δειλὸν ὁ ἔλαφος, […], εὐχάριστον ὁ κύων καὶ πρὸς φιλίαν μνημονικόν." *Basilius von Caesarea: Homilien*, no. 9, ed. de Mendita and Rudberg, 149:13–17.

84 "ἡ δὲ καρδία αὐτοῦ δόλου καὶ ἁρπαγμοῦ πλήρης ὑπάρχει." *Physiologus*, ed. Sbordone, 3rd. Greek redaction, 266.8–9; "Ὁ λέων […] ἔχει καὶ βασιλικῶν χαρίτων." Ibid., 1st red., 2.942–3. See also Aristoteles, *Historia animalium*, 488b.

85 "Κατέχετε τὸν κανόνα ὅτι κύνες οἱ ἀναίσχυντοι, ὅτι λύκοι οἱ ἅρπαγες, ὅτι ὄφεις οἱ δολεροί, […], ἵπποι θηλυμανεῖς […]." *Ps-(?) Ioannes Chrysostomos, ad orationem…*, ed. Migne, PG 59, 513.

86 On Aesop in Byzantium see M. D. Lauxtermann, *Byzantine Poetry from Pisides to Geometres: Texts and Contexts*, 2 vols. (Vienna, 2003 and 2019), vol. 1, 253–60, who highlights the richness in versions and adaptions of his fables and their accessibility to common people. While Lauxtermann stresses the popularity of Aesop in monastic circles, R. Ousterhout and A. Sitz, "Reading Aesop in Cappadocia," in *After the Text: Byzantine Enquiries in Honour of Margaret Mullett*, ed. L. James, O. Nicholson, and R. Scott (London, 2021), 259–72, at 270–72 argue for Aesop's equal popularity in secular milieus. For *Stephanites kai Ichnelates*, see J. Niehoff-Panagiotidis, *Übersetzung und Rezeption: Die byzantinisch neugriechischen und spanischen Adaptionen von Kalīla wa-Dimna* (Wiesbaden, 2003), 34–47. The text is a Greek translation of Indian, middle-Persian, and Arab material. For the Tale of the Quadrupeds and the Poulologos see the chapter by K. Stewart in the present volume.

87 Ousterhout/Sitz, "Reading Aesop," 259.

88 Ibid, 264–67; *Babrii fabulae Aesopicae*, ed. O. Crusius; *Accedunt fabularum dactylicarum et iambicarum reliquiae. Ignatii et aliorum tetrasticha iambica*, ed. C. F. Müller (Leipzig, 1897), 270, 275, 279; *Babrius and Phaedrus*, ed. and trans B. Perry (Cambridge, MA, 1985), 67, 94, 147.

89 See A. Littlewood, "Vegetal and Animal Imagery in the History of Niketas Choniates," in *Theatron. Rhetorische Kultur in Spätantike und Mittelalter*, ed. M. Grünbart. Millennium-Studien 13 (Berlin and New York, 2007), 223–58 and idem, "Imagery in the Chronographia of Michael Psellos," in *Reading Michael Psellos*, ed. C. Barber and D. Jenkins.The Medieval Mediterranean 61 (Leiden and Boston, 2006), 13–56, at 21–22.

90 With a focus on the 8th and 9th c. see N. Koutrakou, "Animal Farm in Byzantium? The Terminology of Animal Imagery in Middle Byzantine Politics and the Eight 'Deadly Sins,' " in *Ζώα και περιβάλλον στο Βυζάντιο*, ed. Anagnostakis, Kolias, and Papadopoulou, 319–77; for the 11th-13th c., see Schmidt, *Tierbildlichkeit*, 353–56.

91 See G. Schmalzbauer, "Regieren mit und ohne Waffen. Die Bienenmetaphorik in byzantinischen Fürstenspiegeln," in *Corona coronaria. Festschrift für Hans-Otto Kröner zum 75. Geburtstag*, ed. S. Harwardt and J. Schwind (Hildesheim, 2005), 305–19.

92 See I. Schaaf (ed.), *Animal Kingdom of Heaven. Anthropozoological Aspects in the Late Antique World*. Millennium-Studien 80 (Berlin and Boston, 2019).

93 See ibid.

94 See Ogden, "Dragon Episodes," 35–58; White, "The Rise of the Dragon," 149–67.

95 For a positive reading of the snake see, for example, *Physiologus*, ed. Sbordone, 3rd red., 272–75.

96 For different interpretations of dogs in Byzantium, see Schmidt, "Hounds," 103–31.

97 The notion of irrational animals was not shared by everyone, and specific texts show remarkable openness to attribute rational capabilities and even *logos* itself to animals. See idem, "Agency – A Core Concept in the Cultural History of Human-Animal Relations," in *Ecologizing Late Ancient and Byzantine Worlds*, ed. T. Arentzen and L. Borghetti (forthcoming).

98 See Maguire, *Earth*, 5–8; 17–20.

99 For animal depictions during iconoclasm see ibid., 6 and S. M. Oberhelman (ed.), *Dreambooks in Byzantium: Six Oneirocritica in Translation, with Commentary and Introduction* (Aldershot, 2008), 9.

100 *Physiologus*, ed. Sbordone, 3rd red., 274:6–18; *Cosma di Gerusalemme. Commentario ai Carmi di Gregorio Nazianzeno*, ed. G. Lozza (Naples, 2000), 153–56; Aristotle, *Historia Animalium*, 601b. See, in a similar manner, the interpretation of the turtle's tendency to leave the water for procreation as a symbol for those who "do not want to have the Lord as father and the baptismal font as mother" in 7th-c.-*Anastasius of Sinai: Hexaemeron*, ed. and trans. Kuehn and Baggarly. Orientalia Christiana Analecta 278 (Rome, 2007), 160–61, and the 12th-c. Annals by Michael Glykas, ed. I. Bekker, *Michaelis Glycae Annales*, CSHB 24 (Bonn, 1836), 104 on the deer's habit to drop its horns, paralleling Adam transgressing divine rule and forfeiting his presence in Paradise.

101 *Physiologus*, ed. Sbordone, 2nd red., 149–50. For the re-shaping of animals' physeis in the *Physiologus* to accommodate Christian teachings, see also the chapter by S. Lazaris in this volume.

102 Augustine, "Commentary on the Psalms," in *Sancti Aurelii Augustini Enarrationes in Psalmos CL-CL*, CI.1.8, ed. D. E. Dekkers and J. Fraipont (Turnhout, 1956), 1431; *Physiologus*, ed. Sbordone, 1st red., 16:1–18:2. On this type of knowledge see N. Henkel, *Studien zum Physiologus im Mittelalter*. Hermaea 38 (Tübingen, 1976), 140–41and P. Beullens, "Like a Book Written by God's Finger: Animals Showing the Path toward God," in *A Cultural History of Animals in the Medieval Age*, ed. B. Resl (Oxford and New York, 2007), 127–52, at 134.

103 Procopius, *Wars*, 7.29, ed. J. Haury and G. Wirth, *Procopii Caesariensis opera omnia*, Vol. 2, *de bellis V–VIII* (Leipzig, 1963), 424–25. On this interpretation, see J. S. Codoñer, "Der Historiker und der Walfisch. Tiersymbolik und Milleniarismus in der Kriegsgeschichte Prokops," in *Zwischen Polis, Provinz und Peripherie. Beiträge zur byzantinischen Geschichte und Kultur*, ed. L. M. Hoffmann and A. Monchizadeh. Mainzer Veröffentlichungen zur Byzantinistik 7 (Wiesbaden, 2005), 37–58.

104 See, for instance, Constantine Manasses, Chronicle, ed. O. Lampsidis, *Constantini Manassis Breviarium Chronicum*, 2 vols. CFHB 36,1–2 (Athens, 1966), vol. 1, 4588 on Leon V's iconoclastic policy figured as a devouring dragon, well known from biblical imagery such as *Bel kai drakon* 23.

105 See, for instance, 12th-c. Theodoros Prodromos, *Historische Gedichte*, no. 64a and 66, ed. Hörandner, 479 and 504 comparing deceased aristocrats to lions sleeping with open eyes. Schmidt, "Father," 983–84; *Physiologus*, ed. Sbordone, 2nd red., 151.

106 On dreams and dream interpretation in Byzantium see M. K. Papathanassiou, "The Occult Sciences in Byzantium," in *A Companion to Byzantine Science*, ed. S. Lazaris. Brill's companions to the Byzantine world 6 (Leiden and Boston, 2020), 464–95, at 468–69. On the Byzantine dreambooks see Oberhelman, *Dreambooks*, 1–58.

107 See Ps.-Nikephoros, *Dreambook*, ed. G. Guidorizzi, *Pseudo-Niceforo: Libro dei sogni*. Koinonia 5 (Naples, 1980), 57:58 and 77:45. On its dating see Oberhelman, *Dreambooks*, 6–9.

108 "ἐὰν ἴδῃ τις, ὅτι εὗρεν ἢ εὐπόρησεν ἀετόν, εἰ μὴν ἐστι βασιλεύς, ἕτερον βασιλέα κατάσχῃ, εἰ δὲ τοῦ κοινοῦ λαοῦ ὁ ἰδών, βασιλεύσαι." Achmet, *Dreambook*, ed. Drexl, 231:9–11.

109 See also the attribution in some manuscripts to "Achmet, the son of Sereim, the dream interpreter of the Caliph Mamoun", probably by Greek copyists (see M. Mavroudi, *A Byzantine Book on Dream Interpretation: The Oneirocriticon of Achmet and its Arabic Sources*. The Medieval Mediterranean 36 (Leiden, Boston and Cologne, 2002), 32–35; for the labels of Persian and Egyptian knowledge that the Greek compiler probably took over from Arabic sources, and his rendering of Muslim religious interpretations as "Indian" to make them more exotic and acceptable to the Byzantines, see ibid., 32–41) as well as the attribution of another book to the "Persian" Astrampsychus (see Oberhelman, *Dreambooks*, 2, 24).

110 See Oberhelman, *Dreambooks*, 50–58; See also B. Krönung, *Gottes Werk und Teufels Wirken. Traum, Vision, Imagination in der frühbyzantinischen monastischen Literatur*. Millennium-Studien 45 (Boston and Berlin, 2014), esp. 49–61.

111 See Palladius, *The Lausiac History*, 18, ed. C. Butler (Hildesheim, 1967), 49–51; *Life of Symeon Stylites the Younger*, ed. and trans. P. Van den Ven, *La vie ancienne de S. Syméon Stylite le jeune*, 2 vols. (Brussels, 1962–70), vol. 1, 161–62; Krönung, *Gottes Werk*, 189, 240. On the boar in God's vineyard see Ps. 79.14LXX.

112 See the list of the items in the imperial travel library, written at the court of Emperor Constantine VII (10th c.) in Constantine Porphyrogennetos, *Three Treatises on Imperial Military Expeditions*, ed. J. Haldon. CFHB 28 (Vienna, 1990), 106. Achmet's dreambook was translated into Latin at the court of Manuel I. See Mavroudi, *Book*, 116.

113 "Φασὶ δὲ ὡς ὄναρ ὁ βασιλεὺς Ἰωάννης θεάσαιτο τὸν νεοστεφῆ υἱὸν τὸν Ἀλέξιον θηρίῳ ἔποχον λέοντι [...], μηδενὸς ἑτέρου ὑπόντος ἐπιτηδείου πρὸς διεξαγωγὴν τοῦ θηρός." Niketas Choniates, *History*, ed. Van Dieten, 17:32–34.

114 "ἐὰν ἴδῃ τις, ὅτι ἐπωχήσατο λέοντι ὑποτασσομένῳ, [...] βασιλεύσει [...]." Achmet, *Dreambook*, ed. Drexl, 219:4–6.

115 Theophanes continuatus, 22, ed. and trans. M. Featherstone and J. S. Codoñer, *Chronographiae quae Theophanis continuati nomine fertur libri I-IV*. CFHB 53 (Boston and Berlin, 2015), 56. For versions of this story see W. Brandes, "Kaiserprophetien und Hochverrat: Apokalyptische Schriften und Kaiservaticinien als Medium antikaiserlicher Propaganda," in *Endzeiten. Eschatologie in den monotheistischen Weltreligionen*, ed. W. Brandes and F. Schmieder. Millennium-Studien 16 (Berlin and New York, 2008), 157–200, at 186–87.

116 See T. Teoteoi, "L'aigle impérial en tant que motif littéraire et source d'histoire comparée," in *Études byzantines et post-byzantines* 4 (2001): 181–97, at 183–84; M. Grünbart, "Traditions and Practices in the Medieval Eastern Christian World," in *Prognostication in the Medieval World: A Handbook*, ed. M. Heiduk, K. Herbers, and H.-C. Lehner, vol. 1 (Berlin and Boston, 2021), 446–52, at 450.

117 See Niketas Choniates, *History*, ed. Van Dieten, 251–52. For the techniques of avimancy see S. Rapisarda, "Traditions and Practices in the Medieval Western Christian World," in *Prognostication in the Medieval World: A Handbook*, ed. M. Heiduk, K. Herbers, and H.-C. Lehner, vol. 1 (Berlin and Boston, 2021), 429–45, at 439. For the association of birds with prognostic capabilities, e.g. to foresee the weather, see also Leontisi, "Οικόσιτα," 290.

118 "Symbolische Ordnungen sind [...] Produkte (Konstrukte) rivalisierender Sinnbedürfnisse." Friedrich, "Ritter," 246.

2

TIMOTHEUS OF GAZA AND THE *ZOOLOGICAL COLLECTION* OF CONSTANTINE VII

Two Byzantine Treatises

Arnaud Zucker

Introduction

Byzantine culture is marked, at least among the intellectual elites, by profound identity tensions. While a major part of the literary environment of formation (*paideia*) is constituted by the pre-Roman heritage and defined by the classical ideal, the spiritual and political framework is that of the Christian 'Roman' empire. This form of schizophrenia affects its intellectual and scientific commitments. As for naturalist and in particular zoological literature, which no longer has anything to do with investigation or with any research program, the least we can say is that it is not brilliant. It seems, basically, disenchanted. The drift, if not the shipwreck, of naturalist discourse between the 6th and 14th centuries in the Byzantine world can be at least partly attributed to the Byzantines' 'loss of faith' in nature, and to a form of civilizational sadness. Immanence, or the completeness of the material world, is no longer appealing. It seems that natural beings have lost their inspiring power and their enigmatic charm. The mystery has left all creatures to reside entirely in the divine miracle.

This impoverished view of the world directs zoological discourse towards a seemingly paradoxical destination: the production of surprise and the reduction of nature to its extraordinary limits. But these two apparently opposite characteristics (disenchantment and thirst for the marvelous) are in fact closely correlated. The natural world, through the lens of Christian doctrine, becomes a page on which is written the divine message: its signifiers are innumerable, but its signifieds are reduced and pre-determined. As a result, zoological discourse seems to walk askew, if not backward, in relation to the ancient path opened by the natural philosophers.

The animal stories, made up of investigations and descriptions, gave birth in Byzantium to theological signs that express, each in its own way, a unique and unified program. The

DOI: 10.4324/9781003055877-4

turning point in the discourse was achieved by the *Physiologus*, from the 2nd–3rd centuries onwards, and the effectiveness of this formula, a cross between fable and evangelical parable, is demonstrated by its overwhelming success.[1] The main purpose of references to animal models in literature is to arouse admiration and amazement at the inventiveness of divine providence. Nature is a fable (*muthos*) about divine action, whose solution (*lusis*) is to be sought in God.

What the surprise opens up, through the diverse and sometimes extravagant natures of the animals, should not be closed by a (mechanical) explanation. This is why animal characteristics and behaviors are not scrutinized and dissected in order to uncover a physiological or psychological explanation but rather are maintained in their puzzling originality and preserved in their mystery, which embodies the divine will. Every creature reveals the action of God, but their very natures are in principle as impenetrable as the ways of the Lord. To expose the details of natural mechanics would risk sapping its divine aspect.

Nature is a cryptic setting whose general key has been 'revealed', and this key (God) allows us to understand it as a whole. This priority given to the meaning/signification over the empirical is a fundamental aspect of the conception and perception of the world during the Christian 'Byzantine' period. It is not only a kind of *filter* or a particular relationship to reality, but both evidence and a principle that impacts the very experience and knowledge of the world.

The regularity of physical phenomena manifests the living permanence of divine omnipotence and intelligence, and these true laws that He imposed on the cosmos are the object of special attention: astronomy is certainly the natural science that most honors this authority, and its place is major in Byzantine science, even though the scientists think they really know enough about the order and laws of the universe; in comparison, zoology exists only in an incidental or illustrative way. For more than a millennium, going back to the beginning of the Christianization of the empire, no theoretical work was written on zoology or on the various problems raised by Aristotle and constituting today its sub-disciplines (anatomy, physiology, genetics, ecology, etc.). Theology, philosophy, and history were the only serious disciplines in the Byzantine period, and Aristotle in his time did not manage to bring zoology into philosophy. The 'animal world', if one can risk this term, is described as a theater, occupied by diverse characters among which permanently popular species stand out, who concentrate the attention of the authors by an original profile. The animal natures are individual and particular, and it is these *physeis* that best manifest the divine commands; and in the general literature, these natures are figures, which signify moral and spiritual principles, and allow us to question the social interrelations of men. In Byzantine times, living beings are thus supposed to be worth knowing only as economic resources, medicine stock, and theological proof of God's providence.

Apart from a few appearances of exotic species in historical texts, and the homiletic tradition, including the *Physiologus*, animals are the subject of almost no scientific-literary texts. Even if many works are lost and our knowledge is limited, the discovery of an unpublished text would be unlikely to alter profoundly our perception. The 'theological' sheath of scholarly discourse and the normative coding of nature is so deep and powerful that it is almost a 'wonder' that some works escape this ideological alignment. This is, however, the case for two documents composed in a Christian milieu, but which make no reference to the theological perspective and are more or less explicitly inscribed in the pre-Christian or non-Christian naturalist tradition.

The only original work we know of for the entire Byzantine period is a treatise by Timotheus of Gaza, which has come down to us in two formats, both incomplete.[2] The other major document for the period is a 10th-century zoological compilation, allegedly commissioned by Constantine VII Porphyrogenitus, but which contains no texts later than the time of Timotheus. Its composition testifies to the absence of a properly Byzantine tradition in zoology. We present here the two most important works of non-physiological and non-catechetical zoological literature of the Byzantine period. Both are steeped in ancient knowledge and make no reference to creation or the creator.

Timotheus' *De Animalibus*

An original work, both literary and scientific, whose title is poorly preserved, and which prudence leads us to render by default simply as *De Animalibus*, is attributed to Timotheus of Gaza (fifth-sixth century). It is partially saved in two recensions: the first contains an epitome in 52 chapters of a part of the work;[3] the second preserves extracts of 18 chapters in an anthology of the tenth century (*Zoological Collection of Constantin VII*).[4] The only information on this author and his work is a brief note in the *Suda*: "Timotheos, of Gaza. Grammarian. He lived under the emperor Anastasius, for whom he wrote a tragedy on the tax that was known as *chrysarguros*. He also wrote with expertise on the four-footed animals in India, Arabia, and Egypt, and on the creatures of Libya; [he wrote] also on unusual foreign birds and snakes (in 4 books)."[5]

The author, who seems to have done important work as a grammarian,[6] lived in the 5th–6th century during the reign of Anastasius I Dicoros, the Silentiary (491–518),[7] and he was probably born in Gaza around 460.[8] Despite the adverb ἐπικῶς (sic) employed by the *Suda*, which sometimes has the meaning of the almost homophonous adverb ἐπιεικῶς (scholarly),[9] this was surely a prose treatise, as suggested by the fact that (1) all the witnesses to the text are in prose and (2) there is an epitome of it (not a paraphrase, as with poetic texts), which shows no trace of poetic rhythm or vocabulary.[10] The stylistic quality of the witnesses preserved in the *Zoological Collection* makes it possible to consider them as true excerpts and corroborates this conclusion. Indeed, we note the revealing presence of numerous figures, forms of redundancy, and certain stylistic tics (such as the disproportionate use of ἀμέλει) as well as an obvious search for rhythmic sequences.[11] The style is entirely consistent with a form of rhythmic prose typical of Gazan literary circles.[12] A manuscript of the epitome (codex Barrocianus 50.2, f.354 r) actually gives the text the following title: Τιμοθέου γραμματικοῦ Γάζης περὶ ζώων τετραπόδων καὶ φυσικῶν ἐνεργειῶν θαυμαζομένων ποιητηκῶς (sic) αὐτοῦ καλλιεποῦντος[13] ("From the grammarian Timotheus of Gaza's *On the quadrupedal animals and their admirable natural powers*, poetically written in a noble style").

The treatise probably dealt only or mainly with animals of the air and the earth. The only categories named in the *Suda* or by indirect witnesses are quadrupeds, snakes, and birds. The two surviving recensions corroborate this limited focus. It is, moreover, unlikely that the treatise was concerned with a restricted geographical area or with wild animals alone. The toponymic references are very rare in the two recensions, and the repertoire of animals described by Timotheus includes familiar animals such as fox, hedgehog, mouse, donkey, weasel, and certain domestic animals such as sheep, dog, and mule. The two recensions, although partial, also support the title given in the epitome, which seems more relevant and reliable than the one given by the *Suda*, especially since the latter erroneously qualifies his lampoon, against the tax named *chrysargyros*, as a 'tragedy'.[14]

The Two Recensions of the Treatise

The first recension (*Epit. Tim.*), which is represented by several similar manuscripts, is an epitome divided into 52 chapters (ca. 5000 words).[15] It contains interpolations, one of which allows us to date the probable time of the writing of the epitome to the middle of the eleventh century, the time of Constantine IX Monomachos (emperor from 1042 to 1055):[16] "This animal has been seen in our day: two animals imported from India were also regularly presented at a spectacle for the people by the emperor [Constantine] Monomachos in the theater of Constantinople." The second recension (*Syll. Const.*) consists of a series of 18 monographic sections, inserted in a Byzantine encyclopedia, the *Zoological Collection of Constantine Porphyrogenitus* (10th century AD).[17] This unique and rich witness (ca. 4800 words), offers long textual units on the wolf, the boar, the mole, etc. and on a number of exotic animals such as the camel, the jackal, the panther, etc. Timotheus's brilliant treatment of certain extraordinary characters and the folkloric and magical uses of these animals perhaps explains the *Suda*'s emphasis on exotic fauna, as well as the fact that his contemporary, John of Gaza, refers to him as "the man who wrote on Indian animals" (ὁ γράψας περὶ ζῴων Ἰνδικῶν).[18]

Each of the chapters in *Epit. Tim.*, titled in the manuscripts with the name of an animal, succinctly lists characteristics (which are systematically and conventionally introduced by the word ὅτι, meaning "[the author says] that" or "[I read] that"), up to 14 items in some chapters. Sixteen animals are common to both reviews, allowing for a precise comparison. There is a perfect correspondence, both for the content and for the order of the information given on each animal, in 11 cases (giraffe, panther, leopard, cheetah, jackal, cat, mouse, weasel, marten, mole, and wild boar), a reversal of two successive notations for three animals (wolf, fox, camel) and a reversal involving several units for two animals (hyena, hedgehog). The fidelity is thus remarkable, and it seems reasonable to consider when there are discrepancies between them, that for each chapter the sequence of remarks proposed by the series of extracts of the *Zoological Collection* is more reliable than the series of the epitome. We can measure the correspondence and the difference between the recensions by an example, about the fox: "The fox has a character that can never be docile or pleasant to deal with, and it can never establish friendly relations with humans: it will always remain unapproachable and never short of deceptions" (*Syll. Const.* B 404) vs "[the author says] that it (sc. the fox) cannot be tamed" (*Epit. Tim.* 5.4). It can be estimated, on the basis of a comparison of the respective volume of the two recensions, that the epitome has reduced the original text by at least 75 percent.[19]

The Composition

As fluid as a poem, Timotheus's treatise, to which his first editor (Christian Matthäus) gives the title Περὶ ζῴων τινῶν ἰδιότητος (modeled on Aelian's title), gathered, for each animal under consideration, a wealth of naturalistic, economic, and medical information. The division into chapters is probably not genuine, and Timotheus's work was probably a more or less continuous text, where the "monographic" sections were probably not indicated by intertitles. Heir to Aelian and Oppian, the Gazan does not seem to follow a systematic order in his treatise either or even seems to neglect the general framework of the great Aristotelian groups (quadrupeds, birds, fish…). The Epitome's chapter 22, for instance, which describes the onager, groups together a few features of the onager (22.1–2) and then adds a series of remarks on the ostrich (22.3–6). The passage from one to the other is "natural", because "the ostrich, the so-called sparrow-camel, does the same" as the wild ass (viz. it throws stones with the feet to defend itself).

We can, therefore, consider that the division into chapters is due to the epitomizer, who is sometimes explicit about the flexible association of animals, titling a section where the tiger (*Epit. Tim.* 9.1–7) and the griffin (9.7–11) are associated: "on the tiger and in the same chapter on the griffin" (περὶ τίγρεως ἐν ταὐτῷ καὶ γρυπός). The scribe, however, detaches from this pair of enemies (9.7) the following "chapter", which deals with the *hippotigris* (i.e. zebra: *Epit. Tim.* 10) and which obviously belonged to the same set. He similarly isolates a chapter (20) that contains only one indication about the squirrel (it is overshadowed by its tail), whereas it is an adventitious remark from chapter 19 that concludes with the turtle's vulnerability to the sun (ἐὰν ἐν τῷ ἡλίῳ ξηρανθῇ). The logic of the transitions is thus not natural but cultural or onomastic, with chap. 6 (hedgehog) following chap. 5 (fox) because of the proverbial pair they form (in Archilochus, fr. 201 West); it may even be anecdotal, as with the transition between wolf (7) and porcupine (8), preserved by the epitomizer: "the porcupine is the size of the wolf…".

A remnant preserved in the *Zoological Collection* confirms, in its own way, this conclusion. Chapter B 363 indeed bears a supertitle: "(According to Timotheus) *On Mice and Shrews* (Ἐκ Τιμοθέου περὶ μυῶν καὶ μυγαλῶν)". This heading does not match the selection given in the *Zoological Collection*, which mentions only mice, but it does fit perfectly with a series in the epitome that includes the chapter "mouse" (μῦς, 38) and the chapter "weasel" (entitled γαλῆ and μυγαλῆ, 39), and thus with a sequence in the original treatise. The excerptor (of the *Zoological Collection*) probably used an intermediate, chaptered version of the Timothean treatise.

Timotheus must, therefore, have favored a cultural articulation based on associations of ideas, like Oppian who moves from the tiger to the boar to the porcupine, then to the ichneumon and the fox (*On hunting* 2). A certain number of blocks are easily distinguished by the explicit relation: tiger/griffin (*Epit. Tim.* 9), pard/leopard/jackal/panther (11–14), onager/ostrich (22), elephant/dragon (25), mouse/weasel/shrewmouse (38–39), crocodile/ Egyptian plover/mongoose (42–43), salamander and catoblepas (52), … There is often an umbrella title for these associated animals, as for the crocodile group under the heading "crocodile" in one ms (Barocc. 354v–355v). This flexibility explains the misunderstanding of some modern scholars who imagined that Timotheus's treatise embraced the entire terrestrial and aquatic macrofauna. Wellmann thus proposed a scheme (in four books) freely inspired by the *Suda*, dealing respectively with quadrupeds, birds, snakes, and fish.[20]

In fact, Timotheus did not follow these broad categories and integrated in each set centered on an animal, more- or less-developed peripheral information. Thus, dealing with the wild goat, Timotheus notes that it feeds its father when he is too old to feed himself, "just as the kingfisher on the sea nourish the *kerylos*, their father. This is indeed the name the bird receives when he grows old. And the storks do the same" (15.2–3). This remark testifies to Timotheus's way of evoking, by association and digression, animals that do not belong to the quadrupeds, such as the crocodile's *trochilus* or the halcyon. The salamander is invoked for its insensitivity to fire and its ability to extinguish it in a section devoted to the *catoblepas* that spits fire through its nostrils. About the fox's mock death ruse (5.1–2), Timotheus adds: "The sea-frog and the torpedo do the same. (That) the basilisk is afraid of the seal's skin." The last remark, concerning the antipathy between a snake and an amphibian, is totally irrelevant, but the two must have been drawn together by some consideration related to the fox,[21] as Timotheus then returns to remarks about the fox.

The Content of Timotheus's Treatise

The redactional flexibility and the attention to animal interactions support the hypothesis of a treatise focused on terrestrial macrofauna, and it can account for the presence of particular remarks about other animals (e.g. marine species) that stood in interaction with the terrestrial animals.[22] This 'grammarian' treatise was an erudite literary work, which brought together traditional materials of what is called zoology. Across the whole of the chapters certain types of data are recurrent: reproduction (with a kind of obsession, likely owing to Timotheus himself and not his sources), for the phenomenon of hybridization, which is at the same time a representation of the fluidity of the animal world and a theme likely to allure the reader,[23] ethological aspects, remedies, antipathies, and local varieties. Some entries suggest a recurring pattern for these headings: mating and birth (ὀχεία, γένεσις) [1], appearance (σῶμα, μορφή) and varieties (εἴδη) [2], behavior and habits (ἤθη, πράξεις, ἐπιτηδεύματα) [3], and medical and technical uses (δυνάμεις) [4]. But the deviations noted from this theoretical pattern are numerous, and physiology and ethology are naturally intertwined.

The chapters, some of which suggest an encyclopedic vocation and the desire to gather the main pieces of the animal file, are presented less as zoological cards than as "portraits", initially physical but above all psycho-ethological, with dominant traits that give unity to the animal character. These portraits often conclude with remarks on the therapeutic or magical use of animal parts, in the Roman tradition, which undoubtedly goes back to the Hellenistic period and to the *Epitome* of Aristophanes of Byzantium. The notations referring to the camel thus present it successively under two temperament "natures" (φύσεις): endurance (B 466–468) and anger (B 469–472), with an appendix on Indian camel-hair rugs. Various peripheral information aggregates to this line.

Similarly, Timotheus describes the deer under two signs: that of its antlers (B 507–510) and that of its race (B 511–515), with a conclusion on the various species or names of cervids. The panther is a diabolical being (with a stinger and which does not bear human gaze: B 261–265) of Dionysian nature: it is born of Pentheus' aunts, it likes to drink wine, and is captured by a wine trick which makes it drunk (B 266–267). Each animal can be expressed in this way by a 'psychological' formula, sometimes delivered in the incipit of the description that gives the *sphragis* of the animal such as: "the boar is a boiling animal with tusks" (B 566), *vel*: Θερμόν τι καὶ ὀξὺ θηρίον ὁ σῦς καὶ ἰταμώτατον τὴν ὁρμήν. The whole record of the boar (B 566–571) fleshes out, in effect, this identity: his tusks are inflammable; though he is so wild that there are none in Libya; he is wild in love, trains hard for his duels, is killed by his own tusks. Timotheus's definition of the wolf is as follows: he has the evil eye (a) and is an enemy of sheep (b). Before the heading devoted to sympathies and antipathies (c), which is almost systematic, all the notations follow on from this basic line, with the exception of a scholarly remark on the origin of the adjective *lykabas*, entailed by the description of the raptor wolf.

The aggregation of peripheral remarks may be important, but the key to organizing Timotheus's zoological portraits is thus less 'thematic' than *Physiologus*-related. The hedgehog in the *Physiologus*, in the last two redactions (of the Sbordone edition), especially in the pseudo-Basilean redaction, is φιλότεκνος: it brings the grapes back to its young, as the proposed digest in the *Zoological Collection* (B 434) explicitly states. This 'physiological' orientation of Timotheus's work manifests the diffusion of a certain zoographic format and a cultural impregnation,[24] but absolutely nothing in his treatise refers to Christian religion. There is only one trace of the Bible in the lexicographical relics of his grammatical production, a reference to *Proverbs*.[25] Contrary to Wellmann's belief, the three missing chapters in the *Epitome Timothei* were not devoted to the angel, the man, and the animal, but to three flagship animals, the last two of which are actually found in the *Zoological Collection* recension, and which are probably the lion, the deer, and the bear.[26]

The scholarly culture of the author, known by lexical glosses and a treatise *On Orthography*, is manifested by philological remarks and by a taste for etymologies: the *kastōr* (beaver) has a white belly (*gastēr*) (*Epit. Tim.* 54); the *boubalos* (antelope or buffalo) is a mixture of bull (*bous*) and deer (*elaphos*) (*Epit. Tim.* 29); the *crocodeilos* (crocodile) is afraid (*deilos*) of the saffron (*crocos*), the *skiouros* (squirrel) makes shade (*skia*) with its tail (*ouros*), etc. It is also evidenced by a strong taste for the myths of metamorphosis which he enjoys dwelling on: the mole as Phineus (*Syll. Const.* B 423),[27] the panthers as nurses of Dionysus (*Syll. Const.* B 266), the wolf as Lycaon (*Syll. Const.* B 237), and the wild beasts in general born of the blood of the Titans during the war between Cronos and Zeus (*Epit. Tim.* 9).

Timotheus also evokes other cultural characteristics, like the manufacture of the lyre by Hermes, starting from a tortoise (*Epit. Tim.* 19), the sacrifices of asses carried out by the Hyperboreans (*Epit. Tim.* 31), or the affinity of Hermes for the dog, of Apollo for the lynxes, and of Dionysus for the monkeys "*sfinges*" (*Epit. Tim.* 46). Timotheus also shows a predilection for precise descriptions of behavior (and comments on the ἦθος),[28] for inventories of local 'varieties' and, like Aelian, for anecdotes of a moral nature. He also gives an important place to the theme of animal sympathy (or antipathy), popularized by the pseudo-democritean tradition (Bolos, Nepualios, Pamphilos …)[29] and by the *Cyranids* as well as to the magical and therapeutic properties and uses of animals.

One can often find parallels to the Timothean developments in the preceding literature, in particular, in Aelian and in Oppian's *Cynegetica*.[30] But this is widely shared knowledge. Timotheus does not follow a model and only once quotes the name of Aristotle; about the fact that "dogs dream, according to Aristotle".[31] If the animal portrait evokes, by its form, the typical focus of the *Physiologus* on salient features, the relation of the treatise to the natures described in the *Physiologus*, including in the pseudo-Basilean version, which is quite close in time, is very distant. Timotheus delivers some quite original notations: he describes the vigil of the marmots ('bear-mice'), reports the making of clothes (καστόρεια ἱμάτια) and shoes of beaver skin (σαπηρινὰ ἐνδύματα), discusses the Indian mugger crocodile, and gives curious hints about a flying rodent, apparently equipped with a patagium, in the chapter on *ictis* (commonly identified as the marten): "but when it falls, it blows itself up and filling its skin with air it is not harmed". Timotheus's work is not a systematic treatise like Aristotle's *Parts of the Animals*, but it offers, in a kind of rhapsody of portraits, a collection of varied information or strands of knowledge expressing remarkable aspects of animal φύσις and synthesizing different zoological facts (biological, ethological, pharmaceutical, technical, philological, …). He approaches the subject of animals as a whole and offers the first wide bestiary in the medieval sense, without the catechism, close to the Western type found in the medieval tradition stemming from the *Physiologus*.

Timotheus is a prose artist, without doctrinal posturing, who dispenses with both physiological exegesis and hexaemeral moralism. The question of the animal *logos*, an obsessive issue of the ancient philosophical debates after Aristotle, is probably for him an idle debate and does not appear in the remainder of his work. He probably has the same flexibility on this question as Aelian: animals (or ἄλογα) do not need the logos to be effectively rational and moral. At any rate, he clearly acknowledges their *phronesis*,[32] but he perhaps did not go as far as the Syrian Tatian (2 AD), who, in a lost work *On animals*, vigorously developed the idea that *aloga* have, like man, intelligence and knowledge (τὰ ἄλογα νοῦ καὶ ἐπιστήμης δεκτικά) with an apologetic but probably also descriptive character, against "the peremptory hucksters".[33] A conventional and commonplace definition of animal is attributed to Timotheus, which does not contradict this general orientation: " 'Animal': Animate and sentient essence (vel) 'the non-speaking' ".[34]

Posterity of Timotheus

The *De Animalibus* enjoyed significant popularity, and the treatise was quickly translated into Arabic (seventh-ninth century),[35] as it was known to Muhammad ibn Ahmed of Jerusalem (tenth century), and probably also into Syriac.[36] In the twelfth century, it was still circulating in the Arab world in its unabridged version and al-Marwazi quotes on numerous occasions "Atmûniyûs the Wise" i.e. Timotheus the Grammarian, by taking literally certain segments present in the *Syll. Const.* such as the chapter on the costume of the giraffes and the bridles and the black men who ride them in the parades. The later tradition of the text, especially in the Arab world, has not yet been systematically studied, and it is likely that parts of his work had been taken up under different names.[37] John Tzetzes testifies to his preservation in the twelfth century and to his Byzantine fame since he cites him among the four principal authors of animal stories, distinguishing him from the imperial trio formed by Aelian, Oppian, and Leonidas of Byzantium: "The earlier stories about animals are written/ by Aelian and Oppian, together with Leonidas,/ and with them Timotheus, the grammarian of Gaza,/ who, in previous times, was coincidental with king Anastasius".[38] It may be added that Timotheus is a probable source for George of Pisidia's *Hexaemeron* (seventh century), and for the *Entertaining Tale of Quadrupeds* (fourteenth century).[39]

The Zoological Collection of Constantine VII

The importance of Timotheus in Byzantine culture, attested also by the preservation of part of his grammatical production, is highlighted by his remarkable place in the zoological collection reported as the *Zoological Collection* of Constantine VII Porphyrogenitus. Its complete "title" appearing in the manuscript, before four dodecasyllabic verses praising the emperor, is *Compendium of knowledge on terrestrial, winged and marine animals, carefully produced for the great king and emperor Constantine*. This subscription is supplemented by a second heading corresponding to the content: *Abridgment made by Aristophanes of Aristotle's work on animals, to which has been added, for each animal, the information given by Aelian, Timotheus, and some others* (Ἀριστοφάνους τῶν Ἀριστοτέλους περὶ ζῴων ἐπιτομή, ὑποτεθέντων ἑκάστῳ ζῴῳ καὶ τῶν Αἰλιανῷ καὶ Τιμοθέῳ ἑτέροις τισὶ περὶ αὐτῶν εἰρημένων).

This work is one of the major witnesses of the Byzantine culture of the tenth century, and it is associated with the name of the emperor, who is also credited with a vast and prestigious encyclopedic program of collection and compilation of useful knowledge known as *Excerpta* or *Collectanea*.[40] The Latin title given by his editor (S. Lambros), namely *Excerptorum Constantini de natura animalium libri duo*, is misleading and seems to affiliate the text to the famous *Excerpta*. But the *Zoological Collection*, no more than the *Geoponica* or the *Hippiatrica*, does not belong to this editorial enterprise in which "neither science, nor the trades and techniques, nor the economy, for example, seem represented".[41] The text does not comprise the invariable and common preamble which constitutes a kind of official seal, present in the four preserved sections; the sources are not divided into Christian authors on the one hand, and pagan authors on the other, as is usual in the *Excerpta*; the title does not match one of the 25 known section titles and, above all, it does not constitute a "great lesson of history" (ἱστορικὴ μεγαλουργία) since zoological knowledge is "luxurious" and useless for a prince. Yet, one of the 53 sections of the *Excerpta* dealt with hunting, and probably compiled excerpts or paraphrases (such as that of Eutecnius, third-fifth century) from the many texts of this tradition.[42] Aristocratic amusement is indeed the only interaction with animals that could be relevant for a prince.

This precious collection, which is known only from a single manuscript, nevertheless offers a Byzantine overview of zoology, the state of the arts of texts in this field in the tenth-eleventh centuries through an organized selection made from the zoological corpus preserved at that time.[43] This assemblage of texts (συλλογή) is a by-product of the work of Aristotle, who is its primary authority and major author, even if his work does not appear directly but rather through various and successive reworkings. It is clearly constituted by the superposition of several textual layers, starting from a structural skeleton provided by the manual of Aristophanes of Byzantium known as the *Epitome of Aristotle's works on animals*, duly quoted at the beginning of the surviving manuscript. Aelian's *On Characteristics of Animals* and Timotheus's *On Animals* provide, together with Aristophanes' work, the bulk of the text which is presented as a combination of summaries and extracts. It is essentially an anthology of texts, but it is not always possible to discern between the two types of literary genre in the second degree (ἐπιτομή and ἐκλογή), not because of the absence of an original, but because the practices of textual abridgment in Byzantium were very free, especially for prose texts, and often combined quotation and condensation.

The treatise consisted of four books, of which we have preserved only the first and most of the second. It followed rigorously the program of Aristophanes' manual. The first book is entirely Aristophanean and is presented as a general and synthetic introduction, proposing successively a list of explained classificatory terms (1–27), a synthesis on the generation of animals (27–97), and an inventory of animal and human singularities and exceptions (97–154). The second book is devoted to mammals (or viviparous quadrupeds) and deals successively with men, fissipeds, bifurcated animals, and solipeds, in monographic chapters devoted individually to different animal species. The end of the book is missing but was intended to deal with cetaceans and dogfish (σελάχια). The two lost books dealt with oviparous animals (reptiles, birds, fish, cephalopods), probably omitting arthropods, in accordance with Aristophanes' framework. The general organization of the treatise indeed faithfully follows the program of Alexandrian scholarship, but the distribution of the material between the two lost books is not certain: either a general introduction (as in book 1) followed by a monographic inventory of oviparous animals; or a book on fish (3) and one on birds (4).

The second arrangement is less likely and is structurally problematic (where to place other categories such as snakes, amphibians? etc.). Aristophanes' treatise, almost certainly distinct from the compendium known as the *Zoica*, was presented as a kind of naturalist guide inspired by Aristotle's works. While it condensed Aristotelian material, it was less like an abridgment than a collection of notes close in form to the loose type of the ὑπομνήματα, and its function was probably that of an ἐγχειρίδιον or an εἰσαγωγή. In the second book of the *Collection*, the Aristophanean sections are perfectly circumscribed and present for the animals' standardized data sheet, describing for each of them in a regular order its name (ὄνομα), anatomy (μόρια), reproduction (ὀχεία), gestation (πόσους κύειν δύναται μῆνας), fertility (ἔκτεξις/ποῖα καὶ πόσα ὑπομένει τίκτειν βρέφη), life (βίος), habit (ἦθος), and lifespan (πόσα δύναται ζῆν ἔτη).

But the treatise has not only absorbed Aristophanes' handbook, which is not preserved anywhere else than in this editorial montage, since he has compiled in book 2, for each animal, many chapters first borrowed from Aelian and Timotheus. To these three levels are occasionally added extracts from other authors, both relevant and available in the libraries, which testify less to the Byzantine zoological canon than to the reference literature. Two texts are systematically exploited: the *Mirabilia* of Ps.-Aristotle and the *Homilies on the Hexaemeron* of Basil. All the passages of these works relating to the animals treated in Aristophanes' list are systematically extracted and literally reproduced after the main block of the three major authors. Three complementary works provide occasional supplements: Ctesias's *History of India* (on the dog, the sheep, the goat, and the pig), Agatharchides' *On the Red Sea* (on the man), and Philostorgius's *Ecclesiastical History* (on the giraffe).

These elements, gleaned from texts not identified as zoological, do not fit the general plan of the work. Their order of appearance, scrupulously respected in the various chapters throughout the book, is as follows: Aristophanes (third-second century BCE), Agatharchides (second century BCE), Aelian (second-third century CE), Timotheus, Ctesias (fifth-fourth century BCE), Basil (fifth century CE), Ps.-Aristotle (fourth century BCE?),[44] Philostorgius (fourth-fifth century CE). Two extracts resist any attribution: a list of dogs (B 193–197) and a list of horses (B 587–609), which describe the qualities of different ethnic morphotypes and their military or hunting use. They are most probably taken from specialized collections, and the hippological excerpt is closely related to a passage in the *Hippiatrica*.[45] The preserved text of the *Zoological Collection* (52,860 words) is distributed by source author in number of words as follows: Aelian (22,557), Aristophanes (16,814), Timotheus (6,846), Agatharchides (3,482), Basil (857), *adespoton* on horses (851), Ps.-Aristotle (747), Ctesias (411), Philostorgius (183), *adespoton* on dogs (112).

The anonymous *Collection*, whose stratification is explicit, shows that the main mutations of zoological discourse and knowledge occur from the Alexandrian period onwards: the distinction between a general and a particular account; the monographic (by animal) and not thematic display; the emphasis on (paradoxical) singularities; the ellipsis of problematic aspects and of critical discussion. Remarkably all the authors it incorporates are prose writers. This selection criterion is implicit, but it probably explains the absence of the work of Oppian of Syria, an author still widely read in medieval Byzantium. This coherence constitutes an additional argument for ruling out the idea that the text of Timotheus could be originally in verse. This patchwork, which has no originality in terms of content and follows an Alexandrian editorial practice, does not show any epistemological reflection. The literary rigor with which the *excerptor* exploits his sources is manifest in the case of Aelian; he proposes from his source two juxtaposed series for each animal: a first batch made up of abridged notices (*epitome*) of *some* of the chapters related to the animal; and a second batch, that completes the first series and proposes in the form of integral extracts (*eclogai*) all the chapters of Aelian which are omitted in the first series, without any overlap. But the criteria for the selection of the texts retained by the *excerptor* are not explicit and are probably partly based on the available texts. Aristophanes is chosen, in preference to Aristotle, for the framework of the treatise for tactical and more precisely economic reasons, but the presence of passages from Ctesias or Philostorgius, rather than from other more specialized authors, is confusing.

The absence of any borrowing from book XI of the *Christian Topography* of Cosmas Indicopleustes (sixth century), which describes Indian fauna, is undoubtedly explained by the limits of the list of animals given by Aristophanes and which does not, apparently, include the Indian bestiary, but it is astonishing that no Christian historian except Philostorgius is exploited, not even George of Pisidia (seventh century), author of a *Commentary on the Hexaemeron* and of a *Universal Chronicle*. The absence of any mention of the *Physiologus*, and of any reference to God, even in the fragments of Basil or Philostorgius, makes it an atypical text in the Christian zoological tradition. This compilation reflects, in fact, a conception and a state of knowledge corresponding to the end of late antiquity rather than to the Byzantine period, and the tenth century is probably only the re-edition of an older collection. This editorial recycling, which promotes through a new preface or a paratextual addition an older literary product, also characterizes another compilation attributed to the time and sometimes to the initiative of Constantine VII: the *Geoponica*.

This collection expanded in the sixth-seventh century by a certain Cassianus Bassus,[46] starting from two works of agriculture composed in the fourth century, of which a *Compendium of Agricultural Practices*[47] was reworked towards the middle of the tenth century by an editor who dedicated it to Constantine VII. The stratigraphy of the text that has come down to us is complex but is made up of three visible layers: that of the two main source authors and a few additional contributors; Cassianus Bassus, who combines, compares, and criticizes them; and the Byzantine editor, whose contribution is undoubtedly almost nothing: a few lexical or even phraseological innovations, and at most a surface polishing. For the dating and attribution of the work, one cannot rely on the occasional and opportunistic preface that tradition has attached to it. The question of dating is essential from a theoretical rather than historical point of view, and the critics maintain a kind of confusion about the status of the work. One can be sure that there was an edition of the *Geoponica* at the time of Constantine VII, but the text, in probably the same shape, had most likely existed for several centuries. The situation of the *Hippiatrica*, another collection attributed to Constantinian encyclopedism, is probably not very different. The richest manuscript in this corpus (*Berolensis*), dated to the tenth century, which offers one of the recensions of the collection, has all the characteristics of an "imperial" manuscript made for the Porphyrogenitus,[48] but the emperor (whose name does not appear anywhere) probably had as little to do with this collection, whose essential part probably dates back to fourth-fifth centuries, as with the tradition and transmission of the veterinary science in the *Hippiatrica*.

Conclusion

Whereas the human world appears entirely determined by God, it seems that the animals, when they escape the hold of the learned Physiologus, who is His minister, play their own naive and secular role. As if they belonged to the old world, that of the physiology and the psychology of the lyric poets and philosophers rather than that of the clerics and their kind of 'natural mythology'. Timotheus thus represents in Byzantium the affirmation (and the victory) of the *polutropos* fox over the univocal hedgehog. The *Zoological Collection*, which makes Timotheus one of the reference authors, and which dates perhaps from shortly after the time of the Gazan, manifests the same spirit and continues the tradition of zoology as profane knowledge. In their own way, the *Geoponica* (Byzantine *Georgics*) and the *Hippiatrica* (veterinary compilation) in their slow and progressive composition also maintain a distant scholarly tradition related to anthropozoological practices and knowledge.[49] But unlike these two applied zootechnical sums, made up of warnings and advice, the *Zoological Collection* probably has no other objective than to disseminate a kind of literary encyclopedia of naturalist knowledge, totally devoid of practical use.

The long poem entitled *On Characteristics of Animals* by the most prolific poet of the final 300 years of the Byzantine empire, Manuel Philes (ca. 1275–ca. 1335), also belongs to this legacy. This poem of more than 2000 iambic dodecasyllables dedicated to Michael Palaeologus, is almost a versified paraphrase of Aelian's comprehensive work, a playful evocation of traditional curiosities, from the resurrection of the drowned and then incinerated fly to the description of the elephant's trunk.[50] For Philes, as for John Tzetzes before him, animals constitute a poetic excuse, the occasion to turn into verse classical *ekphraseis* and to present an anthology of sketches or a gallery of portraits, systematically organized in this case in three parts according to the divisions in flying, terrestrial, and marine animals.

Since Aristophanes of Byzantium, zoology has been a matter for grammarians and poets. During late antiquity, one perceives how much these two forms of expressions converge and, as evidenced by the close relationship of Oppian and Aelian, draw from the same authors. Timotheus's treatise in a mannered and *rhythmical prosa*, as well as the mille-feuille of the compiler of the *Collection*, are part of this heritage. But it is surprising how alien and hermetic to each other remain the two main Byzantine traditions of zoological discourse, that of the *Physiologus* and that of the continuators of Alexandrian encyclopedism and Aelian.

Notes

1 See the chapter by Stavros Lazaris in the present volume.

2 In his monumental history of Byzantine literature, K. Krumbacher barely alludes to some of these texts (Physiologus, Timotheus, treatises of falconry, Late Byzantine commentaries, Manuel Philes), and in less than one page (*Geschichte der byzantinischen Litteratur,* vol. 9 (Munich, 1891), chap. 7, §261, 631–33).

3 The most complete manuscript is the ms. Munich, Bayerische Staatsbibliothek, cod. Grace. 564 (f. 239r–246v) [olim Monacensis gr. 364] of the 13th–14th century, edited by Ch. F. Matthäus (*Brevis historia animalium scriptoris anonymi qui seculo XI sub Constantino Monomacho imperatore Constantinopoli floruit graece* (Moscow 1811)) and then by M. Haupt ("Excerpta ex Timothei Gazaei libris de Animalibus," in *Hermes* 3 (1869): 1–30). A pericope of the epitome is known (cod. Oxford, Oxoniensis Baroccianus 50 of the 11th century, edited by J. A. Cramer). See A. Zucker, "Approche structurelle et phraséologique de l'ouvrage de Timothée de Gaza *Sur les animaux*," in *L'Ecole de Gaza: Espace littéraire et identité culturelle dans l'antiquité tardive,* ed. E. Amato, A. Corcella, and D. Lauritzen. Orientalia Lovaniensia Analecta - Bibliothèque de Byzantion 249 (Leuven, Paris and Bristol, 2017), 367–412.

4 On the manuscript Athous Dionysiou 180 (14th century) see V. Cuomo, "Athos dionysiou 180 + Paris, Suppl. grec 495: Un nuovo manoscritto di Teodosio Principe," *BZ* 98.1 (2005): 23–34. The only edition of the text is *Syll. Const.*, ed. S. Lambros, *Aristophanis Historiae animalium epitome. Excerptorum Constantini de natura animalium Libri duo. Aristophanis Historiae animalium epitome, subjunctis Aeliani Timothei aliorumque eclogis.* Commentaria in Aristotelem Graeca. Supplementum Aristotelicum 1 (Berlin, 1885). There is a pericope of this text with some Timothean snippets (4 lines) in ms. Firenze, Laurentianus gr. Plut. 86.8 (15th century).

5 Suda, ed. A. Adler, *Suidae Lexicon*, 5 vols. (Stuttgart 1967–71), T 621 Adler. Haupt, "Excerpta," 3 considers it likely that Timotheus of Gaza also wrote about rivers: this name appears in a commentary by Cosmas of Jerusalem, as the author of considerations on the role of the rivers that cross Gaul and flow into the Atlantic; but the attribution of a work *On rivers* in 11 books ascribed to an author called Timotheus, in the *De Fluviis* of Pseudo-Plutarch, is not a sufficient guarantee. See R. A. Kaster, *Guardians of Language: The Grammarian and Society in Late Antiquity* (Berkeley and Los Angeles 1988), 368–70 (n°156).

6 See A. Corcella, "Timoteo di Gaza: un grammatico fra tradizione e innovazione," in *L'Ecole de Gaza: Espace littéraire et identité culturelle dans l'antiquité tardive*, ed. E. Amato, A. Corcella, and D. Lauritzen. Orientalia Lovaniensia Analecta - Bibliothèque de Byzantion 249 (Leuven, Paris and Bristol, 2017), 413–53.

7 See Tzetzes, *Chiliades* 4.128, v. 172, ed. P. A. M. Leone, *Ioannis Tzetzae Historiae*. Pubblicazioni dell' Istituto di Filologia Classica. Università degli Studi di Napoli 11 (Naples, 1968), 132; *Opuscula*, ed. M. Haupt, vol. 3 (Leipzig 1875), 276.

8 The treatise is likely of Syrian-Egyptian origin (see M. Wellmann, "Timotheos von Gaza," *Hermes* 62.2 (1927): 179–204, at 189–90). In the *Epitome Tim.*, the mongoose (pharaoh's rat) is also called ὕλλος as in the *Physiologus* (*versio* I, 25); the squirrel bears the same name as in the Syrian Archigenes (σκίουρος).

9 See Suda, ed. Adler, s.v. Ἐπικῶς, E 2412 Adler: <Ἐπικῶς:> ἀντὶ τοῦ λογίως. <Ἐπικῶς> δὲ ἀντὶ τοῦ ποιητικῶς. Cf. Zonaras, ed. J. A. H. Tittmann, *Iohannis Zonarae Lexicon ex tribus codicibus manuscriptis*, vol. 1 (Leipzig, 1808), s.v. Ἐπικῶς, E 855 Tittmann: <Ἐπικῶς>. λογίως, ἀπὸ τοῦ ἔπω, τὸ λέγω. Idem in Suda, ed. Adler, s.v. Οὐκ ἐπιεικῶς, A 877 Adler: <Οὐκ ἐπιεικῶς:> οὐ μετρίως. <Οὐκ ἐπικῶς> δὲ ἀντὶ τοῦ οὐ λογίως. H. Usener, "Commentaria in Aristotelem Graeca," *Göttingische Gelehrte Anzeigen* 26 (1892): 1001-22, at 1019-20), taken up by M. Wellmann ("Timotheos," 180) points out this confusion. Let us add that the corresponding adjective ἐπιεικής occurs with the spelling ἐπεικής (and consequently ἐπεικῶς for ἐπιεικῶς), which makes the two adjectives phonetically indistinguishable.

10 See H. Diels in *Syll. Const.*, ed. Lambros, XIII, n. 1.

11 H. Usener ("Commentaria," 1019-20) notes that sentence members rarely conclude with a double dactyl, and end either with a choriambe (-vv-) or an adonius or adonic (-vv--), but this remark is not based on a systematic analysis and privileges quantitative stress. For a detailed demonstration, see A. Zucker, "Approche structurelle"; M. Haupt, H. Usener, E. Wellmann and A. Steier, "Timotheos (n. 18)," in *Paulys Realencyclopädie der classischen Altertumswissenschaft*, vol. 2.6 (Stuttgart 1937), 1339-41, at 1340.

12 See K. Seitz, *Die Schule von Gaza. Eine litterargeschichtliche Untersuchung ...* (Heidelberg, 1892), 30 ff.

13 See *Anecdota graeca e codd. manuscriptis bibliothecarum oxoniensum*, ed. J. A. Cramer, 4 vols. (Oxford, 1835–37), vol. 3, 263.

14 See A. Steier, "Timotheos (n. 18)," 1340–41. The *Suda*'s entry probably comes from the *Onomatologon* of Hesychius of Miletus (or Hesychius Illustrius) (see Photius, *Library* cod. 69) through a condensed version made in the 9th century.

15 Bodenheimer and Rabinowitz's translation (*Timotheus of Gaza on animals: Peri Zôon: Fragments of a Byzantine Paraphrase of an Animal-Book of the 5th Century AD.* Collection des Travaux de l'Académie internationale d'Histoire des Sciences 3 (Paris, 1949)) is sometimes wrong, and they ignored the existence of the *Zoological Collection*.

16 See Michael Attaleiates, *History*, ed. E. Th. Tsolakis, *Michaelis Attaliatae Historia.* CFHB 50 (Athens, 2011), 39:28-30: Ἐπὶ τούτῳ τῷ ζώῳ καὶ τὴν λεγομένην καμηλοπάρδαλιν ἐξ Αἰγύπτου πεμφθεῖσαν αὐτῷ τοῖς πολίταις ὁ βασιλεὺς (*sc.* ὁ Μονομάχος: p. 50) καθυπέδειξεν. K. Krumbacher

(*Geschichte der byzantinischen Litteratur*, 68) mistakenly thought that another zoological compilation, different from the *Zoological Collection*, was composed at the time of Constantine X.

17 Συλλογὴ τῆς περὶ ζῴων ἱστορίας. *Syll. Const.*, ed. Lambros (see n. 4). See A. Zucker (ed.), *L'encyclopédie zoologique de Constantin VII vel Epitomé d'Aristophane de Byzance*. Rursus, 7 (2012): http://rursus.revues.org/617.

18 In a scholium to the *Palatine Anthology* preserved on fol. 15r of the Suppl. gr. 384, Paris, Bibliothèque Nationale de France (*Epigrammatum Anthologia Palatina...*, ed. F. Dübner. 3 vols. (Paris 1864–90), B 519): ἐλλόγιμοι ταύτης τῆς πόλεως (i.e., Gaza) Ἰωάννης Προκόπιος, Τιμόθεος ὁ γράψας περὶ ζῴων Ἰνδικῶν (quoted by M. Wellmann, "Timotheos," 179). See John of Gaza, ed. P. Friedländer, *Johannes von Gaza, Paulus Silentiarius und Prokopios von Gaza: Kunstbeschreibungen Justinianischer Zeit* (Leipzig, 1912, repr. Hildesheim, 1969), 135; Haupt, "Excerpta," 2.

19 This estimation is a minimum since the *Collection* did not preserve the entire text of Timotheus for the animals mentioned. When comparing each chapter in both recensions, the reduction rate varies in fact between 18% and 90%.

20 M. Wellmann, "Timotheos," 179–80.

21 May be a reversal of remarks with 5.6: "the fox fears the bile of the chameleon" and 5.5: "the fox knows how the wolf fears the squill".

22 Evidence for this is the incipit of the *excerptum* of the *Zoological Collection* on the fox (B 401), probably taken literally from the original: Ἀλώπηξ ἐντεῦθεν εἰσήχθω μεθ' ὧν οἶδε πολλῶν ἐπιτηδευμάτων, "Let us now approach the fox, and the many skills that it possesses..."

23 See T. Buquet, "Les panthères de Timothée de Gaza dans l'encyclopédie zoologique de Constantin VII," *Rursus* 7 (2012). https://doi.org/10.4000/rursus.971.

24 Note that in the ms. Oxford, Bodleian Library, Baroccianus 50 edited by J. A. Cramer (*Anecdota Graeca*, vol. 4, 258–63), the chapters of the epitome of Timotheus follow a 15-chapters pericope of the *Physiologus* with the inscription Φυσιολογήματα περὶ ζῴων.

25 As noted by A. Corcella ("Timoteo di Gaza," 447), the only gloss from the *Scholia to Cyril* relating to an animal is found in the ms. Firenze, Laurentianus 59.49 (fol.164ʳ) and does not correspond to a passage from Timotheus in any of the partial recensions of the περὶ ζῴων: πάνθηρ ὠκύτατον ζῷον ὅπερ πτερωτὸν καὶ μετάρσιον τοῖς ὁρῶσι ἐν τῷ τρέχειν νομιζόμενον. A similar indication appears in the epitome for other animals, viz. the tiger (chap. 9) and the wild goat (chap. 15).

26 See R. Kruk, "Timotheus of Gaza's on Animals in the Arabic Tradition," *Le Museion* 114.3 (2005): 355–87, at 363.

27 Cf. Oppian, *Cyn.* 2.614–16, ed. M. Papathomoulos, *Oppianus Apameensis: Cynegetica, Eutecnius Sophistes: Paraphrasis Metro Soluta* (Leipzig, 2003). For further myths, see *Epit. Tim.* 9, 19, 26, 31 and 46, ed. Haupt, "Excerpta," 9–10, 13, 16–17, 19–20 and 26.

28 For an explicit reference to the *ēthos* category, see *Syll. Const.* B 237 (wolf), B 318 (hyena), B 404 (fox); B 507 (deer).

29 See M. Wellmann, *Die Φυσικά des Bolos, Demokritos und der Magier Anaxilaos aus Larissa*. Abhandlungen der Geistes- und Sozialwissenschaftlichen Klasse-Akademie der Wissenschaften und der Literatur in Berlin 7 (Berlin 1928).

30 See M. Wellmann, "Timotheos," 179.

31 See the cautious conclusion of A. Zumbo ("Timoteo di Gaza. De animalibus," in *Byzantina Mediolanensia. V congresso nazionale di studi bizantini*, ed. Fabrizio Conca (Milan, 1996), 422-29 at 425–26): "Nella maggioranza di casi si deve parlare di conflatio di modelli [...] Timotteo non segue una fonte precisa". According to F. Sbordone (*Physiologi graeci singulas variarum aetatum recensiones codibus fere omnibus tunc primum excussis collatisque* (Naples, 1936), CXIII), the pseudo-Basilean redaction of the *Physiologus* would have been made at the same time as the excerpta of Timotheus at the mount Athos.

32 "Irrational animals apparently think (εἰκὸς φρονεῖν τὰ ἄλογα ζῷα)", *Epit. Tim.* 50, ed Haupt, "Excerpta," 26. The text names in continuation 19 animals that evidence this fact.

33 ὥσπερ οἱ κορακόφωνοι δογματίζουσι. Tatian, *Ad Gr.* 15.1.7, ed. E. J. Goodspeed, "Tatianus. Oratio ad Graecos," in *Die ältesten Apologeten: Texte mit kurzen Einleitungen* (Göttingen, 1915), 266–305, at 282.

34 ζῷον οὐσία ἔμψυχος αἰσθητικὴ τὰ ἄλογα in ms. Roma, Biblioteca Vallicelliana, Vallicellianus E 11 (fol.97ᵛ) and ms. Firenze, Laurentianus 59.49 (fol.91ʳ); see Corcella, "Timoteo di Gaza," 443. The word *alogon* is untranslatable but refers probably more to language than to reason.

35 See M. Wellmann, "Timotheos," 197–99.

36 See Kruk, "Timotheus of Gaza," 360.

37 See L. Kopf, "The Zoological Chapter of the Kitāb al-Imtā'wal-Mu'ānasa of Abū Ḥayyān al-Tauḥīdī (tenth century)," *Osiris* 12 (1956): 390-466, on Abu Hayam (tenth century), *The Book of Enjoyment and Entertainment*; A. Z. Iskandar, "A Doctor's Book on Zoology: al-Marwazī's Ṭabā'iʿ al-ḥayawān (Nature of Animals) Re-Assessed," *Oriens* 27 (1981): 266–312, on al-Marwazi's (twelfth century) *On the nature of animals*; and A. Z. Iskandar, *A descriptive list of Arabic manuscripts on medicine and science at the University of California* (Los Angeles 1984), 21–22. Kruk ("Timotheus of Gaza"), who obviously did not consult the *Syll. Const.* distinguishes three categories of Timothean testimonies: A (explicit quotation from Marwazi, especially in ms. University of California, UCLA Ar. 52), B (notable parallels between the texts); C (presumably Greek or Byzantine notations).

38 Tzetzes, *Chiliades*, 4.128, v. 166–169, ed. Leone, 132.

39 Παιδιόφραστος Διήγησις τῶν Ζῴων τῶν Τετραπόδων. The modern translators of the text are reluctant to refer to sources other than the *Physiologus* (*An Entertaining Tale of Quadrupeds*, trans. N. Nicholas and G. Baloglou (New York and Chichester, West Sussex, 2003), 20), but Timotheus offers the closest text for the elephant (ibid., p. 32).

40 See P. Lemerle, *Le premier humanisme byzantin. Notes et remarques sur enseignement et culture à Byzance des origines au Xème siècle* (Paris, 1971); A. Németh, *The Excerpta Constantiniana and the Byzantine Appropriation of the Past* (Cambridge, 2018).

41 See P. Lemerle, "L'encyclopédisme à Byzance à l'apogée de l'empire et particulièrement sous Constantin VII Porphyrogénète," *Cahiers d'histoire mondiale IX* (1966): 596–616, at 607.

42 See *Leg.* 2, 275, ed. C. G. de Boor, "Excerpta de legationibus. Pt. 2, Excerpta de legationibus gentium ad romanos," in *Excerpta historica iussu Imp Constantini Porphyrogeniti*, ed. Ph. Bossevain, C. De Boor, and Th. Büttner-Wobst, vol. 1 (Berlin, 1903), mentioning a περὶ κυνηγίας; see T. Büttner-Wobst, "Die Anlage der historischen Encyklopädie des Konstantinos Porphyrogennetos," *BZ* 15 (1906): 113–14.

43 The only surviving manuscript of the *Zoological Collection*, dated to the fourteenth century, was probably complete at the time of its production, but it was broken up and cut in half, probably by accident. Book 1 is now in the BnF (ms. Suppl. gr. 495), brought back and probably torn out of the manuscript by Minoïde Mynas in 1843; and book 2, interrupted at folio 387, is preserved in the library of the monastery of St. Dionysios on Mt. Athos (ms. Athous Dion. 180).

44 On the date of the text, see C. Giacomelli, *Ps.-Aristotele, De mirabilibus auscultationibus. Indagini sulla storia della tradizione e ricezione del testo.* Commentaria Aristotelem Graeca et Byzantina 2 (Berlin and New York, 2021).

45 *Corpus hippiatricorum Graecorum*, ed. E. Oder and C. Hoppe. Vol. 2, *Hippiatrica Parisiana, Cantabrigiensia, Londinensia, Lugdunensia, appendix* (Leipzig, 1927), 121–24.

46 This dating is basically suggested by the qualification of *scholastikos*, meaning lawyer, and added to the name of Bassus, whose use disappears after the emperor Heraclius (610–641).

47 The first work (συναγωγή γεωργικῶν ἐπιτηδευμάτων) is attributed to Vindanios Anatolios (and commented on in the codex 163 of Photius's *Library*), and the second (Γεωργικά) in 15 books, to a writer, perhaps a physician, of Alexandria named Didymus.

48 P. Lermerle (*Le premier humanisme byzantin*, 296) devotes no more than four lines to the *Hippiatrica*. The sumptuous ms. Phillipps 1538 of the Staatsbibliothek in Berlin is indeed a product of the imperial Scriptorium of the tenth century.

49 Other zootechnical works are quite rare in Byzantium and concern exclusively treatises on the care of dogs or birds of prey, frequently attributed to Demetrios Pepagomeneos (thirteenth century). See A. Zucker, "Zoology," in *A Companion to Byzantine Science*, ed. S. Lazaris. Brill's Companions to the Byzantine World 6 (Leuven, 2020), 261–301.

50 The poem is divided into 119 chapters of varying length, usually about an animal or a topic (animal antipathies, horn formations...). The author shamelessly touts the weightiness of his poem in his address to Michael Palaeologus, begging for a reward (κύων ἐγὼ σός... καὶ φαγεῖν ζητῶ

βρῶμα). Philes is the author also of several other poems *On silkworm, On vulture, On ostrich…* . The great number of manuscripts of the text (more than 30) proves its success, similar to the Western bestiaries. For a recent study of this text, see A. Vives Cuesta, "Reescritura de las fuentes y sustituciones léxicas en el *De Proprietate Animalium* de Manuel Files: ¿Variación estilística o cambio de nivel lingüístico?", *Studia philologica valentina* 22 (2020): 143–75. Concerning the reception of Philes in the Renaissance, see G. Peers, "Forging Byzantine Animals: Manuel Philes in Renaissance France," *RSBN* 49 (2012): 79–104.

3

CHRISTIANISING ANIMALS
Physiologus and *Hexaemeral* Literature

Stavros Lazaris

In the Middle Ages, nature held a central place in the most diverse literary forms. Its importance was clearly linked to agriculture and animal breeding, which played an essential role in medieval societies. This close relationship between people and nature was also the result of their desire to know God through his creation. We read in Paul's Epistle to the Romans that "For what can be known about God is plain to them [men], because God has shown it to them. Ever since the creation of the world his eternal power and divine nature, invisible though they are, have been understood and seen through the things he has made [...]".[1] This passage is key to understanding the special relationship between humans and nature throughout the Byzantine civilisation. Evagrius Ponticus (346–399),[2] Gregory of Nazianzus (c. 329–390),[3] John Chrysostom (c. 347–407)[4] and Theodoret of Cyrus (c. 393–458/466)[5] have formulated similar ideas.

It is through knowledge, among other things, of the φύσεις of animals that we can access the word of God. George of Pisidia (7th c.) wrote in his *Hexaemeron* that God cloaks his nature in clouds and darkness so that we may learn about it in his creatures.[6] According to some scholars, humans can even learn by following the example of animals. Basil of Caesarea in the 4th century and Theodoret of Cyrus in the 5th century insisted that all animals teach us something. Both scholars and others wrote that God provided the species that were not endowed with reason with some natural advantages, so that humans might reap some benefit.[7]

The early Christians had to gather from earlier (i.e. pagan) literature in the field of natural history that which was most useful. Already from the time of the first apologists, various philosophical doctrines were used to demonstrate the Christian faith. Clement of Alexandria for example, was of the opinion that there was good in every philosophical doctrine,[8] which recalls the words, albeit in an entirely different context, of Paul the Apostle, who wrote "test everything; hold fast to what is good".[9]

Secular knowledge could serve as a propaedeutic tool, the main purpose of which was the elaboration of Christian dogma.[10] For many early Christian thinkers, the usefulness of secular culture was a given. Accepting philosophical, medical and scientific knowledge as a heritage from which to draw that which they considered to be most useful was for some

DOI: 10.4324/9781003055877-5

of them a foregone conclusion. It is through this prism that we can better understand the genesis of works like the *Physiologus* and the *Hexaemera*.

The Greek *Physiologus* was the first Christian work, preserved in its entirety, that highlighted this attitude towards secular natural history texts. This work, still anonymous, is a mixture of ancient pagan naturalist knowledge and Judeo-Christian spirituality, and is of great interest for bridging zoological and exegetical knowledge, but also for the ways in which its author used pagan natural history knowledge in the service of the Christian faith. The *Hexaemera*, or commentaries on the first six days of the Creation of the world as recounted in Genesis, is the other major reference on this topic. The *Hexaemera* are therefore not treatises on Creation as such, but commentaries on Genesis. It should be noted, however, that the inspired exegesis of the first chapters of Genesis often goes beyond the strict character of commentary and takes on various literary forms, such as world chronicles, poetry, homilies, hymns, questions and answers, legendary narratives, monastic literature and *sui generis* forms like that of Cosmas Indicopleustes (*Topographia Christiana*). Although I have grouped all these texts together under the heading "Hexaemeral Literature", so that readers may appreciate their evolution over time, I am aware that, strictly speaking, they do not all belong to the same category.

Both the *Physiologus* and the *Hexaemera* flourished throughout the Middle Ages. This rich production of literature is therefore a very important source to investigate the birth of a Christian science of nature. More specifically, as far as this chapter is concerned, it highlights the place of animals in Byzantine civilisation.

In the following pages I will discuss the Greek *Physiologus* and some *Hexaemera* which provide information about animals. Although this chapter focuses on Greek-language production, I will also mention, where necessary, authors who wrote *Hexaemera* in other languages, but whose reciprocal affinities and influences with the Greek authors, considered in this chapter, are strong. These authors lived in late antiquity, when cultural multilingualism was present all around the Mediterranean.

The *Physiologus*

The *Physiologus* flourished, both in Byzantium and in other medieval cultures. This text was soon translated and adapted into most ancient literary languages (Arabic, Coptic, Syriac, Latin, Armenian, Ethiopian and Georgian) and later into Old English, Slavic and Icelandic, which eventually resulted in the Western medieval bestiaries.[11]

The Greek text has been preserved in four main recensions. The first, the so-called Christian one, consists of 48 or 49 chapters, depending on the edition,[12] and even more in some manuscript sub-families.[13] It was composed in Alexandria, around the 2nd century C.E.[14] Its author, who remains anonymous, describes the properties of animals, plants and minerals in simple and easily understandable Greek.[15] In some manuscripts, the 48 basic chapters are supplemented by a few others. The work contains references to the Church and the faithful; through several translations it wielded a great influence in the Greek-speaking regions and beyond.

The vast majority of scholars place the second, so-called Byzantine, version of this work in the 11th century.[16] The content underwent modifications: plants and stones were removed and nine new animal species appeared.[17] New natural characteristics were attributed to the studied species. It should also be noted that the author seems to have followed an order in the presentation of the chapters that come after the biblical classification of animate beings.[18] The chapters were organised as follows: quadrupeds, fish, large birds, small birds and reptiles. This order was followed scrupulously, except at the end of the work where there is some confusion (chapter 21 on the fox and chapter 25 on the hare).

The third recension, known as the Pseudo-Basil, is generally dated to the 12th century.[19] We find new animals, absent from the other two recensions,[20] and others shared only with the second recension.[21] All hybrid animals (hippocentaurs/onocentaurs, sirens, ant-lions, etc.) are excluded here, without it being possible to know whether this was accidental, even if it is unlikely, or a desire to eliminate them. This act, if it was indeed done consciously, constitutes a very important fact about the conception of this recension and perhaps, more generally, of its environment and its time, about fantastic beings and their distinction from living species. In some chapters King Solomon is identified as the author of the natural history accounts (τοῦ σοφωτάτου Σολομῶντος τὰς τῶν ἀλόγων ζῴων φύσεις).[22] For the second part, however, the text quotes Basil of Caesarea. He is identified as the author of the spiritual interpretations, most often introduced by the expression: Ὁ ἅγιος Βασίλειος εἶπε [...] / Ὁ δὲ μέγας Βασίλειος ἀλληγορικῶς ἡρμήνευσε [...].

The fourth recension appeared between the 14th and 15th centuries and consists of 1,131 unrhymed political verses (15 syllables) in Greek and some prose passages.[23]

Each chapter of this work consists in two parts that follow the διήγησις / ἑρμηνεία model: in the first part, one or more φύσεις (physical aspects, morals, character traits or even supposed qualities and defects)[24] of the studied species[25] are briefly presented and, in the second part, these characteristics are interpreted from a symbolic-allegorical point of view. This Christian hermeneutic section provides a brief theological lesson and moral encouragement to the reader.[26]

There is every reason to believe that the author's main aim was to use the natural world to explain the Bible's parables and teachings. Indeed, like Basil of Caesarea and Ambrose of Milan in their respective *Hexaemera*, but most likely before them, the author of the *Physiologus* produced something new by theocentrically reorienting the classical texts on the φύσεις of animals, plants and minerals. Indeed, this author did not hesitate to reshape the φύσεις according to the needs of the second part.[27] This degree of interventionism should not come as a surprise: the *Physiologus* is linked to the method of interpretation of the Scriptures according to the Alexandrian School. Based on the conviction that nature is the reflection of divine design, the early masters of the Alexandrian School[28] devised a model of symbolic-allegorical exegesis that allowed humans to grasp the divine world through the terrestrial world. By knowing the nature of God's creations on earth, the *Physiologus* comes closer to this symbolic vision. Thus, a Christian typology, whose principle is to juxtapose an image of nature, often fabricated *ex nihilo* with a Christological idea, makes its appearance. In doing so, he adopts, in a way, the definition of the fable given by Aelius Theon (1st c.) in his *Progymnasmata*: "a fictious story picturing reality" (Μῦθός ἐστι λόγος ψευδὴς εἰκονίζων ἀλήθειαν).[29]

The author of the *Physiologus* did not simply borrow the information he needed from the books of the Bible. On the contrary, when reading this work, one soon realises that the books of the Bible are not the source of many of the naturalistic descriptions found in the first part of each chapter. The books of the New Testament were mainly used for their symbolism. The information about animals, plants and minerals often comes from other secular sources which the author of the *Physiologus* had read or was familiar with ἀπὸ φωνῆς. The written works of pagan authors that could be identified are few in number

and almost all were written in Greek. Aristotle, Claudius Aelianus and Pliny the Elder
are among those most often encountered. Then there are authors like Plutarch, Oppian,
Nicander of Colophon, Oribasius and Pedanius Dioscorides; also included in these sources
are the *Cyranides*.[30]

This brief enumeration of the pagan natural history sources used by the author of the
Physiologus[31] is sufficient to demonstrate the difficulty of identifying these sources. The
chapter on the beaver illustrates this point.[32]

Περὶ κάστορος	On the beaver
Ἔστι ζῷον λεγόμενον κάστωρ, ἥπιον πάνυ καὶ ἥσυχον· τὰ δὲ ἀναγκαῖα αὐτοῦ προχωροῦσιν εἰς θεραπείαν. ὅταν δὲ ὑπὸ τῶν κυνηγῶν διώκηται καὶ νοῇ, ὅτι καταλαμβάνεται, τὰ ἀναγκαῖα αὐτοῦ κόψας ῥίπτει τῷ κυνηγῷ· ἐὰν δὲ πάλιν περιπέσῃ ἑτέρῳ κυνηγῷ καὶ διώκηται, ῥίπτει ἑαυτὸν ὕπτιος ὁ κάστωρ, καὶ νοήσας ὁ κυνηγὸς ὅτι ἀναγκαῖα οὐκ ἔχει, ἀναχωρεῖ ἀπ' αὐτοῦ.	There is an animal called beaver, which is perfectly harmless and peaceful. Its testicles have therapeutic properties. When it is chased by hunters and realises that it is caught, it cuts off its testicles and throws them at the hunters. And if it encounters a hunter again and is chased, the beaver lies down on its back and the hunter, noticing that it has no testicles, leaves it and walks away.
Καὶ σὺ οὖν, πολιτευτά, ἀπόδος τὰ τοῦ κυνηγοῦ αὐτῷ· κυνηγός ὁ διάβολός ἐστιν· πορνεία ἐν σοὶ ἢ μοιχεία ἢ φιλαγυρία· ἔκκοψον τὰ τοιαῦτα καὶ δὸς τῷ κυνηγῷ διαβόλῳ, καὶ ἀφήσει σε, ἵνα καὶ σὺ εἴπῃς· "ἡ ψυχὴ ἡμῶν ὡς στρουθίον ἐρρύσθη ἐκ τῆς παγίδος τῶν θηρευόντων".	You too, Christian, give back to the hunter what is his. The hunter is the devil. Prostitution, lust and greed are in you. Remove these vices from you and give them to the devilish hunter and he will leave you alone, so that you too can say: "Our soul has escaped as a sparrow from the snare of the fowlers".[33]

In just a few words (ἥπιον πάνυ καὶ ἥσυχον) the author described the main personality
traits of the beaver, before announcing the therapeutic virtues of its testicles (τὰ δὲ ἀναγκαῖα
αὐτοῦ προχωροῦσιν εἰς θεραπείαν).[34] In general, the scant references to the physical aspects,
personality traits or supposed qualities and flaws of the animals, plants and minerals do not
allow for in-depth analysis and comparison with other works to better understand what
influenced the writer of this text. It is even more difficult to identify those authors and texts
he knew first-hand. Apart from the limited amount of natural history information in the
text and the numerous gaps due to the disappearance of several older works containing
natural history material, this constraint is also due to the author's own silence about the
scientific sources he used. This silence must be related to the primary aim of the author of
this work: indeed, and in all likelihood, he did not so much want to describe the actual
behaviour of the various species, but rather, to present their properties, often marvellous,
as moral or religious symbols. Their φύσεις is often fictitious, and under the false pretence
of a work of natural history, used to explain the parables and teachings of the Bible with a
parenetic objective. Pagan science is harnessed to the Christian faith. It is therefore easier to
understand why the writer of the *Physiologus* omits to mention his pagan scientific sources.

In order to better understand his approach, we must contextualise the time of the genesis of the *Physiologus*. This work, as has already been mentioned, came into being during the first centuries of our era, i.e., at a time when Christians had contrasting positions on pagan knowledge.[35] Whereas "theologians" of the early 2nd century, such as Justin the Martyr, still saw no contradiction between human *logos* (which included pagan knowledge) and divine *Logos*,[36] the end of this century was marked by the emergence of virulent opponents to pagan knowledge. Theophilus of Antioch (2nd c.) contrasted divine knowledge – which was certain – with human knowledge, which was uncertain.[37] He was not alone in opposing pagan knowledge, like Irenaeus (c. 130–c. 202, pseudo-Justin (c. 100–c. 165) or Clement of Alexandria (c. 150–c. 215), for instance. Also worthy of note is the *Didascalia apostolorum*, where we read that the Bible is sufficient to define a Christian culture, and where pagan knowledge is rejected in the name of pure biblical knowledge (2, 6).

Even if one might question whether these (isolated) fragments are truly representative, there certainly was some lingering mistrust. In response, the Alexandrians, mainly in the late 2nd and early 3rd century, were compelled to justify the Christian use of pagan texts by granting them a subordinate position. For many Christian thinkers, ancient Greek *paideia* became a simple, but necessary, Christian *propaideia*. In their eyes, Greek knowledge did not serve to better understand the Bible literally, but could help to better grasp it through allegory. It is precisely in this context that the Greek *Physiologus* was born.[38]

In this work, pagan knowledge of natural history was used to better understand certain aspects of the new religion, thanks to a symbolic-allegorical interpretation of the physical aspects and personality traits of animals together with the supposed qualities and flaws of plants and minerals. This is why so little naturalistic information was provided. The author of the *Physiologus* offered his readers only the necessary facts of natural history, to gain a deeper understanding of Christianity (and in a way, of pagans, Jews and heretics),[39] so that he could adopt an attitude consistent with the life of a good Christian. The natural world was used in this work to explain the parables and teachings of the Bible. As Fr. Sbordone aptly wrote, the author of the *Physiologus* is first and foremost "esegeta della natura secondo i canoni della fede cristiana".[40]

In these stories, nature often loses its objective reality and becomes the background for religious morality; the description of each individual animal, sometimes extended to a variety of characteristics (for instance the chapter on the viper), produced an allegorical interpretation that often differed from one recension to another. The main religious justification for these traits was that animals embodied Christ's passion, his resurrection, chastity, fasting and more generally, his struggle against evil.[41] For the author of the *Physiologus*, secular texts were useful tools to help his reader perfect his Christian education and lead a life in accordance with the principles of the new religion. The structure chosen for each chapter, split into two main parts (one devoted to the διήγησις, the other to the ἑρμηνεία), contributed to this objective.

As we will see in the following pages, animals were used in a similar perspective by the authors of the *Hexaemera*.

The *Hexaemeral* Literature

The title *Hexaemera* refers to the exegetical works on the events associated with the six days of Creation.[42]

The best known non-Christian commentary on Genesis is the *De opificio mundi* by Philo of Alexandria. As early as the 1st century CE, this Hellenised Jew commented on the first chapters of Genesis, which deal with crucial issues concerning the creation of the universe, animals and man.[43] A Pythagorean and Platonist, Philo envisioned Creation on the basis of arithmetical symbolism and a corporeal, sensible world copied from an intelligible, incorporeal world also created by God, to serve as an archetypal model. Philo's aim was to convince pagan philosophers of the validity of the biblical account.

Theophilus, bishop of Antioch after 168, produced the earliest Christian *Hexaemeron* and the Church was clearly interested in this type of work given that we have some testimonials that point to lost versions of the *Hexaemera* in the years that follow until Basil of Caesarea (330–379).[44] Besides the latter, Gregory of Nyssa (c. 335–c. 395), John Chrysostom (c. 347–407), John Philoponus (c. 490–c. 570), George of Pisidia (7th c.) and Anastasius of Sinai (died after 700) are among the best known authors of *Hexaemera* in Greek. We also have Syriac *Hexaemera*,[45] like those by Jacob of Edessa in the 7th century or by Abū-l-Farağ Gamal al-Dīn, known as Bar Hebraeus, in the 13th century. Finally, as early as the 4th century, theologians in Western Christendom wrote commentaries on Genesis in Latin: Ambrose of Milan and Augustine of Hippo were followed by Bede the Venerable, John Scotus Eriugena, Peter Abelard and Henry of Langenstein.

It was in the context of the liturgy that these Christian *Hexaemera* came into being. In the East, in Antioch, Caesarea, Constantinople, as in Milan in the West, Genesis was read during the first week of Lent. The reading was resumed during the Easter Vigil. As I stressed before, the inspired exegesis of the first chapters of Genesis often goes beyond the strict character of commentary and takes on various literary forms. The *Hexaemera*, in all these forms, occupy a prominent place in patristic literature as well as in medieval literature. They constitute a voluminous and rich corpus, even if several *Hexaemera* have only survived in fragments or, indirectly, via the exegetical chains of Genesis, while others have definitively been lost.

Basil the Great

The *Hexaemeron*[46] of Basil the Great (4th century CE) is the earliest Christian document devoted exclusively to the subject of the six days of Creation and is one of the most important in the history of the *Hexaemera*. We have nine homilies on the *Hexaemeron* that Basil preached in Caesarea in the years before his episcopate (370), or even later,[47] which would better explain the apology of his brother, Gregory of Nyssa.[48]

Basil's commentary, which follows the 26 verses of Genesis, is primarily exegetical. However, it is a polemic against the philosophical theses that were deeply entrenched in people's minds at the time. The audience was mixed: it comprised both craftsmen and educated people. The orator, either to instruct or to embellish his remarks, allows himself numerous digressions in which he demonstrates his knowledge of natural history or geography. The fact remains that his aim was essentially to provide a Christian teaching. He warns, from the outset, that he is drawing his inspiration from faith and not from philosophy. His investigation into the constitution of the world, and his contemplation of the universe were not embedded in the wisdom of this world, but came from the teachings which God dispensed to his servant, when he spoke to him visibly, and not by enigmas.[49] His constant concern was to show the coherence and wealth of the sacred text and to draw from it some lessons for living a Christian life.

From the seventh homily onwards, Basil details animal species and seeks moral examples for the lives of Christians. Basil's intention was to depict various animal situations and behaviour as moral *exempla*, maybe as Aesop had once done, to criticise human flaws.[50] Basil encourages Christians to follow these natural models or be wary of the anti-models: "But I wish you would at once rival the crab in cunning and industry, and abstain from harming your neighbour; this animal is the image of him who craftily approaches his brother."[51] More precisely, the last three homilies of his *Hexaemeron* concern the creation of the fish, birds, beasts, herbs and trees. We find for the first time in the *Hexaemeral* tradition, stories about the animal and plant kingdoms, illustrating the idea that Providence always creates things that are useful for humans and beasts, even when they seem to be harmful, and that it endows irrational creatures with the instincts of self-preservation. In truth, most of his examples come from Aelian and Aristotle, but Basil probably found them in the *Physiologus*.[52]

As A. Zucker wrote, these three homilies on Creation testified to Basil's classical knowledge of zoology. Basil writes about "fish" (except crabs and octopuses) in general terms, but distinguished birds (crane, swallow, stork, as well as bats and bees) and quadrupeds in great detail. While he lists different ethological characteristics of animals, which were already documented in earlier literature, he is clearly fascinated by the wonderful anatomical variety of animals and their skills.[53]

Basil is particularly interested in how animals take care of themselves. He reports, for example, that bears, foxes, turtles and even snakes treat themselves with medicinal herbs.[54] George of Pisidia (see below) writes similar things about the falcon, panther and other animals, comparing them to the famous doctors, Hippocrates, Galen and even Asclepius.[55]

In his *Hexaemeron*, Basil summarises the science of his time, which gives his homilies an encyclopaedic character. He draws from various ancient sources, including Aristotle's *Historia Animalium*, which he probably had access to via secondary literature and epitomes; his text presents many parallels and borrowings.[56] He includes an Aristotelian discussion on how fish breathe and is familiar with taxonomic classifications. In this regard, it should be noted that, while initially employing Aristotelian categories, Basil then suggests other possibilities of classification.[57]

This *Hexaemeron* was often imitated in later times. Some of the later *Hexaemera* are largely based on Basil's work, e.g. Ambrose (see below), Augustine,[58] the pseudo-Eustathius (see below), Philoponus (see below), Pisides (see below), John the Exarch,[59] Glykas (see below), and Guillaume de Salluste Du Bartas.[60] Basil's work also served as a model for other works.[61] The Latin translation of his *Hexaemeron* by Eustathius was the main inspiration for the Fathers of the western Church, who knew some of it via Ambrose.

Ambrose of Milan

In his *Hexaemeron*,[62] Ambrose of Milan (c. 339–c. 397) describes Genesis' literal and mystical meaning. His dependence on Basil is evident, especially in its moral applications, when he describes the animals and birds of Creation. However, Ambrose also knew how to distance himself from his illustrious predecessor. He sometimes contradicts or refutes him, but without naming him. In particular, he distances himself from Basil's philosophical concerns. He adapts the geographical allusions to the West and speaks in particular of the Rhone River. The contribution of the bishop of Milan is thus specific to him, as in the case of the partridge.

Indeed, whereas in Basil's *Hexaemeron* it is only mentioned that the partridge, treacherous and jealous, helps the hunters to seize their prey,[63] Ambrose's text is far more complex. The latter's account of the theft of other birds' eggs by the partridge[64] follow the *Physiologus*[65] in sharing the same biblical verse (*Jeremiah* 17, 11) and the same symbolic-allegorical interpretation of an unjust acquisition of property: eggs for the partridge / wealth for an individual. On this basis, it is assumed that the bishop of Milan had knowledge of a copy of the Greek *Physiologus* or a Latin translation, unless it was a common source, a hypothesis that cannot be ruled out in the present state of our knowledge.[66] Moreover, this detail somewhat reinforces the opinion contrary to the traditional idea, which sees Ambrose as a blind imitator of Basil.[67] In general, a careful comparison of their works clearly shows significant differences between them. It is only recently that some scholars, like R. Henke, L. Gosserez and G. Nauroy, have begun to oppose this idea of a servile imitation, pointing out that Ambrose had proceeded to developments, abridgments and other modifications of his model.[68]

Pseudo-Eustathius

The Pseudo-Eustathius' commentary in *Hexaemeron* was ascribed to Eustathius of Antioch in error.[69] In fact, it was probably written in the late 4th century or at the beginning of the 5th. It contains a wealth of zoological data but it is just a compilation of known excerpts;[70] the work builds on various sources (including novels like Achilles Tatius' *Leucippe and Clitophon* when describing a crocodile),[71] from ready-made compilations, particularly in the passages that comment on the 5th and 6th days.

As A. Zucker has already pointed out,[72] the main points of divergence between it and Basil's *Hexaemeron* are the short zoological chapters, with the description of striking characteristics among 67 animals from sawfish to cicada (5th day, PG 18, 724–37) and from lion to flea (6th day, PG 18, 737–50), which are no longer part of theological debates or moralisations but presented in a naturalistic fashion. There is little evidence in the text that this is a theological work, as the author lists zoological curiosities without any explanation. He probably thought that the reader, by reading these colourful stories, would reflect on the infinite greatness of Creation. By combining various parts from the *Physiologus* to Basilean segments, the zoological section of this *Hexaemeron* provides us with a number of sketched portraits, the presence of which is justified by stating that showing animal variety offers us knowledge on divine Providence.[73]

On several occasions the author of this *Hexaemeron* relies on Basil's explanations. This is the case, for example, when he tries to explain why birds and fish, two different groups of animals, are found next to each other on the fifth day of Creation. Paraphrasing Basil, he points out that fish, like birds, move by swimming, some in water, others in the air (PG 18, 728). He also describes where fish live: they can live in seas, in rivers, in swamps and along the coast. He mentions some common characteristics, such as breathing through the gills. He also talks about their diet, saying that all aquatic animals should be considered "fish". He then mentions the different species of fish such as octopuses, crocodiles or hippopotamuses.

This zoological catalogue of sorts, with the exception of one page on God's providence (τὴν Πρόνοια τοῦ Θεοῦ, PG 18, 737A) when introducing the 6th day, is discarded in the Ethiopian and Arabic versions of the text in the *Adam book*.[74]

John Philoponus

John Philoponus (6th century CE), drew on his vast knowledge of classical scholarship.[75] In his *De opificio mundi*,[76] Plato, Aristotle, and the astronomers are his main source of inspiration to defend Basil's position, in particular, against Theodorus of Mopsuestia.[77] Indeed, in his *De opificio mundi*, commentary on the Creation stories in seven books, John Philoponus refutes the Ἑρμηνεία τῆς κτίσεως (*Fragmenta in Genesim*)[78] by Theodore of Mopsuestia and his disciples, among whom Cosmas Indicopleustes (see below),[79] and supports the idea that the biblical "cosmogony" is compatible with the realities of nature (concordance theory). Following Basil, and adopting his linguistic argument from Aquila of Sinope's translation of the Bible, Philoponus argues that matter, space and time were created instantaneously. According to Philoponus, in the sublunary world, Creation followed an order from the simplest to the most complicated, or more precisely, from the least perfect to the most perfect beings.

In his *De opificio mundi*[80] Philoponus claimed to demonstrate the scientific validity of biblical notions on nature, and commented on Gen. 1 through a series of issues like "why land-animals come after fish in the creation", "why water-animals and birds are created together", or "do animal souls perish with the body".[81] He mentions for instance that the octopus and the *adonis* (flying fish) are amphibians and explains the etymology and meaning of ἕρπετα for water animals as well as the old use of θηρία (for animals) by relying on biblical examples.[82]

He was interested in animal intelligence, describing the intellectual gradation from plant to humans and the superiority of land animals over other animals.[83] In the order of creation, the first animate beings created were plants, which although they possess the faculty of feeding, growing and engendering a being similar to themselves,[84] have neither sensibility nor movement. Aquatic animals were the second animate beings to be created, since water is the second element after earth in the Aristotelian hierarchy (earth, water, air, fire). They are more perfect than plants (they have the faculties of movement and sensibility), but less so than aerial beings, because water, being thicker than air, allows less light, sound and odour to pass through, and therefore lessens the sensations of aquatic beings compared to aerial ones. In third place come the birds (air is the third element). They are more perfect than fish, for their senses and imagination are more acute, but they are mostly wild. Fourth are the land animals, which are even more perfect, because some of them, the domestic animals, are reasonable. Last comes the most perfect of all animals, human beings (τὸ τελειότατον ζῷον ἁπάντων).[85]

Cosmas Indicopleustes

Cosmas Indicopleustes (6th century C.E.) is the author of a work known as *Topographia Christiana*.[86] He was probably a spice merchant who travelled a lot, in the Mediterranean, in the Red Sea and in the Persian Gulf, up to Somalia, but he never reached India.[87] A native of Antioch, Cosmas had no formal education, but had read extensively. He regarded Father Mar Abaas as his master, a katholikos (archbishop) of the Oriental Church whom he had met in Alexandria.

Mar Abaas was a follower of the Nestorian School of Nisibia in Persia, an important centre for teaching the doctrines of Theodore of Mopsuestia. Cosmas himself was a strong supporter of Theodore of Mopsuestia. The latter's views include two points that contradict Basil's dogma: that the earth is flat and that there was no pre-creation of an intelligible world with angels. Like John Chrysostom, Theodore did not believe that Hellenic science could contribute to a better understanding of Creation and the organisation of the world.

The zoological section is in the eleventh book of Cosmas' *Topographia Christiana*, whose title is: "Description of Indian animals, Indian trees and Taprobane Island" (Καταγραφὴ περὶ ζώων ἰνδικῶν καὶ περὶ δένδρων ἰνδικῶν καὶ περὶ τῆς Ταπροβάνης νήσου). Eleven species are described. Cosmas often describes these animals in great detail as someone who had seen them with his own eyes during his long journeys.

Unlike Timothy,[88] who only states that the rhinoceros is as big as a hippopotamus, Cosmas describes it in great detail.[89] He points out that the animal's largest horn is on its forehead and is so strong that it can even uproot trees. The rhino's eyes are very low, next to its jaw, and its feet and skin resemble those of the elephant. Its dry skin is so hard that some people use it as a ploughshare. Its greatest enemy is the elephant. Cosmas even gives the explanation of the origin of the Ethiopian name of the animal (Καλοῦσι δὲ αὐτὸ οἱ Αἰθίοπες τῇ ἰδίᾳ διαλέκτῳ ἀρουὴ ἄρισι).[90] He admits that he only saw this animal from a distance, in Ethiopia (presumably because of the peril involved). The next paragraph of his treatise is devoted to the *taurelaphos*, which lives in Ethiopia and India.[91] The species that live in India are tame, able to carry burdens and are used for their milk, from which butter is made. W. Wolska-Conus, in his edition, identified this animal as the buffalo. This may be the Indian *arni* buffalo (*Bubalis arni*) or the African caffar (*Bubalis caffer*) about which Philostorges has already written.

According to Cosmas, the giraffe lives only in Ethiopia ("Ἡ δὲ καμηλοπάρδαλις ἐν τῇ Αἰθιοπίᾳ μόνη εὑρίσκεται").[92] Although it is a wild animal, one or two specimens had been captured at an early age and were kept at court for the king's pleasure. Regarding the habits of the giraffe, the author makes a wise observation: when milk or water is put into a basin in the presence of the king, the animal can only bend down to drink by spreading its front legs, because its legs, torso and neck are too tall. The author proudly adds to this observation that he is relating something he had witnessed himself.

Among the other animals described, Cosmas clearly differentiates the rhinoceros from the *monokeros* (unicorn), which also lives in Ethiopia.[93] He admits that he has not seen this animal, but immediately adds that he has seen four bronze statues of it in the palace of the Ethiopian king. According to him, the animal's solid horn makes it invincible. On the other hand, he claims to have seen the *choirelaphos*, and even to have eaten its meat.[94] This may be the warthog (*Phacochoerus aethiopicus*) with short, strong tusks.

In the *Topographia Christiana*, Ethiopia and India appear together, but this confusion was very common in antiquity and the Middle Ages. On the other hand, we should note an originality in his work. Indeed, although he is not the only one to do so, Cosmas makes sure to mention his personal experience, the *autopsia*, each time.

Jacob of Edessa

Through his exegetical method, Jacob of Edessa (c. 640–708) is linked to the School of Antioch.[95] There is a great affinity between Philoponus' *Hexaemeron* and that of Jacob of Edessa. The latter's work[96] was also the basis for several later essays of the same kind (e.g. on the *Hexaemeron* by Moses bar Kepha).

Jacob of Edessa's *Hexaemeron*, which he wrote towards the end of his life, is the greatest expression of his profound classical learning as well as his Christian background. According to Abbé Martin, this work "c'est moins une œuvre de théologie qu'une œuvre de science. Il y a sans doute de la théologie, comme dans les œuvres des Pères grecs et latins qui portent le même nom, comme dans l'*Hexaméron* de saint Basile ou dans l'*Hexaméron* de saint

Ambroise; mais il y a en plus une revue générale et assez complète des sciences physiques ou naturelles telles qu'on les connaissait en Syrie au VIIe siècle".[97]

Jacob of Edessa divided his *hexaemeron* into seven versified homilies or discourses (*mimre*). It begins with a dialogue between the author and one of his disciples, named Constantine. The fifth and sixth homilies deal with natural history. "Of the animals and creeping things which God brought forth from the waters and of the birds which he also brought forth from the waters" is the title of the fifth discourse (*mimra*). The author delivers a treatise on natural history, speaking of fish, reptiles and birds according to Gen. I, 20–23. The other animals on Earth are the topic of the subsequent discourse (Gen. I, 24–25), entitled: "Of cattle and wild beasts and all reptiles that creep on the earth".

These two parts of the work are therefore a kind of natural history treatise, but Jacob of Edessa relies heavily on earlier treatises.[98] Among other things, the fifth *mimra* deals with bees, silkworms and parrots. With regard to the parrot, Jacob of Edessa notes that it is famous for the songs it can be taught. He adds that some patient people once taught a parrot to say in Greek: "You are holy, God, you are holy, almighty; you are holy, immortal, who was crucified for us; have mercy on us". It should be noted that this anecdote was already known from a poem by Isaac of Antioch.[99]

In his sixth discourse, Jacob deals with domestic and wild animals, he describes the species and their properties. The elephant, the camel, the dromedary, the "river dog", which he says is found in Egypt and Syria,[100] and the "ordinary dog" are among the animals that attract his attention. He reports several stories about dogs, stories that he says he borrowed from Greek poets. On the subject of the elephant, he writes that naturalists (he does not name anyone in particular) claim that it remains in its mother's womb for two years and only reaches full growth after 12 years. He also explains that it serves as a mount and carries burdens commensurate with its strength. According to his account, it is led to war and trained to fight enemies and take revenge on them.

For this purpose it is given wine mixed with incense and in its drunkenness becomes angry and violent against its enemies. He also specifies how the elephant uses its trunk before moving on to a fairly detailed morphological description of the pachyderm. He points out that God gave the elephant feet like columns, straight feet that do not bend, because they have no joints. Consequently, according to Jacob, the elephant does not lie down on all fours, like other animals on earth, because, if it were to lie down, its leg muscles would have difficulty in raising its vast bulk to its feet. He then points out that when the elephant wants to sleep or rest a little, it leans against strong and vigorous trees, or against a wall.[101] Jacob is not the only one to believe that the elephant could not bend its legs. Basil of Caesarea[102] and many others, according to an ancient tradition, thought similarly.

George of Pisidia

George of Pisidia's (7th century CE)[103] *Hexaemeron*,[104] at 1,898 lines is the longest poem, probably elaborated in stages over several years. The preserved version dates from after 632.

Both in its literary form and in the deployment of its content and structure, this work constitutes a decisive transformation of the literary genre of the *Hexaemera* and a distancing from the patristic writings, some of which have been discussed above. The poem does not follow the six-day narrative of Genesis. For George of Pisidia the world is already created. Patristic and biblical exegesis give way to the winged discourse of poetry, a sign of a deeper and more personal interiorisation of Christian truths enshrined in the substance of language. As E. Delli rightly pointed out, "les traités systématiques et souvent polémiques ainsi que les analyses rigoureuses concernant les origines et la fonction du cosmos ne semblent pas indispensables dans la mesure où le christianisme triomphant n'a plus à craindre les doctrines de ses adversaires païens".[105] According to George of Pisidia, scientific reasoning is not enough to unlock the secrets of the universe, for there is a surplus of meaning that resists the measurable, the explicable and the nameable. Where syllogisms prove powerless to capture the true causes of natural phenomena, faith can give courage to speak and write about Creation (πίστις δὲ θαρρεῖ καὶ λαλεῖν καὶ συγγράφειν[106]). One notices in his words a certain distrust of science.

The penultimate part of the poem (v. 899–1609) concerns plants and animals (quadrupeds, fish, insects and birds).[107] George of Pisidia describes their characteristics, their roles and their relationships within the harmony of the cosmos. The author also considers the various plant and animal species as manifestations of divine energy, sometimes transposing their physical qualities to the spiritual plane and thus aiming at the moral edification of the reader. In this section with 47 animals – most of which are not listed in Basil's or in Pseudo-Eustathius' works – the majority of the animal characteristics are drawn from the *Physiologus*. George of Pisidia offers numerous zoological details with a wealth of veterinary terminology. He explains how eels reproduce by rubbing against rocks (v. 1020–1028),[108] for this fish has no womb and produces no semen. He goes on to say that the skin torn from it becomes its offspring. In this way he demonstrates the truth of the Bible regarding Eve's origin from Adam's rib. In fact, he wants more than just to praise Creation's marvels or describe the natural science God expressed through animals – the ibis is wiser than Galen (Ἴβις δὲ ποίων εὐπορεῖ διδασκάλων καὶ τοῦ Γαληνοῦ δεικνύει σοφωτέρα, v. 1117–1118) – he wants to prove the mysteries of the Christian faith. The vulture's parthenogenesis demonstrates that the conception of the Virgin (v. 1124a–r) occurred in nature, and the silkworm which rises from the grave-like cocoon, substantiates the resurrection narrative (v. 1251–1275), just like the swallow that somehow arises in springtime as did Christ (v. 1276–1293).[109] As has been pointed out, in these stories the author repeatedly refers to the lessons that humans can learn from the animal world.

For this section on plants and animals, influences from Claudius Aelianus (*De natura animalium*), Timotheus of Gaza (*De animalibus*) and Theophrastus (*Historia plantarum*) can be identified. Regarding medical themes, he was inspired by Hippocrates and Galen, and for pharmacology and medicinal plants, by Pedanius Dioscorides (*De materia medica*). There are also Stoic echoes (on the presence of God everywhere in the universe, on cosmic sympathy, on the teleological construction of each being, and on the role of Providence) as well as echoes of the *Corpus Hermeticum*. Finally, George of Pisidia draws on the *Cyranides* and the *Physiologus*.[110] In turn, George of Pisidia influenced several authors, even in the post-Byzantine period, with Guillaume de Salluste Du Bartas.[111]

Anastasius of Sinai

The *Hexaemeron*, traditionally ascribed to Anastasius of Sinai (7th–8th centuries), may be one of the most important works of Christian mysticism from the Byzantine era.[112] Unlike other similar treatises, Anastasius of Sinai does not limit himself here to the first chapter of Genesis, but continues his reading until the end of the third chapter, in a resolutely symbolic and mystical exegesis. This does not annul literal exegesis such as can be found in Basil of Caesarea or John Philoponus, but goes beyond it in order to search for "the mystical meaning hidden under the letter".[113]

The author quotes an impressive list of scholars: Philo of Alexandria, Justin, Papias, Irenaeus, Pantaenus, Clement of Alexandria, Origen, Gregory of Nyssa, Gregory of Nazianzus, Epiphanius of Salamis, Cyril of Jerusalem... . In addition to this patristic erudition, he included a wide range of secular sciences (following the example of John Philoponus), evoking the phases of the moon, the phenomenon of the tides, the signs of the zodiac, the symbolism of numbers, and the variations in the level of the Nile. Although the date of composition of this text is still in question, it reflects a seventh-century vision.

Anastasius of Sinai is mainly interested in a Christian perspective and offers simplistic comments on animals; he identifies apostles and sinners with marine monsters and fish respectively.[114] He interprets God's exclusive blessing to fish and humans in Genesis by stating that according to fishing specialists, fish is conceived without semen (τίκτουσιν αὐτομάτως χωρὶς σπορᾶς[115]) from the (spirit of the) wind. In the preface, he explains that he is addressing an audience of initiates, and urges uninitiated listeners not to listen: "the occasion demands that I speak poetically. And if I sing to those that understand, then you that are profane, shut the doors on your ears!" He goes on to explain his intentions: to reveal, in a priestly and ritual context, a secret teaching hidden between the lines. The key to his reading of the origin story lies in the analogy between Creation and the mystery of Christ and the Church: minute details in the biblical text are interpreted as a veiled announcement of this mystery by Moses, the author of the Creation story, who had foreseen the Economy of the Incarnation.

Cosmas of Jerusalem

Cosmas of Jerusalem (vel Cosmas Magnus Melodus, 7th–8th centuries), one of the most important hymnographers of the Byzantine church, was probably born in Jerusalem (or, according to other scholars, in Damascus). As the legend goes, orphaned from an early age, he was adopted by the father of John of Damascus, and the two foster brothers were bound by a friendship which lasted throughout their lives. Cosmas ὁ Μελῳδός became a monk in the famed Great Laura of Saint Sabbas (Kidrov Valley in the Judah Desert, 12 km east of Bethlehem). There, around 740, he was consecrated bishop of Maiouma (a harbor of Gaza).[116]

His work testifies to how naturalistic knowledge spread far beyond *Hexaemeral* commentaries. When discussing some of the works of Gregorius of Nazians and especially the *Carmina*,[117] he expands the allusions to various animals that echo the *Physiologus* and Basil's *Hexaemeron* on sepia, mule, deer, generative stones, turtle, amphibians, asp, phoenix, remora, turtle-dove, crow, burning fish, lodestone, diamond, stone-shining-in-the-water, starling, kite, parrot, bear, lion, elephant, crab, deer, bee, cow, swallow, crow, dog, lioness, dolphin, eagle, viper, cicada, eagle, *dipsas*, flea, chameleon, jackdaw and wasp.[118]

Michael Glykas

Michael Glykas (12th century) is best known for a chronicle (*Annales sive Βίβλος χρονική*),[119] that tells the story of the world from Creation until 1118, the year of the death of Alexios I Komnenos (in four parts: Creation; from Adam to Julius Caesar; from Julius Caesar to Constantine; and the empire after Constantine). This Chronicle must be distinguished from other works in this genre. Michael Glykas elaborates an account of Creation that contains abundant information on animals (and plants and stones)[120] and the long portion on the 5th and 6th days has often been transmitted as a separate text. In one instance, it was even entitled "Physiologus," in the *Paris, BnF, gr.* 1612 (fol.127[r]).[121]

Glykas' first remark in his *Annales* concerns a passage in Genesis about the fifth day, explaining why fish die when removed from water.[122] In fact, this problem is not suggested by the Scriptures: it was something Aristotle wondered about – just like the following remarks on fish denture and the absence of rumination, except for the parrot-wrasse – in *De respiratione*, and Glykas' explanations are directly inspired from Basil's *Hexaemeron*.[123] Most comments on the creation of animals are recycled from the Greek tradition, but Glykas seems to have had access to a wide range of sources, like Herodotus, Aelian, Plutarch, Basil, George of Pisidia, Anastasius of Sinai, the *Physiologus*, Nicander, Pseudo-Alexander's *Problemata*[124] and Theophilaktos Simokattes.

We find in Glykas' text an attempt to explain the virginal conception, according to which Mary conceived Jesus of Nazareth while remaining a virgin, based on the alleged birth customs of vultures.[125] A similar explanation is already used by George of Pisidia. Likewise, in connection with the phoenix, Michael Glykas specifies that it is a bird of Arabia (ὁ φοῖνιξ ὁ Ἀραβικὸς ὄρνις λέγεται[126]), a piece of information already reported by Herodotus (ἐξ Ἀραβίης).[127]

Glykas uses Aristotle in his *Annales*, for he knows his work well, particularly Aristotle's zoological writings. But his re-use of the original texts is not literal: he reduces the data, combines information from different books, supplements the text with passages borrowed from other scientific or Christian sources and even adds his own information or thoughts. His great knowledge of Aristotle's writings reflects a common interest in the ancient author at the time. Interestingly, Glykas' *Annales* has practically no point of convergence with Michael Ephesus' commentaries (*In librum de animalium incessu commentarium*, *In libros de animalium motione commentarium*, *In libros de partibus animalium commentaria*, etc.) which were viewed at the time as the most important work on the topic. The authors draw their information from Aristotle in different ways. In other words, given that they were contemporaries, each represented an independent tradition in the interpretation of Aristotle's works.[128]

Constantine Manasses

Another text of the first six days of creation is the ekphraseis of creation, which opens the verse chronicle written by Constantine Manasses (τοῦ κυροῦ Κωνσταντίνου τοῦ Μανασσῆ Σύνοψις χρονική) in the mid-twelfth century. According to Ingela Nilsson, Manasses also wrote independent ekphraseis, five of which, have survived: Ἔκφρασις γῆς ("Description of the Earth"), Ἔκφρασις κύκλωπος ("Description of a cyclops"), Ἔκφρασις ἀνθρώπου μικροῦ ("Description of a small man"), Ἔκφρασις ἁλώσεως σπίνων καὶ ἀκανθίδων ("Description of the catching of siskins and goldfinches"), and Ἔκφρασις κυνηγεσίου γεράνων ("Description of a crane hunt").[129]

In his Synopsis Chronike, the six days of creation cover about 250 verses and describe, in an extended series of longer and shorter ekphraseis, the creation of Heaven and Earth, flora and fauna, Adam and Eve. The day five contains the creation of the fishes, birds, and animals (v. 144–173). Manasses mentions the eagle, who is the king of the birds (ὀρνεοκράτωρ ἀετός), the falcon, the sparrows, the goldfinches, skylarks, finches, starlings.

He also mentions the lion, the leopard, the tiger, the bear, the wild boar, the broad-breasted elephant (εὐρύστερνος ἐλέφας), the dogs with pointed teeth (κύνες καρχαρόδοντες), and hares on winged feet (πτηνόποδες λαγῖναι). As Henry Maguire explained so well, "the ekphraseis of creation by Constantine Manasses, with its strongly poetic evocation of nature echoing the early Hexaemeron commentaries, was an exception in medieval Byzantine literature".[130]

Neophytos of Cyprus (or Neophytos the Recluse)

Neophytos of Cyprus – or Neophytos the Recluse, 12th–13th centuries – was a Cypriot Orthodox monk and priest whose writings preserved a history of the early crusades. In contrast to what the title suggests (Ἑρμηνεία τῆς ἑξαημέρου), Neophytos' work does not simply comment on the first six days of Creation, but takes all of Genesis into account in the 16 chapters of his work. According to the scientific editors, he wrote his work after 1196.[131]

Reptiles, birds and marine monsters (κήτη)[132] are commented on in Chapter 5. The author lists, among others, the dolphin, the tuna, the swordfish, as well as griffins, vultures, eagles, cranes, geese, partridges and ravens. The next chapter discusses sheep, horses, camels, donkeys, elephants, deer, rams and hares. The author also reviews wild beasts such as lions, leopards, panthers and bears, as well as snakes like the asp. Finally, the author also mentions "mythical" creatures (onocentaur, *monokeros, basiliskos,* and *kerberos*).

Conclusion: Animals in Christian Thought and Spirituality

People in the Christian Middle Ages never ceased to ponder on the subject of animals and, in so doing, constantly brought them to the forefront, introduced them into the churches and gave them an importance in texts and images[133] that they would probably never find again.

The cornerstone of medieval and Byzantine thought and attitude towards animals is the passage in the first chapter of the first book of the Bible, Genesis, of their creation by God – first the sea animals, then the animals of the sky and finally land animals (Gen. I, 20–22) – with the clarification that this creation was entirely good in God's eyes (καὶ εἶδεν ὁ Θεός, ὅτι καλά, and God saw that it was good, Gen. I, 22). This creation by God gives animals an objective value, and implies from the outset that human beings should respect them[134] because by their very existence they bear God's mark.

According to the Fathers of the Church, all of nature's creatures bear witness to the existence of God and the permanent activity of His providence, and each one, to the extent of the possibilities defined by its nature, praises and gives thanks to Him. Animals do this to a greater and better extent than all other creatures of the plant and mineral worlds, not only because of their variety and magnificence (which is also found in other realms), but also because of the greater complexity of their organisms and their capacity for expression, which was considered a sort of language. The Fathers thus often see in the song of birds a form of praise that they unconsciously address to God.[135]

As was pointed out in the introduction to this chapter, in relation to animals (as with all other beings in nature), humans are supposed to see in them the mark of God and through them to ascend in spirit to God. He must also give thanks to God for them. Human beings were created as mediators between nature and God, on the one hand because they gather and combine all the elements of nature in themselves, and on the other hand because, through their reason and intellect (or spirit), they can connect all of nature's creatures to God in their contemplations and prayers. As the king of Creation, gifted with intelligence and speech, the human being is the conscious spokesman of all beings before God, their leader and respondent.[136]

Animals, unlike humans, have no rational soul or intellect. According to Gregory of Nyssa (*Antirrheticus adversus Apollinarium*)[137] for instance, what makes the human soul so special (ὅπερ ἴδιον τῆς ἀνθρωπίνης ἐστὶ ψυχῆς[138]), is the human's intellective faculty (νοητὴν δύναμιν).[139] Animals, on the other hand, are characterised by instinctive behaviour, in which reason and will play no role and are not the result of a conscious, free will. With a few exceptions, many Christian scholars believed that animals were inferior to humans and, unlike humans, were not immortal.[140] In this Christianity differs from Hinduism and all forms of pantheism which consider all beings to be elements of a great undifferentiated whole. Despite this being the case, its vision is not an arrogant and absolute anthropocentrism either, but it considers that there is a hierarchy in Creation that was willed by God himself, and which is also expressed between the creatures of the animal kingdom and those of the plant and mineral realms. This hierarchy also exists between the animal species, the complexity of which varies considerably, as we have previously seen.

Of course, this does not mean that medieval humans were not interested in animals. Far from it. In fact, texts as varied as treatises on agriculture, veterinary medicine and hunting, manuals on war, astrology and oniromancy, even satirical literature, fables and travel accounts, clearly show the extent to which medieval culture was curious about animals. Religious books played no small part, starting with the Bible, the supreme authority and source of all knowledge, considered not only as a sacred book but also as an immense natural history collection. Apart from the *Physiologus* and the *Hexaemera*, briefly discussed in the previous pages, many medieval texts speak of animals or depict them. This abundant literature certainly played an important role in medieval humans' conceptions, perceptions and representations of animals.[141]

It is this close relationship that led some thinkers, particularly in the West, to develop the idea of a true community of living beings and a kinship – not just biological – between humans and animals. This school of thought, which is remarkably modern, is the antithesis of the theories on animals set out above. Its origins are both Aristotelian – see in particular the *De anima* – and Pauline, especially the *Epistle to the Romans* (8, 21). According to M. Pastoureau, the words of Paul the Apostle in this verse have strongly influenced several theologians who have commented on them.[142] Some ask whether Christ really came to save all creatures and whether all animals are really "children of God". The fact that Christ was born in a stable seems to some authors to be proof that the Saviour came down to earth to save the animals as well. Other theologians, especially Protestants and scholastic Catholics, ask about the future life of animals: do they rise after death?[143]

The Byzantine Orthodox position on this subject is more nuanced. Probably with the sole exception of the phoenix, which according to legend is reborn from its own ashes, there is no mention in Byzantine writings of an afterlife for animals. The first reason given by the Fathers is that, unlike humans, they do not possess an immortal soul. A second reason is that the vision of God that the righteous will have in the Kingdom depends not only on the possession of the intellect and the spirit (for it is noetic in nature), but also on the degree of virtue – freely acquired by ascetic effort in synergy with God's grace – and that this notion is not applicable to animals. A rare passage that might have hinted at an afterlife for animals is a verse from the Psalms. Indeed, we read: ἀνθρώπους καὶ κτήνη σώσεις, Κύριε, "You save humans and animals alike, O Lord" (Ps 36 [35], 6). However, Eusebius of Caesarea[144] and Jerome[145] and Athanasius of Alexandria,[146] assign to the word "cattle" (κτήνη) an allegorical meaning and consider that it refers here to those among humans who behave like beasts. Τοὺς κτηνώδεις καὶ ἀνοήτους τῶν ἀνθρώπων, stated Eusebius of Caesarea.

In spite of these nuances regarding how animals were considered, they occupied a predominant place in the Byzantine mentality. Certainly, animals were at the service of a system for which they were not the sole purpose. Medieval thought was, in fact, primarily theocentric and, as a result, its literature was too anthropocentric to make animals its own topic. Animals only appear in this literature in relation to humans or to God, and the two examples, the *Physiologus* on the one hand and the *Hexaemera* on the other, have demonstrated this. This being the case, the questions that people in the Middle Ages asked about animals underline the extent to which Christianity was the vehicle of an extraordinary promotion of animals. Such a promotion had a direct influence on this medieval interest in animals, as I pointed out in the introduction to this chapter and as was highlighted by questions raised by the various authors discussed in this chapter. Their works clearly show how the Byzantine scholars viewed nature with reverence and how much they admired each of its details. Sometimes they saw moral teachings in the behaviour of animals, other times they simply admired the infinite variety of nature.

Notes

1 Τὸ γνωστὸν τοῦ Θεοῦ φανερόν ἐστιν ἐν αὐτοῖς· τὰ γὰρ ἀόρατα αὐτοῦ γὰρ Θεὸς αὐτοῖς ἐφανέρωσε. τὰ γὰρ ἀόρατα αὐτοῦ ἀπὸ κτίσεως κόσμου τοῖς ποιήμασι νοούμενα καθορᾶται [...], Romans 1.19–20.

2 See Evagrius Ponticus, ed. A. Guillaumont and C. Guillaumont, *Évagre le Pontique, Traité pratique ou le Moine*, vol. 2. Sources chrétiennes 171 (Paris, 1971), 92 (τὸ ἐμὸν βιβλίον, φιλόσοφε, ἡ φύσις τῶν γεγονότων ἐστί, καὶ πάρεστιν ὅτε βούλομαι τοὺς λόγους ἀναγινώσκειν τοὺς τοῦ Θεοῦ).

3 See Gregory of Nazianzus, *Orationes XXVII–XLV*, ed. J.-P. Migne, PG 36 (Paris, 1858), col. 57–64 (*oratio XXVIII*). As H. Maguire points out, "a typical ekphraseis of the peacock is contained in the Second Theological Oration of Gregory of Nazianzus. This sermon, which Gregory delivered in Constantinople in the year 380, incorporates a celebration of the created world that resembles the commentaries on the Hexaemeron." H. Maguire, *Nectar and Illusion: Nature in Byzantine Art and Literature*. Onassis series in Hellenic culture (Oxford; New York, N.Y.; Auckland, 2012), 52. On this text, see also the edition (with modern French translation) made by P. Gallay and M. Jourjon, *Grégoire de Nazianze, Discours 27-31*. Sources chrétiennes 250 (Paris, 1978).

4 See John Chrysostomos, *De statuis ad populum Antiochenum* 12.2–3, ed. J.-P. Migne, PG 49 (Paris, 1862), col. 16–222, at 128–32; *In epistulam I ad Thessalonicenses*, ch. 5, homily 10.2–3, ed. J.-P. Migne, PG 62 (Paris, 1860), col. 391–466, at 456–58; *Ad eos qui scandalizati sunt*, 7.30–33, ed. A.-M. Malingrey, *Jean Chrysostome, Sur la providence de Dieu*. Sources chrétiennes 79 (Paris, 2000).

5 Theodoret of Cyrus, *De providentia orationes decem*, *oratio* V, ed. J.-P. Migne, PG 83 (Paris, 1859), col. 556–774, esp. at 623–41.

6 Ὦ τὴν ἀνεξεύρητον ἣν ἔχεις φύσιν κρύψας ὁμίχλῃ καὶ περιστείλας γνόφῳ, ὅπως μάθοιμεν πῶς ἀεὶ κεκρυμμένη ἐκ τῶν ἑαυτῆς φαίνεται ποιημάτων, George of Pisidia, *Hexaemeron sive Cosmopoeia*, v. 81–84, ed. L. Tartaglia, *Georgio di Pisidia, Carmi* (Torino, 1998).

7 See Basil of Caesarea, *Homilies*, 9.4 (197A), ed. S. Giet, *Homélies sur l'Hexaéméron*. Sources chrétiennes 26 (Paris, 1968); Theodoret of Cyrus, *De providentia orationes decem*, *oratio* V, ed. Migne, col. 623–44.

8 See e.g., Clemens Alexandrinus, *Stromata (Buch I–VI)*, 1.9.43, ed. O. von Stählin, L. Früchtel, and U. Treu (Berlin, 1985), 28–29.

9 "πάντα δοκιμάζετε, τὸ καλὸν κατέχετε", 1 Thessalonians 5.21.

10 Basil of Caesarea, in his work *Address to Young Men on Reading Greek Literature*, writes about those who prepare for battle and acquire training through various exercises from which they will benefit at the time of battle (Basil of Cesarea, *De legendis gintilium libris*, ed. F. Boulenger, *Basile de Césarée, Aux jeunes gens sur la manière de tirer profit des lettres helléniques* [Paris, 1965], 2:27–39).

11 See A. Zucker, S. Draelantis, and S. Lazaris, "Introduction," in *RursuSpicae* 2 (2019): https://journals.openedition.org/rursuspicae/1047.

12 Fr. Sbordone lists 48 chapters in his critical edition (*Physiologus*, ed. F. Sbordone [Milan, 1936; repr. Hildesheim and New York, 1976], 1–145). Following this edition, taking into account, among other things, the *New York, Pierpont Morgan Library*, Ms. M 397 (= G), which is the oldest complete example, D. Offermanns proposed a new edition in 49 chapters of the first two subgroups of this list (*Physiologus, nach den Handschriften G und M*, ed. D. Offermanns. Beiträge zur klassischen Philologie 22 [Meisenheim am Glan, 1966]). A few years later, D. Kaimakis published a synoptic edition of the other three subgroups (*Physiologus, nach der ersten Redaktion*, ed. D. Kaimakis. Beiträge zur klassischen Philologie 63 [Meisenheim am Glan, 1974]). Also worthy of mention is the edition of a papyrus, PSI inv. 295, from the 6th century, published by M. Stroppa ("Un papiro inedito del Fisiologo [PSI inv. 295])," in *I papiri letterari cristiani. Atti del convegno internazionale di studi in memoria di Mario Naldini, Firenze, 10–11 giugno 2010*, ed. G. Bastianini and A. Casanova. Studi e Testi di Papirologia N.S. 13 (Firenze, 2011), 173–92 and tav. XXI–XXII). It contains part of chapters 41 and 42 belonging to this first recension.

13 See S. Lazaris, *Le Physiologus grec*. Vol. 1, *La réécriture de l'histoire naturelle antique*. Micrologus Library 77 (Florence, 2016), 69–78.

14 On when and where the Greek *Physiologus* was written, see *Le Physiologus*, 17–30. This date is not accepted by all scholars, some of whom prefer the 3rd or even 4th centuries. Others have adopted it. See most recently in S. Vollenweider, "Der Erlöser im Tarnanzug. Eine Studie zur Christologie des Physiologus, zu seiner Datierung und zur Rezeptionsgeschichte von Psalm 24 (23LXX)," in *Christus in natura. Quellen, Hermeneutik und Rezeption des Physiologus*, ed. Z. Kindschi Garský and R. Hirsch-Luipold. Studies of the Bible and Its Reception 11 (Berlin, 2020), 93–132, at 93, n. 4.

15 One wonders whether the decision of the author of the *Physiologus* to include stones is not related to the ancient idea that animals, plants and minerals possessed the same "force". We read for example in *Orphei Lithica* that from the earth comes the whole race of stones: in them, therefore, is found infinite and diverse forms of power; as much as plants, stones are powerful (ἐκ γαίης δὲ λίθων πάντων γένος, ἐν δ' ἄρα τοῖσι κάρτος ἀπειρέσιον καὶ ποικίλον· ὅσσα δύνανται ῥίζαι, τόσσα λίθοι, *Les Lapidaires grecs: Lapidaire orphique, Kérygmes lapidaires d'Orphée, Socrate de Denys, Lapidaire nautique, Damigéron-Évax*, ed. and trans. R. Halleux and J. Schlamp. Collection des universités de France 151 (Paris, 1985), v. 410–12).

16 Following the analysis of fragments of an unpublished codex, I have recently expressed doubts about this date (see S. Lazaris, "Un nouveau manuscrit illustré du *Physiologus* grec et la date de la deuxième recension: le *Sinaï, Monê tês Hagias Aikaterinês, NE gr. M 103*," in *Annuaire de l'Université de Sofia "St. Kliment Ohridski"* 99 [2017], 233–62).

17 See *Physiologus*, ed. Sbordone, 149–256. On the added animals, see Lazaris, *Le Physiologus grec*. vol. 1, tableau n° 1, 53–65 (third column).

18 See for instance *Genesis* (1.20–25) and *Leviticus* (11.1–30).

19 See *Physiologus*, ed. Sbordone, 257–99.

20 Wolf (chapt. 3), crocodile (chapt. 8), boar (chapt. 12), parrot (chapt. 22), pheasant (chapt. 23) and hydra (chapt. 28).

21 Asp (chapt. 11) and rabbit (chapt. 25). For further details, see Lazaris, *Le Physiologus grec*, vol. 1, tableau n° 1, 53–65 (fourth column).

22 Solomon is also cited as the author in the Ethiopian and Coptic translations of the Greek *Physiologus* and in the Icelandic version of the Latin *Physiologus*.

23 Ed. in C. Gidel and E. Legrand, "Étude sur un poème grec inédit, intitulé ὁ Φυσιολόγος suivie du texte grec édité par Mr. E. Legrand," in *Annuaire de l'association pour l'encouragement des études grecques en France* 7 (1873): 188–296.

24 On the term φύσις (and also φυσιολόγοι and φυσιολογία), see Lazaris, *Le Physiologus grec*, vol. 1, 39–40; J. Sallis, *The Figure of Nature: On Greek Origins. Studies in Continental thought* (Bloomington, Ind., 2016).

25 I use this term not in its taxonomic sense, but in a broader sense. It should also be noted that the notions of genus and species are not the subject of a specific and coherent vocabulary in the *Physiologus*.

26 As far as the composition of the chapters is concerned, other structural elements contribute to the overall design, including a biblical quotation which can be found at the beginning or end of several chapters. These elements are probably arranged by the author according to the requirements of the theme at hand, which explains the great irregularity of their presence in the chapters (see Lazaris, *Le Physiologus grec*, vol. 1, 84–85).

27 For the re-shaping of zoological knowledge in the *Physiologus* to accommodate Christian teachings, see also the chapter by T. Schmidt in the present volume.

28 Philo, without ceasing to defend the literal meaning, adopted the allegorical method advocated by the Greek schools and used by the Alexandrian Jews masterfully. See J. Pépin, *Remarques sur la théorie de l'exégèse allégorique chez Philon* (Paris, 1967); J. Moreau, *Abraham dans l'exégèse de Philon d'Alexandrie: Enjeux herméneutiques de la démarche allégorique*. PhD thesis (Lyon, 2010); F. Calabi et al., *Pouvoir et puissances chez Philon d'Alexandrie* (Tournhout, 2015).

29 See *Rhetores graeci*, 3 vols., ed. L. Spengel (Leipzig, 1853–56), vol. 3.

30 On the author's likely sources, see Lazaris, *Le Physiologus grec*, vol. 1, 37–46; S. Lazaris, *Le Physiologus grec. Vol. 2, Donner à voir la nature*. Micrologus Library 107 (Florence, 2021), 15–18.

31 Unless otherwise indicated, my references to the *Physiologus* concern the first recension.

32 The Greek text comes from D. Offermanns' edition (*Physiologus*, 86).

33 See Psalm 123 (124).7.

34 In reality, it refers to castoreum, the oily and fragrant secretion produced by specific glands (on castoreum in ancient medical literature, see S. Barbara, "Castoréum et basilic: deux substances animales de la pharmacopée ancienne," in *Le médecin initié par l'animal: Animaux et médecine dans l'Antiquité grecque et latine. Actes du colloque international tenu à la Maison de l'orient et de la Méditerranée-Jean Pouilloux les 26 et 27 octobre 2006*, ed. I. Boehm and P. Luccioni. Collection de la Maison de l'Orient méditerranéen Série littéraire et philosophique 12 (Lyon and Paris 2008), 121–48.

35 Many studies have been carried out on the relationship between Christianity and Greek thought, e.g. B. Pouderin and J. Doré, *Les apologistes chrétiens et la culture grecque. Actes du colloque de Paris, 2 et 3 septembre 1996* (Paris, 1998); A. Perrot, *Les chrétiens et l'hellénisme: identités religieuses et culture grecque dans l'Antiquité tardive* (Paris, 2012); S. Morlet, *Les chrétiens et la culture: conversion d'un concept (Ier–VIe siècle)* (Paris, 2016).

36 Justin considered that human reason (as proposed by Plato and other Stoics, poets and writers) was inspired by the divine Word and, since truth was unique, it was necessarily Christian (*Apologia secunda*, 13.4, ed. D. Minns and P. Parvis, *Justin, Philosopher and Martyr, Apologies*. Oxford Early Christian Texts [Oxford, 2009], 320).

37 Theophilus of Antioch, *Libri III at Autolycum*, 3.17, ed. G. Bardy and J. Sender, *Théophile d'Antioche, Trois livres à Autolycus*. Sources chrétiennes 20 bis (Paris, 1948), 238.

38 As I explained in a monograph devoted to this work, and even if some caution is necessary with a work like the *Physiologus*, we would then be very close to early Christianity of the second century. The anonymous writer of our work was therefore already practising, like his colleague in the *Epistle of Barnabas*, Philo of Alexandria's method of allegorical exegesis (see above and Lazaris, *Le Physiologus grec*, vol. 1, 30).

39 In some chapters, there is indeed a plea against the enemies of the new religion.

40 F. Sbordone, *Ricerche sulle fonti e sulla composizione del Physiologus greco* (Naples, 1936), 174.

41 See Lazaris, *Le Physiologus grec*, vol. 1, 89–99

42 In addition to the references cited in E. Delli's bibliography ("Bibliographie secondaire sélective sur les Hexaéméra et les thématiques rattachées," in *Almagest*, 11.1 [2020], 138–56) and those in the following pages, see A. Hamman, "L'enseignement sur la création dans l'Antiquité chrétienne

(suite)," in *Revue des sciences religieuses* 42, 2 (1968): 97–122. Also useful: [https://lacademie.tv/conferences/genese-structure-hexaemeron]

43 On this text, see the critical edition by D. T. Runia, *Philo of Alexandria, On the creation of the cosmos according to Moses.* Philo of Alexandria Commentary Series (Leiden and Boston, 2001). See also D. T. Runia, *Philo of Alexandria: collected studies 1997-2021* (Tübingen, 2023), esp. 315–28.

44 See Theophilus of Antioch, *Libri III at autolycum*, ed. Bardy and Sender. As an example, consider a fragment of a commentary by Origen quoted by Eusebius of Caesarea, *Praeparatio evangelica*, 7.20, ed. K. Mras, *Eusebius Werke.* Vol. 8, *Die Praeparatio Evangelica*, 2 pts. (Berlin, 1982), 402–403 in which Origen postulates and demonstrates the creation of the material world.

45 On the Syriac *Hexaemeral* tradition in general, see T. E. Napel, "Some Remarks on the Hexaemeral Literature in Syriac," in *Orientalia Christiana Analecta* 229 (1987): 57–69; A. Guillaumin, "Genèse I, 1–2, selon les commentateurs syriaques," in *In principio: interprétations des premiers versets de la Genèse.* Collection des études augustiniennes Série Antiquité 51 (Paris, 1973), 115–32.

46 Basil of Caesarea, *Homilies*, ed. Giet.

47 Ibid., 6–7. On his *Hexaemeron*, see also M. Garrido, *Las homilias In hexaemeron de Basilio de Cesarea: una respuesta a la política religiosa del emperador Juliano?* PhD thesis (Louvain-la-Neuve, 1996); J. Bernardi, "La date de 'l'Hexaéméron' de saint Basile," *Studia Patristica* 3 (1962): 165–69; E. Amand de Mendieta, "Les neuf Homélies de Basile de Césarée sur l'Hexaéméron. Recherches sur le genre littéraire, le but et l'élaboration de ces homélies," *Byzantion* 48 (1978): 337–68; T. Arentzen, V. Burrus and G. Peers, *Byzantine Tree Life: Christianity and the Arboreal Imagination.* New Approaches to Byzantine History and Culture (Cham, 2021) esp. 72–73.

48 Two sets of homilies on the creation of man (*De creatione hominis*) and a treatise on the *Hexaemeron* (*Apologia in hexaemeron*) are ascribed to him. In the treatise, written at the request of his brother, Peter of Sebaste, Gregory states that his aim is to defend Basil against certain critics who alleged there was some obscurity in his explanation of how light was created and later the luminaries, as well as in the passages dealing with heaven, the firmament (on Gregory of Nyssa's *Apology*, see, among others, D. Costache, "Approaching An Apology for the Hexaemeron Its Aims, Method and Discourse," in *Cappadocian legacy: a critical appraisal*, ed. D. Costache and P. Kariatlis [Sydney, 2013], 349–71).

49 Ὅτι ἐπειδὴ πρόκειται ἡμῖν εἰς τὴν τοῦ κόσμου σύστασιν ἐξέτασις καὶ θεωρία τοῦ παντός, οὐκ ἐκ τῆς τοῦ κόσμου σοφίας τὰς ἀρχὰς ἔχουσα, ἀλλ' ἐξ ὧν τὸν ἑαυτοῦ θεράποντα ὁ Θεὸς ἐξεπαίδευσεν, ἐν εἴδει λαλήσας πρὸς αὐτόν, καὶ οὐ δι' αἰνιγμάτων [...], Basil of Caesarea, *Homilies*, 6.1 (117B), ed. Giet.

50 On this point, see T. Korhonen, "Anthropomorphism and the Aesopic Animal Fables," in *Animals and their Relation to Gods, Humans and Things in the Ancient World*, ed. R. Mattila, S. Ito, and S. Fink (Wiesbaden, 2019), 211–31. I will revisit this in a study in preparation.

51 Ἐγὼ δέ σε βούλομαι τὸ ποριστικὸν καὶ εὐμήχανον τῶν καρκίνων ζηλοῦντα, τῆς βλάβης τῶν πλησίον ἀπέχεσθαι. Τοιοῦτός ἐστιν ὁ πρὸς τὸν ἀδελφὸν πορευόμενος δόλῳ [...], Basil of Caesarea, *Homilies*, 7.3 (153B), ed. Giet.

52 See B. Rowland, "The relationship of St. Basil's 'Hexameron' to the 'Physiologus'," in *Épopée animale, fable, fabliau. Actes du IVe Colloque de la Société internationale renardienne, Évreux, 7–11 septembre 1981*, éd. G. Bianciotto and M. Salvat. Publications de l'Université de Rouen 83 (Paris, 1984), 489–98.

53 See A. Zucker, "Zoology," in *A Companion to Byzantine Science*, ed. S. Lazaris. Brill's Companions to the Byzantine World 6 (Leiden and Boston, 2020), 261–301, at 276.

54 See Basil of Caesarea, *Homilies*, 9.3 (193A), ed. Giet.

55 See George of Pisidia, *Hexaemeron sive Cosmopoeia*, ed. Tartaglia, esp. v. 931–48.

56 See Zucker, "Zoology," 276.

57 See Basil of Caesarea, *Homilies*, 7.2 (149C–152C), ed. Giet about aquatic animals.

58 Augustine was particularly irritated by the physical sciences and found it useless to discuss such matters. The creation of the birds and fish gave rise to the same topic as in Philo and Basil, i.e., that both fish and birds swim and that both species were created from water, Augustine demonstrates

that they are able to fly only in humid air (closely akin to water). Another one of his theories is that animal life generated by the decay of other animals was not a new creation, but the result of a natural force created during the first creation.

59 This Slavic author and translator was active in Preslav during the reign of Symeon (893–927). On John the Exarch, see M. Capaldo, "Jean l'Exarque en tant que compilateur et traducteur. Problèmes historico-littéraires, textologiques, linguistiques," in *Polata k"nigopis'naja* 3 (1980): 69–89; J. T. Francis, "John the Exarch's Theological Education and Proficiency in Greek as Revealed by His Abridged Translation of John of Damascus' *De fide orthodoxa*," in *Palaeobulgarica / Старобългаристика* 15 (1991): 35–58; L. Sels, "Deux traductions slavonnes du *De hominis opificio* de Grégoire de Nysse (Xe et XIVe ss.)," in *Slavica Gandensia* 29 (2002): 137–64 and ibid., "John the Exarch and his Sources: New Sources of the Sixth Book of the Šestodnev," in *Slavica Gandensia*, 35 (2008): 123–49.

60 *La Semaine ou Creation du Monde* by Guillaume de Salluste Du Bartas (1544–1590) was published in 1578 and has been translated into several languages. It is a long poem in alexandrine verse, the genre of which is uncertain by the author's own admission, who describes his work as epic, panegyric, prophetic and didactic. Regarding Du Bartas, George of Pisidia provided most of the data, mostly from the anecdotes in the *Physiologus*.

61 For his influence on Gregory Palamas, see D. Costache, "Theology and Natural Sciences in St Gregory Palamas," in *God, freedom and nature. Proceedings of the biennial conference in philosophy, religion and culture 2008*, ed. R. S. Laura, R. A. Buchanan and A. K. Chapman (Sydney, 2012), 132–38.

62 Ambrose, *Opera*, ed. C. Schenkl. Corpus scriptorum ecclesiasticorum latinorum 32.1 (Vienna, Prague, and Leipzig, 1896).

63 Δολερὸν ὁ πέρδιξ καὶ ζηλότυπον, κακούργως συμπράττων τοῖς θηραταῖς πρὸς τὴν ἄγραν ("the partridge is treacherous and jealous, wickedly offering its help to the hunters to get their prey"), Basil of Caesarea, *Homilies*, 8.3 (172C), ed. Giet.

64 Ambrose, *Opera*, ed. C. Schenkl, *Pars I. Qua continentur libri exaemeron...*, 6.3.13, col. 246–247.

65 *Physiologus*, ed. Offermanns, 18:5–7.

66 See Lazaris, *Le Physiologus grec*, vol. 1, 18–19.

67 This idea goes back to St. Jerome, according to whom Ambrose of Milan, in his *Hexaemeron*, merely paraphrased Basil, Origen and Hippolytus (*Nuper Ambrosius sic Exaemeron illius [Origenis] compilauit, ut magis Hippolyti sententias Basiliique sequeretur, Ep.* 84.7, ed. J. Labourt, *Jérôme, Lettres*. Vol. 4, *Epistulae LXXI–XCV* [Paris 1954], 134). It has been echoed by several authors. To take just one example, F. E. Robbins does not hesitate to mention that "Ambrose is almost entirely dependent upon Basil, with a few reminiscences of Philo". (F. E. Robbins, *The Hexaemeral Literature. A Study of the Greek and Latin Commentaries on Genesis*. PhD thesis [Chicago, Ill., 1911]).

68 See R. Henke, *Basilius und Ambrosius über das Sechstagewerk: eine vergleichende Studie.* ΧΡΗΣΙΣ - CHRÊSIS 7 (Basel, 2000), esp. 34–35; L. Gosserez, "Sous le signe du phénix (Ambroise de Mila, Exameron, V, 23, 79–80)," in *La Création chez les Pères*, ed. M.-A. Vannier (Berlin and Bruxelles, 2011), 55–75; G. Nauroy, "Ambroise de Milan, émule critique de Basile de Césarée: à propos de Genèse 1, 2," in *La Création chez les Pères*, ed. M.-A. Vannier (Berlin et al., 2011), 77–101.

69 See Ps.-Eustathius, *Commentarius in hexaemeron*, ed. J. P. Migne, PG 18 (Paris, 1857), col. 707–94.

70 See F. Zoepfl, *Der Kommentar des Pseudo-Eustathios zum Hexaëmeron*. Alttestamentliche Abhandlungen 10, 5 (Münster, 1927), 17.

71 *Leucippe and Clitophon*, 4.19, ed. E. Vilborg. Studia Graeca et Latina Gothoburgensia 1 (Stockholm, 1955), 86–87.

72 See Zucker, "Zoology," 277–78.

73 Ps.-Eustathius, *Commentarius in hexaemeron*, ed. Migne, PG 18, col. 757D–760A.

74 On Adam book, see Zucker, "Zoology," 277–78. See also E. Trumpp, *Das Hexaëmeron des Pseudo-Epiphanius. Aethiopischer Text, verglichen mit dem arabischen Originaltext und deutscher Uebersetzung*. Abhandlungen 16, 5 (Munich, 1882). On the use of the Ethiopic and Arabic sources in Trumpp's publication, see the critics of A. Haffner, *Das Hexaëmeron des Pseudo-Ephiphanius*. Analecta Gorgiana 482 (Piscataway, N.J., 2010).

75 On this individual and his work, see John Philoponus, *De opificio mundi*, ed. C. Scholten. Fontes christiani 23 (Freiburg, Basel and Vienna, 1997); M.-E. Allamandy and M. H. Congourdeau, *Les Pères de l'Eglise et l'astrologie. Origène, Méthode, Basile, Grégoire de Nysse, Diodore, Procope de Gaza, Jean Philopon.* Les Pères dans la foi 85 (Paris, 2003); L. S. B. MacCoull, "The historical Context of John Philoponus' *De Opificio Mundi* in the Culture of Byzantine-Coptic Egypt," in *Zeitschrift für Antikes Christentum* 9, 2 (2006): 397–423; M.-H. Congourdeau, "Cosmas Indicopleustès et Jean Philopon. Deux lectures de la Genèse à Alexandrie au VIe siècle," in *Science et exégèse: les interprétations antiques et médiévales du récit biblique de la création des éléments (Genèse 1, 1–8)*, ed. B. Bakhouche. Bibliothèque de l'Ecole des Hautes Etudes 167 (Turnhout, 2016), 147–59; R. Wagner and A. Briggs, *The Penultimate Curiosity: How Science Swims in the Slipstream of Ultimate Questions* (Oxford, 2016), 119–23 ("The Creation of the World"); E. Nicolaidis, "Platon et Hipparque ont lu la Bible: l'Hexaéméron de Jean Philopon," *Almagest*, 11, 1 (2020): 42–55.

76 See John Philoponus, *De opificio mundi libri VII*, ed. G. Reichardt (Leipzig, 1897).

77 The new element of the exegesis of Theodorus and his followers is that the world was not spherical, but oblong and flat. If this was new for the *Hexaemera* it was not new *per se*, and was probably a belief shared by many less educated Christians at the time.

78 On this text, see among others F. Petit, "L'homme créé 'à l'image' de Dieu: quelques fragments grecs inédits de Théodore de Mopsueste," in *Muséon* 100 (1987): 274–80. Photius in his *Bibliotheca* (cod. 38) wrote that Theodorus of Mopsuestia's style is neither polished nor clear.

79 Building on a theological debate about the interpretation of the Bible, two world views clash in the streets of Alexandria in the 550s. On the one hand (Philoponus), a world that we can explain, within the limits imposed by God, by secular knowledge and the natural philosophy of the Ancients, in which the phenomena of nature are due to physical principles laid down during the Creation of the world. On the other hand (Cosmas Indicopleustes), we have a world whose reality is based on theological symbolism and in which the natural philosophy of the Ancients is replaced by the mechanics of the angels, responsible for all the phenomena of nature.

80 John Philoponus, *De opificio mundi*, 5.1, ed. Reichardt, 204–10.

81 See respectively ibid., 5.1; 5.3; 5.13, 204–10; 212–13; 227–28.

82 See ibid., 5.6; 5.12, 217–18; 225–27.

83 Ibid., 5.1, 204–10.

84 On these three faculties, see also in the *De anima* of Alexander of Aphrodisias who wrote τό τε γὰρ τρέφεσθαί τε καὶ αὔξεσθαι καὶ γεννᾶν οἷον αὐτὸ (*Alexandri Aphrodisiensis Praeter commentaria scripta minora: De anima liber cum mantissa*, ed. I. Bruns (Berlin, 1887), 12:12).

85 John Philoponus, *De opificio mundi*, 5.1, ed. Reichardt, 204–10.

86 Cosmas Indicopleustes, "Topographia Christiana," in *Cosmas Indicopleustès, Topographie chrétienne.* Vol. 1, *Livres I–IV*, vol. 2, *Livre V*, vol. 3, *Livres VI–XII and Index*, ed. W. Wolska-Conus. Sources chrétiennes 141; 159; 197 (Paris, 1968–73).

87 See ibid., vol. 1, 16.

88 See "Excerpta ex Timothei Gazaei libris de Animalibus," ed. M. Haupt in *Hermes* 3 (1869): 1–30, at ch. 45, 25–26.

89 See Cosmas Indicopleustes, *Topographia Christiana*, 11.1, ed. Wolska-Conus, vol. 3, 315–17.

90 See ibid., 11.2, 317.

91 See ibid., 11.3, 318–19.

92 See ibid., 11.4, 320–21.

93 See ibid., 11.7, 326–29.

94 See ibid., 11.8, 330–31.

95 On Jacob of Edessa, see R. B. Haar Romeny, *Jacob of Edessa and the Syriac Culture of His Day.* Monographs of the Peshiṭta Institute 18 (Leiden and Boston, 2008) (several contributions). On the *Hexaemeron*, see P. Martin, "L'Hexaéméron de Jacques d'Édesse," in *Journal asiatique*, Sér. 8, t. 11 (1888), 155–219 and 401–90; A. Hjelt, *Etudes sur l'Hexaméron de Jacques d'Edesse: notamment sur ses notions géographiques contenues dans le 3ème traité.* PhD thesis (Helsingfors 1892); M. Wilks, "Jacob of Edessa's Use of Greek Philosophy in His Hexaemeron," in *Jacob of Edessa and the Syriac culture of his day*, ed. R. B. Haar Romeny. Monographs of the Peshiṭta Institute 18 (Leiden and Boston 2008), 223–238.

96 See Jacob of Edessa, "Hexaemeron," ed. J.-B. Chabot, *Iacobi Edesseni Hexaemeron seu in opus creationis libri septem*. Corpus scriptorum christianorum orientalium 56 (Paris, 1928).

97 Martin, "L'Hexaéméron," 165.

98 On these texts, see Hjelt, *Etudes sur l'Hexaméron de Jacques d'Edesse*. Also useful: Martin, "L'Hexaéméron de Jacques d'Édesse."

99 See Isaac of Antioch, ed. G. Bickell, *S. Isaaci Antiocheni doctoris Syrorum, opera omnia: ex omnibus, quotquot extant, codicibus manuscriptis cum varia lectione Syriace Arabiceque primus edidit* (Gießen, 1873), 85–175.

100 This is most likely the otter, also known as a river dog (see Lazaris, *Le Physiologus grec*, vol. 1, n. 173 [pp. 56–58]; Lazaris, *Le Physiologus grec*, vol. 2, 170–73).

101 For a full description of the elephant, see Martin, "L'Hexaéméron," 468–71.

102 See Basil of Caesarea, *Homilies*, 9.5 (201A), ed. Giet.

103 For a full bibliography on this author, see: https://georgios-pisides.univie.ac.at/en/project/bibli ography/. The following titles should be added to this list: N. Radošević, Шестоднев Георгија Писиде и његов словенски предов / *The Hexaemeron of George Pisides and its slavonic translation* (Belgrade, 1979); P. Speck, "Ohne Anfang und Ende: Das Hexaemeron des Georgios Pisides," in *AETOS: Studies in Honour of Cyril Mango presented to him on April 14, 1998*, ed. I. Ševčenko and I. Hutter (Stuttgart and Leipzig, 1998), 314–28.

104 George of Pisidia, *Hexaemeron sive Cosmopoeia*, ed. Tartaglia, 310–425. See also the edition by F. Gonnelli, *Giorgio di Pisidia, Esamerone* (Pisa, 1998), 116–242.

105 E. Delli, "Survie et transformations de la littérature hexaémérique. Entre littérature, théologie et philosophie: Le cas de Georges de Pisidie," in *Almagest*, 11, 1 (2020): 56–79, at 66.

106 George of Pisidia, *Hexaemeron sive Cosmopoeia*, ed. Tartaglia, v. 76.

107 On the zoological part of this work, see L. Tartaglia, "L'excursus zoologico dell'Esamerone di Giorgio di Pisidia," in *Νέα Ρώμη. Rivista di ricerche bizantinistiche* 2 (2005): 41–57.

108 On the same subject, see the passage in Basil's *Hexaemeron* (Basil of Caesarea, *Homilies*, 9.2 (192A), ed. Giet).

109 See Zucker, "Zoology," 279.

110 On the links between the latter two works, see Lazaris, *Le Physiologus grec*, vol. 1, 43.

111 On his work, see above, n. 59.

112 Anastasius of Sinai, *Hexaemeron*, ed. C. A. Kuehn and J. D. Baggarly. Orientalia Christiana analecta 278 (Rome, 2008). On the authenticity of this work, apart from the commentary of the two editors, see D. Zaganas, "The Authenticity of Anastasius Sinaita's Hexaemeron (CPG 7770)," in *REB*, 73 (2015): 189–201; idem, "Encore sur l'authenticité de l'Hexaéméron d'Anastase le Sinaïte," in *BZ*, 110, 3 (2017): 755–78; on the sources used by the author, see ibid., "Anastase le Sinaïte, entre citation et invention: L'*Hexaéméron* et ses sources 'antiques'," in *Academic Journal – Augustinianum* 56, 2 (2016): 391–409 and especially ibid., *L'Hexaemeron d'Anastase le Sinaïte. Son authenticité, ses sources et son exégèse allégorisante*. Vigiliae Christianae, Supplements 172 (Leiden, 2022). On Anastasius' work as a whole, see K.-H., Uthemann, *Anastasios Sinaites: byzantinisches Christentum in den ersten Jahrzehnten unter arabischer Herrschaft*. Arbeiten zur Kirchengeschichte 125 (Berlin and Boston, 2015); ibid., *Studien zu Anastasios Sinaites: mit einem Anhang zu Anastasios I. von Antiochien*. Texte und Untersuchungen zur Geschichte der altchristlichen Literatur 174 (Berlin and Boston, 2017). Also useful: V. Déroche, "La Disputatio adversus Iudaeos d'Anastase le Sinaïte: Authenticité du texte et identification des fragments," in *Ephemerides theologicae Lovanienses* 95 (2019): 427–38 and www.anastasiosofsi nai.org/.

113 *Anastasius of Sinai Hexaemeron*, ed. Kuehn and Baggarly, 1.6.2, 18–20.

114 See ibid., 5.6.2 and 5.7.3, 162 and 166.

115 Ibid., 5.1.3, 148–50.

116 On Cosmas, see K. Krumbacher, *Geschichte der byzantinischen Litteratur von Justinian bis zum Ende des oströmischen Reiches (527–1453)* (Munich, ²1897), 674–676; A. P. Kazhdan and S. Gero, "Kosmas of Jerusalem: A More Critical Approach to his Biography," in *BZ* 82, 1–2 (1989): 122–32; T. Detorakis, Κοσμάς ο Μελωδός, βίος και έργο. Ανάλεκτα Βλατάδων 28 (Thessaloniki, 1979); H.-G. Beck, *Kirche und theologische Literatur im Byzantinischen Reich* (Munich, 1959), 515–16; L. M. Hoffmann, "Cosmas of Jerusalem (Saint)," in Religion Past and Present (2011) [Consulted online on 02 August 2023].

117 See Cosmas of Jerusalem, *Commentarii in Gregorii Nazianzeni carmina*, ed. G. Lozza, *Commentario ai carmi di Gregorio Nazianzeno*. Storie e testi 12 (Naples, 2000).

118 See ibid. On the animal species described in his work, see Zucker, "Zoology," 279.

119 Michael Glykas, *Annales*, ed. I. Bekker and. J. Löwenklau. CSHB 27 (Bonn, 1836).

120 See G. Sarton, *History of Science*, vol. 2.1 (Chicago, 1931), 133.

121 See F. Sbordone, "ΦΥΣΙΟΛΟΓΙΑ Parigina degli Annales di Michael Glycas," in *BZ* 29, 2 (1930): 188–97.

122 Michael Glykas, *Annales*, ed. I. Bekker and J. Löwenklau, 65:5–13. On Glykas, see also Zucker, "Zoology," 280.

123 On this point, see B. Zarkhaia, "Михаил Глика – читатель Аристотеля," in Индоевропейское языкознание и классическая филология 16 (2012): 254–63.

124 On this text, see *Physici et medici Graeci minores*, vol. 1, ed. J. L. Ideler (Berlin 1841), 3–80.

125 See C. Zirkle, "Animals Impregnated by the Wind," in *Isis* 25, 1 (1936): 95–130 at 109.

126 Michael Glykas, *Annales*, ed. I. Bekker and J. Löwenklau, 88:19–20.

127 Herodotus, *Historiae*, 2.73, ed. N. G. Wilson, *Herodoti Historiae*, vol. 1 (Oxford, 2015), 168–69.

128 On this point, see Zarkhaia, "Михаил Глика – читатель Аристотеля."

129 On these texts and their critical editions, see I. Nilsson, "Narrating Images in Byzantine Literature: The Ekphraseis of Konstantinos Manasses," *JÖB*, 55 (2005): 122. On the last ekphrasis, see also S. Lazaris, "Hunting in Byzantium: a case-study in falconry," in *Falconry in the Mediterranean Context during the Pre-Modern Era*, ed. C. Brunet, B. Van den Abeele (Geneva, 2021), 261–76 (Bibliotheca Cynegetica 9), at 264.

130 Maguire, *Nectar and Illusion*, 58. On this text, see Nilsson, "Narrating Images," 121–46. For a modern English translation of the Greek verse chronicle and the prose Bulgarian translation, see L. Yuretich, *The Chronicle of Constantine Manasses*. Translated texts for Byzantinists 6 (Liverpool, 2018), 21–262.

131 Neophytos of Cyprus, "Commentarius in Hexaemeron Genesim," ed. T. Detorakis, in Ἁγίου Νεοφύτοτου τοῦ Ἐγκλείστου Συγγράμματα, vol. 4, ed. D. G. Tsames et. al. (Paphos, 2001), 45–144.

132 On the term κῆτος, see J. Boardmann, " 'Very like a whale' – Classical Sea Monsters," in *Monsters and demons in the ancient and medieval worlds: papers presented in honor of Edith Paroda*, ed. A. E. Farkas, P. O. Harper and E. B. Harrison (Mainz, 1987), 73–84; A. Zucker, "Étude épistémologique du mot κῆτος," in *Les zoonymes. Actes du colloque international tenu à Nice les 23, 24 et 25 janvier 1997*, éd. J.-P. Dalbera (Nice, 1997), 425–54; S. Riccioni, "Dal kētos al sēnmurv? Mutazioni iconografiche e transizioni simboliche del ketos dall'Antichità al Medioevo (secolo XIII)," *Hortus Artium Medievalium* 22 (2016): 130–44. G. Nocchi Macedo's insights may also be useful for the reader (G. Nocchi Macedo, "Réexamen du dessin du "Codex Miscellaneus" de Montserrat (P. Montserrat inv. nr. 154 = MP3 2916.41)," *Aegyptus* 90 (2010): 99–117).

133 Recent and earlier bibliography in Lazaris, *Le Physiologus grec*. vol. 2, esp. n. 294.

134 Man, created last as the crowning glory of Creation, is established by God as its king, having the function of watching over its good order, the preservation of its qualities, its harmonious development, not dominating it as a tyrant, but governing it in the name of God (who remains its only Master and Provider) as a good, wise and faithful steward.

135 On this point, see J.-C. Larchet, *Les animaux dans la spiritualité orthodoxe* (Geneva, 2018), 18.

136 At the same time, animals possessed their own function as mediators between human beings and God, as parts of a signifying kosmos full of divine wisdom.

137 On Gregory of Nyssa and his distinction between humans and animals, see H. Grelier, *L'argumentation de Grégoire de Nysse contre Apolinaire de Laodicée: Etude littéraire et doctrinale de l'Antirrheticus adversus Apolinarium et de l'Ad Theophilum adversus*. PhD thesis (Lyon, 2008), esp. 746.

138 Gregory of Nyssa, *Antirrheticus adversus Apollinarium*, ed. F. Müller and W. Jaeger, *Gregorii Nysseni opera dogmatica minora (Pars I)* (Leiden, 1958), 141.

139 See Aristotle, *De anima*, 414b:18–19. For this problem, see also the chapter by Somma and Vogiatzi in the present volume.

140 On this point, see for example, John Chrysostomos, *Expositiones in Psalmos*, ed. J.-P. Migne, PG 55 (Paris, 1862), col. 35–498, esp. 48.6, col. 232. For an alternative opinion of some Byzantine

scholars about the rationality of animals, see the chapter by Somma and Vogiatzi in the present volume.

141 On this point, see B. Resl (ed.), *A Cultural History of Animals in the Medieval Age*. A Cultural History of Animals 2 (Oxford and New York, 2007); L. Kalof (ed.), A *Cultural History of Animals in Antiquity*. A Cultural History of Animals 1 (Oxford and New York, 2011); Lazaris, *Le Physiologus grec*, vol. 2, esp. 311–43.

142 See M. Pastoureau, "L'animal," in *Le Moyen âge en lumière. Manuscrits enluminés des bibliothèques de France*, ed. J. Dalarun (Paris, 2002), 65–105, at 104–5.

143 On these authors, see ibid.

144 See Eusebius of Caesarea, *Commentaria in Psalmos*, ed. J.-P. Migne, PG 23 (Paris, 1857), col. 320B–D.

145 See Jérôme, *Commentaire sur Jonas*, ed. Y.-M. Duval. Sources chrétiennes 323 (Paris, 1985).

146 Athanasius of Alexandria, *Expositiones in Psalmos*, ed. J.-P. Migne, PG 27 (Paris, 1857), col. 176B–C.

PART II

Literary and Figurative Discourses on Animals

4

ANIMAL RATIONALITY IN BYZANTINE PHILOSOPHY AND ISLAMIC PHILOSOPHY

Bligh Somma and Melpomeni Vogiatzi

A Short Introduction: Animal Psychology and Animal Rationality in Ancient Greek Thought*

The debate regarding the attribution of rationality to animals[1] is an ancient one. Although relevant discussions can be found among the Pre-Socratics and Plato, there is no doubt that Aristotle's examinations of animal rationality[2] have been extremely influential. In fact, Aristotle can be said to be the founder of the field in two respects: first, because of his explicitly stated ambition to study the soul as a whole, including its presence in non-human living beings, and second, because of his pioneering interest in biology, the importance of the study of which he compared to the study of the heavenly bodies. However, despite the uniqueness of his studies and the influence they had on the subsequent thinkers, Aristotle did not deliver an account of animal psychology as such and his discussions of animal rationality are spread throughout his corpus, often even appearing inconsistent.

In particular, in the context of discussing animal psychology, Aristotle distinguished between humans and other animals on the basis of their possession of rational thought and criticized his predecessors for referring to the soul and the mind interchangeably, thus implying that mind can be attributed to all animals.[3] In his view, reason, reasoning faculty, thought and intellect (*logos, logismos, dianoia, nous*) are possessed only by humans[4] and higher beings,[5] whereas animals have only their nutritive and perceptual capacities in common with humans. Sense perception is for Aristotle the psychological capacity that distinguishes animals from plants and that is – either all senses or at least taste and touch – present in all animals. In addition to perceiving through the senses, Aristotle allows some animals the ability to form mental images (*phantasia*) and even recall these images through memory.[6] However, these capacities seem to differ in their presence in humans: in the case of rational animals, *phantasia* and memory are the prerequisite for thinking and learning. Humans are able to reflect upon these images and in this sense their rationality depends upon these images (*phantasiai*),[7] while memory, which depends upon the faculty of *phantasia*, is a prerequisite of learning.[8] Non-rational animals, on the other hand, cannot reflect upon these images, but seem to use them as a compensation for their lack of a capacity for judgement and belief formation. In this sense, *phantasia* is to non-rational animals

DOI: 10.4324/9781003055877-7

what thinking is for rational animals: since they are not able to think and form beliefs, e.g. about what is dangerous, *phantasia* and memory would provide the animal with certain mental images that would enable the animal to avoid danger.[9]

Although this description of animal psychology makes it clear that non-human animals lack reason, in other contexts Aristotle seems to attribute some rationality-related features to such animals. In particular, the *Nicomachean Ethics* (*NE*) briefly discusses the possession of *phronêsis* by non-human animals and states that they have a capacity for forethought (*dynamin pronoetiken*), whereas elsewhere Aristotle calls humans the most prudent (*phronimôtaton*) among animals, thus implying that other animals might be less prudent.[10] In the *NE*, *phronesis* plays a special role among the intellectual virtues by being the proper virtue of the calculative part of the rational soul (*logistikon*), which differs from the scientific part (*epistemonikon*), whose proper virtue is wisdom (*sophia*). Given this, the attribution of *phronêsis* to non-human animals might hint at the possession of a rational faculty by them to some extent. In particular, the calculative part of reason is said to be the one that reflects upon the things that change and for which we deliberate and is distinguished from the scientific part which reflects upon the unchanging things that require understanding (*NE* VI.1 1139a1–16). Although the *NE* does not attribute this latter function to non-human animals, the above-mentioned passages suggest that the former part can, at least to some extent, be found among non-human animals. In favor of this view may also be the fact that the calculative part and its proper virtue, *phronêsis*, are in fact said to be necessary for the acquisition of the virtues, which in other contexts seems to be attributed to non-human animals. Specifically, in the biological writings, Aristotle seems to attribute to non-human animals not only virtues and emotions, such as courage and anger, but also rational capacities, such as intelligence and practical wisdom.[11] Thus, already in Aristotle's discussion, there is a possible disconnect between acknowledging that animals act in apparently reasonable or rational ways, and questioning whether animals are in fact capable of sophisticated cognitive activity.

A passage from his *History of Animals* seems to ease this tension: Aristotle emphasizes the analogy between the human and other animal capacities and states that in some respect humans differ from other animals in degree, namely by having more or less of one quality, but in another respect by analogy, namely when a certain function seen in humans is analogous to a function seen in non-human animals. Examples of the latter criterion of differentiation are technical skill (*technê*), wisdom (*sophia*), and intelligence (*sunesis*). These are properly attributed only to humans, but some animals possess something analogous as a "natural capacity".[12] In a later similar passage we learn that also virtues and emotions, which were mentioned in this current passage as examples, are in animals as "natural capacities",[13] namely as analogical to the human possession of them. Therefore, according to this account, even when we speak of animal virtues, intelligence or their ability to learn, we do not mean the same thing as when we speak of the same features in humans, but we only speak by analogy.

The lack of an explicitly unified account or clear compatibility between the various accounts has already been recognized as problematic from a contemporary point of view. But contemporary readers have not been the first to notice that Aristotle raised issues that he himself left unsolved. Solutions to the questions raised by Aristotle's treatment of animal rationality were offered in antiquity, both in the Byzantine and the Islamic traditions. In this chapter, we will examine salient features of the medieval reception of Aristotle's discussions about animal rationality. As will be shown, both in the Byzantine and the Islamic traditions, Aristotle's discussions were deemed as unsatisfying and in need of alteration. Although the modifications differ, we will find that the two traditions often offered similar answers and made use of the same resources.

Animal Rationality in Byzantium

As with many other topics, the Byzantine accounts on animal rationality are usually left aside in overviews of ancient discussions of the topic. Although the classical and late antique accounts as well as the Western medieval accounts have been fairly well studied, little is said about the Byzantines' views on animal psychology. Since this is a vast topic, we will focus primarily on the reception of Aristotle's zoology and on philosophical views on animal rationality,[14] in particular on views expressed during the 12th-century Byzantine flourishing of Aristotelian studies. At this time we find the first elaborate Byzantine accounts of animals, initiated by Anna Komnene through the establishment of a scholarly group aiming at the production of commentaries on Aristotle's least studied treatises.[15] As a result of this effort, we now possess commentaries on Aristotle's biological writings (*Parts of Animals* (PA), *Movement of Animals* (MA), *Generation of Animals* (GA), *Progression of Animas* (IA)) as well as on treatises on the *Parva Naturalia* (PN) written by the Byzantine scholar Michael of Ephesus.[16] Michael also composed commentaries on selected books of Aristotle's *NE* (books 5, 8–10), whereas other commentaries on the remaining books are also preserved, for instance Eustratius of Nicaea's commentaries on books 1 and 6. In these texts, Aristotle's dichotomy between rational and non-rational animals is adopted, but as we will see, with some serious alterations.

Starting with Michael's account of animal psychology, it becomes clear that Michael follows roughly the Aristotelian hierarchy of the soul functions: the nutritive soul is common to all ensouled beings, the perceptive is distinctive of all animals, among which only humans possess reason. In fact, reason and the ability to engage in scientific reasoning are presented as the most proper feature of humans. By contrast, non-human animals are said to be animals "insofar as they possess the non-rational soul".[17] It is, therefore, the perceptive faculty that is common to all animals, whereas humans are said to be the only animals that can, in addition to perceiving, engage in thinking (*phronein*).[18]

In his commentary on the *Parva Naturalia,* Michael also refers to the animals' possession of *phantasia*. As in Aristotle, non-human animals also possess *phantasia*, and consequently memory (*mnêmê*), whereas reminiscence (*anamnêsis*) can be observed only in humans, since it resembles syllogistic reasoning, which can be seen only among humans.[19] The reason why other animals have memory and *phantasia* is that *phantasia* is an 'affection' of the common perception, through which we perceive the common perceptibles such as size and movement. Following Aristotle, Michael says that if *phantasia* did not belong to perception but to reason, only humans would possess it, who have belief (*doxa*) and *phronêsis*.[20] As Michael explains it, the mental images are properly speaking objects of common perception (*prôtê aisthêsis*) and only accidentally of intellect, when intellect is accompanied by *phantasia*. Hence, animals are able to perceive the proper perceptibles, and through the common and first perception can form mental images. These mental images can then be further used by humans for grasping the intelligibles (*ta noêta*). So far then, Michael seems to follow Aristotle's distinctions.

However, an interesting difference from Aristotle arises with Michael's views on perception, above agreed as being possessed by all animals. In particular, in Aristotle's view, the faculty of perception is responsible for the perception not only of the proper objects of perception, but also for the awareness of perception. In other words, self-awareness, namely the ability to perceive that we perceive, belongs, according to Aristotle, to the faculty of perception and can, therefore, be ascribed to all animals.[21] However, already in antiquity this view was considered problematic due to the fact that, first, the perceptive faculty would be responsible for the perception both of the proper objects and of the activity of perception, and second, the capacity of awareness would be divided between the perceptive faculty (which is aware of our perception) and the rational faculty (which is aware of our thinking).

Among the Neoplatonic solutions to this problem, one is relevant here: later interpreters revised Aristotle's view and added a new power to the rational part of the soul, which would be responsible for self-awareness. As John Philoponus (6th c.) attests,[22] the introduction of the "attentive" (*prosektikon*) power of the rational soul aimed at solving a problem introduced by Aristotle himself. If there were no unifying faculty responsible for awareness, then it would be as if, within one soul, two individuals perceive two different things.[23] According to Philoponus' report, this attentive part has a grasp of all powers of the soul, "the rational, the non-rational and the vegetable", hence accounting for the awareness of the cognitive, perceptive and nutritive functions. Without further explication, Philoponus refers to a further subdivision within this attentive power:

> This attentive part roves over all powers, cognitive and vital. But if it is roving over the cognitive, it is called 'attentive' [...] whereas if it is going through the vital powers it is called conscious (*suneidôs*). This is why the tragedy says 'Conscience (*sunesis*), since I am conscious of myself having done terrible things.'
>
> (Phil., *In DA* 465, 10–15 trans. R. Sorabji)

Turning back to Michael, we can see that he adopts this earlier account of the existence of an 'attentive' part of the soul, which is responsible for self-awareness:

> Thinking and judging, when we are seeing and thinking, that we are seeing and that we are thinking – that which is called the attentive part – is a part of the rational soul and, as it were, the centre of it. For this is what determines that we are seeing when we see, and are reflecting when we are reflecting, and are walking and writing when we are walking and when we are writing. The [phrase] 'we would perceive that we are perceiving and we would think that we are thinking' is the same as 'for we are not unconscious (*anaisthêtos*) or unaware (*anennoêtos*) when we think and perceive that we are thinking and are perceiving, but rather when we are perceiving and thinking we co-perceive (*sunaisthanomai*) and co-understand that we are perceiving and that we are thinking', and these things are nothing other than [the fact that] we exist. For one who co-perceives that he perceives, judges and by-perceives nothing other than that he is living and exists; in the same way, too, one who co-thinks that he is thinking, is thinking this: that since he is thinking, he is living and has not died, and is not a non-existing thing but rather something that exists, an animal (*zôon*) like this one here.
>
> (Michael, *In NE* IX 517, 14–26, trans. D. Konstan modified[24])

Reading the two texts in parallel, it becomes evident that Michael has the same theory in mind when he speaks of an attentive part of the rational soul. Though this is not stated explicitly, it is obvious that also for Michael the attentive part is responsible for reflecting upon both perception and cognition. The two functions are named in the whole passage always side by side: the attentive part allows us to think that we think and to perceive that we perceive, thus being aware of our existence. What then about animals? As we saw, Aristotle's ascription of self-awareness to the perceptive part of the soul would guarantee that animals are able to perceive that they perceive. But if, in this new account, the attentive part is responsible for this function, and it belongs to the rational part of the soul as the "centre of it", and if animals have no rational part, then this should mean that non-human animals would not be able to perceive that they perceive. However, the distinction between the two functions of the attentive part can solve this problem.

In particular, of the two functions of the attentive part, the capacity to rove over the cognitive parts is one that can belong only to humans (who have cognitive parts), whereas we can expect that both humans and non-human animals are able to attend to the capacity of their vital part. If this is so, then humans are able to reflect upon their thinking, perceptual and vital activities, whereas non-human animals can only be aware of their perceiving as well as of their appetites. In fact, Philoponus' text explicitly states that the attentive part reflects not only upon thinking and perceiving, but also upon the lower functions of the soul, such as appetites. We can read the passage from Michael along the same lines: humans can think and judge, when they are thinking and seeing, that they are thinking and seeing, whereas animals can only "judge", when they are seeing, that they are seeing. And this goes not only for eyesight of course, but also for the other kinds of sense-perception (and probably for the other faculties that animals possess, such as imagination and memory). The last line of the passage can provide support for this reading, since Michael suggests that being aware of one's thinking or perceiving means nothing else than being aware of one's existence as an animal. In this sense, non-human animals can be also aware of their existence through their perceiving.

Support for this reading can be drawn from the language used in Philoponus' text for referring to the reflection upon the vital activities. In particular, the term *sunesis* was often used in the ancient debate regarding animal rationality as indicating animal reason. For instance, one of the first supporters of animal rationality in ancient thought, Plutarch, argued that all animals have a share in reason, which he signalled with the term *sunesis*, above translated as consciousness. Animals were said by Plutarch to have consciousness or awareness and to differ from humans only with respect to the degree in which they exhibit their intellect.[25] In fact, it has been argued that Michael's discussions of animals bear similarities with Plutarch's argumentation, although the two accounts differ in content.[26] Therefore, it would be no surprise if Michael adopted the Neoplatonic introduction of the attentive part together with the view that animals might have a share in it – even to a smaller degree than humans, since they share only some functions of the attentive faculty.[27] We will see below that this is exactly what Michael thinks regarding other rational functions, namely that animals possess them to a lesser degree than humans.

We can here note that the introduction of such a faculty in order to account for awareness has certain advantages for Aristotle's psychology in general. First, by having one faculty that is responsible for the awareness both of perception and cognition, one secures that there are no overlaps between the various faculties, but each has its proper object: sense-perception is responsible for the perception of the sensible things (perceived each through the proper sense organ), whereas attention is responsible for awareness across an entire spectrum of functions, from the vital to the cognitive. In this sense, the soul becomes more unified in that each faculty rests on the subordinate faculties without sharing their object with them. Second, the attentive power contributes to a clearer distinction between humans and non-human animals. As mentioned above, Aristotle's account leaves much room for doubt regarding the possession of higher soul functions by humans and other animals, for instance, imagination and desire. The introduction of the attentive faculty and especially the distinction of functions into attentive and conscious that we analysed above can solve this problem: up to the conscious function of the attentive part, all soul functions are shared by humans and non-human animals, whereas the awareness of thinking and the other rational capacities belong only to humans.

Animals' share in the rational part of the soul can be further supported by an examination of the ontological status of animals. It has been convincingly argued[28] that, in his commentary on the *Parts of Animals*, Michael presents animals as supreme objects of inquiry, though not in the same way as humans.

> [...] there are things that need a brief survey, so that nothing should be left unexamined. Then, this is said about 'the exchange as to the philosophy of divine matters'; as if the animals and the plants were saying to us: 'men, although the heavenly bodies are noble and most divine, there are still things of sublimity (*thaumasia*) about us so that you should take us into account [make a rational inquiry about us] and do not despise us in every respect'. 'Those that do not please the senses', he [sc. Aristotle] said, are the most disgusting and aversion-provoking [creatues], such as the snail and many others.[29] And 'not to provoke childish aversion' means that we should not avoid like children those animals that are not pleasant to the eye but approach them for the sublimity [*thaumasion*] that there is in them.
>
> The story about Heraclitus is the following: Heraclitus of Ephesus was sitting inside the *ipnos* (and *ipnos* means the bread oven in a house where we bake the bread and thus we speak about ipnites bread); sitting then in the *ipnos* and feeling hot he asked the strangers who came to see him to enter; 'even here, he said there are gods'. Because, the phrase 'all is full of gods' is a Heraclitean doctrine. And because in the works of nature there is above all the final cause, and everything is or becomes because of the final cause; and as finality, he [sc. Aristotle] considered the realm of the good (because everything that is to become is becoming because of something that is taken as its good); and because it is like that, it is imperative that we investigate it.
>
> (*In PA* 22,20 ff. trans. G. Arabatzis)

This passage exhibits a shift from Aristotle's text: animals and plants are given voice, through which they ask humans to engage in rational inquiry about them and to recognize their sublimity also in relation towards other beings. We can already note that the fact that animals are given speech, might also be Michael addressing the Stoic rejection of rationality to animals on the basis of their inability to speak.[30] At any rate, as Arabatzis has shown, this shift from Aristotle's text entails an interesting feature: under the usage of the 'as if' language, Michael attributes a belief to animals, that is, 'belief in their own value'. In other words, animals not only exhort humans to a new investigation, but they also justify this exhortation with an appeal to their own value.

However, this passage not only emphasizes the need for scientific investigation, but also makes a claim about "the ontological status of the knowing object". In particular, the personification of animals and plants, through which animals and plants turn to humans and ask them to proceed to a rational inquiry, is Michael's own initiative, employed in order to show animals as sublime objects of nature. All animals are said to be worthy of investigation not only because of the importance of the humans' understanding of them, but because they themselves partake in sublimity. This participation in sublimity is further explicated: every natural thing has a final cause and accordingly tends towards its good as the final cause. This passage is philosophically dense, and here we do not have room for a detailed analysis. However, what is most pertinent is that animals are positioned within the sphere of nature and the need for their investigation is justified with reference to their having the good as their final cause in the same way that other natural things do, including the heavenly bodies. Michael's use of the anecdote about Heraclitus aims to confirm this sublimity of animals: the anecdote, which Michael enriches with the addition of the phrase 'all is full of gods' (which he falsely attributes to Heraclitus instead of Thales), shows all living beings as partaking in sublimity, although not in the same way: animals and plants, though having a share of sublimity, are not comparable to humans and the divine heavenly bodies. As Arabatzis argues, in Michael's understanding, humans are necessary in this process, since "animals and plants are in need of human perception in order for their value to be formally recognized".[31] In this sense, humans' priority over other living beings is not threatened.

Interestingly, all of Michael's accounts that we have examined so far entail this two-level description of animals. As in Aristotle, human excellence is ensured and rationality in its highest form is attributed only to them. However, in Michael's texts animals are said to have a share in sublimity by themselves, even though lower than humans; and they are able, contrary to Aristotle's account, to partake in the lowest part of the rational part that grants them self-awareness through perception. Therefore, so far, we can say that Michael consistently adheres to a 'lesser degree' thesis, according to which animal souls are inferior to humans with respect to the degree of their possession of rational attributes. More specifically, keeping in mind the Aristotelian distinction between human and animal soul attributes differing in degree or by analogy, it seems that Michael takes a different view than Aristotle. Contrary to Aristotle's insistence that references to animal rationality, virtues and emotions are only said by analogy, Michael seems to take the view that animals have rational attributes to a lesser degree than humans and not simply by analogy.

This is even explicitly stated in a more ethical context, in particular in Michael's description of animals' natural pursuit of the good:

But a decent person, who lives in accord with the mind, does just those things that one should in truth do. For every mind, provided it has not been maimed by pleasure or sickness, by its own nature chooses and pursues what is best and advantageous to itself. This is obvious also from irrational animals, for all these have a certain illumination of mind – some more, some less, as Aristotle himself says elsewhere – and through this illumination they seek and find, by their own nature, the things that benefit them. For a snake, when it is sick in its eyes, rubs fennel on [them], and a bear, when it has been feverish, eats ants and cures itself [...].

(*In NE* IX 506,1–14, trans. D. Konstan modified)

As mentioned above, for Aristotle, any reference to animal rationality is only analogous to human rationality – we speak of animal intelligence only metaphorically. By contrast, Michael here states that animals possess a glimmer or illumination (*ellampsis*) of mind, which makes them pursue what is good and advantageous for them. A similar thought can be found also in Eustratius of Nicaea's commentary on *NE* VI. In particular, when commenting on Aristotle's statement that beasts can be said to be in a way prudent because they seem to have foresight (1141a26), Eustratius gives similar examples as Michael that show how non-human animals seek by nature what will contribute to their survival. He adds that though there is no reason and foresight in them properly speaking, there are echoes of reason (*apechemata logou*) in them by nature.[32]

Besides the moral-psychological implications of this, for instance that animals in this view must hold a kind of belief as to what is advantageous, the reference to illumination[33] of mind can grasp what we argued for above: humans are rational to a fuller extent, whereas non-human animals partake to rationality to a lesser degree. Before turning to the meaning of the illumination language, we want to point to the examples given as illustrating the animal understanding of the good. As mentioned above, both Byzantine commentators make use of similar examples that indicate the animals' natural grasp of their own good: animals use appropriate herbs in order to heal their wounds and overcome their illnesses. As we will see in the next section, the Islamic philosophers also make extensive use of such anecdotes, similar in content to the examples used by Michael and Eustratius in order to indicate that animals possess an understanding of their own good.[34]

Turning now to the meaning of the notion of illumination, Michael refers to its origin from the divine as a "filling with the immaculate light", and similar expressions are also found in Eustratius, who also speaks of the divine origin of illumination.[35] The origin of the Byzantine use of the notion of illumination has been already traced back to the Neoplatonic tradition.[36] In the next section, we will see that the same Neoplatonic background was shared by the Islamic philosophers and used in discussions of animal rationality. As for Michael, he does not speak in this context about animals, but about the contact of the contemplative human with the divine. Nevertheless, we can confidently affirm that the illumination of reason to animals – though it is in a lesser degree than in humans – can also be of divine origin. For the sentence of the commentary immediately following indicates such a gradation of illumination and states that "the clearer illumination of the light of its knowledge and the more perfect union to it" are God's rewards for those who cultivate their intellect.[37] According to this passage, humans can be subjects to a different degree of divine illumination depending on their choice of a life of contemplation, virtue or pleasure.

Elsewhere, Michael stated that humans can be also characterized as *alogoi* just as animals are, insofar as they do not make use of their reason, but live in accordance with their emotions.[38] Life itself can have two forms – it can be either in accordance with perception or with reason, both of which are possible for humans to choose. Living in accordance with perception means, for Michael, to live following one's impulses, a life which only the bad people choose (*phaulos*).[39] That animals also participate in this gradation is also made clear: when commenting on Aristotle's statement that even among the *phauloi* there is something naturally good (1173a4), Michael speaks of the possibility that with this Aristotle refers to non-human animals, who would be *phaula* due to their lack of reason and due to their pursuing of pleasure. Even so, says Michael, animals can still be said to pursue by nature what is good:

For there is in all of them – to a greater or lesser degree – something divine, i.e. an illumination of intellect, by means of which [each of them] by nature seeks and discovers its own proper good, that is, its own proper end. For each attains this [end] by nature, and accordingly [each] welcomes pleasure, too, as conducive to the attainment of this [end].

(*In NE* X 538, 27–31, trans. J. Wilberding/J. Trompeter/A.Rigolio[40])

The similarity of this passage with the previously cited passage is evident. Animals are again said to pursue by nature what is good and to partake – to a lesser degree than humans – in what is divine by having an illumination of mind. What is new in this passage is that Michael emphasizes the natural criteria for the goodness or badness of animals. Animals can be said to be naturally bad because they seek pleasure without restraint,[41] but are also naturally good in pursuing their divine end.

The ascription of natural virtue and vice to animals is confirmed by Michael's account of happiness and its attainment by non-human animals, which will be our last point of investigation into Michael's "lesser degree" thesis. In his comment on Aristotle's statement that other living beings do not participate in happiness because they cannot contemplate (1178b24), Michael says:

By 'this sort of activity' Aristotle means contemplation, that is, the intellectual life and knowledge of the best and divine things. [...] Aristotle is saying according to the assumptions about happiness that other philosophers – Epicureans and the later Stoics – have made, one can grant a share of happiness to the non-rational living things, too; yet according to me and Plato and all of us who set up happiness in the intellectual life, it is impossible for the non-rational living things to be happy **on this conception**, since they are deprived of intellect and rational life. [...]

But what is the end of natural desire? Clearly, it is the zenith, which does <not>belong to the development or the deterioration. Both of these – the development and the deterioration – are motions (*kinêseis*) and are opposed to each other [...]. And this state of rest which is in between development and deterioration is said to be the zenith of natural creatures and the end of nature, and once the nature responsible for producing the natural [creatures] has gotten hold of this, it attempts to preserve and hold onto it as its proper good. [...] **Of this sort of happiness**, then, as we say, even the non-rational living things partake, but in no way do they partake of the happiness that is determined to lie in intellectual life and the knowledge of the most divine things.

(*In NE* X 598,16–599,29 trans. J. Wilberding/J. Trompeter/ A.Rigolio)

In an earlier part of his commentary, Michael had distinguished between two types of virtue, the practical or political ones and the theoretical ones.[42] This division of virtues was introduced in order to account for his division of two types of happiness. Theoretical happiness includes both practical/political and theoretical virtues, whereas political happiness includes only the political virtues and the external goods. In the current passage, Michael focuses primarily on theoretical happiness, from which all non-human animals are excluded. Unlike the teachings of the Stoics and the Epicureans, he says, he follows Aristotle and Plato in not attributing the theoretical *eudaimonia* to all animals, but only to those that can participate in the intellectual life. We can expect that he would exclude non-human animals also from political happiness, which consists in the acquisition of external goods as well as on the exercise of the ethical virtues.[43]

However, contrary to Aristotle's account of happiness, Michael does not completely deny happiness to non-human animals, who are said to participate in a more basic type of happiness. In particular, by referring here to natural desire, which he describes as a state of being at rest, which is intermediary between being in development and being in deterioration, Michael repeats his account of the animals' natural pursuit of the good: non-human animals reach their end and proper good by attaining this state of being at rest. This state is described here as a sort of happiness, that is, "natural happiness of achieving one's zenith by becoming a healthy, fully matured adult".[44] This is in agreement with what we have seen above: non-human animals that have an illumination of intellect and seek by nature what is good for them possess natural virtues and can attain a natural type of happiness, both being more basic than the virtues and happiness attained by humans, which would require full rationality.

Animal Rationality in Islamic Philosophy

After the advent of Islam, animal rationality received sustained attention within the development of Islamic philosophy and science in the 9th–13th centuries CE. Like debates on animal rationality in Byzantium, Aristotelian philosophy, and Greek thought in general, was highly influential in the Islamic world. Aristotle's *History of Animals*, *Parts of Animals*, and *Generation of Animals* were translated into Arabic as a unified whole called the *Book of Animals* (*Kitāb al-ḥayawān*). The records of apparently intelligent animal behavior contained therein became the starting point for discussions of animal rationality. Translations of the Neoplatonic texts of Plotinus and Proclus were also widely influential. Nevertheless, these were not the only source texts and influences regarding theorizing the rational capacities of animals. The Qur'ān itself also contains many remarks on the apparently rational behavior and capacities of animals, frequently portraying animals as intelligent, capable creatures. Islamic thinkers thus took these remarks into consideration in their assessment of animal powers. From the beginning, Islamic philosophers were tasked with navigating sometimes competing explanations of rational behavior in animals and with explaining any rational or cognitive capacities that account for apparently rational behavior in animals. The philosophical debates on animal rationality proceeded, we will argue, broadly in three waves. Each articulated the connection of apparently rational behavior in animals to the possession (or not) of a rational power or rationality per se differently, though these accounts were not altogether unrelated to one another.

The first wave follows upon the translations of Aristotelian and Plotinian texts during the translation movement of the 8th–10th centuries. In these texts, rational behavior in animals was either minimized through careful translation choices or explained by an active principle outside of animals completely. In the *Book of Animals*, descriptions of animals as rational (*phronima*) are usually rendered by ascribing them forbearance (*ḥilm*), which, while suggestive of intellectual capacity, does not necessarily imply it.[45] In the above-discussed passage from *HA* VIII.1 588a19–30, the analogous relation between human and animal intelligence is lost completely, and it is not clear that humans and animals share even character traits in the way the original text suggests. Rather, all animal characters are reduced to traces (*al-āṯār*) of their human correlates.[46]

The use of traces in this context calls to mind the prevalence of the trace structure within the imagistic metaphysics of Arabic Neoplatonism, where we find that animals are relegated to the level of nature separate from intellect. In the pseudo-*Theology of Aristotle* we read that the soul "originates an image (*tabtabi'u ṣanaman*), which is sensation and nature (*al-ṭabī'a*) in simple bodies and plants and animals and every substance", which complicates a straightforward relation between animals and soul.[47] In the collection of Plotinian material called *Sayings of the Greek Sage*, we also find discussion of the metaphysical hierarchy of intellect-soul-nature that specifies the relation of natural things to intellect, reading, "Nature is the last of the intellectual things, [and] as for things which are posterior to nature, they are images and traces of the intellectual things."[48]

The text further explains that bodies are generated by nature and farthest from intellect, which makes them "a trace of the trace of its trace, namely, nature".[49] (That is, bodies are produced by nature, being themselves a trace of nature, which is a trace of soul, which is intellect's trace.) Thus, animals stand in a rather precarious position, separated from intellect and even soul in the Arabic Plotinian texts, and appear to be pretty firmly separate in Aristotle's *Book of Animals* as well, depending on how one interprets the changes in the text.

The influence of Neoplatonism is discernible in many originally non-Neoplatonic texts as well, for example the translation of Aristotle's *Parva Naturalia*, called in Arabic after the translation of the title of the first treatise therein, *Kitāb al-ḥiss* (*Sense and Sensibilia*). We find here that not only animal rationality but even sensation is explained by the higher soul, since animals "do not have the ability to perceive (*adraka*) things in their true reality", but perceive obscurely "since they have come to have some of the light of the rational soul within them".[50] Even more strongly here, it is due to the influence of the rational soul outside of animals that they can perceive aspects of things at all. We saw above that Byzantine thinkers also appealed to illumination in order to explain rational behavior in animals. While such an interpretation will not become mainstream in Islamic philosophy, it was nevertheless one possible explanation of rational behavior in animals in Islamic philosophy. For example, Miskawayh (d. 1030), a Neoplatonic philosopher active within the Būyid dynasty, discusses animals and animal behavior throughout his influential *Refinement of Character* (*Tahḏīb al-aḫlāq*).

Each one of these ranks includes many grades, for some animals are nobler than others because they are apt to receive training. The horse is nobler than the donkey only for this reason. Similarly, the falcon is superior to the raven. If you observe the animal kingdom as a whole, you will find that the animal which is apt to receive training—this being a trace of rationality (*āṯār al-nuṭq*), i.e. the rational soul (*al-nafs al-nāṭiqa*)—is better than the rest.[51]

The trace of rationality and of the rational soul is the explanation of the rational behavior animals are able to learn. It is not their own capacity per se, but something they obtain through a certain relation to an outside, rational cause (e.g. the human intellect capable of training).[52] The fact that their rationality is only a trace does not render it null, but grounds a genuine likeness between rational and non-rational animals. Nevertheless, here in Miskawayh the status of the trace and its epistemic standing is left open, and it is unclear whether this trace constitutes a power of animal soul or is merely a behavioral feature useful to human beings. In part, the lack of clarity within this paradigm will motivate the development of the next wave of explanation of animal behavior.

Further, the many anecdotal reports of animal behavior included throughout the *Book of Animals* proved influential. They were an impetus for the development of accounts of rational behavior in animals and accorded well with the presentation of rational behavior in animals in the Qur'ān. A popular example is the discussion of animals from the Brethren of Purity's (fl. 10th c.) famous animal epistle, the 22nd of their encyclopedic collection. Most of the epistle comprises an allegorical narrative in which animals take human beings to court before the king of the *jinn* in order to hold them accountable for their harmful treatment of animals and to challenge the human beings' assumption of their superiority. In Chapter 27, the bee argues against claiming human superiority based on reason, saying, "We, too, have knowledge and understanding, awareness, discernment, thought, judgment, and governance, subtler, wiser, and finer than theirs."[53] There follows standard examples of rational behavior, such as bees' construction of hives and ants' group organization. But the Brethren also quote several verses from the Qur'ān and present animals as capable of worshiping and serving God in the way that human beings can, which is implicitly taken as evidence of their rational capacities, even though it is also said that the animals are guided by God through inspirations (*ilhām*).[54] The Brethren's combined philosophical-theological approach here will later become standard: the attribution of discernment and reason to animals is based on empirical evidence *and* revealed text. But here, that behavior is not explained by or connected to any particular power of the soul.

We find similar treatment of animal rationality in Abū Bakr al-Rāzī (d. 925), who also acknowledges some degree of rationality in animals in explanation of their behavior. His discussion of animal treatment was heavily influenced by tenets from the theological school of the Muʿtazilīs, wherein there was concern for justice for animals in light of animal suffering, and in Abū Bakr's arguments for just treatment of animals he relies upon positions similar to those of this school.[55] His arguments regarding animal rationality, by contrast, appear to be a development of his own. The most generous allowance of rationality as a feature of animal life occurs in his *Doubts on Galen*, where he asks how one can deny reason in animals, given the rational behavior they exhibit, such as the mouse using his tail to extract oil from a narrow-mouthed bottle.[56]

This text presents the basic issue we are highlighting here well. Animals act in ways that appear rational, and, for Abū Bakr, that is enough to conclude there is some sort of rational activity in them. Specifically, he concludes that one cannot deny that animals would be unable to accomplish their tasks were they bereft of conception (*taṣawwur*), reflection (*rawiyya*), and thinking (*fikr*).[57] However, he does not offer an account for the way that activity is possible; he does not identify a specifically rational capacity in animals' souls. In this respect, his portrayal aligns with the others we have seen so far: animals seem to *act* rationally, and while Abū Bakr goes further than other thinkers by claiming that animals are in fact rational, he offers no developed account of animal psychology.

One feature of Abū Bakr's account is telling for the development for which we are arguing. In his analysis of rational behavior in the *Doubts on Galen* passage, he begins by highlighting that *both* animals and human infants act in ways that, without rational capacities, they should not be able to act. Specifically, both know to flee from what is harmful and to cling to what is beneficial. But any account of animal rationality that proposes a rigorous explanation of rational behavior in animals must identify the structure by which this behavior is possible. Anyone looking for a philosophical account of rational behavior or capacities in animals is left, again, without a satisfactory answer. The capacity that accounts for animals' behavior is exactly what Avicenna will aim to identify.

The second wave of philosophical discussion of animal rationality occurs with Avicenna's (d. 1037) paradigm-shifting development of the internal powers of animal soul, specifically, of the power of estimation (*wahm*). His theory of internal psychological powers is a development of Aristotle's psychological powers, which Avicenna formalizes into the five external and five internal powers of the soul.[58] (His notion of estimation was also influential in medieval Latin discussion of animal rationality, where *wahm* was translated as *aestimatio* and hence is called 'estimation' in English.[59]) The activity of the power of estimation is to extract non-sensible features (*ma'ānin*) from the sensible forms in the world. Estimation's apprehension of meanings functions as a pseudo-rational activity that accounts for apparently rational behavior in animals. That means that with Avicenna we have the first account of rationality in animals that identifies a psychological power proper to animals that is the source of apparently rational behavior.

The canonical discussion of estimation is in Books I and IV of Avicenna's *Book of the Soul* (*al-Nafs*) from his *Healing* (*al-Šifā'*). Here the existence of estimation is determined by its unique function: it is responsible for extracting certain features from sensible objects.[60] While common sense apprehends the form of a sensible object that is in the sense organs, estimation apprehends the non-sensible features of otherwise sensible objects, such as "the thing that the soul perceives of the sensible thing without the external sense perceiving it first, like the sheep's perception of the hostile meaning in the wolf".[61] Thus, the sheep flees the wolf "without the [external] sense perceiving it at all". This then, is estimation's function, and we can see immediately the life ensuring function it has in animal living.

Later in Book IV, Avicenna returns to estimation to discuss its inner working more explicitly, specifically, regarding its structural possibility outside of the influence of intellect. He initially explains the inner working of estimation in the case of human infants. "One of [the ways of estimation] is the emanative inspirations (*ilhām*) upon the whole from divine mercy", and when an infant grabs hold of a table as he falls, it is owing to "something innate in the soul, [which] divine inspiration places in it".[62] The example of the infant emphasizes estimation's power to guide without experience, since the infant can be guided neither by a rational capacity nor by experience, having none yet available. Avicenna turns then to non-human animals, explaining that the same process is at work in their use of estimation as well.

> Similarly animals have innate inspirations [...] and by these inspirations estimation comes to the features mixed with the sensibles about what harms and is beneficial, such that every sheep is wary of the wolf, even if he has never seen [a wolf], nor had an affliction from a wolf.[63]

As shown in the infant's case, a key aspect of estimation's explanatory power is that with it Avicenna can explain why animals reliably act in reasonable ways in new situations. Aristotle's explanation of this sort of animal action relied on memory, which the sheep would use to gain experience, which in turn explains rational behavior.[64] The major problem with this account is that memory in animals, according to Aristotle and Avicenna, is unreliable. Thus, to explain the sheep's flight at sight of the wolf on Aristotle's account, one must have recourse to (1) the sheep's past experiences of the wolf, and (2) the chance that the sheep in fact remembers the wolf when he sees it.[65] Also on Avicenna's account of memory, animals cannot recollect – they cannot recall things at will – they can only remember, which is irregular.[66] With the power of estimation, Avicenna explains the sheep's immediate flight the very first time he sees the wolf.

In addition to its apprehending function, estimation works in animals as the power responsible for making judgments. In service of this function, estimation utilizes the material and function of other powers, especially common sense and imagination.[67]

> [Estimation is] the one that makes judgement in animals, a judgement that is not a distinction, like the intellectual judgement; rather, it is an imaginative judgement connected with the particular, and with the sensible form, and from which many animal activities derive.[68]

The qualification of estimation as capable of judging particulars on the one hand accounts for apparent animal judgements, but on the other hand would seem also to leave Avicenna open to the objection that he is merely rebranding intellect into animal form without fully acknowledging the consequences. Estimation is *not* per se a rational power – neither the internal sense power of cogitation (*mufakkira*), which is the imaginative power in human beings, nor the intellectual power of the rational soul proper to human animals alone and responsible for their capacity to intellect universals.[69] The features within estimation are only particular in content, and thus do not rise to the level of the universal that is, on Avicenna's (and Aristotle's) account, necessary for qualifying as rational activity in the richest sense. Nevertheless, the activity of estimation explains apparently rational *behavior* in animals by providing a mechanism by which animals have access to information that would otherwise require rational thought. In turn, the sheep need not make an inference based on the wolf's behavior and appearance to conclude that he is dangerous. Instead, estimation provides the sheep immediately with the feature "hostility", which allows the sheep to respond in a reasonable way, i.e. by fleeing, reliably and *without* experience. Thus, estimation is central in animals' ability to pursue what is in their interest.[70]

However, even here we find that at crucial points of Avicenna's account, he has recourse to explanations reminiscent of the Neoplatonic-inspired first wave theory of apparent animal rationality. Specifically, when he pinpoints the source of estimation's grasp of features, he cites the innate inspirations animals share with human infants. However, the appeal to innate inspirations (or, in human infants, an emanative effect) is not explained. Avicenna does not say where these inspirations come from or why they are reliable – whether they come from an animal's nature, from a source higher up in the emanative structure, or from somewhere else entirely (for example, from God). Thus, although he has given us the power of estimation in explanation of apparent animal rationality, we are left to wonder whether Avicenna has really settled the issue or whether he has begged the question by moving the crucial explanandum back a step.[71] We saw above an appeal to inspiration in the explanation of animal rationality by the Brethren of Purity, and Avicenna's use of this concept in this context recalls their account. Although the psychological power of estimation has far-reaching philosophical and historical influence, without a thorough explanation of the inner working of the estimative power in animals, his account runs the risk of being *ignotum per ignotius*. We will see below that at least one response to Avicenna's account of animal pseudo-rationality seizes the issue at its core and acknowledges that universals are necessary if we are to provide a philosophical account for rational animal behavior.

The final wave we would like to present here is the work of an early respondent to and active critic of Avicenna, Faḫr al-Dīn al-Rāzī (d.1210).[72] A highly innovative and controversial thinker, he challenged Avicenna's philosophy on many fronts, but a consistent point of dispute was the psychology of animals. He particularly challenged the claim that animals have souls consisting only of powers proper to or seated in the body, and his considerations of animal behavior frequently led him to question whether animals do, in fact, have separable souls. In these short remarks, we will restrict ourselves to the remarks he presents in his *Eastern Investigations* (*al-Mabāḥiṯ al-mašriqiyya*), *Epitome on Philosophy and Logic* (*al-Mulaḫḫaṣ fī l-ḥikma wa-l-manṭiq*), and *Exalted Topics of Inquiry* (*al-Maṭālib al-ʿāliya*). One of the crucial points on which Faḫr al-Dīn criticizes Avicenna's psychological account is the power of estimation. He was not the only one to do so, and more well-known (to contemporary scholarship) critiques were leveled by Abū Ḥāmid al-Ġazālī (d.1111) and Averroes (d.1198).[73]

Faḫr al-Dīn's critiques focused on two points: first, whether estimation was indeed necessary as a distinct power of the soul, and second, whether estimation was in fact just a type of intellectual power (something that Avicenna in fact denied of animals). In his criticism of the internal senses, Faḫr al-Dīn targets Avicenna's account of estimation on just these grounds. For example, in his *Eastern Investigations*, he rejects estimation as a distinct power of the soul, presenting a dilemma aimed at proving the superfluity of estimation. Using the case of the apprehension of hostility in a given individual as an example, he argues that one of two things must be the case. Either the hostility of a given individual inheres in the individual itself, and if this is this case, the common sense could apprehend the hostility, since it apprehends particulars.[74] Or, the hostility of a given individual does *not* inhere in that individual, in which case the apprehender grasps a universal, and in this case the power responsible for apprehension is the intellect, not estimation.[75] In either case, it looks like estimation is not really needed after all.

Faḫr al-Dīn is committed to a unified soul capable of its acts without distinct powers.[76] A fundamental disagreement he has with the Aristotelian tradition is on the nature of sensation, specifically, whether two different powers are responsible for apprehending particulars and universals. Faḫr al-Dīn argues that one thing must apprehend *both*, since, on his account, sensation requires both a particular and a universal to do its work.

When we apprehend some individual person, we know that person is a particular falling under the universal *human*, and not a particular falling under the universal *horse*. What judges the particular human to be a particular falling under the universal *human* and not a particular falling under the universal *horse* must surely be one and the same as what apprehends the individual human, the universal *human*, and the universal *horse*. So what apprehends particulars is one and the same as what apprehends universals.[77]

Since animals are capable of sensation – and since on the Aristotelian model animals are characterized by sensation – Faḫr al-Dīn's objection here aims to refute Aristotelians on their own terms and to show that they have underestimated the psychological complexity required for animals to live their lives.[78] If animals are capable of sensation, it looks like they must have access to universals, although in the *Eastern Investigations* he aporetically leaves this conclusion open.

But later in his *Exalted Topics of Inquiry*, Faḫr al-Dīn argues at length that animals must be capable of some rational activity by appealing to their apparently rational behavior.[79] He offers proofs of two different sorts, those on rational grounds and those on authoritative grounds. The former proofs include the above-mentioned example of the mouse extracting oil from a narrow-mouthed bottle with his tail, the prudent behavior of ants, and the orientation skills of camels. Many of these examples go back to the Aristotelian tradition, and while he likely found many in Avicenna's own *Book of Animals* (*Kitāb al-ḥayawān*), many are found already in Aristotle's zoological texts.[80] The arguments on authoritative grounds include the Qur'ānic account of Solomon's conference with the birds in 27:16 and of birds' ability to praise God in 24:41. Faḥr al-Dīn takes these examples to provide evidence that animals are capable of the sort of actions that require rational capacity.

As part of the groundwork for his argument, Faḥr al-Dīn makes the crucial observation that acknowledging that animals have rational capacities close to those of human beings does not thereby entail that non-human and human animals are one in essence. As he argues, the fact that souls share a privative quality, in this case being *in*corporeal (something entailed by the knowledge of universals), does not entail that they share all essential features. To complete this observation, at the end of his discussion Faḥr al-Dīn differentiates between types of rationality or cognitive activity, saying that, "there are many degrees of knowledge and intelligence", and "perhaps each of [the degrees] is specific to a [kind of] soul, which can only have a particular type of reason and a specific degree of intelligence".[81] Thus, he concludes the entire discussion as follows:

> So if what is meant by 'reason (*'aql*)' is all the forms of knowledge (*'ulūm*) that belong to the human, then it would be true to say that [animals] are not 'rational.' But if what is meant by 'reason' is any of the types of intelligence, then it would evidently be the case that this intelligence should be attributed to them.[82]

Although he adds the proviso that God alone knows the truth of this matter, he introduces a new level of sophistication to theories on animal intelligence. The rationality proper to human beings is not the only type of rationality, and concluding that animals are rational does not identify their essences with the human essence. Faḥr al-Dīn's argument points out a crucial mistake that philosophers have made regarding animal rationality. They have assumed that the only true cognitive capacities are *human* cognitive capacities, and for that reason they also assume that if one grants that animals have functionally sophisticated cognitive capacities then one also grants that animals are, more or less, the same as human beings. Not only does that assumption beg the question by defining rationality as human rationality alone (a criterion animals will always fail to satisfy because they are simply *not* human beings). It misses the complexity of rational activity in all animals, human and non-human.

Finally, in his *Epitome*, he argues explicitly that in order to explain animal behavior we must acknowledge that they have the capacity to know universals. He begins by highlighting, "We find [that] each species (*naw'*) of animal seeks a certain species of food and detests everything else. If the [the animal] could not distinguish the species it seeks from the species it detests, this would not be the case".[83] In response to an objection that an animal just desires whatever particular vegetation they happen to come across, Faḥr al-Dīn replies that such a counter explanation is insufficient to account for the basic animal behavior of searching for food.

> For we say: this is false, since the beast, whenever he senses the smell of vegetation, either has that given vegetation as his goal – which is absurd, for if [the animal] has not seen it, how could it be [the animal's] goal? – or [the animal's goal] is the vegetation insofar as it is vegetation, which is what was to be shown.[84]

Here Faḫr al-Dīn's response rests on the fact animals tend to use smells while searching for food, and *smells* are universal. I do not smell a particular pizza, I smell *pizza*. The deer does not smell one patch of grass, she smells *grass*. Thus, saying that an animal desires whatever food they happen come across requires that the animal has apprehended, that is, *seen* a particular patch of vegetation. Without that apprehension, the animal cannot desire it, since without properly apprehending it, the animal does not actually have a particular object of desire, because (and this is the crucial part) smells are connected to the knowledge of universals. But the interlocutor is trying to deny that animals know universals. Thus, Faḫr al-Dīn concludes, if an animal seeks any sort of food whatsoever, they must know the universal of that food determining their desire from the beginning. In this discussion as well, Faḫr al-Dīn emphasizes that the incorporeality of animal souls does not identify them with human beings. Here in what we have designated the third wave of theorizing on animal rationality, we find the strongest account yet, which argues that the rational behavior displayed by animals must in fact be due to their own rational capacity.

Conclusion

Our examination reveals primarily two parallels between the Byzantine and Islamic discussions on animal rationality. Both traditions incorporate a new power of the soul that works, at least in part, to account for apparent animal rationality – the attentive power in the Byzantine tradition and the estimative power in the Islamic tradition. And both traditions incorporate an appeal to Neoplatonic emanation or illumination in explanation of rational behavior in animals, although we have noted that this feature may have been elided or replaced with theological inspiration in the Islamic tradition. A primary difference is that in the Islamic tradition, consideration of our topic is disconnected from Aristotle's texts rather quickly, and in general, Avicenna's own thought replaces Aristotle as the focal point of philosophical discourse. This change does not mean that Aristotle's theory of animal minds was set aside. Rather, it was at work in the background of all the discussions we considered.

The Byzantine tradition is unique to medieval philosophy insofar as its debates on animal rationality developed without the Avicennan power of estimation. The power of estimation was crucial in debates on animal minds in the medieval Latin tradition, especially in the flurry of philosophical activity on animal rationality spurred by the translations of Aristotle's work by Michael Scot (from Arabic) and William of Moerbeke (from Greek) into Latin in the first half of the 13th century.[85] (Avicenna's *Book on the Soul* was translated into Latin earlier in the 12th century.[86]) As an internal sense power, estimation allowed thinkers to explain the appearance of rational behavior from animals while having recourse only to the sensitive parts of the soul, that is, those parts of the soul that are never separate from the body. Thus do figures like Albert the Great, Thomas Aquinas, and John Duns Scotus rely in part on the labor of estimation to limit the realm of the rational to the human animal.[87] While the incorporation of estimation sophisticates accounts of animal cognition, those accounts maintain the strong division between the human (rational) soul and the animal (sensory) soul present in early texts influential in Latin Christendom, such as Augustine's *On the Magnitude of the Soul*.[88] To our knowledge, within these debates, it is only in the Islamic tradition in the 13th century that this dichotomy loses its force, when Faḫr al-Dīn al-Rāzī rejects the restriction of rationality to human beings alone by adding the crucial premise that even if animals have separate, rational souls, they are still not human beings.

Nevertheless, all three traditions use the philosophical tools developed in the earlier Aristotelian and Neoplatonic traditions to interpret and to refine the original Aristotelian position in an effort to account for what may seem to be rational behavior or activity in animals. A marked feature of the Byzantine tradition is its utilization of Neoplatonic theory, which the tradition seems to integrate more extensively than others. The attentive power developed in Philoponus's thought was used to organize the internal powers of the animal soul, and the illumination from reason occupied a more central position in Byzantine thought on animal rationality. The centrality of the latter may be due to the need to explain the potentially rational aspect of animal behavior. The attentive power does not explain all aspects of rationality, only self-awareness. Without the power of estimation that was available to the Islamic and Latin tradition alike, Byzantine thinkers like Michael of Ephesus had to develop a different explanation of the rationality animals appear to display. As a result, he relies on the Neoplatonic motif of an illumination of reason in an attempt to account for the rational features of animals.

Notes

1 This article was written under the aegis of the project "Animals in the Philosophy of the Islamic World", which has received funding from the European Research Council (ERC), under the European Union's Horizon 2020 research and innovation programme (grant agreement No. 786762). We thank Peter Adamson, our referee, and the editors for their feedback on our paper. We are also very grateful to Christof Rapp and all the participants of the Munich School of Ancient Philosophy Research Seminar for the useful input on an earlier draft of this paper.

   Unless otherwise indicated, with "animals" we denote non-human animals.

2 As will become clear in what follows (and unless otherwise explicitly defined by the original authors), with 'rationality' we refer to any kind of end-oriented reflective thinking process, including reflection and combination of mental images included in memory, cognition and self-awareness (e.g. see below on attention). Hence, we aim to investigate whether and to what extent non-human animals possess rationality in any of these senses, using each feature as a sufficient condition of rationality.

3 *DA* I.2 404b1–6; 405a8–16; III.3 427a19–29.

4 *DA* I.2 404b4–6; III.3 428a24; III.10 433a11–12; *NE* I.7 1098a3–4; *EE* II.8 1224a27; *PA* I.1 641b7.

5 *DA* II.3 414b18–19.

6 *Met.* A 980a27–b28; *DA* III.3 428b26; III.8 432a11–13; III.10, 433a9–13. See D. Henry, "Aristotle on Animals," in *Animals: A History*, ed. P. Adamson and G.F. Edwards (Oxford, 2018), 9-26 on Aristotle's views on animal psychology.

7 *DA* III. 431a14–9, 431b2–10; *De Mem.* 449b31–450a1.

8 *Met.* A 980a27–b28.

9 See R. Sorabji, *Animal Minds and Human Morals: The Origins of the Western Debate* (Ithaca, N.Y., 1993), 35 ff. on the relation between perceptual *phantasia* and belief or thought in Aristotle. See also Henry, "Aristotle on Animals," and S.M. Connell, "Animal Cognition in Aristotle," in *The Cambridge Companion to Aristotle's Biology*, ed. S.M. Connell (Cambridge, 2021), 200.

10 *NE* VI.7 1141a22–28. *DA* II.9 421a30–32. See also Connell, "Animal Cognition in Aristotle," 196–199.

11 *HA* 612b19–26; 614b19–21; 618a25–27; 623a8; *PA* 650b19–20.

12 *HA* VIII.1 588a18–31; IX.1 608a11–21. See W. Fortenbaugh, "Aristotle: Animals, Emotion and Moral Virtue," *Arethusa* 4 (1971) 137–65, esp. 153–57; Henry, "Aristotle on Animals," 15–16.

13 *HA* IX.1 608a1121. It is not clear whether Aristotle intends the two options of quantitatively different and analogously different capacities in human and non-human animals to be mutually exclusive. It is possible that he meant that with respect to some capacities, humans and animals differ in quantity, while with respect to others, they differ by exclusion, i.e. the capacities belong only to humans strictly speaking and it is only by analogy when we attribute them to animals.

However, none of the examples given here can correspond to the quantitative difference. The rational capacities as well as the virtues and emotions are said to belong to animals as analogical 'natural capacities'. It seems that only tameness (*hêmerotês*) and wildness (*agriotês*) can be examples of attributes being in a higher or lesser degree in humans and animals, although *hêmerotês* can be also said to be analogously used with reference to humans and to mean gentleness. For this reason, we take all references to such animal capacities to be analogical and not to indicate any form of gradualism in the thought of Aristotle. But see again Connell's conclusion in "Animal Cognition in Aristotle," at 198–99.

14 Thus, we will leave aside medical texts and other non-philosophical literature. See A. Zucker, "Zoology," in *A Companion to Byzantine Science*, ed. S. Lazaris (Leiden, Boston, 2020) for the current state of the art of Byzantine zoology.

15 See George Tornikes, *Funeral Oration to Anna Comnene* (J. Darrouzes, *Georges et Demetrios Tornikes: lettres et discours* (Paris 1970), 283, 4–7). See also R. Browning, "An Unpublished Funeral Oration to Anna Comnena," in *Aristotle Transformed: The Ancient Commentators and their Influence*, ed. R. Sorabji (Ithaca, N.Y., 1990), 423–38.

16 See M. Trizio, "The Byzantine Reception of Aristotle's Parva Naturalia (and the zoological works) in eleventh- and twelfth-century Byzantium: an overview," in *The Parva Naturalia in Greek, Arabic and Latin Aristotelianism: Supplementing the Science of the Soul*, ed. B. Bydén and F. Radovic (Gotheburg, 2018), 155–68.

17 Michael of Ephesus, *In PA*, ed. M. Hayduck, *Michaelis Ephesii in libros de partibus animalium, de animalium motione, de animalium incessu commentaria*. CAG 22, 2 (Berlin, 1904), 6:25: εἰ δὴ εἶδος ἀνθρώπου ψυχὴ ἢ μέρος ψυχῆς, τοῦ φυσικοῦ περὶ ψυχῆς ἂν εἴη λέγειν, καὶ εἰ μὴ περὶ πάσης, τῆς τε λογικῆς καὶ ἀλόγου, ἀλλὰ τέως περὶ τῆς ἀλόγου, καθ' ἣν τὰ ζῷα τὸ ζῷα εἶναι ἔσχηκεν. Cf. ibid., 71:2 ff.; Michael of Ephesus, *In SE*, ed. M. Wallies, *Alexander, quod fertur Michael Ephesius, In Aristotelis sophisticos elenchos commentarium*. CAG 2, 3 (Berlin, 1898), 133:20); Michael of Ephesus, *In NE*, ed. G. Heilbut, *Eustratii et Michaelis et Anonyma in ethica Nicomachea commentaria*. CAG 20 (Berlin, 1892), 482:18-19.

18 Michael of Ephesus, *In PN*, ed. P. Wendland, *Michaelis Ephesii in parva naturalia commentaria*. CAG 22, 1 (Berlin, 1903), 49:29.

19 Michael of Ephesus, *In PN*, ed. Wendland, 6:3). Cf. Arist., *De mem.* 453a6–10.

20 Michael of Ephesus, *In PN*, ed. Wendland, 12:5–13:24). Cf. Arist., *De mem.* 450a9.

21 *DA* III.2 425b12–25. Aristotle here discusses the possibility that we perceive that we see through another sense besides sight and rejects it. Each sense is responsible for its self-perception.

22 J. Philoponus, *In DA*, ed. M. Hayduck, *Ioannis Philoponi in Aristotelis de anima libros commentaria*. CAG 15 (Berlin 1897), 464:24–465:31. See R. Sorabji, *The Philosophy of the Commentators, 200-600 AD: a Sourcebook in Three Volumes*, vol. 1 (London, 2004), 152–53.

23 *DA* III.2 426b17–21.

24 We modified the translation of the two terms *anaisthêtos* and *anennoêtos* in order to indicate the parallel to what Philoponus says about conscience: conscience is the function of the attentive part that has to do with the vital powers, whereas the attentive function (not to be confused with the attentive part) is responsible for the cognitive functions. We also added the transliteration of some Greek terms for clarification. See D. Konstan (trans.), *Aspasius, Anonymous, Michael of Ephesus: On Aristotle Nicomachean Ethics 8-9* (London, New York, 2001).

25 Plutarch, *On the Cleverness* 960a; 968a–c; 975e–976d; 979a; 985c; *Are Beasts Rational* 992a–d.

26 G. Arabatzis, "Animal Rights in Byzantine Thought," in *Animal Ethics: Past and Present Perspectives*, ed. E.D. Protopapakis (Berlin, 2012), 103–11.

27 Animals do not share in the function of the attentive faculty that is related to cognition. The attentive faculty in animals is, we take it, responsible for self-awareness as well as for the unification of the animals' sensation and desire. The attentive faculty in humans also unifies sensation with thought.

28 G. Arabatzis, "Michael of Ephesus and the Philosophy of Living Things: *In De Partibus Animalium*, 22.25–23.9," in *The Many Faces of Byzantine Philosophy. Papers and monographs from the Norwegian Institute at Athens*, series 4, 1, ed. B. Bydén and K. Ierodiakonou (Athens, 2012), 51–78.

29 In this sentence we have modified the translation.

30 Chrysippus was allegedly the first Stoic who supported the view that animals lack rationality due to their soul's imperfection, which he connected to the fact that their utterances – unlike those of humans – have no significant meaning. This view was in turn used as an argument for the humans' natural superiority over animals and the impossibility of affinity (*oikeiosis*) between humans and animals. Diog. Laert., *Vit. Phil.* VII.55, 1650; VII.129, 234–35. See a similar thought in Aristotle at *NE* 1161a30–b2.

31 G. Arabatzis, "Michael of Ephesus and the Philosophy of Living Things," 58.

32 Eustratius of Nicaea, *In NE*, ed. G. Heilbut, *Eustratii et Michaelis et Anonyma in ethica Nicomachea commentaria.* CAG 20 (Berlin, 1892), 327:25–328:15.

33 We translate *ellampsis* as illumination and not as glimmer only in order to be consistent with the next passage's reference to *ellampsis*.

34 The examples have, of course, Aristotle's *HA* as a common source.

35 Michael of Ephesus, *In NE*, ed. Heilbut, 603:16–31; Eustratius of Nicaea, *In NE*, ed. Heilbut, 294:19–25. On Eustratius' commentary and its Neoplatonic sources see M. Trizio, "Neoplatonic Source-Material in Eustratios of Nicaea's Commentary on Nicomachean Ethics VI," in *Medieval Greek Commentaries on the Nicomachean Ethics*, ed. Ch. Barber and D. Jenkins (Leiden, Boston, 2009), 71–109.

36 K. Ierodiakonou, "Some Observations on Michael of Ephesus' Comments on Nicomachean Ethics X," in *Medieval Greek Commentaries on the Nicomachean Ethics*, ed. Ch. Barber and D. Jenkins (Leiden/Boston 2009), 197-98; M. Trizio, "Neoplatonic Source-Material in Eustratios of Nicaea's Commentary on Nicomachean Ethics VI"; G. Arabatzis, "Michael of Ephesus on the Empirical Man, the Scientist and the Educated Man (In Ethica Nicomachea X and the De Partibus Animalium I)," in *Medieval Greek Commentaries on the Nicomachean Ethics*, ed. Ch. Barber and D. Jenkins (Leiden/Boston 2009), 163–84.

37 Michael of Ephesus, *In NE*, ed. Heilbut, 603:31–35.

38 Ibid., 481:4–6.

39 Ibid., 514:25–31.

40 J. Wilberding/J. Trompeter/A. Rigolio (trans.), *Michael of Ephesus on Aristotle Nicomachean Ethics 10 with Themistius On Virtue* (London and New York, 2019).

41 That is, animals' lack of self-control makes them seek pleasure without restriction, and in this sense, they are worse than humans who are able to control themselves. However, even this point is not accepted by Michael who states that, since the prudent beings, that is, humans, also seek pleasure, then the argument against the view that all things pursue what is good is false. See Michael of Ephesus, *In NE*, ed. Heilbut, 538:3 ff.

42 Michael of Ephesus, *In NE*, ed. Heilbut, 571:30–572:5. As Ierodiakonou, "Some Observations on Michael of Ephesus," 194–95 has argued, Michael's distinction between practical/political virtues and theoretical virtues is not to be confused with Aristotle's division between ethical and intellectual virtues, but is rather of Neoplatonic origin.

43 On the differences of this view from Aristotle's views see Ierodiakonou, "Some Observations on Michael of Ephesus," 194–95.

44 Wilberding/Trompeter/A.Rigolio, *Michael of Ephesus on Aristotle's Nicomachean Ethics 10*, 148, n. 466.

45 See *HA* 611a15–17; *HA* 612a1–3, b1, b18–22; *PA* 648a5–8.

46 Aristotle, *History of Animals* 588a2–3, ed. L. Filius, *The Arabic Version of Aristotle's* Historia Animalium. *Book I–X of the* Kitāb al-Hayawān (Leiden, 2019), 162.

47 *Theologia Aristotelis* X.16, ed. A. Badawī, *Aflūṭīn ʿinda l-ʿArab. Plotinus apud Arabes. Theologia Aristotelis et fragmenta quae supersunt* (Cairo, 1966), 136; trans. G. Lewis, *Plotiniana Arabica* (Paris, 1959), 293, altered. On this topic, see B. Somma, "The Causal Efficacy of Nature in the *Neoplatonica Arabica*" in *Reading Proclus and the* Book of Causes, *Volume 3: On Causes and the Noetic Triad*, ed. D. Calma (Leiden, 2022) 281–302. On the nature of the surviving Arabic translation of Plotinus, see M. Aouad, "La Théologie d'Aristote et autres textes du Plotinus Arabus," in *Dictionnaire des philosophes antiques* I, ed. R. Goulet (Paris, 1989) 541–90.

48 *Sayings of the Greek Sage* II.67–68, ed. and trans. E. Wakelnig, *A Philosophy Reader from the Circle of Miskawayh* (Cambridge, 2014), 138–39.

49 *Sayings of the Greek Sage* II.69, ed. and trans. Wakelnig, 140–41.

50 *Kitāb al-ḥiss*, trans. R. Hansberger in "Averroes and the Internal Senses," in *Interpreting Averroes: Critical Essays*, ed. P. Adamson and M. Di Giovanni (Cambridge, 2016), 151–52.

51 Miskawayh, *The Refinement of Character*, ed. C.K. Zurayk, *Tahḏīb al-aḫlāq* (Beirut, 1966), 46; trans. idem, *Miskawayh: The Refinement of Character* (Chicago, 2012), 42.

52 We find this explanation in Jewish philosophy as well. We find a similar account of animal rationality: "When the souls of animals [...] incline towards the rational soul, some cognition will be found in them according to the degree of their inclination towards the rational soul, be it large or small," at §6 of Isaac Israeli's *Mantua* text, trans. A. Altmann and S.M. Stern, *Isaac Israeli* (Chicago, 2009), 125.

53 Brethren of Purity, *Epistle on Animals*, trans. L.E. Goodman and R. McGregor, *Brethren of Purity: The Case of the Animals versus Man Before the King of the Jinn* (Oxford, 2009), 243, with the general discussion from 242–47.

54 The use of inspiration for animal behavior is ultimately a theological notion. In philosophical reflection on animals, it is also found as early as *adīb* and Muʿtazilī theologian Abū ʿUthmān al-Jāḥiẓ (d. 868) in his *Kitāb al-ḥayawān* 2.147–48.

55 P. Adamson, "Abū Bakr al-Rāzī on Animals," *Archiv für Geschichte der Philosophie* 94 (2012): 268–72. The classic reference work for concern for suffering within this school is M.T. Heemskerk, *Suffering in Muʿtazilite Thought* (Leiden, 2000).

56 P. Adamson, "Abū Bakr al-Rāzī on Animals," 261. On the example of the mouse using his tail for oil, see S.M. Virgi, "The Mouse's Tale: Al-Jāḥiẓ, Abū Bakr al-Rāzī, and Fakhr al-Dīn al-Rāzī on Animal Thinking," *British Journal for the History of Philosophy* 30, 5 (2022): 751–72.

57 Abū Bakr al-Rāzī, *Doubts on Galen*, ed. and trans. P. Koetschet, *Doutes sur Galien* (Berlin, 2019), 76–79.

58 T. Alpina, "Avicenna's Account of Memory," in *Memory and Recollection in the Aristotelian Tradition: Essays on the Reception of Aristotle's De memoria et reminiscentia*, ed. V. Decaix and C. Thomsen Thörnqvist (Turnhout, 2021), 67–92, at 72–75 .

59 See A. Oelze, trans., *Animal Minds in Medieval Latin Thought* (Cham, 2021), Part I, 31–142.

60 On the three distinctions by which Avicenna determines the various powers of the soul, see T. Alpina, "Avicenna's Account of Memory," 70–73. Although we do not have space for it here, there is another aspect of estimation's activity in animals, namely, self-awareness. On this topic, we refer you to L.X. López-Farjeat, "Self-awareness (*al-shuʿūr bi-l-dhāt*) in human and non-human animals," in A.G. Vigo (ed.), Oikeiosis *and the Natural Bases of Morality: From Classical Stoicism to Modern Philosophy* (Hildesheim, 2012), 121–40; idem, "Avicenna on Non-conceptual Content and Self-Awareness in Non-human Animals," in J. Kaukua and T. Ekenberg (eds.), *Subjectivity and Selfhood in Medieval and Early Modern Philosophy* (Cham, 2016), 61–73; and A. Alwishah, "Avicenna on Animal Self-Awareness, Cognition, and Identity," *Arabic Sciences and Philosophy* 26 (2016): 73–96.

61 Avicenna, *Book of the Soul*, I.5, ed. Rahman, *Avicenna's De Anima* (London, 1959), 43; trans. T. Alpina, in *Subject, Activity, Definition: Framing Avicenna's Science of the Soul* (Berlin, 2021), 43.

62 Avicenna, *Book of the Soul*, IV.3, ed. Rahman, 183–84, my translation as found in B. Somma, "Avicenna on Animal Goods," *Journal of Islamic Ethics* 6.1 (2022): 20.

63 Avicenna, *Book of the Soul*, IV.3, ed. Rahman, 184; Somma, "Avicenna on Animal Goods," *Journal of Islamic Ethics* 6.21 (2022).

64 See S.M. Connell, "Animal Cognition in Aristotle," 200–201.

65 See ibid., 200.

66 See T. Alpina, "Avicenna's Account of Memory," 83.

67 D. Black, "Estimation (*wahm*) in Avicenna: The Logical and Psychological Dimensions," *Dialogue* XXXII (1993): 227.

68 Avicenna, *Book of the Soul*, IV.1, ed. Rahman, 167; T. Alpina, "Avicenna's Account of Memory," 90.

69 T. Alpina, *Subject, Definition, Activity*, 130–57.

70 For a discussion of just this topic, see B. Somma, "Avicenna on Animal Goods," 20–23.

71 This issue is distinct from another concern about the role of emanation in psychological activity, namely, the so-called emanationist and abstractionist interpretations of human intellection

according to Avicenna. For a recent overview and critical discussion, see J. Kaukua, "Avicenna's Outsourced Rationalism," *Journal of the History of Philosophy* 58, 2 (2020): 215–40.

72 On Faḫr al-Dīn as a paragon example of post-Classical thought, see F. Griffel, *The Formation of Post-Classical Philosophy in Islam* (Oxford, 2021), 264–303.

73 See the overview at D. Black, "Estimation (*wahm*) in Avicenna," 221–24.

74 Faḫr al-Dīn feels entitled to make this claim because he rejects a basic Avicennan principle—that from only one power can come one effect, which means that each unique activity of the soul must have a unique power responsible for it. See W.M. Amin, "From the One, Only One Proceeds," *Oriens* 48 (2020): 123–55. As Adamson points out, Faḫr al-Dīn elsewhere argues that estimation would in fact need to know universals. See P. Adamson, "Faḫr al-Dīn al-Rāzī on Animal Intelligence," fn 24.

75 Faḫr al-Dīn al-Rāzī, *al-Mabāḥiṯ al-mašriqiyya*, ed. M.M. al-Baġdādī, *Faḫr al-Dīn al-Rāzī: al-Mabāḥiṯ al-mašriqiyya*, 2 vols (Beirut, 1990), vol. 2, 342–43.

76 J. Janssens, "Fakhr al-Dīn al-Rāzī on the Soul: A Critical Approach to Ibn Sīnā," *The Muslim World* 102 (2012): 569–70; M. İskenderoğlu, "Fakhr al-Din al-Razi on the Immateriality of the Human Soul," *Journal of Oriental and African Studies* 14 (2005), 121–36.

77 Faḫr al-Dīn al-Rāzī, *al-Mabāḥiṯ al-mašriqiyya*, ed. al-Baġdādī, vol. 2, 255; trans. P. Adamson and B. Somma in "Faḫr al-Dīn al-Rāzī on Animal Cognition and Immortality," *Archiv für Geschichte der Philosophie* (2022): 7.

78 Cf. C. Freeland's discussion of the role of perception in animals according to Aristotle in "The Science of Perception in Aristotle," in *Cambridge Companion to Aristotle's Biology* ed. by S. Connell (Cambridge, 2021), 159–75.

79 This passage has been translated and discussed by Peter Adamson at P. Adamson, "Faḫr al-Dīn al-Rāzī on Animal Intelligence," forthcoming. Thus, we will highlight only a few examples here and otherwise refer you to his analysis.

80 See P. Adamson, "Faḫr al-Dīn al-Rāzī on Animal Intelligence," 4–5.

81 Faḫr al-Dīn al-Rāzī, *al-Maṭālib al-ʿāliya*, ed. A.Ḥ. al-Saqqā, *al-Maṭālib al-ʿāliya min al-ʿilm al-ilāhī*, 9 vols (Beirut, 1987), VII, 311; trans. P. Adamson, "Faḫr al-Dīn al-Rāzī on Animal Intelligence."

82 Faḫr al-Dīn al-Rāzī, *al-Maṭālib al-ʿāliya*, ed. al- Saqqā, *al-Maṭālib al-ʿāliya.*, VII, 311; trans. Adamson "Faḫr al-Dīn al-Rāzī on Animal Intelligence".

83 Faḫr al-Dīn al-Rāzī, *al-Mulaḫḫaṣ fī-l-ḥikma wa-l-manṭiq* (trans. Adamson and Somma, *al-Mulaḫḫaṣ fī-l-ḥikma wa-l-manṭiq*, 239v). References are to Berlin, Staatsbibliothek, or. oct. 623, with Leiden, Universiteit Leiden, or. 132 also consulted.

84 Faḫr al-Dīn al-Rāzī, *al-Mulaḫḫaṣ fī-l-ḥikma wa-l-manṭiq* (trans. Adamson and Somma, *al-Mulaḫḫaṣ fī-l-ḥikma wa-l-manṭiq*, 239v–240r).

85 A. Oelze, *Animal Rationality: Later Medieval Theories 1250–1350* (Leiden, 2018), 34–35 and 12–15, respectively.

86 See J. Janssens, "Ibn Sīnā, Latin translations of," in H. Lagerlund (ed.), *Encyclopedia of Medieval Philosophy: Philosophy between 500–1500* (Dordrecht, 2011), vol. 2, 522–27.

87 All found in Oelze, *Animal Minds in Medieval Latin Thought*, 77–88; 119–23; and 127–30, respectively.

88 Augustine, *On the Magnitude of the Soul*, trans. Oelze, *Animal Minds*, 31–39, especially Augustine's conclusion at 38–39.

5

UNSUNG HEROES OF BYZANTINE HAGIOGRAPHY

The Role of Animals in Martyrs' *Passions*

Christodoulos Papavarnavas

Byzantine hagiography is particularly rich in scenes with animals. Animals appear at crucial moments in hagiographical stories and make strategic contributions to the development of the narrative by taking on a positive or a negative role in the spiritual progression of the holy protagonist. In this regard, hagiographical stories often contain multiple heroes: the holy persons and the animals, both good and bad. The form, symbolism, and agency of animals are well-represented aspects in this kind of narrative. For example, two of the most popular hagiographical stories throughout the Byzantine periods – and well-known even today – provide conflicting images of animals by revealing the ill or virtuous intentions behind their actions: in the *Life of Antony* (*BHG* 140/*CPG* 2101), written by Athanasios of Alexandria in the fourth century, demons appear in the form of wild animals such as lions, bears, leopards, bulls, wolves, poisonous snakes, and scorpions, and strive in vain to discourage the holy man from continuing his hard *ascesis* in the cemetery where he sought isolation in a tomb.[1] By contrast, in the seventh-century *Life of Mary of Egypt* (*BHG* 1042), a huge lion appears in the desert on the other side of the river Jordan to help the monk Zosimas dig a pit and bury Mary, the repentant prostitute who had led a solitary life there for forty-eight years up to her death.[2] While in the first case lions and other wild animals embody evil forces wishing to hinder Antony's spiritual progress, in the second case the lion appears to be a well-intentioned animal sent by God in recognition of the sanctity of Mary's dead body. This twofold function of animals, as delineated in the hagiographical texts mentioned above, would have been quite familiar to the Byzantines. This is because the relevant *Lives* were broadly circulated both during the early period when they were first composed and subsequently as part of the popular tenth-century *Menologion* of Symeon Metaphrastes, which incorporated these two texts without alterations and thereby ensured their continued dissemination.[3]

Descriptions of animals and their actions can be found not only in saints' *Lives* but also in all other hagiographical subgenres.[4] So far, only one systematic study, a collective volume, has been devoted to animals in Byzantium; it considers saints' *Lives* among other

DOI: 10.4324/9781003055877-8

sources, but largely excludes literary material from other hagiographical subgenres.[5] In an effort to fill – at least partially – this gap in research, the present chapter will delve into the most prolific hagiographical subgenre throughout the Byzantine centuries, the martyrs' *Passions* or martyrdom narratives.[6] Specifically, the aim of this chapter is to focus on the structure of these texts and to investigate the role of animals in the four basic stages of martyrdom: the protagonists' interrogation, torture, imprisonment, and execution along with burial.[7]

The Greek *Passions* recount stories of Christian men and women who, following their arrest, were subjected to several trials (i.e., interrogations, tortures, and imprisonments) and were finally killed by the Roman (or other non-Christian) authorities because they categorically refused to offer sacrifice to pagan gods and to convert to paganism. Although such martyrdom stories are set primarily during the persecutions of Christians in the Roman Empire (41–313 CE),[8] most of these texts, often anonymous, were composed by Byzantine authors much later than the events they describe (approximately between the fourth and the tenth century), and are thus characterised by limited historicity.[9] Due to their restricted historical character, they were unable to meet the needs of modern scholarship, which for many years sought to identify historical aspects in hagiographical texts.[10] Therefore, Greek hagiographical *Passions* were largely neglected until recent decades, when scholars began paying attention to their literary merits.[11] Most of the *Passions* are still understudied, and their literary aspects remain to be thoroughly explored. By examining the interaction between martyrs and animals from a literary-narrative vantage point, the present survey is intended to contribute to further exploration of the *Passions'* literariness.

For the purposes of this study, the following four early and middle Byzantine martyrdom narratives were selected: the *Passion of Zosimos (BHG 2476)*, the *Passion of Panteleimon (BHG 1414)*, the *Passion of Marina of Antioch (BHG 1165)*, and the *Passion of Sergios and Bakchos (BHG 1624 and 1625)*. These narratives, taken together, enable an expanded investigation of the different key phases of martyrdom, in which animals intervene in various ways – either to the benefit of the martyrs or to their detriment. In both cases, animals prove to be important hagiographical heroes or antagonists. The martyrdom accounts examined here fall into the category of "epic *Passions*" characterised by a certain detachment from historical reality and by several extravagant and fanciful features.[12] That the holy protagonists of these accounts were popular among the Byzantines is indicated by the fact that brief summaries of their martyrdom stories were also included in the tenth-century *Synaxarion of Constantinople*,[13] meaning that they were commemorated in the churches of the Byzantine capital and beyond.

Before we take a closer look at the selected texts, providing their summaries and dates of composition will be helpful. According to François Halkin, the editor of the *Passion of Zosimos*, this text must have been written in the fifth century.[14] Zosimos, an ascetic living in the mountains in the company of wild animals, is arrested by pagan soldiers and brought before the governor Dometianos for interrogation. A speaking lion suddenly appears in the courtroom and manages to persuade the pagan persecutor of the truth of the Christian God. Although both Zosimos and the lion are about to be executed, they ultimately do not die a martyr's death. Instead, they disappear along with a recently converted official into a chasm in a rock.

The version of the *Passion of Panteleimon (BHG 1414)* discussed below was composed by Symeon Metaphrastes in the tenth century on the basis of an earlier account.[15] During the reign of Maximian (286–305), a young man named Pantoleon (later Panteleimon) is instructed in the Christian doctrine, first by his Christian mother and then by the priest Hermolaos who also baptises him. At the wish of his pagan father, Pantoleon begins studying medicine and soon he gains a reputation as a good physician. All sick people turn to him for healing, which they receive by divine intervention. Other physicians, consumed with envy, reveal to the emperor that Pantoleon is an advocate of the Christians and the Christian faith. Pantoleon is immediately brought before the emperor and is commanded to sacrifice to the pagan gods. Despite Maximian's exhortations and threats, the Christian man remains steadfast in his faith. Then the martyr is subjected to several tortures, from which he emerges unscathed. God appears to the martyr in the form of his teacher Hermolaos and helps him endure all of the ordeals bravely. As part of his tortures, Pantoleon is also thrown to the wild beasts in the arena. Yet, instead of devouring him, the animals behave like right-minded humans and pay reverence to the holy martyr. His teacher Hermolaos, along with two other Christian men, is decapitated, and Pantoleon is sent to prison. After a further series of trials and tortures, the emperor decides to behead him. Before being executed by the sword, the martyr pleads with God to forgive his executioners, who have recently converted to Christianity. God accepts his request and calls him by his new name, Panteleimon, as many people will be benefitting from his mercy (*eleos*). At his own instigation, the executioners carry out the beheading, and the Christians bury his dead body.

The *Passion of Marina of Antioch* (*BHG* 1165) was composed by the seventh century at the latest.[16] Marina, the daughter of a pagan priest, converts to Christianity under the influence of her nanny. Impressed by her beauty, the prefect Olybrios wishes to marry her. However, once her Christian identity is revealed, she is brought before Olybrios, who interrogates her and orders her to sacrifice to the pagan gods. Her refusal to comply with this command results in her cruel torture and imprisonment. In prison, she is threatened first by a dragon and then by a demon, but the martyr emerges victorious from her combat with these evil creatures. Christ appears in jail in the form of a dove and announces that Marina will gain entry into Paradise. A further sequence of interrogations and tortures follow the martyr's imprisonment up to the point when she is finally baptised by the same dove and decapitated by a pagan soldier outside the city of Antioch.

The *Passion of Sergios and Bakchos* (pre-Metaphrastic version [*BHG* 1624]; the tenth-century Metaphrastic version [*BHG* 1625] is used here only occasionally for comparisons) was presumably written in the mid-fifth century.[17] The protagonists, Sergios and Bakchos, are two army officers and close friends of the emperor Maximian (286–305). Nevertheless, when the emperor learns of their Christian faith, he commands them to offer sacrifices at the temple of Zeus and eat from the sacrificial meat. Their refusal to obey this imperial command leads to their public punishment. Afterwards the emperor sends both men to Antiochos, the *dux* of the province of Augusto-Euphrates, with orders to sentence them to death if they do not convert to paganism. Bakchos is tortured and killed first, while Sergios is executed by the sword at a later stage. Antiochos orders that the body of the first martyr be thrown out of the camp to be devoured by wild animals. Instead, wild animals of the land and of the air protect the saint's body until ascetics come to the spot and take the remains to bury in their caves. When Sergios is led to his place of execution, he is accompanied not only by people but also by beasts mourning for the martyr's imminent death. Some of the eyewitnesses bury the martyr's body on the spot, where a monument [*mnēma*] of stones and clay is then erected. Sometime later, fifteen bishops construct a shrine [*martyrion*] at a nearby castle to accommodate the saint's remains. Many miraculous healings take place not only at the shrine but also at the location of the first tomb. Every year, on the day of the martyr's death, animals and humans gather together to pay reverence to the holy martyr.

Interrogation

The interrogation process is an indispensable part of martyrs' *Passions* and is usually the first step of the sequence towards martyrdom, which begins after the arrest of the Christian protagonist. In these scenes the pagan judge vainly strives to convince the martyr to relinquish his or her Christian faith. The *Passion of Zosimos* (*BHG* 2476), a considerably neglected text in modern scholarship,[18] offers a unique narrative because of two peculiarities: firstly, the story is characterised by an unusual structure, as it is almost entirely devoted to an interrogation; tortures are succinctly presented, and there is no execution. Secondly and most importantly, the long interrogation scene, which serves as the core of the entire story, centres on Zosimos' acquaintance with wild animals and the vivid presence of a speaking lion in the courtroom. These peculiarities are further discussed below. The *Passion* reveals Zosimos' role models in his way of life and his relationship with wild animals, which go back to the Old and New Testaments: he was an ascetic living in the mountains like John the Baptist, who had camel's hair as his clothes and wild honey as his food (Mk. 1.6), and Elijah, the prophet who received his food from ravens (3 Kings 17.4).[19] Of course, Christ also spent a period of his life with wild animals in the desert (Mk. 1.12–13). According to the text's editor, the mythical poet Orpheus, who was able to charm and calm the savage animals with his singing and the sound of his lyre, may have also served as an inspiration for the figure of Zosimos.[20]

Already in the first lines, the hagiographer indicates that animals will be critical actors in the story by mentioning that Zosimos "was not only living with them [the wild animals] but also conversing with them, and they [Zosimos and the animals] had everything [i.e., their habits] in common" (οὐ μόνον σύνοικος αὐτῶν, ἀλλὰ καὶ συνόμιλος αὐτῶν· καὶ ἦν αὐτοῖς ἅπαντα κοινά).[21] This phrasing gives the impression that Zosimos was living in a kind of *koinobion* where the animals served as his fellow monks, equally pursuing the same way of life.[22] Zosimos is even said to have grown spiritually along with the wild animals, with whom he sang psalms and addressed prayers to God every day. The hagiographer clarifies that the protagonist was living with the wild animals not for pleasure but because he aspired to mitigate their savagery and to show one of them to be a martyr of Christ (τὴν ἀγριότητα κατασβέσαι βουλόμενος καὶ ἀναδεῖξαι καὶ ἐξ αὐτῶν τῶν θηρίων μάρτυρα τοῦ Χριστοῦ).[23] These words would have surprised the audience at the time, who, based on their acquaintance with the conventional structure and contents of a martyrdom narrative, would certainly not have expected an animal in the role of a Christian martyr. In this way, the hagiographer presumably tried to catch the attention of the intended audience and encourage them to hear or read the rest of this extraordinary story attentively. In any case, the hagiographer's prologue functions as a foreshadowing of the whole story.

Zosimos is arrested (apparently in the mountains where he was dwelling with wild animals) and brought before the pagan judge Dometianos in the city of Anazarbus in Cilicia. The dialogue between the pagan persecutor and the Christian man begins as follows:

Καὶ εἰσαχθέντος αὐτοῦ ὁ ἄρχων εἶπεν· "Τίς καλεῖ;" Ζώσιμος εἶπεν· "Χριστιανός." Ὁ ἄρχων εἶπεν· "Τῶν πρὸ σοῦ μηδὲν ὠφελησάντων ἐκ τοῦ ὀνόματος τούτου, λέγε τὸ ὄνομά σου." Ὁ δὲ εἶπεν· "Ζώσιμος καλοῦμαι." Ὁ ἄρχων εἶπεν· "Λέγε οὖν, Ζώσιμε, πῶς εἶ σύνοικος μετὰ τῶν θηρίων;" Ζώσιμος εἶπεν· "Πῶς οὖν δοξάζεται Χριστός; Τοίνυν· μετηλλάγη γὰρ τὸ γένος τῶν ἀνθρώπων οὐ τῇ φύσει ἀλλὰ τῷ τρόπῳ· καὶ γεγόνατε θηριωδέστατοι, οἵτινες λατρεύετε τῇ κτίσει παρὰ τὸν κτίσαντα (cf. Rom. 1.25). Τὰ δὲ ἄγρια ζῷα ἐπιγινώσκοντα τὸν Θεὸν τῷ Θεῷ λατρεύουσιν, οὐ τῇ κτίσει ἀλλὰ τῷ κτίσαντι· καὶ γὰρ προφητεύεται αὐτοῖς ἐκ τῆς θείας γραφῆς, τὰ θηρία καὶ πάντα τὰ κτήνη δοξάζειν τὸν Θεόν (cf. Ps. 148.10). Ἕνεκα οὖν ταύτης τῆς ὑποθέσεως σύνοικός εἰμι μετὰ τῶν θηρίων." Ὁ ἄρχων εἶπεν· "Πάλιν σοι λέγω· παυσάμενος ταύτης τῆς μωρᾶς λέξεως, προσελθὼν θῦσον τοῖς θεοῖς."[24]

And when he [Zosimos] was brought in, the governor said: "What is your name?" Zosimos said: "[I am a] Christian." The governor said: "Your predecessors have not benefited from this name at all. Tell me your name." He said: "I am called Zosimos." The governor said: "Tell me now, Zosimos, how does it happen that you live with beasts?" Zosimos said: "How does it happen that Christ is praised? Well, there is no doubt that human beings have been changed not in regard to their nature but to their way of life, and you behave most bestially in that you worship creation instead of the Creator [cf. Rom. 1.25]. Wild animals, in contrast, are aware of God and they worship Him – not the creation but the Creator. For, according to the prophecy of the Holy Scriptures, the beasts and all animals glorify God [cf. Ps. 148.10]. So, in light of this reasoning, I am living with beasts." The governor said: "I command you again to stop saying these foolish words and come forward to sacrifice to the [pagan] gods."

The pagan governor takes on the role of judge, while the martyr appears in the courtroom as the accused. As in the majority of *Passions*, the interrogation begins with the pagan judge asking the name of the accused, and the martyr first confesses to being a Christian and then reveals his name. The judge's reaction when he learns the religion and name of Zosimos is highly interesting. Instead of focusing on the religious confession of the accused, which is actually the main reason for his arrest, Dometianos hastens to ask him about his relationship with the beasts (θηρία). On a narrative level, this question is posed to satisfy not only the curiosity of the pagan judge and the bystanders, but also that of the text's recipients. Zosimos goes so far as to characterise the pagans as more bestial than the beasts themselves (θηριωδέστατοι) since, and unlike actual beasts, they are not able to recognise and glorify God. Although the pagans are humans in nature (φύσις), their way of life (τρόπος) is not the befitting one. Zosimos' words imply that he decided to abandon the city to dwell with wild animals in the mountains because there he could share with them the same – ascetic – way of life (τρόπος). In this manner, the text again demonstrates wild animals' central role in the life of the holy hero and, by extension, in the whole story. The hagiographer seems to underscore the importance of animals on a linguistic level as well, namely through the play on words between θηρία (the Christianised beasts) and θηριωδέστατοι (the bestial pagans), as outlined above. Only after the relationship between Zosimos and the animals has been clarified does Dometianos proceed to the accusations against the martyr.

Because of Zosimos' offensive words and implicit refusal to convert to paganism, Dometianos orders that he be punished through bodily torture. Zosimos' ears are thus pierced with red-hot irons, which the martyr endures uncomplainingly. Zosimos is then thrown into a boiling cauldron, but God brings forth rain from a cloud, and the cauldron becomes colder than snow. The martyr emerges from this ordeal unscathed. These short episodes of torture and divine intervention take up only twenty lines in total (not even one-eighth of the whole text). Then the interrogation continues with Zosimos promising to invite "one of the beasts" (ἐν τῶν θηρίων) to come to the courtroom and contribute to the interrogation in an effort to persuade the pagan judge that animals truly live in fear of God.[25] Here the animals' piety is obliquely juxtaposed once again with the pagans' impiety. The judge commands that Zosimos be hung upside down with a rock around his neck, implying that he intended to kill him by strangling. However, at that very moment, the martyr begins to pray to God, asking him to send a lion with a "spirit of speech" (πνεῦμα λαλοῦν) so that the promise he gave to the pagan judge will be fulfilled.[26]

Indeed, a lion enters the courtroom and declares itself ready to participate in the interrogation process. The relevant passage reads as follows:

Ἔτι δὲ αὐτοῦ εὐχομένου ἦλθεν λέων μαινόμενος καὶ βρύχων τοῖς ὀδοῦσιν ἐν τῷ δικαστηρίῳ, καθεζομένου τοῦ ἄρχοντος καὶ παντὸς τοῦ ὄχλου περιεστῶτος, μηδένα καταβλάπτων. Καὶ ἰδὼν ὁ λέων τὸν ἅγιον Ζώσιμον κρεμάμενον ἐπὶ κεφαλὴν ἅμα τῷ λίθῳ, ἀπελθὼν πρὸς αὐτὸν καὶ ἑαυτὸν ὀρθώσας ὑπεβάσταζεν τὸν λίθον. Ἡ δὲ τάξις ἰδοῦσα τὸν λέοντα ἐν τῷ δικαστηρίῳ ἔφυγεν· ὁμοίως καὶ πᾶς ὁ δῆμος τῆς πόλεως ἔφυγεν· καὶ κατελείφθη ὁ ἄρχων μόνος· ὅθεν καὶ αὐτὸς ἐπειρᾶτο φυγεῖν καὶ οὐκ ἠδυνήθη· ὁ γὰρ λέων πρὸς βραχυτάτην ὥραν καταλείψας τὸν ἅγιον Ζώσιμον ἐπεκατέλαβε τὸν ἄρχοντα ἐπὶ τοῦ πυλεῶνος φεύγοντα καὶ ἐκράτησεν αὐτὸν ἐπὶ τοῦ θρόνου αὐτοῦ. Ἐλάλησεν δὲ ὁ λέων φωνῇ ἀνθρωπίνῃ λέγων τῷ ἄρχοντι· "Μὴ φοβοῦ, δικαστά· ἦλθον γὰρ πρὸς ἀπολογίαν σου· οὐ μόνον σοῦ, ἀλλὰ καὶ τοῖς ἠλπικόσιν ἐπὶ τὸν κύριον ἡμῶν Ἰησοῦν Χριστόν." Καὶ καταλείψας αὐτὸν ἐπὶ τοῦ θρόνου αὐτοῦ, πάλιν ἀπελθὼν καὶ ὀρθώσας ἑαυτὸν ὑπεκούφιζεν τὸν λίθον ἐκ τοῦ τραχήλου τοῦ ἁγίου Ζωσίμου.[27]

While he [Zosimos] was praying, a lion appeared in the courtroom, raging and roaring with sharp teeth, but it did not harm anyone – neither the governor who was seated [there] nor the entire crowd standing around. And when the lion saw Zosimos hung with his head down and a rock [tied around his neck], it went to him and placed itself under him, supporting the weight of the rock. When the group of officials saw the lion, they ran away from the court. Likewise, all the citizens in attendance ran away; and the governor was left alone [in the courtroom] from which he also tried to run away, but he did not manage. For, the lion left saint Zosimos for a little while and caught him [the judge Dometianos] at the gateway as he was about to flee, and it forced him to remain seated on his [judge's] throne. The lion spoke in a human voice, saying to the governor: "Don't be afraid, Judge! For, I came to [my faith's] defence on account of you; [in fact,] not only on account of you but also on account of those who expect [the coming of] our Lord Jesus Christ." And the lion left him on his throne and went and placed itself again [under Zosimos], alleviating [the weight of] the rock around saint Zosimos' neck.

At the sight of the lion in the courtroom, the audience (officials and bystanders) immediately flees to safety. However, the lion hastens to place itself under Zosimos to help him bear the weight of the rock around his neck. In doing so, the lion not only makes known its intention to support Zosimos (literally!) and the Christian faith before the court, but by its own volition it also takes on the role of the accused next to the Christian man. When the pagan governor also tries to abandon the courtroom, the lion forces him to stay in his place and carry out his task as judge. Although at first glance the lion seems to disrupt the interrogation process, it is in fact the lion that helps this proceeding come to an end by delivering a speech (ἀπολογία) in defence of the Christian faith before the judge. In the meantime, the audience realises that the lion has no ill intentions, and they return to the courtroom, completing the standard picture of a trial and its main actors (i.e., the judge, the accused, and the audience). Then the pagan judge orders that the lion be slaughtered (σφάξατε τὸν λέοντα τοῦτον). The animal expresses its willingness to spill its blood for the sake of Christ along with the Lord's servant Zosimos (ὑπὲρ Χριστοῦ τὸ αἷμά μου ἐκχέειν ἅμα τῷ θεράποντι τοῦ κυρίου Ζωσίμῳ),[28] but it wishes first to complete its confession of faith before the court.

It is interesting to observe how the text draws a parallel between the lion and Zosimos. First, both of them appear in the role of the accused and are called on to confess their Christian faith. Second, they are both about to be killed by having their throats cut – in martyrs' *Passions*, beheading is the most common method of execution. And third, the lion describes itself as a co-martyr who wishes to suffer martyrdom, together with Zosimos, for the sake of Christ. In the end, neither of them dies a martyr's death because, thanks to the gift of human speech given to it by God, the lion explains Christ's stages of life and brings the pagan judge to believe its words. In this context, the lion characterises itself as a preacher of God's words: "I am a beast by nature, but a preacher in my behaviour" (τῇ φύσει θηρίον εἰμί, τῷ δὲ τρόπῳ κῆρυξ).[29] Here the opposition between φύσις and τρόπος can be observed once again. Similarly to the first passage cited above, in which Zosimos states that humans have been changed not in regard to their nature but in regard to their way of life as they behave savagely, the lion indicates that, even though it remains a beast, its behaviour as a Christian preacher is more appropriate than that of the pagans who support a false religion. Through the vivid example of the lion, Zosimos' words are verified: wild animals live in fear of God. In this sense, the lion's intervention in the interrogation and, by extension, in the story proves to be of paramount importance.

Overall, the narrative starts out by declaring the close relationship between Zosimos and wild animals, the background of which the protagonist himself explains in the courtroom during his interrogation. The narrative reaches its climax when this relationship is corroborated through the appearance of a speaking lion, which also actively participates in the interrogation. As Halkin has already noted, this martyrdom story with the speaking lion is presumably based on the *Acta Pauli*, according to which another talking lion, which Paul had baptised during a journey, appeared in the arena and was expected to devour Paul who had been condemned to death.[30] The lion from the *Acta Pauli* appears in one more – similar – story, that of the *Passion of Sebastiane* (*BHG* 1619), who was a disciple of Paul and, like her teacher, was also thrown to the wild animals in the arena.[31] In these stories, the lion recognised the Christian protagonists (Paul and Sebastiane) and did not harm them. Although the speaking lion as an image and a concept may go back to the *Acta Pauli*, the martyrdom narrative examined here remains unique for its structure and contents: the anonymous hagiographer of Zosimos' *Passion* managed to effectively incorporate the lion into the story by vesting a leading role, specifically that of a martyr, in this wild animal. The lion thus appears to be a main hero of the story – almost equal to Zosimos. The story comes to a close when a chasm opens in a certain rock to receive into it Zosimos, Athanasios (a pagan official who at the end converted to Christianity), and the lion. Even though the image of a rock that opens to receive a saint is a recurring motif in martyrdom accounts,[32] there is no other example of an animal accompanying the saint at that very last moment. The inclusion of the lion may be interpreted as an indication that not only Zosimos and the converted official but also the lion gained sanctity and entry into Paradise. This final scene of the story with the lion disappearing into the chasm in a rock and its implications for the animal's sanctity may have shocked the compiler of the *Synaxarion of Constantinople* who, although mentioning the lion in the courtroom, silently omitted its presence from the closing scene in the *Synaxarion*.[33] This "holy lion" may serve as further evidence of the examined *Passion*'s extreme originality.

Torture

According to martyrs' *Passions*, the Christian men and women, who during their interrogation did not follow the Roman judge's order to offer sacrifices to gods and convert to paganism, were tortured in various ways. One well-known and spectacular method of public punishment was being thrown into the arena to fight wild animals (θηριομαχεῖν).[34] There, martyrs were attacked by the beasts and severely injured, or even fatally wounded. Those who survived this ordeal were subjected to further tortures and finally decapitation. The Greek *Passion of Perpetua and Felicity* (BHG 1482), which was translated from the third-century Latin original and is considered a historical account,[35] describes vividly this kind of spectacle in the amphitheatre: during the reign of Valerian and Gallienus (253–260),[36] Perpetua and Felicity, along with their Christian companions, Revokatos, Saturninos, and Satyros, are sentenced to contend with wild animals. Revokatos and Saturninos are killed by a leopard and a bear, while Perpetua, Felicity, and Satyros are seriously injured by a wild cow and a leopard and are afterwards beheaded by the sword.[37] Such descriptions of wild animals attacking martyrs in the arena can be found in the "epic *Passions*" as well. A case in point is the *Passion of Euphemia* (BHG 619–619a): during the reign of Diocletian (284–305), Euphemia, after suffering several tortures because of her steadfastness in faith, is finally thrown into the arena to be devoured by lions and other wild beasts. Indeed, one of these beasts bites her badly. Then a voice is miraculously heard from heaven, announcing that Euphemia will now gain a celestial reward for her efforts, namely entry into Paradise, and she dies on the spot.[38] In the above cases, the animals harm the holy protagonists but paradoxically contribute at the same time to their path to Heaven by expediting their death for the sake of Christ – the martyrs' ardent desire.

The *Passion of Panteleimon* (BHG 1414) offers a striking variant of this kind of torture involving wild animals in the arena and is thus worthy of a more detailed discussion. As indicated from the summary given above, this narrative is uncommonly long and rich in its descriptions. Yet the following analysis will focus solely on the episode with the wild animals and their interaction with the holy protagonist.[39] After a series of trials and corporal punishments, the emperor Maximian orders that the martyr Pantoleon (he receives the name Panteleimon from God later in the narrative) be thrown to "wild beasts of every kind" (θῆρας [...] παντοδαπούς) because of his continued refusal to convert to paganism.[40] The pagan emperor expects the martyr to be horrified at the sight of the beasts and change his mind. However, Pantoleon feels confident that God will tame the feral animals on his account, making them even more harmless than sheep.

Encouraged by Christ who appears to him, Pantoleon proceeds to this punishment with the confidence and bravery of a lion (ὡς λέων πεποιθώς).[41] Obviously, his name

Pantoleon – meaning 'the man who possesses all the attributes of a lion' – plays on the courage of his actions and anticipates his encounter with the wild animals. Interestingly, the martyr is positively connected with the features of a wild animal, while at the same time the emperor is depicted as "a tyrant [who is] far more ferocious than the beasts" (τοῦ πολλῷ τούτων [sc. θηρίων] ἀγριωτέρου τυράννου).[42] As in the case of Zosimos' *Passion*, in which the governor Dometianos and his pagan subjects are deemed more bestial than the animals because of their impiety towards God, the pagan emperor Maximian is characterised here as unfit to wield power (*tyrannos*) since he cannot control his emotions and acts instead more savagely than the beasts.[43] Characteristics of wild animals (i.e., bravery and savagery) thus become a rhetorical device in the hands of the hagiographer to illustrate his characters, both the Christian protagonist and his pagan opponent.

Pantoleon's firm belief in God's help is finally vindicated when the wild animals are set free in the arena:

Τοσοῦτον γὰρ ἀπέσχον οἱ θῆρες τοῦ κακῶς διαθέσθαι τὸν ἅγιον, ὡς μικροῦ καὶ τὸ μὴ θῆρας εἶναι ὅλως αὐτοὺς νομισθῆναι, θῆρας, ἔφην, μηδ᾽ ἁπλῶς ἄλογα καὶ τοῦ φρονεῖν ἀπεστερημένα, ἀλλ᾽ οἷόν τινας ἔμφρονας καὶ τῆς τῶν λογικῶν ὄντας μερίδας, μετὰ πολλῆς τῷ μάρτυρι προσιόντας εὐλαβείας, σαίνοντας δὲ ἡδέως καὶ φιλανθρώπῳ γλώσσῃ καὶ θεραπευτικῇ τῶν ποδῶν παραψαύοντας, ἄλλως τε δὲ καὶ ἀλλήλοις διαμιλλώμενοι, τίς πρῶτος αὐτῶν προσέλθοι, οὐ πρότερον ἀφιστάμενοι, πρὶν ἂν ὁ μάρτυς αὐτοῖς ἐπιθῇ τὴν χεῖρα καὶ εὐλογήσειεν. Ἦν οὖν ἰδεῖν πρᾶγμα κομιδῇ καινότατον· ἐπεὶ γὰρ τοιαύτη τοὺς ἀνθρώπους ὑπῆλθε κακία, ὥστε ὁμοῦ καὶ ἀλόγων αὐτοὺς ἄνοιαν καὶ ὠμότητα θηρίων ὑπερβαλεῖν, τὴν μὲν ἐν τῷ μὴ σέβεσθαι τὸν πεποιηκότα, τὴν δὲ ἐν τῷ τιμωρεῖσθαι τοὺς αὐτὸν σεβομένους, ᾠκονόμησεν ὁ πάντα μετασκευάζων Θεός, ἔμπαλιν τοὺς θῆρας οἷα λογικοὺς ὀφθῆναι καὶ ἀνθρώπων ἡμερότητα μιμουμένους, κήρυκας ἀψευδεῖς ὄντας τῆς ἐκείνων κακίας καὶ τῆς αὐτοῦ ἀφράστου μακαριότητος. Ὅθεν καὶ οἱ παρόντες ἐπὶ τῷ γενομένῳ· "Μέγας ὁ τῶν Χριστιανῶν Θεός ἐστιν, ὁ μόνος καὶ ἀψευδής," ἐβόων, καί· "Ἀφιέσθω ὁ δίκαιος." Τί οὖν ὁ ἀληθῶς θὴρ καὶ ὠμότατος βασιλεύς; Ἐπάγει τὸν θυμὸν κατὰ τῶν θηρίων καὶ πάντα εὐθὺς ἀναιρεῖ, ὥσπερ διαφθονήσας αὐτοῖς τῆς συνέσεως καὶ τὸν ὑπ᾽ ἐκείνων οὐκ ἐνεγκὼν ἔλεγχον. Τὰ δὲ καὶ οὕτως ἀναιρεθέντα ἐπὶ πολλαῖς ἔμενε ταῖς ἡμέραις, οὐδενὶ βρῶσις τῶν σαρκοβόρων γινόμενα, τιμῶντος κἂν τούτῳ ἀθλητὴν τοῦ Θεοῦ καὶ πρὸς εὐσέβειαν τοὺς ἄλλους παρακαλοῦντος, ὥστε τὸν ἀσύνετον βασιλέα τοῦτο μαθόντα, προστάξαι τὰ θηρία κατορυγῆναι τῇ γῇ, πολλοῖς οὕτω τὴν πίστιν ἐπιστηρίζοντα.[44]

For the beasts refrained from harming the saint to such an extent that they were almost not considered beasts at all – I say beasts, yet, in fact, they could be not even considered unreasoning [creatures] lacking prudence, but instead a kind of prudent and rational [being]. They moved towards the martyr with great piety, pleasantly wagging their tails, and they harmlessly touched his feet with their friendly and mellow tongues. They even vied with each other over which of them would approach [the martyr] first, and they did not go away until the martyr placed his hand on them and blessed them. So, one was able to see an absolutely strange thing. For, people were overwhelmed by such a great malevolence that they surpassed both the folly and the savagery of irrational animals, because, on the one hand, they did not feel fear before the Creator and, on the other, they punished those who lived in fear of Him. By contrast, God, who is able to transform everything, arranged [things in such a way] that the beasts were proved to be somehow rational, and adopting human gentleness, they became true preachers of [the pagan] people's malevolence and His [i.e., God's] indescribable bliss. For this reason, the bystanders, struck by the happenings, exclaimed: "Great is the God of the Christians; this is the only and true one!" and [added]: "This righteous man [Pantoleon] should be released." What did the real beast and savage emperor do then? He vented his anger on the animals and killed them all immediately as though he was envious of them for their sagacity and could not bear their reproach. And although the animals that were thus killed remained [there] for many days, they were not devoured by any flesh-eating [beasts]. In this manner, God wished to pay honour to His athlete [i.e., martyr] and prompt the others to piety. As a result, when the witless emperor was informed of that, he ordered the animals to be buried in the earth, thereby providing many [people] with further proof of the [Christian] faith.

According to the passage above, the wild animals react not as the emperor expected but as the martyr wished. They become tame and harmless. It is in fact a literary *topos* that wild animals behave tamely in the presence of saints, which indicates that the saints live in close touch with nature.[45] Yet, what distinguishes this episode from other similar stories are the hagiographer's assessments and further detailed descriptions that enrich this *topos*. The hagiographer addresses his considerable awkwardness in trying to properly characterise the animals surrounding Pantoleon. He acknowledges that the term "beasts" (θῆρες) does not fit in this case since these animals act like prudent and rational beings. Indeed, the animals appear to possess simultaneously both zoological and human characteristics: while they approach the martyr like domesticated animals, happily wagging their tails and licking his feet, they consciously wish to receive his blessing, which is, in fact, a typical behaviour of humans meeting a holy person.[46] Thus, the wild animals function as gentle and pious humans by recognising the sanctity of the martyr and paying reverence to him. On the other hand, by comparing the behaviour of the animals with that of the pagan people, the hagiographer puts the pagan people to shame for acting impiously towards God and the holy martyr. Through their actions and reactions, the wild animals "preach" God's truth and convince the audience of both the existence of the Christian God and the sanctity of the martyr who has been unjustly sentenced to martyrdom. In this way, the wild animals seem to decisively contribute to the protagonist's God-pleasing efforts to lead more people to Christianity through his martyrdom.

The animals' significant role in the events is also corroborated by the emotional reactions of the emperor who is seized mainly with anger because of the pagan bystanders' conversion to Christianity. According to the hagiographer, the animals' intelligence even arouses the emperor's envy, while their critical attitude towards his deeds, expressed through their disobedience (they do not attack the martyr), is the underlying reason for his decision to kill them. Maximian's actions are thus determined by his emotions, and he again proves ineffective as a ruler since he cannot in fact influence the course of things. The wild animals are ultimately executed before the very eyes of the spectators and are abandoned in the arena for several days. However, their carcasses are wondrously not devoured by any other carnivorous beasts. Here two groups of wild animals can be observed: those whose piety led them to a martyr's death in the arena (in a great human-animal role-reversal), and those whose piety kept them from devouring the carcasses of the martyred animals. In this way, all other creatures of nature seem to pay reverence to the wild animals of the arena for their self-sacrifice to the Creator of all things and to the Christian faith.[47]

This wonder of the wild animals' carcasses occurs during Pantoleon's martyrdom to underline, first and foremost, the martyr's importance as a paradigm of Christian virtue. In this manner, God chooses to honour the martyr, who is about to die for his sake, and concurrently demonstrates the great piety of the wild animals, who do not hesitate to participate in the holy protagonist's ordeals by suffering actual martyrdom. In doing so, as in the case of the lion in Zosimos' *Passion*, the wild animals create the impression that they also attain (some kind of) sanctity like the martyr. This impression is reinforced by the fact that the emperor eventually orders the remains of the dead wild animals – like a saint's relics – to be buried to prevent more pagans from witnessing this wonder and converting to Christianity. But, in fact, according to the hagiographer, this deed has the opposite effect, as it gives further evidence of the power of the Christian faith. By acquiring human traits and even a kind of holy status, the wild animals offer to the recipients of the text an ideal example for imitation.[48] To put it another way, the text's audience would probably have recognised the pious behaviour of the wild animals and aspired to follow their example.

The wild animals are thus killed and buried, but the martyr is still alive. After surviving the "combat with the beasts", which was intended as a form of torture, Pantoleon is imprisoned and then subjected to further tortures until he is finally decapitated. Shortly before his decapitation, the martyr implores God to forgive his executioners, and a divine voice responds, giving him a new name, Panteleimon (meaning 'the man who shows mercy [*eleos*] on all people'), and explaining to him that from that point on, he will act as an intercessor between men and God. In this way, following the recognition of his holiness first by the wild animals and then directly by God, Pantoleon/Panteleimon's martyrdom comes to an end. It should be noted here that not only Pantoleon/Panteleimon as a holy figure but also the wild animals appear to act as intercessors between man and God during the martyrdom as they become the reason for many to believe in God and for others to strengthen their Christian faith.

In sum, the episode with the wild animals in the arena has an important function in the narrative due to its symbolic character; it seems to foreshadow the outcome of the martyrdom and the martyr's entry into Paradise. In general, animals seem to be strongly connected with the interrelated concepts of Paradise and the Last Judgement. Being at the threshold of death, martyrs enjoy "the dominion he [God] had given to Adam over the animals in Paradise, a dominion that had been lost through the fall".[49] Moreover, according to the Byzantine conception of the Last Judgement as visualised on an icon at Mount Sinai, wild beasts, both terrestrial and marine, which have devoured their human prey, appear to then disgorge the body parts since, on the Day of Judgement, both good and bad souls will have their bodies restored to reach their final destination (Heaven or Hades); at the same time, some intact figures are represented in Heaven.[50] In a similar way to this depiction of the Last Judgement, in which the animals pay respect to the dead bodies by returning them to God, in the case of Pantoleon/Panteleimon, the animals function as agents of God's will by displaying the respect due to the body of the Christian protagonist who was about to die a martyr's death. The animals' reactions betoken the steadfast and courageous attitude that the martyr will maintain until the end of his martyrdom and the celestial reward he will gain thereupon, namely entry into Heaven. In this respect, the martyr's encounter with the wild animals augurs his encounter with God.

Imprisonment

After interrogations and tortures, the Christian martyrs are most often sent to prison. On a narrative level, imprisonment is intended to give the martyrs space and time to change their minds and convert to paganism while they await further corporal punishments and trials or execution. The imprisoned martyrs, however, are encouraged by God and strengthened in their resolve to suffer torment and death for their faith. Prison scenes punctuate the narrative flow by accommodating additional episodes with new characters and events, thus achieving a delay in the martyr's death and suspense for both characters and actual readers or listeners. Hagiographers systematically present the imprisonment of martyrs as an independent stage of martyrdom rather than as an additional form of torture, and in several cases prison scenes even occupy the largest part of the narrative. This is affirmed by art historical evidence as well: Byzantine manuscript miniatures depict the four stages of martyrdom (i.e., interrogation, torture, imprisonment, and execution) distinct from one another or sometimes they exclusively focus on the prison scene.[51] This evidence demonstrates that prison was an integral part of the concept of martyrdom and a crucial step in attaining holiness.

In the *Passion of Marina of Antioch*, to which I turn now, the episode that takes place in prison occupies more than one-third of the whole text and captures a confrontation between the forces of evil and the divine, both materialised in the form of animals or animal-like beings.[52] Marina, having devoted her virginity to God, rejects the sexual advances of the prefect Olybrios and reveals her Christian identity. Olybrios interrogates her and orders that she be stripped naked and subjected to several tortures. In a prayer, she asks God to help her defeat the enemy and preserve her virginity. At that very moment, she is led to prison. After a heavy earthquake, "a big and extremely terrifying dragon" (δράκων μέγας καὶ φοβερὸς σφόδρα) suddenly emerges from one of the corners of the prison.[53] Its appearance and behaviour are described in detail:

ποικίλος τῇ χρόᾳ· ἡ δὲ θρὶξ αὐτοῦ καὶ τὸ γένειον ὥσπερ χρυσὸς ὑπῆρχεν· οἱ δὲ ὀδόντες αὐτοῦ ἐξήστραπτον, καὶ οἱ ὀφθαλμοὶ αὐτοῦ ὅμοιοι μαργαρίτῃ ὑπῆρχον· ἐκ δὲ τῶν μυκτήρων αὐτοῦ ἐξεπορεύοντο φλὸξ πυρὸς καὶ καπνὸς πολύς· ἡ δὲ γλῶσσα αὐτοῦ ἦν ὡς αἷμα· περὶ δὲ τὸν τράχηλον αὐτοῦ ὄφεις εἱλιγμένοι· οἱ δὲ κανθοὶ τῶν ὀφθαλμῶν αὐτοῦ ἦσαν ὡς ἄργυρος. καὶ οὗτος ἔστη ἐν μέσῳ τῆς φυλακῆς κράζων καὶ συρίζων· ἔτρεχεν δὲ κύκλῳ Μαρίνης μετὰ γυμνῆς ῥομφαίας, καὶ ὁ συριγμὸς αὐτοῦ ἐποίησεν δυσωδίαν δεινὴν ἐν τῇ φυλακῇ·[54]

It [the dragon] had mottled skin, its hair and beard looked as if they were golden, its teeth were flashing, and its eyes resembled pearls; its nostrils emitted flames and much smoke; its tongue was blood-red; snakes coiled around its neck; the corners of its eyes looked as if they were made of silver. And it stood in the middle of the prison, screaming and hissing. It moved swiftly in a circle around Marina holding a large, unsheathed sword, and its hissing caused a terrible stench in the prison.

Although dragons often appear in hagiographical texts, the dragon's form, as depicted in the *Passion of Marina*, is unique.[55] By describing its skin, coat, beard, teeth, tongue, eyes, and corners of its eyes, the hagiographer provides the text's intended audience with the outline of the strange creature. At the same time, its hissing, that is, the sound it produces, along with the hagiographer's endeavour to connect this creature with snakes (by giving the detail of the snakes twisting around its neck), conveys the impression that it could be a giant serpent or another reptilian beast, such as a lizard.[56] At any rate, this animal or animal-like beast is invested with supernatural characteristics, including the ability to exude flames from its nostrils.

Such strange creatures were well-established in the Byzantine supernatural imagination and appear to be dangerous foes of holy persons.[57] In the case of Marina, for example, the overall appearance and behaviour of the dragon leave no doubt that it is ill-disposed towards the female protagonist. At the sight of the dragon, the imprisoned martyr begins to tremble and turn pale from her great fear. Marina addresses a prayer to God and implores him to prevent the dragon from "harming" her (ἀδικηθῆναι) and to help her vanquish its "fire" (φλόγα).[58] Being terribly frightened, the holy protagonist does not react at all when the dragon drags her towards itself, draws her up and swallows her into its belly (ἔσυρεν πρὸς ἑαυτὸν τὴν ἁγίαν κόρην [...] ἀνιμήσατο αὐτὴν καὶ κατέπιεν εἰς τὴν κοιλίαν αὐτοῦ).[59] But as soon as she makes the sign of the cross, the dragon's innards burst open and she comes out of its belly "unharmed" ([lit. without being harmed at all] μηδὲν ἀδικηθεῖσα).[60]

Immediately after Marina's victory over the dragon, a demon enters her cell from another corner of the prison. Marina now acts bravely and confidently. She is ready to overcome this new challenge by fighting the demon both verbally and physically and eventually triumphs over him. She violently removes his beard and his right eye, kicks him in the neck and hits him on the head with a brass hammer. The demon squalls in pain. It is interesting to note the demon's words, which come to complete the dragon's profile. He discloses that he had dispatched his kinsman named Rufus in the form of a dragon to kill the martyr (ἐγὼ γὰρ τὸν συγγενέα μου Ῥοῦφον ἀπέστειλα ἐν σχήματι δράκοντος, ὅπως σε ἀποκτείνῃ).[61] Thus, as in the case of Antony discussed at the very outset of this study, the animal-like being that appears to Marina is, in fact, a demon who seeks to discourage the holy protagonist from continuing her spiritual struggle by frightening her. The chief demon, whose name is Beelzebul or Satan,[62] has even assigned this wild creature to kill the imprisoned martyr because of her steadfastness in faith. According to late antique and Byzantine belief, at important stages of life, demons lie in wait for their victims and attack them in moments or places of transition (e.g., doorsteps, baptism, death).[63] In the case of martyrs, imprisonment represents an important transitional phase in the attainment of holiness, and its crucial importance is emphasised by the presence of demons. Marina is on the threshold of acquiring her new identity as a saint, and her combat with the evil forces in prison proves to be a prerequisite for her spiritual perfection. In fact, her victory over the dragon and the demon prefigures the successful accomplishment of her martyrdom.

Yet, in her confrontation with the evil forces, Marina is not alone; she is supported by Christ who visits her in prison in the form of a winged animal. According to the text, the prison suddenly lights up, and a huge cross that reaches to the sky appears in the middle of the cell, while "a dove sits on the cross" (περιστερὰ ἦν ἐπάνω τοῦ σταυροῦ).[64] The dove tells the martyr in a human voice that after her victorious battle with the demon and her prayer to receive "the sweet-smelling oil" (τὸ ἔλαιον τῆς εὐωδίας) she is prepared to follow the path to Paradise.[65] The sweet-smelling oil serves here as an allusion to the chrism and especially the baptism Marina will receive by divine intervention at the end of her martyrdom.[66] The huge cross implies the passion and crucifixion of Christ, while also foreboding Marina's martyrdom for the sake of Christ. The most important element in this scene is, however, the dove, which influences the course of action.

The hagiographer himself indicates later in the narrative what the dove incarnates and symbolises: the martyr asks Christ to send "the holy dove of His [Your] Spirit" (ἡ ἁγία περιστερὰ σου τοῦ πνεύματός σου) to bless the water of her baptism, and right before her decapitation, "the Lord [Jesus Christ] comes from the heavens in the form of a dove" (ἐλθὼν ὁ κύριος ἐκ τῶν οὐρανῶν ἐν εἴδει περιστερᾶς) to promise her the status of a saint and the power of intercession after death.[67] The scene of Marina's baptism is strongly reminiscent of Christ's baptism in the New Testament, according to which the Spirit of God or the Holy Spirit descends from heaven and approaches Christ in the form of a dove (Mt. 3.16; Mk. 1.10; Lk. 3.22; Jn. 1.32–33). Therefore, in the *Passion of Marina*, the talking dove, which first appears in jail, incarnates Jesus Christ and, by extension, His Father, and the Holy Spirit. The dove-Christ encourages the imprisoned martyr to continue her fight against the demon until he confesses all his evil deeds. In this manner, Christ not only recognises Marina's spiritual struggles but also participates, in the form of a dove, in her confrontation with the evil forces. Immediately after the divine apparition, the martyr is healed of her physical wounds and regains her strength. Even the demon notices the change in the martyr's appearance and behaviour caused by the renewed inner strength that Christ had given her. Overall, Marina's encounter with the dove-Christ in prison played a decisive role not only in her victory over the demon, but also in her courageous attitude towards the final stages of her martyrdom, including execution by the sword.

What makes this *Passion* particularly interesting within the framework of this study is certainly the apparition of the animal-like creature, namely the dragon, which opens the scene of the protagonist's imprisonment. The above examination, focusing on the highlights of the martyrdom story as a whole, permits a further interpretation of Marina's combat with the dragon in prison. As I have thoroughly investigated elsewhere and will try to summarise here, in late antique and Byzantine martyrdom literature, demons (including dragons) that enter the prison cell of a female martyr symbolise a sexual threat or temptation as well as the female martyr's weak self and impure thoughts.[68]

In the *Passion of Marina*, the confrontation of the Christian protagonist with sexual threat and temptation is introduced even before the prison scene, but it becomes central with the presence of the dragon and the demon in jail. More precisely, Marina's martyrdom begins after she has rejected the sexual advances of the prefect Olybrios by stressing that he will not succeed in "raping" her (ἀδικεῖν).[69] In this context, the verb ἀδικῶ (lit. harm) is to be understood as meaning "to sexually abuse". This, in fact, applies to the other two cases mentioned above, where the same verb is used: in her combat with the dragon, the female martyr asks God to help her be not sexually abused by this creature (ἀδικηθῆναι), and when the dragon eventually dies the hagiographer clarifies that the martyr comes out of its belly unharmed (μηδὲν ἀδικηθεῖσα), namely without losing her virginity, which was her primary concern. The dragon's fire (φλόγα) also symbolises sexual longing and hence sexual threat or temptation.[70] Likewise, the "large unsheathed sword" with which the dragon is ready to leap at the female martyr can be understood as a metaphor for an erect phallus.[71] The dragon's desire to swallow Marina comes across as an active effort to take possession of her body and, by extension, as attempted rape, while the martyr is portrayed as remaining silent and passive during this attack. Such a juxtaposition of submission and dominance clearly refers – also in the view of the Byzantines themselves – to the sexual act between a passive and an active person.[72] The dragon is thus depicted as a lustful creature.[73] However, in the end, the female protagonist not only kills the dragon through her prayer but also becomes more confident so as to actively face the next adversary, i.e., the demon. In this way, Marina and her demonic foes swap the roles of victim and tormentor, with Marina taking on the active role of the tormentor. Furthermore, while the dragon and the demon are characterised by their stench (δυσωδία), which reflects their impure character and inappropriate intentions towards the female protagonist,[74] Marina seeks to protect herself from those foul odours and, at the same time, turns to the God-pleasing sweet smell (εὐωδία). Beyond its association with the sacrament of baptism mentioned above, the sweet smell serves as a symbol for the martyr's virginity that she successfully maintains to the end and thereby gains sanctity.[75]

For the depiction of Marina's combat with the dragon and the demon, the hagiographer seems to draw on the Bible. The dragon's attempt to swallow the martyr must have been inspired by the Old and New Testament story of Jonah, according to which the prophet Jonah spent three days and three nights in the belly of a sea monster, thereby prefiguring Christ's sojourn in the tomb, his descent into the underworld, and ultimately his resurrection (Jn. 2.1–11; Mt. 12.39–41; Lk. 11.29–30).[76] In contrast to the relevant episode from the *Passion of Marina*, the biblical passages have no sexual connotations, but the events they describe can be interpreted as a victory over the Devil and death and, by extension, as a kind of rebirth. In this sense, Marina's triumphant exit from the dragon's belly symbolises her spiritual rebirth or transformation, which enables her to assume a more active role in overcoming new sexual temptations and challenges, especially the demon's attack.

The Old Testament story of Adam and Eve must have also provided background information for Marina's combat with the evil forces in prison. This episode from Marina's *Passion* seems to continue the biblical tradition that Eve, as being of the weaker sex, was deceived by the demon in the form of a serpent and then misled Adam, too (Gen. 3.1–24). The dragon (probably a giant serpent, as explained above) and the demon appear in prison to try to bend the martyr Marina (as Eve had back then) to their will and thus achieve what the pagan persecutor was not able to do in the public torture, namely, to seduce her ideologically and sexually. However, through the victory of the female martyr over these evil forces, a woman's position in late antique society is inverted. In this manner, the *Passion of Marina* strives to put a positive spin on the female sex after Eve's fall in the eyes of the Byzantines. The martyr takes on the role of Eve, who retrospectively defeats the evil serpent through her steadfastness in faith and martyrdom. The Christian martyr as a "new Eve" thus follows the example of the Virgin Mary, who atoned for the fall of Eve by giving birth to Christ.[77]

Finally, although the dragon and the demon are depicted as entering the prison in a physical form, they would likely also have been understood by the text's audience at the time as imaginary creatures symbolising the female martyr's weakness and impure thoughts (*logismoi*). In this regard, prison was the place where the female martyr wrestled with her own sexuality, overcame it, and advanced towards holiness. Indeed, already in the early Christian period, passions/emotions (*pathê*) were often considered forces that could be triggered by the sensual world, thus approaching people from outside; hence negative emotions/bad passions, such as sexual desire and love of lust (*philêdonia*), are usually portrayed as demons attacking a virtuous person.[78] Negative emotions and temptations, however, could conquer someone not only externally by controlling his or her body, but also internally by controlling his or her thoughts. In this way, the dragon and the demon in the *Passion of Marina* can be identified with the protagonist's inner weaker self, against which she fights tooth and nail and ultimately prevails. In his comments on this text, Methodios I, later Patriarch of Constantinople (843–847), writes: "How could one believe that, by the grace [of God], the Saint [Marina] defeated the enemy *mentally* if he [the demon] had not appeared and if she had not conquered him *in a perceptible* [esp.

visible] way?" (πῶς δὲ καὶ εἶχε πιστεύεσθαι ἡ ἁγία διὰ τῆς χάριτος νικῶσα τὸν ἐχθρὸν *νοητῶς*, εἰ μὴ καὶ φανταζόμενον ἐξενεύρου *αἰσθητῶς*; emphasis added).[79] In this sense, the depiction of these evil figures as external threats functions as a metaphor for the inner world of the protagonist and her concerns; by assuming a concrete form in the narrative, these become more easily comprehensible to the readers and listeners at the time. Thus, beyond physical martyrdom, the female protagonist also faces inner challenges. Even though the presence of the dragon and the demon can be regarded as a state of mind, it is fascinating that the hagiographer chose to depict at least one of the female martyr's antagonists as an animal-like being. The animal-like form adopted by the demon-dragon highlights the opposition between (animalistic) desire and (human) virtue, represented by the dragon and the martyr respectively. Marina's victory over the dragon, namely her own weaknesses, leads to her exaltation. Therefore, not only Christ who appeared as a dove but also the dragon was eventually conducive to the growth of the protagonist's spirituality and her path to sanctity.

Execution and Burial

As a rule, the "combat with the beasts" investigated above was intended to torture and injure the martyrs, not to kill them.[80] However, as we have seen, *Passions* also describe cases in which martyrs are fatally wounded by the wild animals in the arena, thereby turning this punishment into a method of execution. In these cases, wild animals seem, indeed, to act "rather like a squad of official executioners".[81] Yet, the most usual method of execution was beheading, a process in which animals were not involved. Although there are a few exceptions of saints who are deemed martyrs because they were tortured but did not die a martyr's death, such as Thekla the Protomartyr or Zosimos mentioned above, a martyr's identity as a saint is, in fact, established only after the completion of his or her martyrdom, including execution.[82] Byzantine illustrations and narrative depictions mostly focus on this very last – and almost unavoidable – phase of martyrdom,[83] which is often followed by the burial of the martyr's remains. In the *Passion of Sergios and Bakchos*, which will be discussed below, wild animals appear to witness – at least parts of – the martyrdom and assume human behaviour during the execution and burial of the two male martyrs.

Bakchos and Sergios refuse to offer sacrifices to the pagan gods, and on command of the emperor Maximian, they are sent to the *dux* Antiochos in the province of Augusto-Euphrates. Antiochos decides to separate the two men by imprisoning Sergios and torturing Bakchos. Bakchos ultimately dies from the harsh whipping and beating, and the governor orders that the martyr's remains not be buried but thrown out of the camp to be eaten by dogs, wild beasts, and birds of prey. But that is not what subsequently happened:

τοῦ δὲ λειψάνου ῥιφέντος μήκοθεν τοῦ κάστρου, συναχθὲν πλῆθος θηρίων περιεκύκλωσαν αὐτό. Τὰ δὲ ὄρνεα ἄνωθεν ἱπτάμενα οὐκ εἴων τὰ αἱμοβόρα θηρία ἅπτεσθαι αὐτοῦ. Καὶ παρέμειναν φυλάττοντα αὐτὸ ἄχρις ἑσπέρας βαθείας. Ὀψίας δὲ γενομένης, κατελθόντες τινὲς τῶν ἐκεῖσε οἰκούντων ἀδελφῶν ἐν τοῖς σπηλαίοις ἐπῆραν τὸ λείψανον τοῦ ἁγίου προπεμπόμενον ὑπὸ τῶν θηρίων ὥσπερ ὑπό τινων λογικῶν ἀνθρώπων. Καὶ ἔθαψαν αὐτὸ ἐν ἑνὶ τῶν σπηλαίων αὐτῶν.[84]

When the dead body was tossed some distance from the camp, a crowd of animals gathered around it. The birds flying above would not allow the bloodthirsty beasts to touch it. And they [the animals, including the birds] kept guard throughout the night. In the evening, some of the monks who lived nearby in caves came and collected the dead body, which the animals – as if they were rational human beings – accompanied first [in the procession]. They buried it in one of their caves.[85]

Contrary to the expectations of the pagan governor, two groups of animals, one on the ground and one in the air, undertake to protect the martyr's dead body from other bloodthirsty beasts.[86] In his tenth-century version, Symeon Metaphrastes stresses that the wild animals become tame by divine intervention.[87] Both groups of animals seem to recognise the holy status of the dead man and by acting as guards try to preserve his body intact as a venerable relic. But the animals' action is not limited to their careful guarding. As in the case of Pantoleon and the beasts in the arena examined above, the animals here behave like rational human beings, who even participate in the funeral procession and the martyr's burial performed by some Christian ascetics late in the evening. In this way, the wild animals are integrated into the crowd of Christians and support the Christian faith.

After imprisonment and torture, the second martyr, Sergios, is sentenced to decapitation. The wild animals are also present at Sergios' death and burial, and like humans, they feel sympathy for Sergios at his execution:

Πολύ τε πλῆθος ἀνδρῶν τε καὶ γυναικῶν καὶ παιδίων ἐπηκολούθει ἰδεῖν τὸ μακάριον αὐτοῦ τέλος. Ὁρῶντες δὲ τὸ ἐπαρθοῦν τῇ ὄψει αὐτοῦ κάλλος, καὶ τὸ παμμεγεθὲς καὶ νεανικὸν τῆς ἡλικίας ἐδάκρυον πικρῶς ἐπ' αὐτῷ στενάζοντες. Τὰ δὲ θηρία τῶν ἐκεῖσε τόπων καταλειπόντα τὰς ἑαυτῶν μάνδρας ἀνάμιξ ἅπαντα συνεληλύθει, οὐδένα μὲν τῶν ἀνθρώπων ἀδικοῦντα, ἀλόγῳ δὲ βοῇ τὴν τοῦ ἁγίου μάρτυρος ἀναίρεσιν ὀδυρόμενα.[88]

A great crowd of men, women, and children followed, to see his [Sergios'] blissful end. Seeing the beauty blooming in his face, and the grandeur and nobility of his youth, they wept bitterly over him and bemoaned him. The beasts of the region left their lairs and gathered together indiscriminately, without harming any people, and bewailed with inarticulate sounds the passing of the holy martyr.[89]

According to the passage above, not only humans of every age but also wild animals of every kind participate in this procession, accompanying the martyr to the place of execution. The wild animals surpass the limits of their nature: they gather together, without attacking each other or the people, and, like the humans who are present there, mourn for the imminent execution of the young martyr. The wild animals are thus depicted as possessing human characteristics while simultaneously recognising, as in the case of Bakchos, the sanctity of the protagonist. In his version, Symeon Metaphrastes goes even so far as to describe the connection between the martyr Sergios and the wild animals as strongly emotional by mentioning that the wild animals hastened to follow the martyr "as if they also felt love [lit. amorous desire] for the athlete [the martyr]" (καθάπερ τι ἐρωτικὸν καὶ οὗτοι [sc. θῆρες ἄγριοι] πρὸς τὸς ἀθλητὴν παθόντες).[90] This strong affection undoubtedly emerges from the common connection to the divine that Sergios and the wild animals pursue. Symeon

Metaphrastes also explains that this procession of humans and animals aimed at reproving the pagans for their inappropriate behaviour, as they were not able to follow the example of the wild animals which showed respect towards the martyr. The wild animals are thus depicted as pious examples to imitate.

Unlike the tenth-century version by Symeon Metaphrastes, the pre-Metaphrastic one includes significant references to the wild animals at two more crucial points of the text: in Sergios' final prayer to God and in the text's epilogue. When Sergios arrives at the place of execution, he asks permission to pray one last time. In his prayer, he implores God to forgive his executioners and gives thanks for sending "the beasts of the field and the birds of the sky" (Τὰ θηρία τοῦ ἀγροῦ καὶ τὰ πετεινὰ τοῦ οὐρανοῦ) "to turn through their unreason the reason of humans to knowledge of Him [You]" (ὅπως τὴν τῶν λογικῶν ἀνθρώπων φύσιν διὰ τῆς ἑαυτῶν ἀλογίας ἐπιστρέψωσι πρὸς τὴν σὴν ἐπίγνωσιν)".[91] It is important that right before his death, the martyr underscores the function that the animals were expected to fulfil in his martyrdom and recognises that, unlike the pagan people, they possess the knowledge of God. Interestingly, although, in the case of Bakchos, the animals are described as rather "rational human beings", the hagiographer refers here to their "unreason" in contradistinction to "the reason of humans". It is, in fact, an ironic comment that, in the end, is indirectly reversed: it is the pagan people who are not able to reason, and not the pious animals.

Immediately after the completion of his prayer, Sergios is beheaded. Some of the eyewitnesses bury his body on the spot, while later it is transferred to a nearby shrine which has been erected in honour of the saint. Beyond many miraculous healings which the saint performed at that shrine and at his first tomb, he was also able to tame the wild animals. In fact, the whole text closes by stressing that the wild animals venerate the saint every year:

ὁ ἅγιος [...] καὶ τὰ θηρία τὰ ἄγρια εἰς πολλὴν ἡμερότητα μεταβάλλει, τὴν γὰρ τῆς τελειώσεως αὐτοῦ ἡμέραν κατ' ἔτος φυλάττοντα ὥσπέρ τινα νόμον τὰ ἄλογα συντρέχοντα ἐκ τῆς πέριξ ἐρήμου καὶ συναναστρεφόμενα τοῖς ἀνθρώποις οὐδένα τὸ σύνολον ἀδικοῦσιν, οὔτε μὴν τῇ τῆς ἀγριότητος ὁρμῇ εἰς βλάβην τῶν συντρεχόντων ἀποκέχρηνται, ἡπιότητι δὲ μᾶλλον τὸν ἅγιον μάρτυρα τιμῶντα προσεδρεύουσι τῷ τόπῳ προστάγματι τοῦ Θεοῦ, ὅτι αὐτῷ πρέπει δόξα, τιμή, κράτος, νῦν καὶ ἀεὶ καὶ εἰς τοὺς αἰῶνας τῶν αἰώνων. Ἀμήν.[92]

The saint is also able [...] to render savage beasts completely tame. The animals, in fact, observe the day of his death every year as if it were law, coming in from the surrounding desert and mingling with the humans without doing them any harm, nor do their savage impulses move them to any violence against the humans who come there. Rather, they come to the place in gentleness out of reverence for the holy martyr, at the command of God, to whom is due glory, honour, and power, now and forever and to the ages of ages. Amen.[93]

Once again, the wild animals act like human beings, but now their behaviour is attributed to the saint Sergios who can turn their savagery into tameness. Yet, what is particularly interesting here is that they are also able to remember the date of the martyr's death (7 October) and attend the yearly commemoration held for him. In this way, they confer honour not only on the saint but also on God himself. In this last section of the present survey, all the text's passages about wild animals affirm their prudence and even their ability to participate in ceremonies organised by Christian people: first, in the funeral procession and burial of Bakchos, then in the procession leading Sergios to his place of execution, and finally in the annual memorial services. It becomes apparent that the hagiographer's intention was to present the wild animals not only with human prudence but also with human – ceremonial – habits of thought and behaviour. In the *Passion of Sergios and Bakchos*, the animals thus appear only in scenes connected to the saints' deaths, specifically their executions, burials, and commemoration, and, as discussed in the case of Pantoleon/ Panteleimon above, may also prefigure the saints' transition to Heaven. Wild animals occupy a special place in this *Passion*'s epilogue as they remind the recipients one last time of the appropriate spiritual behaviour towards the martyr and God.

Conclusions

The present chapter offers an investigation of Greek *Passions* of martyrs, the most fertile hagiographical subgenre in Byzantium; it explores the four fundamental phases of martyrdom – the holy protagonists' interrogation, torture, imprisonment, and execution including subsequent burial – by focusing, from a literary-narrative angle, on the role of animals in these phases. Animals embrace active roles and drastically influence the plot, either as main and secondary heroes or as antagonists: in the *Passion of Zosimos*, the speaking lion that appears in the courtroom and participates in the interrogation takes on a leading role in the narrative, while in all other *Passions* examined here the animals assume a secondary, yet critical role, in the course of action. In the *Passion of Pantoleon/ Panteleimon*, the wild animals in the arena contribute to the recognition of the protagonist's holiness and suffer actual martyrdom themselves. In the *Passion of Marina of Antioch*, the demonic dragon, probably a giant serpent, which functions as an antagonist of the female martyr, comes into indirect conflict with Christ, who appears during the martyr's imprisonment in the form of a dove. In the *Passion of Sergios and Bakchos*, the wild animals witness the martyrs' deaths and pay reverence to their holy remains. As a whole, this investigation has uncovered a wide range of roles taken on by animals in hagiographical *Passions*: they appear as the accused in the courtroom, preachers of the Christian faith, co-martyrs, witnesses to the martyrdom, protectors/guards of the martyrs' dead bodies, but also as demons/enemies of the holy protagonists. With the exception of the dragon that appears to Marina, the examined animals serve as virtuous examples to be imitated by the spectators within the narrative and the texts' recipients.

More specifically, in three of these *Passions* (Zosimos, Pantoleon/Panteleimon, and Sergios and Bakchos), the wild animals' piety and prudence is juxtaposed with the pagan people's impiety and unreason. In these texts, the wild animals seem even to acquire a kind of holy status because of their participation in the martyrdom or they command the respect of the humans around them for their bravery and wisdom. Due to their close relationship with wild animals, Zosimos (who invites a lion to confess its Christian faith before the pagan judge) and Pantoleon (who inspires the wild animals to be martyred in the arena) span both the animal and the human worlds, and their intermediary role is attested on two levels: they intermediate in the relationship between animals and God, and at the same time they become a bridge of communication between humans and animals with God. During the martyrs' ordeals, these animals exemplify appropriate Christian behaviour and lead many pagans to become Christians or to recognise the power of the Christian God, as in the case of the pagan governor in Zosimos' *Passion*. Unlike Pantoleon/Panteleimon, who is sentenced to contend with wild animals in the arena, Marina fights the wild dragon and the demon in prison. Apparently, the hagiographer of Marina's *Passion* tried to reinvent the punishment known as "combat with the beasts". However, as discussed above, this variation bears a wealth of deep, especially sexual, connotations, hidden behind the symbolic character of the dragon as the female martyr's weak self. Marina must first wrestle with and overcome her own sexuality in order to successfully accomplish her purpose, namely suffering martyrdom for Christ's sake.

By and large, this investigation of animals based on their interaction with the holy protagonists enables a better understanding of their role both in the spiritual progression of the martyrs and in the storylines as a whole. On a narrative level, animals, both good and bad, real and imaginary, seem to contribute to the martyrs' efforts along their path to holiness, either by helping them surmount obstacles or, on the contrary, by creating obstacles for them to strengthen themselves and fortify their will. At any rate, the scenes featuring animals examined here foreshadow the successful completion of the martyrdom and the martyr's entry into Paradise. Regardless of whether they assume a primary or secondary role in the story and occupy more or fewer lines of text in the *Passions*, animals prove to be important elements of these hagiographical narratives – the unsung heroes who deserve closer attention.

Notes

1 Athanasios of Alexandria, *Life of Antony*, chs. 9–10, ed. G. J. M. Bartelink, *Athanase d'Alexandrie, Vie d'Antoine: introduction, texte critique, traduction, notes et index.* Sources Chrétiennes 400 (Paris, 1994), 124–377, esp. 158–64.

2 *Life of Mary of Egypt*, chs. 39–40, ed. J.-P. Migne, PG 87 (Paris, 1865), 3697–726, esp. 3724–25. This text was presumably written in the seventh century, and later Byzantines erroneously attributed it to Sophronios, Patriarch of Jerusalem (634–38). For a discussion of the date of composition and Sophronios' authorship of the text, see M. Kouli, "Life of St. Mary of Egypt," in *Holy Women of Byzantium: Ten Saints' Lives in English Translation*, ed. A.-M. Talbot, Byzantine Saints' Lives in Translation 1 (Washington, DC, 1996), 65–93, esp. 66 (with further references). A similar scene with two lions digging a hermit's grave can be found in the *Life of Paul of Thebes*, chs. 16 and 17 (pre-Metaphrastic [*BHG* 1466] and Metaphrastic version [*BHG* 1468]): K.S.T. Corey (ed.), "The Greek Versions of Jerome's *Vita Sancti Pauli*," in *Studies in the Text Tradition of St. Jerome's* Vitae Patrum, ed. W.A. Oldfather et al. (Urbana, 1943), 158–72, esp. 170–71 and 217–33, esp. 231–32).

3 The *Metaphrastic Menologion* constitutes the most extensive and widely copied collection of full-length Greek hagiographical texts that has come down to us. It consists of 148 hagiographical works and is preserved in over 800 manuscripts (including fragments) dating from the first half of the eleventh century onwards. See C. Høgel, *Symeon Metaphrastes: Rewriting and Canonization* (Copenhagen, 2002), esp. 11, 124, 130, 138 (for the insertion of the *Life of Antony* and that *of Mary* into that *Menologion* without changes, see 92, incl. n. 16, 141, 196, 201); idem, "Symeon Metaphrastes and the Metaphrastic Movement," in *The Ashgate Research Companion to Byzantine Hagiography*. Vol. 2: *Genres and Contexts*, ed. S. Efthymiadis (Farnham and Burlington, VT, 2014), 181–96, esp. 183. Høgel's work builds upon the remarks and the list of manuscripts compiled by A. Ehrhard, *Überlieferung und Bestand der hagiographischen und homiletischen Literatur der griechischen Kirche von den Anfängen bis zum Ende des 16. Jahrhunderts*, 3 vols. Texte und Untersuchungen zur Geschichte der altchristlichen Literatur 50–52/2 (Leipzig, 1936–1952), 2:306–659. For the manuscripts containing the texts in question, see also Bartelink, *Athanase d'Alexandrie*, 77–95; Kouli, "Life of St. Mary of Egypt," 67, incl. n. 12.

4 Some of these are martyrs' *Passions*, miracle accounts, translations of saints' relics, edifying stories, etc. For a detailed overview of all hagiographical subgenres, see M. Hinterberger, "Byzantine Hagiography and Its Literary Genres: Some Critical Observations," in *The Ashgate Research Companion*, ed. Efthymiadis, 25–60.

5 See I. Anagnostakis, T. G. Kolias, and E. Papadopoulou (eds.), *Animals and Environment in Byzantium (7th-12th c.)* (Athens, 2011). Thomas Pratsch's brief list of *topoi* concerning the interaction between saints and animals is also solely based on passages from saints' *Lives*, see T. Pratsch, *Der hagiographische Topos: Griechische Heiligenviten in mittelbyzantinischer Zeit.* Millennium-Studien 6 (Berlin and New York, 2005), 286–89.

6 *Passions* constitute by far the largest percentage of Greek hagiographical works. For their extensive manuscript transmission, see Ehrhard, *Überlieferung und Bestand.*

7 These phases have been discussed in recent scholarship, but not in regard to the role of animals: S. Constantinou, *Female Corporeal Performances: Reading the Body in Byzantine Passions and Lives of Holy Women.* Studia Byzantina Upsaliensia 9 (Uppsala, 2005), esp. 33–35, 48–49, with a focus on interrogation and torture; and C. Papavarnavas, *Gefängnis als Schwellenraum in der byzantinischen Hagiographie: Eine Untersuchung früh- und mittelbyzantinischer Märtyrerakten.* Millennium-Studien 90 (Berlin and Boston, 2021), esp. 24–29, 50–53, with a focus on imprisonment and its relationship with interrogation, torture, and execution.

8 For the historical circumstances in which the ideal of martyrdom developed among Christians, see, e.g., W. H. C. Frend, *Martyrdom and Persecution: A Study of a Conflict from the Maccabees to Donatus* (Oxford, 1965); R. Lane Fox, *Pagans and Christians. Religion and the Religious Life from the Second to the Fourth Century A.D. when the Gods of Olympus lost their Dominion and Christianity, with the Conversion of Constantine, triumphed in the Mediterranean World* (San Francisco, 1986); G. W. Bowersock, *Martyrdom and Rome* (Cambridge, 1995).

9 Cf. H. Delehaye, *Les Passions des martyrs et les genres littéraires*, 2nd ed. Subsidia Hagiographica 13B (Brussels, 1966), 171 (on historicity) and 223–24 (on their anonymity and dating): "Ces récits anonymes si nombreux, si semblables, si souvent retouchés [...] se laissent difficilement situer dans le temps."

10 Cf. Hinterberger, "Byzantine Hagiography and Its Literary Genres," 26; M. Detoraki, "Greek *Passions* of the Martyrs in Byzantium," in *The Ashgate Research Companion*, ed. Efthymiadis, 61–101, esp. 61.

11 It should be noted here that the very first attempt to examine literary aspects of martyrs' *Passions* was made in 1921 by Hippolyte Delehaye, see Delehaye, *Les Passions des martyrs.* However, this work largely aimed to divide these texts into different categories (historical, panegyrical, and fictional) and to identify similar scenes in various *Passions*, rather than to analyse their contents in depth. Beyond a few relevant shorter studies, the first monograph to offer a comprehensive literary analysis of mainly Greek *Passions* was published only in 2011 by Anne P. Alwis, *Celibate Marriages in Late Antique and Byzantine Hagiography: The Lives of Saints Julian and Basilissa, Andronikos and Athanasia, and Galaktion and Episteme* (London and New York, 2011). Previous relevant works include C. Høgel, "The Redaction of Symeon Metaphrastes: Literary Aspects of the Metaphrastic Martyria," in *Metaphrasis: Redactions and Audiences in Middle Byzantine Hagiography*, ed. C. Høgel. KULTs skriftserie 59 (Oslo, 1996), 7–21; and Constantinou, *Female Corporeal Performances*, 19–58.

12 See Delehaye, *Les Passions des martyrs*, 171–226.

13 *Synaxarion of Constantinople*, cols. 115–16 (feast day of Sergios and Bakchos [7 October]); 369–70 (Zosimos [4 January]); 825 (Marina [17 July]); 847–48 (Panteleimon [27 July]), ed. H. Delehaye, *Synaxarium ecclesiae Constantinopolitanae: Propylaeum ad Acta Sanctorum Novembris* (Brussels, 1902).

14 F. Halkin (ed.), "Un émule d'Orphée: la légende grecque inédite de S. Zosime, martyr d'Anazarbe en Cilicie," *AB* 70 (1952): 252–53.

15 Cf. Ehrhard, *Überlieferung und Bestand*, 2, 642. The scene with the animals is more elaborated in the Metaphrastic reworking (ed. J.-P. Migne, PG 115 [Paris, 1899], 448–77) than in the pre-Metaphrastic version of the *Passion* (*BHG* 1412z); for this reason, the pre-Metaphrastic version is not part of the present analysis.

16 P. Boulhol, "Hagiographie antique et démonologie: notes sur quelques Passions grecques (*BHG* 962z, 964 et 1165–66)," *AB* 112 (1994): 255–303, esp. 259–61.

17 E. K. Fowden, *The Barbarian Plain: Saint Sergius between Rome and Iran.* The Transformation of the Classical Heritage 28 (Berkeley, Los Angeles, and London, 1999), 26–28. Cf. J. Boswell, *Same-Sex Unions in Premodern Europe* (New York, 1994), 147, incl. n. 172.

18 To the best of my knowledge, besides the edition of this *Passion* (Halkin, "Un émule d'Orphée," 249–61) that is accompanied by a brief introduction, only one article touches on the text in question in the context of Byzantine popular traditions transmitted in hagiographical texts, see M. G. Varvounis, "Les traditions populaires byzantines dans les textes hagiographiques," *Byzantiaka* 18 (1998): 153–70, esp. 154, 160.

19 Cf. *Passion of Zosimos*, ch. 1, ed. Halkin, 255. Some other male martyrs, such as Mamas, Eleutherios, and Blasios, are also narratively depicted as living with wild animals in the mountains. For the *Passion of Mamas* (*BHG* 1019), see A. Berger, "Die alten Viten des heiligen Mamas von Kaisareia. Mit einer Edition der Vita *BHG* 1019," *AB* 120 (2002): 241–310 (edition, German translation, and commentary). For a brief discussion of the *Passion of Eleutherios* (*BHG* 568–70) and that *of Blasios* (*BHG* 276 and 277), along with references to their editions, see C. Papavarnavas, "Imprisoned Martyrs on the Move: Reading Holiness in Byzantine Martyrdom Accounts," *BZ* 114.3 (2021): 1241–62, esp. 1253–56.

20 Halkin, "Un émule d'Orphée," 249–50, 254 n. 3.

21 *Passion of Zosimos*, ch.1, ed. Halkin, 254. Translations from the Greek are my own unless otherwise indicated.

22 In a *koinobion*, all the monks of the monastery must lead a common and egalitarian way of life. For further information, see A.-M. Talbot, "Koinobion," ODB 2, 1136. Another similar case can be found in John Moschos' *Spiritual Meadow* (*BHG* 1441-1442; *CPG* 7376) dating from the sixth/seventh century: one day, a monk named Gerasimos (living in a *lavra*) healed a wounded lion, and the lion stayed with him, eating and behaving like a monk until its death, see John Moschos, *Spiritual Meadow* ch. 107 (ed. J.-P. Migne, *PG* 87/3 (Paris, 1865), 2852–3112, esp. 2965–2969).

23 *Passion of Zosimos*, ch. 1, ed. Halkin, 255.

24 Ibid., ch. 2, ed. Halkin, 255–56.

25 Ibid., 257.

26 Ibid., ch. 3, ed. Halkin, 257–58.

27 Ibid., ch. 2, ed Halkin, 257.

28 Ibid., ch. 4, ed. Halkin, 258.

29 Ibid., 259.

30 Halkin, "Un émule d'Orphée," 251, incl. n. 2. Cf. C. Schmidt and W. Schubart, Πράξεις Παύλου: *Acta Pauli nach dem Papyrus der Hamburger Staats- und Universitätsbibliothek* (Glückstadt and Hamburg, 1936), 38–40, 86–87. The story of the speaking and baptised lion was very popular among Christians, see T. Adamik, "The Baptized Lion in the Acts of Paul," in *The Apocryphal Acts of Paul and Thecla*, ed. J. N. Bremmer. Studies on the Apocryphal Acts of the Apostles 2 (Kampen, 1996), 60–74, esp. 62–68.

31 *Passion of Sebastiane*, chs. 19–21 (*AASS* Iun. VI [1715], 60–70, esp. 68–69).

32 See Halkin, "Un émule d'Orphée," 251 (with references to the relevant examples).

33 See *Synaxarion of Constantinople*, ed. Delehaye, cols. 369–70 (4 January). Cf. Halkin, "Un émule d'Orphée," 251, incl. n. 11.

34 See, e.g., D. Potter, "Martyrdom as Spectacle," in *Theater and Society in the Classical World*, ed. R. Scodel (Ann Arbor, 1993), 53–88, esp. 66; Bowersock, *Martyrdom and Rome*, 18. For further aspects of this kind of punishment (*damnatio ad bestias*) and some relevant manuscript miniatures, see N. Patterson Ševčenko, "Wild Animals in the Byzantine Park," in *Byzantine Garden Culture*, ed. A. Littlewood, H. Maguire, and J. Wolschke-Bulmann (Washington, D.C., 2002), 69–86, esp. 78–79; eadem, "Eaten Alive: Animal Attacks in the Venice *Cynegetica*" in *Animals and Environment*, ed. Anagnostakis, Kolias, and Papadopoulou, 115–35, esp. 122–26.

35 For the historical dimensions of this *Passion*, see Delehaye, *Les Passions des martyrs*, 49–55. The Greek version of the text was most probably composed during the pre-iconoclastic era, though it is also possible that it was contemporary with the Latin original, see M. White, "The Rise of the Dragon in Middle Byzantine Hagiography," *BMGS* 32.2 (2008): 149–67, esp. 155, incl. n. 23 (with further references).

36 According to the Latin original, the martyrdom took place under Geta (198–209). For a discussion of the date of the martyrdom, see т. J. Heffernan, *The Passion of Perpetua and Felicity* (Oxford, 2012), 62–78, 138.

37 *Passion of Perpetua and Felicity* chs. 19–21, ed. C. I. M. I. van Beek, *Passio sanctarum Perpetuae et Felicitatis: textum graecum et latinum ad fidem codicum Mss.* (Nijmegen, 1936), 5–53, esp. 45–53.

38 *Passion of Euphemia*, ch. 17, ed. F. Halkin, *Euphémie de Chalcedone*, Subsidia Hagiographica 41 (Brussels, 1965), 56–79, esp. 77–78.

39 The relevant episode takes up around one and a half columns in the *Patrologia Graeca*, making it a relatively small part of the entire text (15 columns in total), see *Passion of Panteleimon* by Symeon Metaphrastes, chs. 19–21, ed. J.-P. Migne, PG 115 (Paris, 1899), 448–477, esp. 468–69).

40 *Passion of Panteleimon* by Symeon Metaphrastes, ch. 19, ed Migne, PG 115, 468.

41 *Passion of Panteleimon* by Symeon Metaphrastes, ch. 20, ed Migne, PG 115, 469. The lion was mainly considered a mighty beast throughout the Byzantine centuries (and beyond), cf. A. Karpozilos, A. Kazhdan, and A. Cutler, "Lions," ODB 2, 1231–32, esp. 1232; T. Schmidt, *Politische Tierbildlichkeit in Byzanz: Spätes 11. bis frühes 13. Jahrhundert*, Mainzer Veröffentlichungen zur Byzantinistik 16 (Wiesbaden, 2020), esp. 75–95.

42 *Passion of Panteleimon* by Symeon Metaphrastes, ch. 20, ed. Migne, *PG 115*, 469.

43 For the term τύραννος as a characterisation of the pagan persecutor in the martyrs' *Passions*, see also M. Hinterberger, "A Neglected Tool of Orthodox Propaganda? The Image of the Latins in Byzantine Hagiography," in *Greeks, Latins, and Intellectual History: 1204–1500*, ed. M. Hinterberger and C. Schabel, Recherches de théologie et philosophie médiévales. Bibliotheca 11 (Leuven, Paris, and Walpole, MA, 2011), 129–49, esp. 140, incl. n. 42; Papavarnavas, *Gefängnis als Schwellenraum*, 68.

44 *Passion of Panteleimon* by Symeon Metaphrastes, chs. 20–21, ed. Migne, PG 115, 469.

45 A. P. Kazhdan, "Holy and Unholy Miracle Workers," in *Byzantine Magic*, ed. H. Maguire (Washington, DC, 1995), 73–82, esp. 75–76; Pratsch, *Der hagiographische Topos*, 287–288. Another example of a martyr who tames wild beasts (lions) in the arena is Mokios. For the *Passion of Mokios* (*BHG* 1298–1298c), see H. Delehaye, "Saints de Thrace et de Mésie," *AB* 31 (1912), 163–176.

46 In martyr's *Passions*, people are even depicted as visiting the imprisoned martyrs to receive physical healing and spiritual support, see C. Papavarnavas, "Miracle and Martyrdom: Performing and Experiencing Healing Miracles in Prisons and Shrines," in *Le rôle des miracles et des recueils des miracles*, ed. V. Déroche and S. Efthymiadis (Paris, forthcoming).

47 Often, it is the dead martyrs whose bodies are abandoned unburied at the place of execution and are not touched by wild animals, see, for instance, the case of Sergios and Bakchos analysed below.

48 Note that in hagiographical texts, not only the protagonists but also the secondary characters (in this case, the animals) may serve as appropriate examples for Christians to imitate, see C. Papavarnavas, "The Role of the Audience in the Pre-Metaphrastic Passions," *AB* 134.1 (2016): 66–82.

49 N. Patterson Ševčenko, "The Hermit as Stranger in the Desert," in *Strangers to Themselves: The Byzantine Outsider (Papers from the Thirty-second Spring Symposium of Byzantine Studies, University of Sussex, Brighton, March 1998)*, ed. D. C. Smythe. Society for the Promotion of Byzantine Studies 8 (Aldershot and Burlington, VT), 75–86, esp. 85. Nancy Patterson Ševčenko's observation cited above was made on the basis of hagiographical descriptions of desert ascetics, but it seems to apply to the holy martyrs, too.

50 Icon of the Last Judgement, eleventh/twelfth century, 62.2 x 45.8 cm, Holy Monastery of St. Catherine, Sinai (Princeton work number 98, Michigan Inventory number 107–108); image available at: www.sinaiarchive.org/s/mpa/item/6950#?c=&m=&s=&cv=&xywh=-33%2C2140%2C3411%2C2143. See Patterson Ševčenko, "Eaten Alive," 129.

51 For a thorough analysis of the narrative functions of imprisonment in martyrs' *Passions* as well as a presentation of relevant manuscript illustrations, see Papavarnavas, *Gefängnis als Schwellenraum*, esp. 50–60, 155–56.

52 The relevant episode takes up 13 out of a total of 33 pages in the edition, see *Passion of Marina of Antioch*, 23:36–36:6, ed. H. Usener, "Acta S. Marinae et S. Christophori," in *Festschrift zur fünften Säcularfeier der Carl-Ruprechts-Universität zu Heidelberg*, ed. idem (Bonn, 1886), 15–47, esp. 23–36.

53 *Passion of Marina of Antioch*, 25:23–24, ed. Usener.

54 Ibid., 25:24–33, ed. Usener.

55 This dragon's appearance seems to amalgamate three different literary traditions, the ancient, the biblical, and the hagiographical, see Boulhol, "Hagiographie antique et démonologie," 262–67. For other episodes with dragons in Byzantine hagiographical literature, see White, "The Rise of the Dragon," 149–67; Papavarnavas, *Gefängnis als Schwellenraum*, 90–115 (with further references). On dragons in the early Christian world, see in general D. Ogden, *Dragons, Serpents, and Slayers in the Classical and Early Christian Worlds: A Sourcebook* (New York, 2013); idem, *Drakōn: Dragon Myth and Serpent Cult in the Greek and Roman Worlds* (Oxford, 2013); idem, "The Function of Dragon Episodes in Early Hagiography," in *Animal Kingdom of Heaven: Anthropozoological Aspects in the Late Antique World*, ed. I. Schaaf. Millennium-Studien 80 (Berlin, 2019), 35–57.

56 In Byzantium, dragons were mostly thought to be giant serpents or other reptilians, sometimes with two or more heads, see P. Koukoules, Βυζαντινῶν βίος καὶ πολιτισμός, vol. 1.2. Collection de l'Institut Français d'Athènes 11 (Athens, 1948), 252–53; A. Karpozilos and A. Cutler, "Snakes," ODB 3:1920; C. Walter, *The Warrior Saints in Byzantine Art and Tradition* (Aldershot, 2003), 128; White, "The Rise of the Dragon," 149–50. Notably, in early hagiographical texts, dragons are even identified with the serpent of Eden and the Devil, see Ogden, "The Function of Dragon Episodes," 41–46.

57 White, "The Rise of the Dragon," esp. 149–50; Papavarnavas, *Gefängnis als Schwellenraum*, esp. 90–93.

58 *Passion of Marina of Antioch*, 26:6–8, ed. Usener.

59 Ibid., 27:7–8, ed. Usener.

60 Ibid., 27:13–14, ed. Usener.

61 *Passion of Marina of Antioch*, 29:20–21 [135v] and 29:1 [136r], ed. Usener. Cf. the scholarly discussion about the connection between dragons and demons: contrary to Monica White, who tends to believe that dragons and demons are two distinct evil forces, Pascal Boulhol argues that these figures can be directly or indirectly identified with each other, see White, "The Rise of the Dragon," esp. 149–50 and Boulhol, "Hagiographie antique et démonologie," 264, incl. n. 44 and 45. Cf. Ogden, "The Function of Dragon Episodes," 43–46.

62 *Passion of Marina of Antioch*, 31:9 and 32:24, ed. Usener.

63 C. Rapp, "Heilige, Teufel und Dämonen: Frömmigkeit im Alltagsleben," in *Das goldene Byzanz und der Orient*, ed. F. Daim, C. Gastgeber, and D. Heher (Schallaburg, 2012), 105–17, esp. 108–9.

64 *Passion of Marina of Antioch*, 30:29, ed. Usener.

65 Ibid., 30:35.

66 When the pagan adversary orders the martyr to be drowned in a large vessel full of water, she prays to God to convert this torture into her "sacrament of baptism" (μυστήριον βαπτίσματος). Marina's wish is instantaneously fulfilled, see *Passion of Marina of Antioch*, 38:36–40:29, ed. Usener. For the association between a sweet smell and baptism, see B. Caseau, "The Senses in Religion: Liturgy, Devotion, and Deprivation," in *A Cultural History of the Senses in the Middle Ages*, ed. R. Newhauser (London, New Delhi, New York, and Sydney, 2014), 89–110, esp. 92–94.

67 *Passion of Marina of Antioch*, 38:47–48 and 41:25–26, ed. Usener. Following her baptism, Marina addresses a final prayer to Christ, expressing her special requests for posthumous abilities, which Christ affirms. For an interpretation of this scene, see Constantinou, *Female Corporeal Performances*, 56–58; B. Flusin, "Le contrat de Marina: passions épiques et culte des saints," in *Culte des saints et littérature hagiographique: accords et désaccords*, ed. V. Déroche, B. Ward-Perkins, and R. Wiśniewski. Monographies du Centre de Recherche d'Histoire et Civilisation de Byzance - Collège de France 55 (Leuven, Paris, and Bristol, CT, 2019), 39–53, esp. 47–53 (with further references). Cf. Papavarnavas, "Imprisoned Martyrs on the Move," 1249–51, incl. n. 22.

68 See Papavarnavas, *Gefängnis als Schwellenraum*, 87–119, esp. 99–104, 112–15.

69 *Passion of Marina of Antioch*, 20:36, ed. Usener

70 Note that the metonymy of "fire" for sexual temptation appears again later in the narrative: the demon reveals that by instilling a "burning fire" in humans (πῦρ καιόμενον), he endeavours to lead them into (sexual) "sin" (ἁμαρτία), see *Passion of Marina of Antioch*, 33:11–12, ed. Usener. The imagery of fire (passionate love, sexual desire) goes back to ancient Greek literature, especially to the myth of Eros, see B. S. Thornton, *Eros: The Myth of Ancient Greek Sexuality* (Colorado and Oxford, 1997), esp. 15–16, 28–33, 217.

71 For a discussion of similar scenes and symbolism in other *Passions* of female martyrs, see Papavarnavas, *Gefängnis als Schwellenraum*, 93–94 [*Passion of Juliana of Nicomedia*], 105 and 110 [*Passion of Perpetua*]. Daniel Odgen's opinion that in Marina's *Passion*, the "sword" (ῥομφαία) is used metonymically for the dragon's tongue is, in my view, not convincing, cf. Ogden, *Dragons, Serpents, and Slayers*, 226, 244–46. In the relevant passage cited above, the author of Marina's *Passion* explicitly refers to various body parts of the dragon, including its tongue. Thus, in this context, a metonymy for the tongue, namely for something that has already been explicitly mentioned a few lines earlier in the narrative, and is also mentioned by name later in the narrative when the dragon swallows the female martyr, would be completely unnecessary. Moreover, the tongue itself is not necessarily a threatening body part warranting comparison with a sword, unlike the phallus, which, as I suggested above, would be so to a sworn virgin.

72 Cf. e.g., P. Brown, *The Body and Society: Men, Women and Sexual Renunciation in Early Christianity*. Lectures on the History of Religions 13 (New York, 1988), esp. 9–17; Boswell, *Same-Sex Unions*, esp. 12, 58, 79, 154, 243–45.

73 The scene of the dragon and its symbolism can also be found in the medieval French version of the relevant *Passion* (twelfth century) written in verse, in which the protagonist appears with the name Margaret, see S. Kay, "The Sublime Body of the Martyr: Violence in Early Romance Saints' Lives," in *Violence in Medieval Society*, ed. R. Kaeuper (London, 2000), 3–20, esp. 15–16; eadem, *Courtly Contradictions: The Emergence of the Literary Object in the Twelfth Century*. Figurae: Reading Medieval Culture (Stanford, CA, 2001), esp. 227–28. Another case of a lustful dragon is provided by the late Byzantine romance *Kallimachos and Chrysorrhoe*, 84:415–94:584, ed. C. Cupane, *Romanzi cavallereschi bizantini: Callimaco e Crisorroe, Beltandro e Crisanza, Storia di Achille, Florio e Plaziaflore, Storia di Apollonio di Tiro, Favola consolatoria sulla Cattiva e la Buona Sorte*. Classici greci: Autori della tarda antichità e dell'età bizantina (Turin, 1995), 58–212, esp. 84–94. See also Koukoules, Βυζαντινῶν βίος, vol. 1.2, 252–53.

74 The text connects the concepts of stench and impurity to both of the protagonist's enemies, the dragon and the demon, see *Passion of Marina of Antioch*, 25:33; 26:12–14; 28:31 [136v]; 28:2 [135v]; 33:9–12, ed. Usener.

75 For the relationship between the sense of smell and sanctity or salvation, see, e.g., J.-P. Albert, *Odeurs de sainteté: la mythologie chrétienne des aromates*, 2nd ed., Recherches d'histoire et de sciences sociales 42 (Paris, 1996); S. Evans, "The Scent of a Martyr," *Numen* 49.2 (2001): 193–211; S. A. Harvey, *Scenting Salvation: Ancient Christianity and the Olfactory Imagination*. The Transformation of the Classical Heritage 42 (Berkeley, Los Angeles, and London, 2006).

76 Cf. E. Gamillscheg, "Die griechischen Texte über die heilige Marina" (PhD diss., Universität Wien, 1974), 30; Boulhol, "Hagiographie antique et démonologie," 267.

77 For the Virgin Mary as "the new Eve," see, e.g., G. Podskalsky, "Virgin Mary," ODB 3:2173–74; M. Evangelatou, "Krater of Nectar and Altar of the Bread of Life: The Theotokos as Provider of the Eucharist in Byzantine Culture," in *The Reception of the Virgin in Byzantium: Marian Narratives in Texts and Images*, ed. T. Arentzen and M. B. Cunningham (Cambridge and New York, 2019), 77–119, esp. 89, incl. n. 66 (with further references). On a comparison between the progenitor Eve and Mary the Mother of God as well as the latter's contribution to the salvation of humanity, see, e.g., A.-M. Talbot, "Women," in *The Byzantines*, ed. G. Cavallo, trans. T. Dunlap, T. L. Fagan, and C. Lambert (Chicago and London, 1997), 117–43, esp. 117, 142. Repr. in: A.-M. Talbot, ed., *Women and Religious Life in Byzantium*. Variorum Collected Studies 733 (Aldershot and Burlington, VT 2001), I, 117–43, esp. 117, 142; eadem, "Female Sanctity in Byzantium," in *Women and Religious Life*, ed. eadem, VI, 1–16, esp. 1.

78 M. Hinterberger, "Emotions in Byzantium," in *A Companion to Byzantium*, ed. L. James (Malden, MA, Oxford, and Chichester, West Sussex, 2010), 123–34, esp. 128, 131.

79 *Scholia on the Passion of Marina of Antioch by Methodios I* 50:12–15, ed. Usener, "Acta S. Marinae," 48–53, esp. 50. Cf. W. R. Larson, "The Role of Patronage and Audience in the Cults of Sts Margaret and Marina of Antioch," in *Gender and Holiness: Men, Women and Saints in Late Medieval Europe*, ed. S. J. E. Riches and S. Salih. Routledge Studies in Medieval Religion and Culture 1 (London and New York, 2002), 23–35, esp. 27. See also P. Brown, *The Making of Late Antiquity* (Cambridge, MA, 1978), 90: "The demonic stood not merely for all that was hostile *to* man; the demons summed up all that was anomalous and incomplete *in* man."

80 Potter, "Martyrdom as Spectacle," 66.

81 Patterson Ševčenko, "Eaten Alive," 122.
82 Note the comment by M. Kaplan: "Par définition, un saint martyr ne devient saint qu'au jour de sa mort," see M. Kaplan, "Le miracle est-il nécessaire au saint byzantin?" in *Miracle et karāma: hagiographies médiévales comparées* 2, ed. D. Aigle, Bibliothèque de l'École des Hautes Études, Sciences Religieuses 109 (Turnhout, 2000), 167–96, esp. 172. Cf. R. Bartlett, *Why Can the Dead Do Such Great Things? Saints and Worshippers from the Martyrs to the Reformation* (Princeton, NJ, 2013), esp. 3–7.
83 Beyond the martyrdom narratives which almost always refer to the execution of their protagonist(s), there are also various relevant miniatures in the illustrated *Menologia* of Symeon Metaphrastes and Basil II. On this matter, see N. Patterson Ševčenko, *Illustrated Manuscripts of the Metaphrastian Menologion.* Studies in Medieval Manuscript Illumination. Chicago Visual Library Text-Fiche Series 54 (Chicago and London, 1990), 11–180, 187, 190; eadem, "Synaxaria and Menologia," in *A Companion to Byzantine Illustrated Manuscripts*, ed. V. Tsamakda, Brill's Companions to the Byzantine World 2 (Leiden and Boston, 2017), 319–27, esp. 320–21, 323; Papavarnavas, *Gefängnis als Schwellenraum*, 27.
84 *Passion of Sergios and Bakchos*, ch. 19, ed. I. van Den Gheyn, "Passio antiquior SS. Sergii et Bacchi graece nunc primum edita," *AB* 14 (1895): 375–95, esp. 389.
85 Trans. Boswell, *Same-Sex Unions*, 385, modified.
86 Although the pre-Metaphrastic text does not clarify the intentions of the first group of animals gathering around the holy body, they should be understood as well-intentioned animals since, in the following lines, they appear to participate in the funeral. Symeon Metaphrastes' version clearly mentions the good intentions of the wild animals. Another relevant example is provided in the *Passion of Lukianos of Antioch* (*BHG* 997), according to which the martyr's dead body, which had been thrown into the sea on the order of the pagan persecutor, was found and transferred to the shore by a dolphin, see *Passion of Lukianos* by Symeon Metaphrastes, ch. 16, ed. J. Bidez and F. Winkelmann, *Philostorgius Kirchengeschichte mit dem Leben des Lucian von Antiochien und den Fragmenten eines arianischen Historiographen*, 2nd ed. Die griechischen christlichen Schriftsteller der ersten Jahrhunderte 21 (Berlin, 1972), 184–201, esp. 197.
87 *Passion of Sergios and Bakchos* by Symeon Metaphrastes, ch. 14, ed. J.-P. Migne, PG 115 (Paris, 1899), 1005–32, esp. 1024.
88 *Passion of Sergios and Bakchos*, ch. 27, ed. van Den Gheyn, 393.
89 Trans. Boswell, *Same-Sex Unions*, 385, modified.
90 *Passion of Sergios and Bakchos* by Symeon Metaphrastes, ch. 19, ed. Migne, PG 115, 1028.
91 *Passion of Sergios and Bakchos*, ch. 27, ed. van Den Gheyn, 393. Trans. Boswell, *Same-Sex Unions*, 388.
92 Ibid., ch. 30, ed. van Den Gheyn, 395.
93 Trans. Boswell, *Same-Sex Unions*, 390, slightly modified.

6

ANIMALS IN BYZANTINE HISTORICAL WRITING

Stephanos Efthymiadis

For José
My greathearted little cat!

When, in his long account of the campaign of King Xerxes against the Greeks, Herodotus inserted a report about the odd fact that lions attacked and devoured only the camels of the Persian army, "seizing nothing else, [neither] man or beast of burden", he set the tone for analogous intrusions of animals into other narratives of Greek historiography or, rather, historiography in Greek. As the Father of History records in Book Seven of his *History*, many lions were to be seen in the area bordered by the two rivers, Nestus in Macedonia and Achelous in Western Greece. Herodotus wondered "what was the reason that constrained the lions to touch nothing else but attack only the camels, creatures whereof till then they had no sight or knowledge" (VII,125–126).[1] Had he not posed this precise question, the entire animal digression would have been just one more exciting story among many – good to entertain his listeners with an intermission in his account of the Persian expedition. In other words, even though the story of the lions as dainty eaters might not be entirely relied upon, it could nevertheless give this particular chapter a symbolic meaning and effect. Was this idiosyncratic behavior a sign prefiguring the ultimate defeat of the Persians by the brave Greeks? As animals lent themselves to a variety of symbolic personifications and interpretations, did the native lions' exclusive taste for Asian camels function as another omen pointing to the malign developments that would affect the Persians in the first decades of the fifth century BCE?

The genres of Byzantine historical writing and the figurative use of animals

Herodotus's attraction to mythography and fondness for paradox found no echo in either the rationalist Thucydides or the realist Xenophon. But reports of animals coming into contact with humans are common among later Roman and Byzantine historians. As a literary genre focused on recording the business of warfare and politics, classical and classicizing historiography developed rather a strong distaste for the realm of 'everyday life' and its

DOI: 10.4324/9781003055877-9

anonymous heroes and other 'ordinary' living beings, turning their attention instead to emperors, kings, and high-ranking officials. Byzantine historiography did not deviate from this elitist narrative orientation and, as a rule, it would foster interest chiefly in animals in the context of warfare. Populist distractions, including animal related stories, were rather left for world chronicles or works of a similar kind that covered large periods of human history. In revisiting biblical history, these texts – the second group of historiographical writing in Byzantium – turned to animals as God's creatures and beings that interacted with humans from the time of creation onwards; also, in their coverage of later history, recent or otherwise, these authors show a keen and widespread interest in episodes and expressions featuring animals. In complete contrast, by virtue of its principal interest in persons and events that shaped the history of the Church from the fourth to the sixth century, ecclesiastical history, the third but short-lived form of historical writing in Byzantium, paid very little attention to the animal world.

To be sure, we should not presume that, by virtue of their primary thematic orientation (warfare, politics, and Church affairs), any of the three types of Byzantine historical writing was impervious to the world of animals with its wide range of literary uses, dramatic, figurative, or proverbial, let alone its potential in terms of characterizing the *dramatis personae*. Putting the multitude of Byzantine historical accounts together and following their course from the Church historians of the fifth century to the 'secular' Greek historiographers and chroniclers who recorded the Fall of Constantinople in 1453, their student will come up with a rich number and variety of references to the animal world. Furthermore, some authors can be singled out as providing multiple references to animals, in most cases fully aware of the range of meanings inherent in animal life. One of them, even though he is not the most typical representative of the historiographical genre, was Constantine Manasses (twelfth c.). His world history in verse, noted for its baroque imagery, integrates a surprisingly rich catalog of animals, some of them paraded one after another in quick succession, to the extent that we may call his account 'a zoological park'.[2]

To begin with, animals are part and parcel of fables, parables, metaphors, allegories, and the rest of the gamut of figurative language and speech.[3] Historiographical narrative is teeming with references to all kinds of animals that, in many cases, cannot be taken as real creatures, but their names and characters must be read in a non-literal way. For instance, the introductory chapters to book I of Prokopios's *History of the Wars*, devoted to the Persian kings of the fifth century responsible for the decay of their empire, contain two fables that can puzzle readers. The first is about a lion lured into a trap by some hunters using a tethered goat (I,3.13);[4] the second tells the story of a "swimming pearl oyster that was persistently pursued by a shark and caught by a fisherman at the cost of his life" (I,4.18–31).[5] The reasons why these fanciful vignettes are embedded in the narrative at particular points or how they tie in with the whole account is open to interpretation.[6] At any rate, they fulfill certain basic goals of fables such as to express thoughts in a concealed way and/or prefigure an upcoming development on the battlefront or in the political arena. In another passage (VIII,19.9–14), Prokopios allusively criticizes Justinian by making Sardil, king of the Utigurs, through the mouths of his envoys addressing the Byzantine emperor, distinguish between dogs that are grateful to their masters and wolves that are able to hide their beastly character. The Utigur leader thus built an argument designed to elicit the sympathy of Justinian and discourage him from supporting the Kutrigurs, the Utigurs' enemies.[7]

Animal-related proverbs in Byzantine historiography

Whether as an echo of some biblical story, a reminiscence of Aesop, or evoking some other literary source, using the animal kingdom in proverbial constructions was the order of the day. Naturally, this assumed a stereotypical image for each animal, deriving, on the one hand, from the distinction between domesticated and wild fauna, and extending, on the other, to their moral qualities as perceived by humans. As a genre chiefly interested in warfare and politics, historiography did not dabble in small animals but showed a natural preference for the more ferocious ones such as the lion, the wolf, the fox, or the eagle.[8] Thus, wolves were routinely contrasted with sheep and, to a lesser extent, eagles with sparrows, suggesting that some groups of people were easy prey to a violent enemy; or, more rarely, that ruthless characters could play the innocent or simulate virtue to conceal their vicious nature.[9]

A similar opposition between ferocious and vulnerable is found in the quote by the acclaimed Athenian general Chabrias: "*An army of deer commanded by a lion is more to be feared than an army of lions commanded by a deer.*" The emperor-historiographer John Kantakouzenos (r. 1347–54) cites it as pronounced by the *basileus*, i.e., himself, in a long speech addressed to a certain Synadenos, envoy of the aristocratic statesman Alexios Apokaukos in 1343 (III,59).[10] The same saying, but more succinctly phrased, was previously cited in the *Continuation of the Chronicle of John Skylitzes*, a work dating from the first decade of the twelfth century.[11] More interestingly, the tenth-century historiographer Leo the Deacon has his central hero, Emperor Nikephoros Phokas (r. 963–969), addressing his troops at the siege of Antioch in 969 using a proverb that ridicules the sheep dogs that had been charged with scaring off wolves, but that ended up tearing the flock apart themselves. His recourse to this proverbial fable, first found in Plato's *Republic* (416a), was prompted by Nikephoros's concern to prevent his exhausted army from ravaging and damaging what was thought to be a Roman country.[12] The historian Niketas Choniates (ca. 1155–1217) in turn evokes a daring image when he says that the Emperor Manuel I Komnenos's futile attempt to claim back from the Cilician Armenians the castles once taken by his father were as if a shepherd put his hand down a wolf's throat to remove a piece of liver (IV,2.2).[13]

What looked to be a common favorite proverb among such eloquent historians as Anna Komnene, Niketas Choniates, and Nikephoros Gregoras was the one contrasting or associating a lion's skin with a fox pelt.[14] As a metaphor, it was meant to criticize changes in behavior and/or instances of hypocrisy in *dramatis personae*, for example, in the case of the Emperor Theophilos (r. 829–42) who, when confronting the Graptoi brothers, defenders of icon-worship, was said to have been divested of his lion's skin and invested with a fox skin;[15] or that of the conspirator Nikephoros Diogenes who, according to Anna Komnene, "clothed himself in humility as if it were a fox skin, and then again shewed his courage like a lion" (*Alexiad*, IX,6.5);[16] or, the second thoughts of Manuel I Komnenos about the army of Conrad, the German crusader, whom, as Niketas Choniates puts it, the Byzantine emperor mistrusted lest he wear the fox skin together with the lion's skin; in similar vein, Andronikos I Komnenos, who, when he fell from power in 1185, was unable to recover the lion's skin, so put on the fox skin.[17] Then again, in his *Roman History*, the fourteenth-century historiographer Nikephoros Gregoras considered flattery incompatible with wisdom, just as Achilles would not like being Thersites and a lion would refuse to change his skin for a fox skin (XIX,3.5).[18]

Thus, although each animal proverb acquires its didactic and stereotyped meaning in historical discourse, it is not always clear whether an author is evoking a particular proverbial saying or is simply playing with a proverbial concept. "Trying to escape a wolf, they fell in with a lion" is, for instance, the venomous comment that Anna Komnene reserves for those of the Norman invader Bohemond's "Celts", who, in their efforts to avoid a clash with the army of the Byzantine general Kantakouzenos, fell into the eddies of a river and drowned (XIII,6.5);[19] and again Nikephoros Gregoras (1290/4–1358/61), reprimanding Emperor John VI Kantakouzenos as a supporter of his personal opponent Gregory Palamas during the council of 1351, claimed that he exchanged the fox for the bear skin, referring to his furious reaction once he felt that he had been beaten by Gregoras's anti-Palamite argument.[20] Earlier in his extensive *Roman History*, the same learned author likens Yathatini (Gr. Ἰαθατίνης), the Persian (i.e., Seljuk) sultan Kay-Kusraw I, to a lion who, despite losing nails and teeth in the fray, was always ready to attack bears and wolves (I,3).[21] Prior to him, in his *History* George Pachymeres (1242–*post* 1310) introduces the word ἀλωπεκισμός, a *hapax legomenon*, to designate the sly acts and behavior of Emperor Andronikos II's entourage when he visited Thessaloniki (X,3).[22]

Animal names calling for puns and characterization

Animal names, themselves borne by humans or names recalling animals, invited comparisons and characterizations. For instance, Leo, a name of Latin origin and the only animal-name given to male humans in Greek, lent itself to attributing an unpleasant role in political and religious affairs to those so named. For two of the six Byzantine emperors who go by this name, this figurative use comes as no surprise. For the ninth-century chronographer Theophanes the Confessor, Leo III was the loathsome emperor who introduced Iconoclasm and, as such, he was the "fitly named wild beast" whom the holy patriarch Germanos had to fight in ca. 730.[23] As another chronographer, the so-called *Scriptor Incertus*, has it, Leo V the Armenian, the emperor who restored Iconoclasm in 815 after betraying his former declaration of orthodoxy, was said to be nicknamed Chameleon, an animal commonly used as a paradigm for human fickleness.[24]

The iconophile chronographer George the Monk (or the Sinner) weaves a long and composite 'monstrous' metaphor referring to the same emperor as the abominable monster born of an Assyrian lioness and an Armenian leopard; on the one hand, being a lion by name, he possessed a ravaging nature, on the other, he was a chameleon in terms of the varying forms of his face and soul, due to the devious nature of his race.[25] More than a century later, the historian Joseph Genesios, though acknowledging Leo V's military prowess and skills as a politician, persists with the same metaphor, calling him "beastly Leo" (I,18: θηριότροπον Λέοντα), and ends by criticizing his obscure origins, saying that his parents gave birth to a wild beast (I,24: ἀνήμερον θῆρα).[26]

Names of other wild animals inspired wordplay and nicknames when they were borne by persons for whom negative sentiments were reserved. Again for Theophanes the Confessor, the Arian heretic Loukios was an obvious *Lykos*, i.e., a wolf;[27] and for Niketas Choniates, John Lagos, the head of police in Constantinople, "disappeared surpassing a hare in racing away" once he saw the mob assembled before the praetorium to catch him (XVI,11.2).[28] In the *vita Basilii*, the Life of the emperor Basil I (867–886) and the fifth book of the historical work known as *Theophanis Continuatus*, we hear of the patrician Imerios (a name sounding like a cognate of the adjective ἥμερος, tame), whom "the emperor himself [i.e.,

Michael III] called 'the Hog' [Χοῖρος] on account of the fierce expression of his counten-
ance, <although> in fact, he deserved such a designation more on account of his swinish
and sordid way of life".[29] Much later chroniclers derided the same emperor, Michael III, for
being a χοιρόβιος, i.e., someone living like a hog.[30] They also reserve the same unflattering
characterization for the Emperor Alexander (912–13) some forty years later, best known
for his short reign and inglorious death.[31]

Animal similes and comparisons pointing out moral behavior

In the tenth century, the illiterate emperor Michael II (r. 821–29), a man from a humble
background and of suspect origins in terms of religious beliefs, would be sarcastically
described as 'knowledgeable' only as regards how to behave toward animals and how
to discern their vigor, velocity, and breeding.[32] Such descriptions, intended as they were
to construct a character metaphorically, could revolve around a person as comparisons
(*parabolai*) without necessarily completely blackening their name.[33] In the *History* of Leo
the Deacon, the reluctant conspirator Marianos Argyros compares himself unfavorably
with the emperor Nikephoros Phokas when replying to Joseph Bringas, the instigator
of the plot: "Stop provoking and inciting a monkey to fight with a fully armed giant!"
(III,2).[34] In Michael Psellos's *Chronographia*, the spouse of Michael IV (1034–41), Zoe
Porphyrogennita, appears as having inspired fear as if she were some lioness (IV,17).[35]
According to the historiographer Anna Komnene, her father Alexios I marched against his
opponent and pretender to the imperial throne, Nikephoros Basilakios, "like a grim lion
against a long tusked boar" (I,7.3).[36]

For all their ferocity, such expressions, inserted as occasional similes and metaphors in
the narrative, are quite unlike the 'depersonalized' characterizations of iconoclast emperors
that are typical of earlier chronographers like Theophanes the Confessor and George the
Monk mentioned above. For George, the iconoclast Theophilos is a chameleon who "as
word has it, can easily become anyone and can take on any color apart from white".[37]
Conversely, for Anna Komnene, the metaphor that contrasts her father to his ambitious
opponent has a temporary, impressionistic effect, and betrays no intention to assign an identity to either of her subjects. We may assign such a change of approach to the attractions for
Byzantine historians after the tenth century of the post-classical historiographical model,
as chiefly represented by Polybios and Plutarch.[38] In line with this 'humanistic trend', as
Romilly Jenkins called it, humans would be portrayed with their lights and shades, that is,
as real persons, not as unidimensional, caricature figures, positive or negative.[39] Indeed, in
the biographical model of writing that historiographers adopted from the tenth century
onwards, animal imagery would exhaust its literary function in a light metaphorical association with persons and their actions. Comparisons of humans with animals would no
longer be interpreted as characterizations, i.e., as irrevocably showcasing the character of
an individual.

This change did not, however, diminish the application of animal imagery in Byzantine historiography of the following centuries. On the contrary, the ironic gaze with which Niketas
Choniates (late twelfth/ early thirteenth c.) looks at both people and political situations
looms large behind his attraction to animal imagery. Sometimes, this imagery employs a
vegetal alternative: in describing the rivalry developed between Maria Porphyrogennita, the
half-sister of the young emperor Alexios II Komnenos (r. 1080–83), and the *protosebastos*
Alexios (who was a regent for Alexios), Choniates has her insisting that the latter be ousted

from the palace where he had grown like a weed alongside the noble plant and was choking the emperor like wheat. However, he clung to the imperial chambers like an octopus that had clamped its suckers onto rocks once and for all (IX,3.4).[40] All in all, Choniates's use of metaphors and similes where animals (as well as plants!) appear is much more varied and creative than that of any of his predecessors, and it forms a forceful part of his rhetorical and ironic weaponry.[41] The expressive power of the animal metaphor was also explored by George Pachymeres (thirteenth and early fourteenth c.), in relation to an incident that occurred in 1265. After clashing verbally with Michael VIII Palaiologos in Hagia Sophia, Patriarch Arsenios left the church, flying away like an eagle, passing through the iron gate of the *triklinos* and leaving the emperor ashamed (IV.5).[42]

Animals as real and concrete beings in Byzantine historical narrative

No doubt, as in other genres of literature, though perhaps to a lesser extent, the metaphorical use of animals in Byzantine historical writing far overshadowed their mentions in episodes and scenes from real life.[43] Mention of wild animals as real creatures crop up in different contexts and milieus, whether in their natural environment or in an urban setting. A violent incident briefly narrated in the *Chronicon Paschale* (early seventh c.) reminds us of the wild aspect of stray dogs wandering around in a city like Constantinople. Such dogs ate the dead body of Komentiolos, patrician and *magister militum*, who, along with other high-ranking officials, was executed near the church of St. Conon as a supporter of the deposed emperor Maurikios after the usurper Phokas's coup in 602.[44] The appearance of a solitary wolf in a park in Constantinople, an episode recorded in the *vita Basilii* that we shall discuss later, must have been an even less welcome sight.

In Prokopios's *Wars* we hear of a wolf appearing before children minding flocks in the Italian countryside while the general Belisarios was campaigning, causing the children to flee, which was taken as a bad omen for the ultimate fate of the Samnites, a local people (V,20.2–4).[45] Further on in his account of the Byzantines' Italian campaign, Prokopios cites the report of a guard on the walls of Rome who thought that a wolf was lurking in an empty aqueduct. However, this proved to be a contrivance devised by the Goths who were besieging the city (VI,9.3–5).[46]

Quite unexpectedly, John Skylitzes (twelfth c.) attests to the presence of lions in the imperial palace, apparently kept for the sole purpose of perpetuating the Roman tradition of *damnatio ad bestias*, saying: "an imperial chamberlain condemned of having tried to poison the emperor was fed to them" (*Basileios and Konstantinos*, 45).[47] If Anna Komnene had not provided us with another mention of a lion kept in the palace, this violent incident would have been hardly credible. In a long excursus on astrology and astrologers, she states that the lion in question, though reported dead after suffering four days of fever, disproved the claims of those who said that its death would go along with the death of Alexios Komnenos (VI,7.5).[48] It is worth noting that the detail about the lion's illness might be interpreted as implying that someone was entrusted with the daunting task of taking care of the beast's health.

Starting with Theodosios II, who in 450 fell off a horse and died two days later,[49] and ending with John II Komnenos (1118–1143), who fell victim to an attack by a boar, several Roman emperors suffered a death caused by a wild animal. It is perhaps in such descriptions of violent incidents that we must seek the most realistic appearances of non-domesticated animals in the historiographical narrative. They do not refer just to Roman emperors but may also occur in relation to a foreign ruler or a simple soldier. The first Byzantine historian to give a lengthy and dramatic account of a fatal confrontation between a man and an animal is Agathias in his *Histories*, a work that covers the years 552–559. The victim is not a Roman, but Theudebert the Merovingian, killed while hunting in Thrace or Illyria in 547/548. He was attacked by a huge bull with gigantic horns. "Not one of the domesticated kind that draws the plow", Agathias adds, "but a creature of the woods and the mountains that deals death with its horns to its adversaries. I believe they are called 'buffaloes'" (I,4.5–6).[50]

Theophylaktos Simokattes pays similar attention to the accidents his main hero, the emperor Maurice, faced when leaving the city and abandoning the security of the imperial palace. Since an 'ominous message' is assigned to these accidents, they will be discussed further below. In later centuries, not all historians and chroniclers paid the same attention to these accidents that befell prominent humans, whether wild animals were involved or not. The death of Emperor Basil I in August 867, caused by an attack by a stag and narrated in great detail in the *Life of Patriarch Euthymios*,[51] is reduced to a simple oblique mention in the conclusion of his laudatory biography, the *Vita Basilii*, which says: "the emperor succumbed to a consumptive illness closely following upon gastric diarrhea, which in its turn owed its origin to some hunting accident" (ch. 102).[52] It is likewise summarily treated in the *Chronicles* of Symeon Magister and Pseudo-Symeon,[53] whereas Genesios records a different version of the event. We are told that, while out hunting, Basil threw his club, hitting the hind legs of a large deer and causing it to bolt out of the forest; he did the same on another occasion with a wolf. His unheroic death was the result of an attack of diarrhea, which had followed his falling from a horse while out hunting (IV,40 and 42).[54]

In keeping with the tendency of later historiographers to give detailed descriptions of imperial deaths,[55] John Kinnamos and Niketas Choniates relate the bad hunting experience John II Komnenos had in Cilicia in 1143 when a large boar rushed ferociously at him. In the encounter, the emperor scratched his arm on one of the (poisoned?) arrows he carried with him. His injury was not immediately lethal but led to further illness, and the two historians discuss at some length the ensuing but fruitless efforts of physicians to save the emperor's life.[56] Finally, a much shorter narration of a hunting accident, again due to a boar, is found in Doukas's *Chronography*. The victim was Mehmed I (1413–1421) and the accident came about while he was hunting on horseback outside Adrianople in Thrace. While trying to pierce the beast with his spear, the Ottoman ruler fell off his horse, probably because of a stroke. They took him to the palace, brought in the best physicians, and presented him before his army as if he was fit. Despite their acclamations, he was confined to bed thereafter, where he died the next day (XXII 8).[57]

Horses as commonplace animals

When it comes to literal rather than figurative usage, it is a domesticated animal that contributes the largest share of references to animals in Byzantine historiography and chronography alike.[58] In their multiple and varied occurrences in either of these genres, horses are the carriers of several symbols yet are almost never cited in metaphorical terms. An exception is found in Anna Komnene's *Alexiad*: the course of her historical narrative is likened by the learned princess to a galloping horse that has run off the highway as a result of her digressions. In implicitly assessing the epic character of her account Anna excuses herself and urges the horse of History to find its way back onto the main thoroughfare and return to its previous course, despite having been let loose (I,16.7).[59]

Naturally, the overwhelming majority of references to horses in the historiographical record are associated with their playing a crucial or at least an auxiliary role on the battlefield. Mentions of warhorses and of cavalry (heavy or light) occur in all works of historiography and chronography known from the Byzantine period. They are not even absent as martial animals from Michael Psellos's *Chronographia*, a work little, if at all, concerned with warfare.[60] All in all, the plethora of references to the horse as a domestic animal in the service chiefly of male humans reflect the diversity of its functions and uses in a premodern society. Authors such as Prokopios, Leo the Deacon, Anna Komnene, or Michael Critoboulos betray a predilection for descriptions of horses in relation to military strategy, being transported by boat, or lavishly ornamented when carrying emperors.[61]

A distinction, however, has to be made here. The increasing attention paid to horses, as noted in the historiography produced from the tenth century onwards, illustrates the priorities of Byzantine society and the aspirations of a rising aristocracy. Indeed, the animal so closely associated with martial arts and virtues is no longer treated in a dry, taken-for-granted sort of way but is portrayed in human terms and endowed with human sentiments. On the one hand, it became the key symbol of someone rising to and/or falling from power and, on the other, it lent glamor to any occasion.

By the same token, horses fit well with personal heroic action just as they contribute to distinguishing men for their martial and other skills. Thus, according to Leo the Deacon, an author who by and large promotes the image of the emperor-warrior and was keen on such descriptions, Emperor John Tzimiskes was able to throw the javelin, artfully draw and shoot the bow, and, best of all, leap over three riding horses in a row and land on the last one like a bird (VI.3).[62] "Vaulting nimbly onto a horse" was part of the military training to which Nikephoros Phokas subjected his servants and household retainers when he was on the throne of Constantinople (III.9).[63]

Fully aware of the multivalent imagery and symbolism of the horse, Michael Psellos in his *Chronographia* animates scenes of this type of rise to power, followed by social prominence, personal decline, and fall in which the horse becomes an essential element of the narrative or, to put it another way, contributes to sketching a person's character or supports the dramatization of narrated events. Interestingly, Psellos's positive treatment of Basil II extends to a portrayal of him standing on horseback. Although he was a rather short man, Basil II would as such afford an impressive sight. Moreover, "when he gave rein to his horse and rode in the assault, he was erect and firm in the saddle, riding uphill and downhill alike, and when he checked his steed, reining it in, he would leap on high as though he had wings, and he mounted or dismounted alike with equal grace" (I,36).[64]

Conversely, Romanos III, finding himself abandoned by his army on the frontline on one occasion in Syrian Antioch, would have risked being captured by the enemy "if someone had not helped him on to his horse, given him the reins, and counselled him to escape" (III,9).[65] The epilepsy of Romanos's successor, Michael IV, sometimes caused him to roll off the horse's saddle and be seen by the mob lying on the ground having a seizure (IV,18–19).[66] The zeal and obsession of his brother, John Orphanotrophos, a figure whom Psellos treats ambivalently, to have everything under his control also resulted in his suddenly riding off on his horse at nights and inspecting every corner of the city, passing through like a flash of lightning (IV,12).[67] By the same token, after dethroning Michael V in 1042, the Constantinopolitan mob dragged Zoe Porphyrogennita's sister, the nun Theodora, from her convent and, forming a circle around her, led her to Hagia Sophia, clothed in a magnificent robe and seated on a horse (V,39).[68] Significantly, these two elements, the one a symbol of secular status, the other a symbol of secular power, marked Theodora's passing from the nunnery to her imperial coronation.

The same episode with Theodora entering Hagia Sophia on horseback is similarly narrated by Psellos's younger contemporary, Michael Attaleiates (IV,7).[69] For such a 'functional' use of horses, both realistic and symbolic, his *History* is a case in point. As the historian and judge by profession states in his introductory speech addressing the emperor Nikephoros III Botaneiates, he embellished his narrative by extending it "with various signs that occurred unexpectedly"; in these, he includes some topics touching upon physical sciences and the appearances of animals as witnessed in those years.[70] Attaleiates's *History* covers the period 1034–1079/1080, the first twenty-three years (1034–1057) rather summarily.

It is specifically in this opening section that Attaleiates repeatedly highlights images from the civil life of Constantinople, whether this refers to emperors appearing in public or the populace's reaction and provocation of political turmoil. Seen as an intruder in the reigning Macedonian dynasty, Emperor Michael V is finally overthrown by the mob, blinded, and confined to a monastery. Formerly presented as triumphantly parading on horseback on the occasion of the Easter celebration (IV,4), he is later set on a wretched mule to be exposed to public ridicule (IV,9).[71] The rebel Leo Tornikios, a relative of Constantine XI Monomachos and another pretender to the throne, escaped from Constantinople riding a relay of horses and reached Adrianople in Thrace "as if of winged feet" (VI,2).[72]

In the same *History*, Michael Attaleiates transmits a personal experience that he had with his horse while accompanying emperor Romanos IV Diogenes (1067–1071) on his long, and finally unsuccessful, Anatolian campaign. The accident occurred on the narrow road across Mt. Tauros in Cilicia and was caused by the horse being afflicted with something called 'lykoenteritis'. Whereas the horse itself lurched uncontrollably and fell over the cliff, Attaleiates managed to remain safe, which he took to be a miracle (XVII,21).[73] By his recurring mentions of animals on public display, Attaleiates aimed at a theatrical reconstruction of characters and past events that would create a strong impression, if not even arouse some emotions, for the reader. Unlike historiographers of subsequent generations, such as the married couple Nikephoros Bryennios and Anna Komnene, horses in his narrative are not invested with any epic associations or role in the plot. If they are associated with a certain person, this is not to support the idea of chivalry or of the heroic warrior but to symbolically represent their fate, whether good or ill, on the stage of public action.

Naturally enough, in Anna Komnene's *Alexiad*, a work replete with episodes from the battlefield or other instances featuring horses and horsemen and which aims at resuscitating Homeric values through an epic representation of contemporary warriors, the obsession with horses comes as no surprise. Any brave act is part and parcel of a particular performance of that particular animal. Both warrior and horse share a common fate whether marching under the sun in a dry season (I,11.6)[74] or on the battlefield. In the lively description of the battle where her father and his adversary Nikephoros Bryennios, a skillful warrior, confront each other, and "both horses and men are sorely wounded" (πᾶς κατετέτρωτο ἵππος καὶ ἄνθρωπος), Anna paints the scene of Bryennios's capture in dramatic tones, by highlighting the elegant joust he performed while standing on horseback against two high-born Turks (I,6.6). One of them leaped off his horse and attacked him like a leopard whereas Bryennios, showing his martial prowess, "kept twisting round like an animal in his endeavors to stab him with his sword".[75] In a different, rather unheroic context, the plotters who managed to break into the city walls at Blachernai and enter the imperial stables were careful to disable the emperor's stabled horses by slitting their hind-legs from the flank down (II.6,1).[76]

We hear of other horses suffering a sorry fate because of serious sores in the *History* of John Kinnamos, who bears witness to a disease that, generally speaking, afflicts the hooves of the equine species. These were the horses of the army stationed in Attaleia in 1158.[77] Kinnamos was a historian keen on bringing into his prose scenes featuring horses either in a heroic or unheroic context that was usually meant to reflect a person's character. Manuel I Komnenos was so brave that he hunted bears and boars holding a spear and dismounted from his horse (III,17).[78] In a subsequent scene, Andronikos Komnenos, who wanted to overthrow his cousin, emperor Manuel I, took advantage of Manuel's night-long hunting expedition. He took the swiftest of his horses, but as he approached the emperor's tent, he asked his Isaurian followers to hold his horse, taking a mule with him. When he dismounted, he donned an Italian cloak in order not to be recognized. Andronikos was noticed by those surrounding the emperor and, in order to avoid troubles, he bent down pretending that he was to make an evacuation of his belly. In sum, the plot failed; what is of interest here is the gradual fall of Andronikos from the heroic horse to the unheroic mule, the ridiculous form of dress he assumed as a disguise and the comical situation in which he was finally involved (III,18).[79]

Interestingly, in a subtle play on manliness and gender, George Akropolites contrasts the army of Michael Palaiologos campaigning in the vicinity of Vodena in Macedonia in 1257 to that of a certain Manuel Lapardas, the commander of a rabble army, "most of whom were riding mares and had them loaded down with provisions". As Akropolites adds, when Michael Palaiologos's "brave soldiers, who rode stately horses and were clad in full armor, encountered manikins who were without arms, low-born, and riding mares, they defeated them all instantly" (ch. 71).[80]

Elephants and camels as unusual animals

Notably, in late antique historical writing horses feature alongside other animals used in war such as elephants and camels, whose exotic provenance accounts for the attention they receive.[81] Elephants in particular are mentioned in the context of military tactics, siege warfare, and other combat roles. Apart from their physical qualities, they had a psychological impact on the enemy, another good reason, especially for the Sassanian Persians, to deploy them in various military assignments.[82] In his *Wars*, Prokopios refers to eight such elephants joining the cavalry of the Persian general Mermeroes (Mihr-Mehroe) who campaigned in Lazica in 551: "upon them the Persians were to stand and shoot down on the heads of the enemy just as if from towers" (VIII,13.4).[83] Later on, during the Persian army's clash with the Romans, it so happened that a wounded elephant became so excited that it ran amok in the Persian ranks. Prokopios then proceeds to recall a similarly fierce reaction by an elephant during the siege of Edessa by Khusro in 544. The animal alone, "an engine for the capture of cities" (μηχανὴ ἑλέπολις), as Prokopios characterizes it, was on the verge of overpowering the men defending the city walls until they upset it by dangling a squealing pig from the tower and it retreated (VIII,14.32–37).[84]

In a similar vein, Agathias allots considerable space in his *Histories* to the description of a battle between the Romans and the Persians in the mid-550s where he highlights the terrifying action of an elephant in naturalistic detail. The bulky animal, struck with a spear above its brow, was enraged by the blow and, driven wild, turned his tusk around, thereby causing terror and confusion in the Persian camp (III,27).[85] Notably, not unlike the example in Prokopios, this is a rare case of 'personalizing' a war animal other than a horse and valorizing its action.

Simokattes, in turn, records the interesting case of an Indian elephant sent from Emperor Maurikios to the khagan of the Huns, i.e., the Avars, as a diplomatic gift.[86] The animal was described as "the most outstanding of the beasts bred by the Roman emperor", and, as such, it was arrogantly requested by the foreign ruler.[87] Nonetheless, once the khagan saw the elephant, whether from sheer terror or caprice, he commanded it be returned to the emperor, asking for it to be replaced by a gold couch that the Romans were to fashion themselves (I,3.8–13). Such a reconstruction of the event raises the question of whether the elephant was indeed bred by the Byzantines as the Avar khagan believed or was merely a recycled gift. At any rate, in later centuries, historiography did not lose interest in recording animals dispatched to foreign rulers as diplomatic gifts, yet these were no longer elephants but thoroughbred horses.[88]

Compared to elephants, the role of camels in warfare must have been more modest and discreet, i.e., less likely to play a decisive part in the issue of the battle. This might have seemed to Prokopios to underestimate their effectiveness. In *Wars* (III.8.27), he reports their 'apotropaic' presence in the ranks of the Moors in terms of their effect on the Vandals' horses; the latter, terrified at the sight of them, refused to engage in the battle.[89] In the context of the same Vandal war, the sixth-century historian records the terrible episode of camels being slaughtered en masse. The killing of two hundred camels by the Roman army in Libya was ordered by the eunuch Solomon, one of Belisarios's senior officers, and helped win the war against the Moors (IV,11.51–52).[90]

All in all, in the context of late antique historical writing, wherever they are present, camels are associated with peoples of the desert such as the Moors and Arabs. Camel cavalry had a role in desert warfare, yet camels accomplished other services too. They may appear as mere beasts of burden, supporting a military campaign or being integrated as exotic animals into the public life of the city. If a military triumph involved proudly parading on horseback, mounting a camel, a mule, or an ass, and being led through the streets of the city meant public derision. The record of people subjected to this humiliating process includes pagans and Christians alike. In Emperor Julian's time, pagans inflicted this penalty on the dead body of George, bishop of Alexandria, which they put on a camel and carried through the whole city.[91] In his *Wars*, Prokopios makes an extensive reference to a certain Christian Armenian by the name of Arsakes whom, though he was accused of treason, Justinian mildly punished with a flogging and being paraded on a camel's back (VII,32.1–3).[92] Prokopios's contemporary, the chronicler John Malalas, summarily reports that, following a Justinianic decree prohibiting gambling in any city, gamblers had their hands cut off and were paraded on camels (XVIII, 47).[93]

Not surprisingly, the military roles assigned to elephants and camels cease with the end of late antique historiography. In later historiographies, elephants acquire the function we associate them with today, i.e., as exotic, awesome animals attracting curiosity for their size, appearance, and behavior. As such, they were still worthy of being exhibited in the menagerie. By contrast, camels retained their 'supporting role' in the scenery of public derision, and the episode which epitomizes this pertains to none other than Andronikos I Komnenos and is narrated by Niketas Choniates in terms of brutal realism.[94]

Animals in ethnographic digressions: the imposing shade of the Herodotean tradition

Interest in ethnographic and anecdotal stories is a basic narrative element and practice that Byzantine historiographers and, to some extent, chronographers inherited from the Father of History.[95] In the sixth century, Prokopios, embedding in his narrative several anecdotes about animals, was the Byzantine historian par excellence who cherished ethnographic memories from Herodotus. His attention to the entertaining and didactic function of such stories is further revealed in that most of them are set and transferred in the exotic quarters of the enemy's court or some barbarian homeland. In a similar fashion, in his final book of the *Wars*, in order to explain the migration of a hyperborean people southwards, Prokopios recollects the story of a doe encountered by "some Cimmerian youths who were out hunting" (VIII.5.7–8). The doe leaped into the waters of a nearby river, and they gave chase, but once they caught up with her on the opposite bank, she disappeared from sight. Yet this, we are told, was taken by the barbarians as a clear sign that they could cross the river, i.e., the Danube, then fight and plunder. However, they ran into the Goths who slew many of them and turned the rest to flight. In conformity with the Herodotean tradition, Prokopios's use of animals was ironic and reserved exclusively for the 'other', the barbarian and the uncivilized.

Such a perspective explains why stories about and comparisons with animals occur in the ethnographic sections in late antique and later historiographers, which are usually inserted in the form of digressions from descriptions of warfare and events of political interest. For instance, in book IV (6.5–13) of *Wars*, Prokopios treats the soft life lived by the Vandals in a derogatory fashion and compares it unfavorably with the eating habits and other customs of the, in his idea, rather primitive Moors, concluding with the comment that the habits of the latter did not differ from those of animals. The underlying argument of this comparison is that, unlike the primitive Moors, the decadent Vandals, besieged as they were by the army of Belisarios, were unable to endure harsh conditions.[96]

This sort of distinction, emerging from comparisons of the *mores* of humans with those of animals, implicitly bears out the assumption of the cultural superiority of the Romans over their barbarian enemies, a perception shared with much later historiographers. The seminomad and pagan Pechenegs are typified by their beastly conduct, as Michael Psellos contends in his *Chronographia* (VII,67–70). These Mysians, as he calls them, when they run short of water, quench their thirst by drinking the blood of their horses. After that they will light a fire, and having lightly cooked the horses' amputated limbs, they devour their meat, blood and all.[97]

Niketas Choniates in his *History* implicitly likens the Serbs, despite them being Christians, to "herds of cattle grazing the green, waiting to be slaughtered". They were abandoned by their leader, who reached the neighboring mountain to find refuge and protection from the Byzantine army. Nonetheless, the *basileus*, i.e., Manuel I Komnenos, "like a lion trusting in its prowess, cut down the barbarian regiments as if they were herds of cattle or flocks of goats" (III,9.1).[98] Moreover, according to the same historiographer, Conrad III of Germany addressed his troops in 1148, criticizing the Byzantines for "rearing wolf-cubs [i.e., the Seljuk Turks] for themselves as sacrificial beasts and fattening them ingloriously on their own blood, when they should have been driving them from their lands and cities like beasts from their flocks" (II,8.6).[99]

At least two generations younger than Prokopios, Theophylaktos Simokattes shared his ethnographical interest in Greek historiography. He lived in the age of the emperor

Herakleios (610–641) but he chose to write the history of the reign of Maurice (582–602), the tragically murdered emperor who was overthrown by the tyrant Phokas. It seems that animal references intrigued Theophylaktos far more than his predecessors Prokopios and Agathias, hence their use in a variety of contexts in his work; more significantly, the way in which he used them was altogether more historical and less legendary. To begin with, for Simokattes animals and their treatment represent a cultural element, a kind of distinctive feature that can be used to counterpose Romans to Asians, Persians, and other foreign peoples. Thus, accused of high treason as the instigator of a rebellion, the Armenian Smbat was sentenced to be thrown to the beasts; but as the theater was full of people and the culprit was about to become easy prey, "upon the acclamations of the people the emperor displayed clemency"; Smbat "was separated from the beasts and reaped unforeseen salvation" (III,8.6–8).[100]

As a kind of contrasting sequel, where cruelty prevails over clemency, Simokattes goes on to relate the story of a prominent Persian by the name of Sarames, who was sent by the Persian king Hormisdas to capture the rebel Bahram and present him at once in disgrace at the palace. Yet, by a reversal of fate, it was Bahram who overpowered Sarames and handed him over for punishment by one of the largest elephants (III,8.9–12).[101] Taking delight in this particular method of execution, the same Persian usurper also had his other would-be assassins trampled by elephants. To underscore Bahram's cumulative cruelty, the scene was then described in naturalistic detail (IV,14).[102]

Asian barbarity and brutality, yet this time reflecting opulence and debauchery, is also implied in the detailed description of exotic animal dishes presented at the Persian royal table: "choice cuts of antelope, gazelle, and wild ass" (IV,7.2).[103] In similar vein, Theophanes the Confessor, in his *Chronographia*, details what Herakleios's army found in Dastagerd after they defeated the Persians of Khusro II. His palaces hosted "an infinite number of ostriches, gazelles, wild asses, peacocks, and pheasant, and in the hunting park, huge live lions and tigers".[104] He speaks in a derogatory fashion of the Arabs roaming around the Sea of Marmara in 716/717 for having eaten all their dead animals, namely horses, asses, and camels, even though he knew they were starving.[105] Oriental opulence was mirrored in the gold and precious stones with which the oxen drawing royal ladies' carriages were decorated in the far-off city of Taugast, somewhere in what is now Western China (VII,8.7).[106]

Animals for pleasure and entertainment

In his *Life of Pericles* (1.12) Plutarch quoted Julius Caesar saying that human love and affection should be granted to fellow human beings rather than the dogs and monkeys that he saw being carried as pets in the arms of some rich men in Rome. In Byzantine historiography, we get quite a few glimpses of men in the palace being closely attached to animals either for entertainment or for sentimental reasons. For instance, Honorius, emperor of the Roman West, considered a rooster he called Rome a pet. While he was in Ravenna, one of the palace's eunuchs, "evidently a keeper of the poultry", informed him that Rome had perished. The stupid emperor, as Prokopios contends, exclaimed "Yet he just ate from my hands!" Then, the eunuch, comprehending his words, said that it was the *city* of Rome that had been conquered by Alaric, only to hear the emperor say with a sigh of relief: "But I, my good fellow, thought that my fowl Rome had perished" (III,2.25–26).[107]

Centuries later, the eunuch patriarch of Constantinople, Theophylaktos (933–56), the second son of Emperor Romanos I Lekapenos, who hardly deserved a clerical office, developed a passion for acquiring horses. Apart from ordaining clerics to take care of his favorite animals, this also resulted in the construction of a stable close behind Hagia Sophia.[108] John Skylitzes, who provides us with the lengthiest account of the patriarch's passion, recounts an anecdote about an episode that occurred while Theophylaktos celebrated the office of the Thursday of the Holy Week. Delighted to hear from his deacon that his mare was foaling, he rushed out of the Great Church (i.e., Hagia Sophia) to reach Kosmidion (a region close to the Theodosian walls) to see the newborn foal.[109]

This same *Chronicle* of Skylitzes provides us with a rare attestation of a name given to an animal. In the context of the hand-to-hand fighting between the two usurpers to the throne, Bardas Skleros and Bardas Phokas, on the battlefield, we hear of Aegyptios, the former's horse, which escaped its handlers and ran away, thereby causing confusion in the ranks and having a negative effect on the outcome of the fight.[110] So, in addition to 'Rome', the name of the rooster owned by the late Roman Emperor Honorius, this Aegyptios, a name perhaps referring to the dark color or the provenance of the animal, provides a second instance of an animal identified by name in Byzantine historiography.

We glimpse another sentimental attachment to horses in the *Chronicle of George Sphrantzes*, which is by and large made up of his memoirs. Narrating the vicissitudes that befell him while in the service of Constantine Palaiologos in the Peloponnese, the chronicler turns his attention to this master's horse, wounded by an arrow shot by one of the inhabitants of Patras. Palaiologos was able to escape once he was disentangled from his horse, but the chronicler himself was wounded, as was his horse. Weakened to the point of collapse, it fell, unseating him. After stating that it had been a perfect horse (ἄλογον ἄριστον), Sphrantzes sets down all its changes of ownership before it was passed on to him by his own brother (ch. XVII).[111]

Animals contributed to the Byzantines' leisure pursuits in terms of hunting and other sporting activities.[112] In his *Chronographia*, Michael Psellos gives information in this respect about Emperor Isaakios I Komnenos (1057–1059), furthermore about Andronikos, brother of Emperor Michael Doukas, and about John Doukas, a member of the Constantinopolitan aristocracy (VII,72–73, VII,178 and VII,180–81). They were all passionately devoted to hunting animals and birds, displaying all the qualities an experienced hunter might have.[113]

Apart from supporting imperial hunting, horses stand out as animals deployed in the service of human entertainment and vanity, whether for racing in the hippodrome, processing through Constantinople on the occasion of a victory parade or taking part in the equestrian sport of *tzykanizein*, a ball game played on horseback in the precincts of the Great Palace and elsewhere. Several historians make mention of this sporting habit, but we owe its lengthiest description to Manuel I Komnenos's historian, John Kinnamos.[114]

It is in this context of animals used for pleasure or rather for satisfying the curiosity of, above all, the urban populace that we should interpret two digressions devoted to exotic animals brought into Constantinople in Byzantine historiographers.

In concluding the opening section of his *History*, Michael Attaleiates evaluates the character of Constantine IX Monomachos. Among this emperor's generous acts, appropriate to his imperial status, he reckons on the exhibition of an elephant and a giraffe brought from Egypt to Constantinople in 1053.[115] Attaleiates reports that the emperor's subjects took delight in such a spectacle and, by giving a detailed physical description of each animal introduced into his narrative, he delights his readers likewise. Couched as they are in rhetorical language, these descriptions give the impression that Attaleiates's inspiration came from literary sources rather than personal witness.

Judging by a number of earlier references, exotic animals had always held a fascination for the Byzantines and had been displayed in the Hippodrome or elsewhere. Yet Attaleiates's description, which is somewhat exceptional in terms of historiographical narrative, finds only one later echo and that is in the *History* of George Pachymeres, which relates the presence of a giraffe on the streets of Constantinople not long after the restoration of Byzantine rule in 1261. The exotic animal was a diplomatic gift from Baybars, the Mameluk sultan in Egypt, to Michael VIII Palaiologos (III,4).[116] Pachymeres gives the description of this "unusual and admirable" animal in the context of an ethnographic digression, which betrays his contempt for those he calls "Ethiopians".[117]

Animals as portentous creatures

It is chiefly wild animals that cast a shadow over the ups and downs of human history and provide auspicious or, more commonly, inauspicious signs regarding the fate of historically notable individuals.[118] It was not only Byzantine chronographers but also historiographers who indulged in integrating anecdotes about animals into their narratives, stories by and large mixing history and legend.

Theophylaktos Simokattes's *Ecumenical History* exemplifies the 'functional' and impressionistic use of animals as creatures of good or bad omen. Such use is reflected in the historian's critical attitude towards the earlier reign of Maurice (582–602) and, most probably, his concern about a turn for the worse at the end of the rule of Herakleios.[119] When his narration reaches certain dramatic events in the empire, the events that led to the murder of the emperor Maurice and his family, Theophylaktos scatters his account with omens of an apocalyptic kind. Interestingly, here, reality replaces animal imagery.

Much like his fifth- and sixth-century predecessors, Maurice was an emperor who never campaigned in person but rather resided permanently in Constantinople. When, however, at some point, the situation on the Balkan frontier proved critical, he decided to lead a military expedition against the Avars. We are told that in the ninth year of his reign (in 590), he went out "a distance of one and a half parasangs from the capital" and set up camp at Hebdomon. As the author goes on to say, not only did he witness an eclipse of the sun, but he was also struck by "violent gusts of wind, a fierce southerly, so that even the pebbles of the deep were virtually churned up by the turbulence of the swell". Forced to return to the capital, he spent the night in Hagia Sophia, but, since no vision came to him in his dreams, he visited the shrine in the monastery of the Source (Pege), which lay outside the walls of the capital. From there he set out again for the European hinterland of Constantinople to resume his initial plans. But, at some point, while marching with his escort early in the morning, he was attacked by a boar which spooked his horse. Even though Maurice was not unseated by this incident, Simokattes concludes: "the beast [the wild boar] was not attacked by anyone and in its irresistible might had an unpunished passage" (V,16.1–12).[120]

Nonetheless, accidents and mishaps with animals did not always end like that. During the same campaign, the most distinguished of the imperial horses collapsed and died. A herd of deer attacked the emperor and his attendants turned to hunting them, resulting in further confusion (VI,2.1–12).[121] Clearly, all these episodes were portents of the ill-fated

Maurice's being overthrown, which would come to pass soon thereafter. However, portents aside, they had a more general implication: confronted with nature, out of his urban comfort zone, even a man of power could become weak and vulnerable.

The case of Theophylaktos Simokattes is noteworthy not only because, in many instances, his way of doing things is at odds with the typology of animal imagery established in the classical historiographical tradition. It also points to a kind of innovation in that he brought a Christian dimension into his account, infusing it with an ethnographic element. In book VII, Simokattes gives a long description of his native Egypt and the Nile, which then witnessed the strange epiphany of anthropomorphic animals, constituting glaring examples of teratogenesis, i.e., birth of deformed creatures. This is the last omen inserted into the narrative, in this case to convey an entirely apocalyptic agenda (VII,16).[122] It is perhaps no accident that the portents of disaster reached Constantinople from a province which, as the end of Theophylakt's life approached, was lost to the insurgent Arabs, never to be reconquered.

When classicizing historiography was revived in the imperial court of Constantine VII Porphyrogennitos, after almost three centuries in eclipse, the interest in anecdotes of this sort was rekindled. In the biography of the emperor's grandfather Basil I, a man of humble origins who violently usurped the throne, the symbolic use of animals greatly contributes to building up a positive portrayal of him. In particular, wild animals cast a shadow over the beginning and, as we have seen, the end of Basil's life. A story about the king of birds standing over and shielding the baby Basil placed in a cradle by his mother while harvesting in Thrace brought about the legendary rehabilitation of the founder of the dynasty's name. The fact that Basil was repeatedly shielded by an eagle was a clear sign that the offspring of a family of poor peasants was destined to become a glorious emperor (ch. 5).[123]

Such a story, associating an eagle with an imperial legend, was not without precedent in Byzantine historical writing. Some four centuries earlier, Markianos, another emperor from a humble background, was said to have been given portents by Providence of his rise to imperial power through being shaded by an eagle. When captured by the Vandal Geizerich/ Genzerich in Africa, Markianos was shielded from the mid-day sun by an eagle, leading to his release on the condition that he signed a peace treaty with his former enemies. This episode is variously recorded in historiography and chronography with some versions placing the incident not in Libya, but in Lycia![124]

It seems that the eagle's association with the imperial throne was so deeply impressed on the collective unconscious that even its appearance in a dream could incur sanctions. During the reign of Tiberios III (698–705), the ambitious official Philippikos was exiled to the island of Kephalonia "for he claimed to have seen in a dream that his head was shadowed by an eagle".[125] Nevertheless, being shaded by eagles might not always be a good omen. In the *Continuation of Theophanes* (reign of Romanos I, ch. 16) and in the *Chronicle* of Symeon Logothete (ch. 136,37) we hear of two eagles flying over Emperor Romanos I Lakapenos and the Bulgarian tsar Symeon as they engaged in peace talks. The two birds screeched, clashed with one another, and flew away in opposite directions. The experts who could interpret the flight and the cries of birds saw this as a bad sign, pointing to the dissent that was to separate the two rulers.[126]

Another example of Basil I's rise to power being anticipated by some auspicious omen involving animals can be found in the *vita Basilii*. Engaged in early manhood in the service of the emperor Michael III, who had a passion for horses and hunting, Basil was able on one occasion to get hold of the emperor's runaway horse. The animal is described as arrogant and insubordinate, yet brave and beautiful and the scene is indicative of what its tamer would one day become (ch. 13). Symbolically enough, Basil managed to leap from his own horse onto the emperor's riderless horse and this sufficed for him to be chosen as the *protostrator* in charge of the imperial stables. It was in that capacity that he saved his patron's life when a wolf unexpectedly attacked him, this time within the city's walls, in the imperial gardens of Constantinople known as the *Philopation*. And it was on that occasion that Caesar Bardas ominously uttered the words "I think this man will prove the undoing of our whole race" (ch. 14).[127]

Signs and omens could also be provided by harmless, gentle animals. The final pages of the same *vita Basilii* (ch. 101) attest to the presence in the imperial palace of "a mimicking and talkative" parrot who kept repeating "ay, ay Sir Leo!". This elicited tears and distress from Basil's guests at a banquet, who took the bird's repeated plaintive squawks as a request to release the emperor's son Leo from custody. Basil revoked his former decision and restored Leo to his ranks and honors.[128] Michael Attaleiates's account of Emperor Romanos IV Diogenes (r. 1068–71) embarking on his fatal Anatolian campaign makes less light reading. While sailing across the Bosporus from Constantinople to Chalcedon, a dove, not completely white but chiefly dark, was noted to fly over the imperial ship. It came to rest right in the hands of the emperor, who sent it to the empress Eudokia Makrembolitissa who had chosen to remain in the palace. At the time it could not be determined whether this was an ill omen or a good one (XX,2).[129]

Animals also foretell the future in the *Chronike Diegesis* of Niketas Choniates, an author who managed to mix historiography with literature. Two episodes, both related to the most obscure and negative figure of the entire *History*, namely the tyrant-to-be Andronikos Komnenos, are worthy of attention. After lurking in the shadows, waiting to seize power throughout the reign of his cousin Manuel I Komnenos, Andronikos saw his chance when the child Alexios II ascended to the throne. One of the first victims in his step-by-step plan to get rid of all of his opponents was the ambitious *protosebastos* Alexios Komnenos; driven out of the church where he was imprisoned at dawn and forced to sit on a very small horse with a flag lodged in a piece of reed, he was put into a small boat and then blinded, an act which was approved by the senate, manipulated by then by the wily Andronikos (IX,5.12).[130] Although the humble ass of the Gospels is replaced by a small horse, there is every reason to believe that Niketas is somehow parodying here the scene of Christ's triumphal entry into Jerusalem. However, by introducing into the scene an animal situated somewhere between the glorious and the humble, he takes a rather ironic stance toward Alexios. Albeit a victim of the cruel Andronikos, the *protosebastos* was by no means to be seen as a martyr.

The second episode harks back to the omens seen in Theophylaktos Simokattes. The awe generated by the appearance of a comet in the sky above Constantinople is enhanced by a white falcon seen flying inside the church of Hagia Sophia.[131] As some men unsuccessfully tried to catch it, the bird flew off over a nearby palace where newly enthroned emperors used to receive acclamation from the populace, and then entered Hagia Sophia again. Once this flight back and forth had been repeated three times, the white falcon was apprehended and taken to the young emperor Alexios II Komnenos. Most people, Choniates says, took it as a sign that the white-haired Andronikos would be arrested shortly, but a few predicted that, before being arrested, he would rule over Constantinople for three years.[132]

Finally, a special category that should be considered in this chapter is the gnomic utterances of horses, i.e., when their neighing is interpreted as offering portents of the future. Prophetic *chremetismos* as a literary theme originates in the *Iliad* (17.426 and 19.405) where Achilles's horses, endowed with speech, foretell the death of their master. A clear Byzantine parallel, albeit in reverse, is found in the *Alexiad* in the description of the battle of Larissa, a victory for Anna's father. The astonishing fact that all the horses of the army were heard to neigh was interpreted as a good omen (V,5.7).[133] The neighing of the Arabian horse which Manuel I Komnenos rode as he entered the palace was likewise interpreted by the experts in such things as a good sign for his longevity, to which the historian Niketas Choniates did not give full credit (II,2).[134] Nevertheless, despite his skepticism, the same historiographer does not fail to cite more examples of hippomancy, namely cases featuring horses neighing that are taken as divinations as to the positive or negative turns in the lives of future emperors. The similarly rich crop of examples found in later Byzantium in the *Roman History* of Nikephoros Gregoras confirms the common belief that horses were well acquainted with the divine will.[135]

Conclusions

This survey of Byzantine historiography and chronography or, more accurately, of its chief exponents, has shown that animals regularly occupy a secondary role in narratives that are chiefly about humans or, rather, the human social elite. The picture of animals we get fails to represent animal life as such; we are only allowed glimpses through the looking glass of moral teaching and concessions made to anecdotal material by means of which the dry sequence of political and military events is embellished and enriched. Shaped by the Greco-Roman tradition of the anthropocentric selection of facts, Byzantine historiographers turned their attention to recording scenes with real animals when these filled in, crept into, or disrupted the human landscape.

Significantly, as writers began to depart from the classical models prevailing in late antiquity, the treatment and documentation of animals became more realistic, pluralistic, and gentle. The more historiography adopted a biographical model of narrative making humans predominate over events, the more information was imparted on the relation of humans with animals. There is no doubt that the place of animals in the narrative seems to acquire more significance and occupy more space by the eleventh and twelfth centuries when this 'biographical model of writing' prevailed.[136] Also, the intrusion of more and more persons and characters in the narrative that we observe in historians such as Anna Komnene, Niketas Choniates, and their peers of the Palaiologan period (1261–1453) played a role in that respect. Their characterization called for comparisons with the world of animals and their portrayal in heroic or unheroic terms involved anecdotal stories especially featuring martial animals.

All in all, in terms of selection and presentation of facts, the presence of animals was not accidental. Animals were indicators of an author's views, their reconstruction of the past, their priorities and underlying purposes; in short, what we call their 'system of values' today. Historiography, a major branch of literature since its classical beginnings, is a blend of reason and imagination, both features of the human mind. To integrate the animal world into such a narrative was to demonstrate an author's literary skills, their concern for enlarging the focus of their audience's expectations, and their ability to make explicit or implicit comparisons between 'civilized' humanity and wild nature. Despite critical and occasional divergences between individual historiographers, they all agreed on one thing: the treatment of animals required far more imagination than reason. Even when recorded in 'real life' circumstances, they were mostly interpreted in a figurative way, and almost never to be seen only as creatures of flesh and blood.

Notes

1 See W.W. How and J. Wells, *A Commentary on Herodotus* (Oxford, ⁵1957), 174. Note that Aristotle makes a similar report in his *Historia Animalium* (VIII,28).

2 See *The Chronicle of Constantine Manasses*, trans. and comm. L. Yuretich (Liverpool, 2018), 9–10.

3 On these aspects see the preliminary remarks of N. Koutrakou, " 'Animal Farm' in Byzantium? The Terminology of Animal Imagery in Middle Byzantine Politics and the eight 'Deadly Sins'," in *Ζώα και περιβάλλον στο Βυζάντιο*, ed. I. Anagnostakis, T. Kolias, and E. Papadopoulou (Athens, 2011), 319–21 (with references to previous bibliography).

4 Prokopios, *Wars*, ed. J. Haury, *Procopii Caesarensis Opera Omnia*, vol. 1, *De bellis libri I–IV* (Leipzig, 1905), and vol. 2, *De bellis libri V–VIII* (Leipzig, 1905); vol 1, 12–13. On this fable, see G.-J. van Dijk, "The Lion and the He-Goat: A New Fable in Procopius," *Hermes* 122 (1994): 376–79.

5 Prokopios, *Wars*, ed. Haury, vol. 1, 17–19.

6 For one such interpretation, see A. Kaldellis, *Procopius of Caesarea. Tyranny, History, and Philosophy at the End of Antiquity* (Philadelphia, 2004), 69–70 and 75–80.

7 Prokopios, *Wars*, ed. Haury, vol. 2, 587–88; tr. E.B. Dewing, rev. A. Kaldellis, *Prokopios, The Wars of Justinian* (Indianapolis and Cambridge, 2014), 504–5.

8 For the emblematic cases of the lion and the eagle, see T. Schmidt, *Politische Tierbildlichkeit in Byzanz. Spätes 11. bis frühes 13. Jahrhundert. Mainzer Veröffentlichungen zur Byzantinistik 16* (Wiesbaden, 2020), 75–163; also idem, "Father and Son like Eagle and Eaglet: Concepts of Animal Species and Human Families in Byzantine Court Oration (11th-12th c.)," *BZ* 112 (2019): 968–83.

9 E.g., Niketas Choniates, *Chronike Diegesis*, II,7.4, ed. L. van Dieten, *Nicetae Choniatae Historia. CFHB* 11.1 (Berlin and New York, 1975), 61; George Pachymeres, *History*, XI.16, ed. A. Failler, *Georges Pachymérès. Relations Historiques*, 5 vols. CFHB 34.1–5 (Paris, 1984–2000), vol. 4, 443:14; Nicephoros Gregoras, *Roman History*, ed. L. Schopen, *Nicephori Gregorae. Byzantina Historia*, 3 vols. CSHB (Bonn, 1829–55), vol. 2, 1018; Ioannes Kantakouzenos, *Historiae*, III, 27, ed. L. Schopen, *Ioannis Cantacuzeni eximperatoris historiarum … .*, 3 vols. CSHB (Bonn, 1828–32), vol. 2, 170; Doukas, *Chronographia*, chs. XXVI, 4, XXXIII, 11 and XL, 5, ed. D. R. Reinsch, *Dukas, Chronographia. Byzantiner und Osmanen im Kampf um die Macht und das Überleben (1341-1462)* (Berlin and Boston, 2020), 320, 410, 526; Laonikos Chalkokondyles, *Historiae Demonstrationes*, IX, 53, ed. E. Darkó (Budapest, 1922), 234.

10 Io. Kantakouzenos, *Historiae*, ed. Schopen, vol. 2, 365. On the central role that Kantakouzenos occupies in his historical narrative, see A. P. Kazhdan, "L'*Histoire* de Cantacuzène en tant qu'œuvre littéraire," *Byz* 50 (1980): 279–335.

11 *John Skylitzes continuatus*, ed. E. Th. Tsolakis, *Ἡ Συνέχεια τῆς Χρονογραφίας τοῦ Ἰωάννῃ Σκυλίτζῃ (Ioannes Scylitzes Continuatus)* (Thessaloniki, 1968), 128: "it is rightly said in an ancient maxim that it is better that a lion commands deer than a deer leads lions".

12 Cf. A. Kaldellis, *Streams of Gold, Rivers of Gold. The Rise and Fall of Byzantium, 955 A.D. to the First Crusade* (Oxford, 2017), 63–64.

13 Niketas Choniates, *Chronike Diegesis*, ed. van Dieten, 103.

14 Cf. the paroemiographers Zenobius 1.93; Apostolius 4,19a, etc. See *Corpus Paroemiographorum Graecorum*, ed. E. L. Leutsch and F. G. Schneidewin, 2 vols. (Göttingen, 1849–51).

15 John Skylitzes, *Synopsis Historiarum*, ch. 10, ed. H. Thurn. CFHB 5 (Berlin and New York, 1973), 61. In the relevant passage *Theophanis Continuatus* is using quite the opposite expression for Theophilos; see bk III,14, in *Chronographiae quae Theophanis Continuati nomine fertur libri I–IV*, ed. M. Featherstone and J. Signes Codoñer. CFHB 53.1 (Berlin, 2015), 150–51: "Theophilus threw off the forbearance he had feigned until then and revealed the beast."

16 Anna Komnene, *Alexias*, ed. Reinsch and Kambylis, 272.

17 Niketas Choniates, *Chronike Diegesis*, II, 7.4 and VII, 3.10, ed. van Dieten, 61 and 286.

18 See Nikephoros Gregoras, *Roman History*, ed. Schopen, vol. 2, 938. On Gregoras's 'menagerie', which is not as extensive as that of Niketas Choniates, see the remarks of A. Kazhdan (in collaboration with Simon Franklin), *Studies on Byzantine Literature of the Eleventh and Twelfth Centuries* (Cambridge, 2009), 268–73.

19 See Anna Komnene, *Alexias*, ed. D. R. Reinsch and A. Kambylis. CFHB 40.1 (Berlin and New York, 2001), 402. On this Kantakouzenos, see B. Skoulatos, *Les personnages byzantins de l'Alexiade* (Louvain la Neuve and Louvain, 1980), 49–52.

20 Nikephoros Gregoras, *Roman History* XVIII,6.5, ed. Schopen, vol. 2, 899.

21 Ibid., ed. Schopen, vol. 1, 19.

22 George Pachymeres, *History*, ed. Failler, vol. 4, 312–13. The word occurs only in the *De pace ecclesiastica*, a treatise by pro-Latin Patriarch John Beccos, an older contemporary of Pachymeres; see J. Darrouzès and V. Laurent (eds.), *Dossier grec de l'union de Lyon (1273–1277)*. Archives de l'Orient Chrétien 16 (Paris, 1976), 435.

23 Theophanes, *Chronicle*, ed. C. de Boor, *Theophanis Chronographia*, 2 vols. (Leipzig, 1883–85), vol. 1, 408, 20; tr. C. Mango and R. Scott, *The Chronicle of Theophanes Confessor: Byzantine Near Eastern History, 284–813* (Oxford and New York, 1997), 564.

24 See Fr. Iadevaia (ed.), *Scriptor Incertus* (Messina, 1997), 39, 53, 56.

25 See George the Monk, *Chronicle*, ed. C. de Boor, *Georgii Monachi Chronicon*, with corrections by P. Wirth, vol. 2 (Stuttgart, 1978), 781–82.

26 *Joseph Genesios*, ed. A. Lesmueller-Werner and I. Thurn, *Iosephi Genesii regum libri quattuor*, CFHB 14 (Berlin, 1978), 16–17 and 21.

27 Theophanes, *Chronographia*, ed. de Boor, 59,22–23; tr. Mango and Scott, 92.

28 See Niketas Choniates, *Chronike Diegesis*, ed. van Dieten, 525.

29 *Vita Basilii*, Ch. 27,17–20, ed. and tr. I. Ševčenko, *Chronographiae quae Theophanis Continuati nomine fertur liber quo vita Basilii imperatoris amplectitur*, CFHB 42 (Berlin, 2011), 104–5.

30 See Konstantinos Manasses, Σύνοψις χρονική, ed. O. Lampsidis, *Constantini Manassis Breviarium Chronicum* (Athens, 1996), v. 5002. Cf. V. Vlysidou, "Ὁ χοίρος ὡς σύμβολο ευδαιμονίας του βυζαντινού ανθρώπου," in Ζώα και περιβάλλον στο Βυζάντιο, ed. I. Anagnostakis, T. Kolias, and E. Papadopoulou (Athens, 2011), 42.

31 See Ioannes Zonaras, Ἐπιτομὴ Ἱστοριῶν, ed. T. Büttner-Wobst, *Ioannis Zonarae Epitomae historiarum*, CSHB (Bonn, 1897), vol. 3, 456–57.

32 See bk II,4 of *Theophanis Continuatus*, ed. Featherstone and Signes Codoñer, 68–69.

33 For a thorough discussion of animals in Byzantine abusive language with examples drawn from historiography and chronography, see P. Eliopoulos, "Τα ζώα στον προσβλητικό λόγο των Βυζαντινών," *Byzantina Symmeikta* 31 (2021): 51–120.

34 Leon Diakonos, *History*, ed. C. B. Hase, *Leonis Diaconi Caloensis Historia… CSHB* (Bonn, 1828), 37; tr. A.-M. Talbot and D. F. Sullivan, *The History of Leo the Deacon. Byzantine Military Expansion in the Tenth Century*. Dumbarton Oaks Studies XLI (Washington, D.C., 2005), 88.

35 Michael Psellos, *Chronographia*, ed. D. R. Reinsch, *Michaelis Pselli Chronographia*, Millennium-Studien 51 (Berlin, 2014), 58–59. Though keen on vegetal metaphors, Psellos has little interest in animal imagery: see A. Littlewood, "Imagery in the *Chronographia* of Michael Psellos," in *Reading Michael Psellos*, ed. C. Barber and D. Jenkins. The Medieval Mediterranean 61 (Leiden and Boston, 2006), 21–22.

36 See Anna Komnene, *Alexias*, ed. Reinsch and Kambylis, 28. On Basilakios, as Anna calls him, see Skoulatos, *Les personnages byzantins de l'Alexiade*, 35–39.

37 See George the Monk, *Chronicle*, ed. C. de Boor, 799.

38 On this development in Byzantine historiography, see the seminal study of R. Jenkins, "The Classical Background of the *Scriptores post Theophanem*," *DOP* 8 (1954): 13–30 (= idem, *Studies on Byzantine History of the 9th and 10th Centuries*. Variorum Reprints, London 1970, IV).

39 See ibid., 14–15.

40 See Niketas Choniates, *Chronike Diegesis*, ed. van Dieten, 233. Allusion is made here to *Odyssey* 5.432–434. The simile also turns up in I.5, ibid., 8. Another metaphor related to the octopus is encountered in XVIII,I.4, ibid., 566.

41 For an exhaustive list of his extensive collection of such imagery, see A. Littlewood, "Vegetal and Animal Imagery in the History of Niketas Choniates," in *Theatron. Rhetorische Kultur*

in Spätantike und Mittelalter/Rhetorical Culture in Late Antiquity and the Middle Ages, ed.
M. Grünbart. Millennium Studies 13 (Berlin, 2007), 223–58. As part of an analysis of Choniates's
literary identity, his animal imagery was first discussed by A. Kazhdan, *Studies on Byzantine
Literature of the Eleventh and Twelfth Centuries* (Cambridge, 1984), 256–86. A later analysis was
provided by L. Bossina, "La bestia e l'enigma: Tradizione classica e cristiana in Niceta Coniata,"
Medioevo Graeco 0 (2000): 35–68.

42 See George Pachymeres, *History*, ed. Failler, vol. 2, 343.

43 For the semiotics of animals, see also the chapter by T. Schmidt in this volume.

44 See *Chronicon Paschale*, ed. L. Dindorf, 2 vols. CSHB (Bonn, 1832), vol. 1, 694.

45 See Prokopios, *Wars*, ed. Haury, vol. 2, 101; tr., Dewing, rev. Kaldellis, *Prokopios, The Wars of
Justinian*, 298.

46 See Prokopios, *Wars*, ed. Haury, vol. 2, 189; tr. Dewing, rev. Kaldellis, *Prokopios, The Wars of
Justinian*, 337–38.

47 John Skylitzes, *Synopsis Historiarum*, ed. Thurn, 367; tr. J. Wortley, *John Skylitzes. A Synopsis of
Byzantine History, 811–1057* (Cambridge, 2010), 347. On this episode, see N. Ševčenko, "Wild
Animals in the Byzantine Park," in *Byzantine Garden Culture*, ed. A. Littlewood, H. Maguire, and
J. Wolschke-Bulmahn (Washington, D.C., 2002), 78–79.

48 See Anna Komnene, *Alexias*, ed. Reinsch and Kambylis, 182. On the whole passage relating to
astrologers, see P. Magdalino, *L'Orthodoxie des astrologues. La science entre le dogme et la divination à Byzance (VIIᵉ–XIVᵉ siècle)* (Paris, 2006), 96–107.

49 The fatal event is recorded in the *Chronicle* of John Malalas, XIV,27, ed. I. Thurn, *Ioannis Malalae
Chronographia*. CFHB 35 (Berlin and New York, 2000), 288 among other sources.

50 Agathias, *History*, ed. R. Keydell, *Agathiae Myrinaei Historiarum libri quinque* (Berlin, 1967),
14–15; tr. J. D. Frendo, *Agathias, The Histories* (Berlin and New York, 1975), 12–13.

51 *Life of Patriarch Euthymios*, in *Vita Euthymii Patriarchae CP.*, ed. P. Karlin-Hayter. Bibliothèque
de Byzantion 3 (Brussels, 1970), 2–5.

52 *Vita Basilii*, ed. and tr. Ševčenko, 334–35.

53 See respectively Symeon Magistros, *Chronicle*, ed. S. Wahlgren, *Symeonis Magistrae et Logothetae
Chronico*. CFHB 44.1 (Berlin and New York, 2006), 270 and ed. I. Bekker, *Theophanis
Continuatus Chronographia*. CSHB (Bonn, 1938), 699 (ch. 23).

54 *Joseph Genesios*, ed. Lesmueller-Werner and Thurn, 89–90; tr. A. Kaldellis, *Genesios, On the
Reigns of the Emperors*. Byzantina Australiensia 11 (Canberra, 1998), 111–112.

55 Cf. D. R. Reinsch, "Der Tod des Kaisers: Beobachtungen zu literarischen Darstellungen des
Sterbens byzantinischer Herrscher," *Rechtshistorisches Journal* 13 (1994): 247–70.

56 See John Kinnamos, *Epitome*, ed. A. Meineke, *Ioannis Cinnami Epitome rerum ab Ioanne et
Alexio Comnenis gestarum*. CSHB (Bonn, 1836), 24–29; tr. Ch. Brand, *Deeds of John and Manuel
Komnenos by John Kinnamos* (New York, 1976), 27–31. For Choniates, see Niketas Choniates,
Chronike Diegesis, ed. van Dieten, 40–41.

57 Ed. Reinsch, *Dukas, Chronographia*, 234–36.

58 For a survey, see T. Kolias, "The Horse in the Byzantine World," in *Le cheval dans les sociétés
antiques et médiévales. Actes des Journées internationales d'étude* (Strasbourg, 6–7 novembre
2009), ed. S. Lazaris. Bibliothèque de l'Antiquité tardive 22 (Turnhout, 2012), 87–97.

59 See Anna Komnene, *Alexias*, ed. Reinsch and Kambylis, 53.

60 For such war scenes in the *Chronographia*, see e.g., chs. III,9, VI,111 and VII,22, ed. Reinsch,
35–36, 156–57, and 216.

61 See Kolias, "The Horse in the Byzantine World," 90–91.

62 See Leon Diakonos, *History*, ed. Hase, 97; tr. Talbot, Sullivan, 146. Leo must have been inspired
here by the description of Ajax in the *Iliad* (15.679–684) portrayed as a man who could tie four
horses together and leap from one to another.

63 See Leon Diakonos, *History*, ed. Hase, 50–51; tr. Talbot, Sullivan, 100–1.

64 Michael Psellos, *Chronographia*, ed. Reinsch, 22–23, tr. E. R. A. Sewter, *Michael Psellus. Fourteen
Byzantine Rulers* (London, ²1966), 28.

65 Michael Psellos, *Chronographia*, ed. Reinsch, 36.

66 See Michael Psellos, *Chronographia*, ed. Reinsch, 59–60. On Michael IV's illness see G. Makris,
"Zur Epilepsie in Byzanz," *BZ* 88 (1995): 381–82.

67 See Michael Psellos, *Chronographia*, ed. Reinsch, 56. On this figure's portrayal by Psellos, see Ch. Messis, *Les eunuques à Byzance. Entre réalité et imaginaire*. Dossiers byzantins 14 (Paris, 2014), 282–86.

68 See ibid., 99–100.

69 Michael Attaleiates, *History*, ed. I. Pérez-Martín, *Miguel Ataliates. Historia*. Nueva Roma 15 (Madrid, 2002), 13; *Michaelis Attaliatae Historia*, ed. E. Th. Tsolakis. CFHB 50 (Athens, 2011), 13; tr. A. Kaldellis and D. Krallis, *Michael Attaleiates, The History*. DOML 16 (Cambridge, Mass. and London, 2012), 24–26.

70 Michael Attaleiates, *History*, ed. Pérez-Martín, 3; ed. Tsolakis, 4; tr., Kaldellis, Krallis, 6.

71 See Michael Attaleiates, *History*, ed. Pérez-Martín, *Historia*, 10–14; ed. Tsolakis, 10 and 13–14; tr. Kaldellis, Krallis, 18–19 and 26–27.

72 Michael Attaleiates, *History*, ed. Pérez-Martín, 18–19; ed. Tsolakis, 18–19; tr. Kaldellis, Krallis, 38–39.

73 See Michael Attaleiates, *History*, ed. Pérez-Martín, 91–92; ed. Tsolakis, 94–95; tr. Kaldellis, Krallis, 220. On this autobiographical reference see L.R. Cresci, "Categorie autobiografiche in storici bizantini," in *Categorie linguistiche e concettuali della storiografia bizantina*. Atti della V Giornata di studi bizantini. Napoli, 23–24 aprile 1998, ed. U. Criscuolo and R. Maisano (Naples, 2000), 132–34. For Attaleiates's participation in military campaigns see the chapter "The Judge on Horseback: The Empire at War," in D. Krallis, *Serving Byzantium's Emperors: The Courtly Life and Career of Michael Attaleiates* (Cham, 2019), 161–87.

74 See Anna Komnene, *Alexias*, ed. Reinsch and Kambylis, 38.

75 See ibid., 25–26.

76 See Anna Komnene, *Alexias*, ed. Reinsch and Kambylis, 69.

77 See John Kinnamos, *Epitome*, ed. Meineke, 179; tr., Brand, *Deeds of John*, 137.

78 See John Kinnamos, *Epitome*, ed. Meincke, 127; tr., Brand, *Deeds of John*, 100.

79 See John Kinnamos, *Epitome*, ed. Meineke, 129; tr., Brand, *Deeds of John*, 101–2.

80 George Akropolites, *History*, ed. A. Heisenberg, *Georgii Acropolitae Opera*, vol. 1, (Leipzig, 1903), 146–47; tr. R. Macrides, *George Acropolites. The History, Introduction, Translation, and Commentary* (Oxford, 2007), 330.

81 For animals employed in Greek-Roman warfare, see A. Mayor's chapter in *The Oxford Handbook of Animals in Classical Thought and Life*, ed. G. L. Campbell (Oxford, 2014), 282–93.

82 On the use of elephants especially by the Sassanids, see Ph. Rance, "Elephants in Warfare in Late Antiquity," in *Acta Antiqua Hungarica* 43.3–4 (2003): 355–84; and M. B. Charles, "The Rise of the Sassanian Elephant Corps: Elephants and the Later Roman Empire," *Iranica Antiqua* 42 (2007): 355–84.

83 See Prokopios, *Wars*, ed. Haury, vol. 2, 553; tr. Dewing, rev. Kaldellis, *Prokopios, The Wars of Justinian*, 490.

84 See Prokopios, *Wars*, ed. Haury, vol. 2, 562–63; tr. Dewing, rev. Kaldellis, *Prokopios, The Wars of Justinian*, 494–95.

85 See Agathias, *History*, ed. Keydell, 119–120; tr. Frendo, 97–98.

86 On this episode, see also the chapter by N. Drocourt in this volume, p. 317.

87 Theophylaktos Simokattes, *History*, ed. C. de Boor (Stuttgart, 1972), 45–46.

88 See N. Drocourt, "Les animaux comme cadeaux d'ambassade entre Byzance et ses voisins (VIIᵉXIIᵉ siecle)," in *Byzance et ses périphéries. Hommage à Alain Ducellier*, ed. B. Doumerc and Chr. Picard (Toulouse, 2004), 67–93; idem, "Le cheval animal diplomatique entre Byzance et l'Occident (IXᵉ-XIIIᵉ s.)," in *Byzance et l'Occident VI, Vestigia philologica*, ed. E. Egedi-Kóvacs (Budapest, 2021), 113–30; and his chapter in this volume.

89 See Prokopios, *Wars*, ed. Haury, vol. 1, 350.

90 See ibid., vol. 1, 469–70.

91 See Socrates Scholastikos, *Ecclesiastical History*, III,2.8, ed. G. Chr. Hansen, *Sokrates Kirchengeschichte* (Berlin, 1995), 194; and *Chronicon Paschale*, ed. Dindorf, 546.

92 See Prokopios, *Wars*, ed. Haury, vol. 2, 433–34.

93 See John Malalas, *Chronicle*, ed. Thurn, 379.

94 See Niketas Choniates, *Chronike Diegesis*, ed. van Dieten, 349–51. Discussion of this episode by P. Magdalino, "Tourner en derision à Byzance," in *La dérision au Moyen Âge. De la pratique sociale au rituel politique*, ed. É. Crouzet-Pavan and J. Verger (Paris, 2007), 64–67; and, more generally, M. Perisanidi, "Byzantine Parades of Infamy through an Animal Lens," *History Workshop Journal* 90 (2020): 1–25.

95 See generally A. Kaldellis, *Ethnography after Antiquity. Foreign Lands and Peoples in Byzantine Literature* (Philadelphia, 2013); and Schmidt, *Politische Tierbildlichkeit in Byzanz*, 63–71.

96 On this episode see Kaldellis, *Ethnography after Antiquity*, 19–20.

97 See Michael Psellos, *Chronographia*, ed. Reinsch, 241.

98 Niketas Choniates, *Chronike Diegesis*, ed. van Dieten, 90; tr. H.J. Magoulias, *O City of Byzantium. Annals of Niketas Choniates* (Detroit, 1984), 53.

99 See Niketas Choniates, *Chronike Diegesis*, ed. van Dieten, 70; commentary and tr. by Littlewood, "Vegetal and Animal Imagery," 238. More generally, for this comparison of Turks with lions and wolves see A. Papageorgiou, "Οἱ δὲ λύκοι ὡς Πέρσαι: the image of the Turks in the reign of John II Komnenos (1118–1143)," in *BSl* 69 (2011): 149–61; and Schmidt, *Politische Tierbildlichkeit in Byzanz*, 312–29.

100 Theophylaktos Simokattes, *History*, ed. de Boor, 125–26; tr. Michael and Mary Whitby, *The History of Theophylact Simocatta. An English Translation with Introduction and Notes* (Oxford, ²1997), 84.

101 See Theophylaktos Simokattes, *History*, ed. de Boor, 127; tr. M. and M. Whitby, 84.

102 See Theophylaktos Simokattes, *History*, ed. de Boor, 180–181; tr. M. and M. Whitby, 125.

103 See Theophylaktos Simokattes, *History*, ed. de Boor, 162; tr. M. and M. Whitby, 147.

104 Theophanes, *Chronographia*, ed. de Boor, 322,11–14; tr. Mango, Scott, 451.

105 See Theophanes, *Chronographia*, ed. de Boor, 397, 23–25; tr. Mango, Scott, 546.

106 See Theophylaktos Simokattes, *History*, ed. de Boor, 261; tr. M. and M. Whitby, 227.

107 See Prokopios, *Wars*, ed. Haury, vol. 1, 314–15; tr. Dewing, rev. Kaldellis, *Prokopios, The Wars of Justinian*, 147.

108 See *Theophanis Continuatus Chronographia*, ed. I. Bekker, 444 and 449.

109 See John Skylitzes, *Synopsis Historiarum*, ed. Thurn, 242–43; tr. J. Wortley, 234–35.

110 See John Skylitzes, *Synopsis Historiarum*, ed. Thurn, 326–27; tr. J. Wortley, 310.

111 Georgios Sphrantzes, *Chronicle*, ed. R. Maisano, *Giorgio Franze. Cronaca.* CFHB 29 (Rome, 1990), 44–47.

112 Cf. Ch. Messis and I. Nilsson's chapter in this volume (with earlier bibliography); and Schmidt, *Politische Tierbildlichkeit in Byzanz*, 196–238.

113 See Michael Psellos, *Chronographia*, ed. Reinsch, 243–44, 292, and 294–95. On these personages and their keen interest in hunting see Messis and Nilsson in this volume.

114 See John Kinnamos, *Epitome*, ed. Meineke, 263–64; tr. in Brand, *Deeds of John*, 198. Other references in Leon Diakonos, *History*, VI,3, ed. Hase, 97; tr. Talbot, Sullivan, 146–47 (about John Tzimiskes's skill at this game); Anna Komnene, *Alexias*, IX,6.5 and XIV,4.2, ed. Reinsch and Kambylis, 273–74 and 439.

115 See Michael Attaleiates, *History*, ed. Pérez-Martín, 37–38; ed. Tsolakis, 38–40; tr. Kaldellis, Krallis, 86–89. The event is also referred to by John Skylitzes, ed. Thurn, 475, and Michael Glykas, *Annales*, ed. I. Bekker (Bonn, 1836), 597. For other sources, see Ševčenko, "Wild Animals in the Byzantine Park," 77–78. On the ethnographic interest of the passage see Kaldellis, *Ethnography*, 99. See also Drocourt's and Schmidt's chapters in this volume.

116 See George Pachymeres, *History*, ed. Failler, vol. 1, 238–41.

117 On this ethnographic digression, see A. K. Petrides, "Georgios Pachymeres between Ethnography and Narrative: Συγγραφικαὶ Ἱστορίαι 3.3–5," *GRBS* 49 (2009): 295–318.

118 For animals as related to divination in ancient times see P. Struck's ch. in *The Oxford Handbook of Animals in Classical Thought and Life*, 310–23.

119 On the question of the implicit targets of his *History*, see S. Efthymiadis, "A Historian and His Tragic Hero: A Literary Reading of Theophylaktos Simokattes' *Ecumenical History*," in *Byzantine History as Literature*, ed. R. Macrides (Farnham and Burlington, 2010), 167–83.

120 Theophylaktos Simokattes, *History*, ed. de Boor, 218–20; tr. M. and M. Whitby 156–57.

121 See Theophylaktos Simokattes, *History*, ed. de Boor, 222.

122 See ibid., 273–74. These apparitions are also recorded at some length and in much more terrifying terms in Theophanes's *Chronographia*, ed. de Boor, 280–81 and the *Chronicle* of George the Monk, ed. de Boor, 656–58. For a discussion of this scene, see Ch. Messis, "L'impureté corporelle suprême: la monstruosité à Byzance, ses perceptions et ses élaborations littéraires," in *Il corpo impuro e le sue rappresentazioni nelle letterature medievali*, ed. Fr. Mosetti Casaretto (Alessandria, 2008), 178–80.

123 See *Vita Basilii*, ed. and tr. Ševčenko, 23–26. Before it began to receive widespread scholarly attention, the episode was first discussed at length by G. Moravscik, "Sagen und Legenden über Kaiser Basileios I," *DOP* 15 (1961): 83–88.

124 See Prokopios, *Wars*, III, 4, ed. Haury, vol. 1, 325; Evagrius, *Ecclesiastical History*, II,1, in *Histoire ecclésiastique* ed. J. Bidez and L. Parmentier (London, 1898), 37–38; Symeon Magistros, *Chronicle*, ch. 98, ed. Wahlgren, 129. The original source used by all must have been Priskos of Panium, see P. Allen, *Evagrius Scholasticus. The Church Historian*, Spicilegium Sacrum Lovaniense, 41 (Leuven, 1981), 96. For the 'Lycian' version see Theophanes, *Chronographia*, ed. de Boor, 104 and Zonaras, Ἐπιτομὴ Ἱστοριῶν, XIII, 24, 4–11, ed. Büttner-Wobst, vol. 3, 113–14.

125 See Theophanes, *Chronographia*, ed. de Boor, 372; tr. Mango, Scott, 519. Also, the *Chronicle* of Symeon Magistros/the Logothete, ch. 116, ed. Wahlgren, 174.

126 See Symeon Magistros, *Chronicle*, ed. Bekker, 409 and ed. Wahlgren, 324.

127 See *Vita Basilii*, ed. and tr. Ševčenko, 50–56. For an analysis of this passage see Ch. Messis, "Est-elle possible une lecture subversive de la *Vie de Basile*? Stratégies narratives et objectifs politiques dans la cour de Constantin VII Porphyrogénète," in *Storytelling in Byzantium: Narratological Approaches to Byzantine Texts and Images*, ed. C. Messis, M. Mullett, and I. Nilsson (Uppsala, 2018), 201–22. On the Philopation, see Ševčenko, "Wild Animals in the Byzantine Park," 69–71; and H. Maguire, "The Philopation as a setting for imperial ceremonial and display," *BF* 30 (2011): 71–82.

128 *Vita Basilii*, ed. and tr. Ševčenko, 331–32. On this story, see M. Leontsini, "Οικόσιτα, ωδικά και εξωτικά πτηνά: Αισθητική πρόσληψη και χρηστικές όψεις (7ος–11ος αι.)," in *Ζώα και περιβάλλον στο Βυζάντιο*, ed. I. Anagnostakis, T. Kolias, and E. Papadopoulou (Athens, 2011), 295–97.

129 See Michael Attaleiates, *History*, ed. Pérez-Martín, 107; ed. Tsolakis, 111; tr. Kaldellis, Krallis, 260–63.

130 See Niketas Choniates, *Chronike Diegesis*, ed. van Dieten, 249.

131 See also T. Schmidt's chapter in this volume.

132 See Niketas Choniates, *Chronike Diegesis*, ed. van Dieten, 251–52.

133 See Anna Komnene, *Alexias*, ed. Reinsch and Kambylis, 156–57.

134 See Niketas Choniates, *Chronike Diegesis*, ed. van Dieten, 51.

135 Evidence collected and analyzed by S. Costanza, "Nitriti come segni profetici. Cavalli fatidizi a Bisanzio (XI-XIV sec.)," *BZ* 102 (2009): 1–24; also idem, "Wiehernde Pferde und westlicher Einfluss auf die Divination der Komnenen- und Palaiologenzeit," in *Byzanz und das Abendland V*, ed. E. Juhász. Studia Byzantino-Occidentalia (= Antiquitas, Byzantium, Renascentia 32) (Budapest, 2018), 99–113.

136 Cf. the analysis provided by Schmidt, *Politische Tierbildlichkeit in Byzanz*, based on a wider spectrum of literature.

7
ANIMALS IN LEGAL SOURCES

Johannes Koder

Special laws on agriculture, animal livestock breeding (and generally on animal related issues) did not exist in premodern times, with few exceptions. In the Byzantine legislation these issues are scattered throughout general bodies of laws, and in no Byzantine legislative text does there exist a systematic overview of animal-related regulations. The specific bibliography reflects these assumptions.[1]

The sources

Between the 4th and the 15th centuries more than twenty legislative sources contain information about animals. Of interest are: *Edictum Diocletiani* (301), Tarif of Anazarbos (5th/6th centuries), *Novellae et Chrysobulla Imperatorum post Justinianum* (after 565), *Nomokanōn* (6th/7th centuries), *Ecloga* (circa 741), *Nomos Geōrgikos* (probably after the *Ecloga*), *Nomos Mōsaïkos* (mid-8th century), *Ecloga Privata Aucta* (8th–12th centuries), *Eisagōgē* ("*Epanagōgē*", 886), *Basilika* (circa 888, often with references to the Justinianic *Codex*, *Digest* and *Novellae*), *Eparchikon Biblion* (circa 900), Leōn VI, *Novellae* (circa 900), Leōn VI, *Tactica* (circa 900), *Procheiron* (907), *Eisagōgē Aucta* (circa 920/930), *Procheiron Auctum* (circa 1330), *Procheiron Calabriae* (9th–14th centuries), *Peira* (10th/11th centuries), *Tipukeitos* (12th century), Harmenopoulos, *Hexabiblos* (in particular *App.* 3, 1345), Harmenopoulos, *Hexabiblos aucta* (late 14th century), and Harmenopoulos, *Hexabiblos Aucta* (after 14th century).[2]

In this chapter references to *Procheiron, Eisagōgē Aucta, Procheiron Auctum, Peira, Tipukeitos* and Harmenopoulos, *Hexabiblos* are restricted to those paragraphs that do not exist already in their basic source text *Basilika*, except when they differ from them. *Scholia Basilicorum, Synopsis maior* and *Synopsis minor* are not cited either, because their content is identical with the *Basilika*. Reference to *Hexabiblos*, App. 3, is only made if it did not adopt the respective paragraph from the *Nomos Geōrgikos*. Finally, *charters* for monasteries are not cited, as most of them refer to specific cases. Hence, they are not legal sources in the proper sense; where appropriate, references to the LBG for further evidence in the sources are made.[3]

DOI: 10.4324/9781003055877-10

In addition to the legal texts, a small selection of other texts is mentioned paradigmatically, in order to better understand the importance of livestock breeding in Late Antiquity and the Middle Ages, be it in a nomadic or an agricultural environment:[4] Hēsychios (5th century?), Physiologus (5th/6th centuries), *Geōponika* (10th century), and Kekaumenos' *Stratēgikon* (circa 1078).

Three legal text corpora are of special significance:

1. The "Farmer's Law" (*Nomos Geōrgikos*): The particular importance of this text results from its express designation as legal text with regard to agriculture and livestock keeping. It offers comprehensive information and is perhaps the oldest formally published special law in this sector.[5] A reason that could have prompted its publication was the occurrence of a famine in Constantinople, dated to the autumn of 743, hence a few years after the *Ecloga*: This famine occurred when Constantine V besieged Constantinople, which at this time was under control of the usurpator Artabasdos – Theophanēs recounts that this siege implied extreme supply shortages and included high price increases for cereals, pulses, olive oil and wine.[6] Another possible cause for the publication of the *Nomos Geōrgikos* are the tax reforms by Constantine V, approximately at the same time.[7]

2. The "Imperial Laws" (*Basilika*): They mention a high number of animal names, and regulate, obviously in each case occasion-related, livestock breeding and pasture farming. Often they refer to the corresponding texts in the Justinianic Corpus (therefore they are spread over the entire text).

3. The "Book of the Eparch" (*Eparchikon Biblion*): It contains legal provisions for craft professions and guilds related to the import and processing of animals, within the scope of responsibility of the prefect of Constantinople. Similar regulations probably existed also for other cities and regions.

Animals mentioned in the legal sources

The regulations in the legal sources pertain to animals in agriculture and livestock, mainly working and breeding animals, but also grazing and some other animals. These animals can roughly be divided into the following four groups (some names are mentioned in more than one group; marking with an *asterisk signifies a collective meaning):

Work-, pack- and riding-animals:

ass → donkey. – beast of burden, carrying on the back: *nōtoforon* scil. **zōon* (*Basilika* 44.3.81.2–5 (D 32.81), *Peira* 42.19). – beast for the yoke, beast of burden: **hypozygion* (N. *Mōsaïkos* 2.4,10, 2.30,39, 12.1, 12.153, 13.1,13.171, 13.1, *Basilika* 60.3.2 et passim, *Basilika* 19.10.43 *hypopzygia thoryboumena*, *Eisagōgē* 21.6, 40.71, 42.67, *Procheiron* 39.53, Harmenopoulos, *Hexabiblos* 3.3.63f., 6.5.9). – brood mare: *phorada, phoras* (*Basilika* 60.3.27,39, 60.12,48,52, *Tipukeitos* 60.3.39). – camel: *kamēlos* (*Basilika* 54.4.1, 60.3.2, Leōn, *Tact.* 18.106, *Eisagōgē* 23.5, *Nov. Imp.* 3.19, *Procheiron* 7.13.9, *Tipukeitos* 60.3.2, *Geōponika* 16.22). – donkey, ass: *onos* (N. *Mōsaïkos* 15, 19, App.3, N. *Geōrgikos* 36, 38, 39, 41, 43, [45, 50], 51, *Basilika* 60.3.2, *Eisagōgē* 23.5, *Tipukeitos*, Harmenopoulos, *Hexabiblos* app3.1.19, 2.6,6, 6.1,3, *Geōponika* 16.21). – elephant: *elephas* (*Basilika* 60.3, Leōn, *Tact.*, *Tipukeitos* 60.3, Physiologus 43, *Geōponika* 3, 5). – four-footed animals, tetrapod: **tetrapoda* scil. *zōa* (*Basilika* 10.3.7, 42.4.7, 44.3.65, 60.2.1,3–5, 3.2, 3.5, 5.tit,

Eisagōgē A. 43.18, *Nov. Imp.* 4.30.125 (Alexios I. Komnēnos), *Tipukeitos*, Harmenopoulos, *Hexabiblos* 6.1.3). – horse 1, also → animal: *alogon* (*Ecloga* 17.10, *Basilika* 20.1.30, 58.3.3, 60.3.2, 60.3.39, *Eparch. B.* 21.9, Leōn, *Tact.* 10.13 et passim, *Eisagōgē* 24.21, *Procheiron* 17.21, *Procheiron A.* passim, *Peira* 49.9, Harmenopoulos, *Hexabiblos* 1.1.9, 3.3.61, 4.9.1, 4.15.2, app1.2.55, app3.4.2), horse 2: *hippos Ecloga* 17.7, 17.54, *Basilika* 60.3.2, *Basilika* 60.25.3–4, Leōn, *Tact.* 1.7, 5.6–7, 9, 11–14, 17–18,20, *Eisagōgē* 23.5, 40.78, 52.43, *Procheiron* 39.50, Peira 9.8, 62.2, 63.4, 66.27, Harmenopoulos, *Hexabiblos* 6.14.11, *Geōponika* 16.1–20). – kicking (sc. animal): *laktizousa, laktistēs, laktistikos* (Basilika 19.10.43, 60.2.1.5–8, 60.2.5, *Tipukeitos* 60.2.1,5, 60.39.17, Harmenopoulos, *Hexabiblos* 3.3.63f., 6.1.3, 23.3.63, *Peira* 38.23). – mare, grazing animal: *phorbas* (Nov. var.4, *Basilika* 2.2.24, 15.1.5, 44.3.65, *Tipukeitos* 15.1, 60.3, 12). – mare mule: *moula* (*Basilika* 19.10.43, 20.1.60, 44.3.62, 60.2.5, 3.2, 12.67, *Tipukeitos* passim, *Hexabiblos* 3.3.63f., 6.1.4). – mule 1: *bordōn(ion)* (*Basilika* 44.3.62, *Tipukeitos* 44.3.261), mule 2: *hēmionos* (*Nov. Imp.* 13.2, *Procheiron A.* 39.154), mule 3: *moularion* (*Nov. Imp.* var.4.68, *Nov. Imp.* 4.30.121).

Grazing animals:[8]

animal, beast, pl. also herds: *ktēnos* (*Nov. Imp.* 13.2, *N. Mōsaïkos* 2.10, 13.1, 14, 18, *N. Geōrgikos* 36, 40, 47, 71, 73, 74, 78, 79, *EclogaA* 17.10, *Peira* 37.2, *Tipukeitos* 44.10.12, Harmenopoulos, *Hexabiblos* 3.3.60, app3.1.16, Kekaumenos 3.88, → pastureland, → wild animal). – calf, young bull *moschos* (*N. Mōsaïkos* 2.153, 13.172, 15.185f., 191, 19,217, App. 3.17,24, *Basilika* 19.11.41f., *Peira* 38.66, *Tipukeitos* 19.11.41, → cattle). – cattle: *bous* (Edictum Diocl. 4.1.2,14,16,49, Tarif Anazarbos 181 (*boïdion*), *N. Mōsaïkos* 2.4,10, 12.1, 13.1, 15, 19, *N. Geōrgikos* passim, *Basilika* 19.10.43, 25.5.26, 58.3.3, 60.3.2, 60.25.3–4 et passim, *Procheiron* 25.2, *Tipukeitos* 44.3.420, → calf). – cow: *agelas* (*Nov. Imp.* 4.30, → cattle). – flock, herd: *poimnē* and *poimnion*, esp. of sheep (*Nov. Imp.* 3.19, 4.12, 4.19, 4.41, 4.42, *N. Geōrgikos* [46],47,66,75 *kyōn poimnēs*, *Basilika* 50.1.43, *Tipukeitos* passim, Harmenopoulos, *Hexabiblos* app1.4.8, app3.6.10f., 9.3). – four-footed → 1. Work animals. – goat: *aix* (Edictum Diocl. 4.1.3,48, *Nov. Imp.* 4.4, *Nov. Imp.* 4.30, *Basilika* 60.3.2). – grazing mare → livestock, → 1. Work animals: mare. – grazing animals, livestock: *thremma*, often pl. *thremmata*, mostly of tame animals, esp. sheep, but also goats (*Ecloga* 17.8,53, *N. Geōrgikos* 46, 72, *Basilika* 2.2.28, 58.3.6 et passim, *Eisagōgē* 18.2, 24.16, 26.14 et passim, *Eparch. B.* 15.1,5, *Procheiron* 8.2 et passim, *Peira* 19.41, 406, Harmenopoulos, *Hexabiblos* passim). – herd: *agelē* (*Eparch. B.* 15.3, *Basilika* 60.25.3, *Procheiron Cal.* 35bis.17, Harmenopoulos, *Hexabiblos* 3.8.14, 5.11.6, 6.5.13).[9] – herd of cattle: *boukolion* (*Basilika* 2.2.86, 16.1.68, 16.70.2, 44.3.81, 60.25.1, *Procheiron* 35bis.16). – herd of horses: *hippophorbion* (*Basilika* 23.3.39, *Peira* 19.47, *Tipukeitos* 23.3.39, → animal, → flock). – herd of pigs: *choiragelē* (*Eparch. B.* 16.3). – lamb → sheep. – livestock, grazing animal, in pl. fatted beasts, cattle: *boskēma* (*Nomokanōn Coll.* XXV cap. 23.243, *N. Geōrgikos* 34, *Basilika* 4.1.23, 6.21.1, *Procheiron* 12.31, *Eisagōgē* 21.6, 39.43, *Peira* 61.4.23, *Tipukeitos* 23.3.318). – ram *krios* (*N. Geōrgikos* [45], *Ecloga* 17.9, *Ecloga Priv. A.* 17.55, *Eisagōgē* 40.79, 52.45, *Basilika* 15.1.23, 44.3.81, 60.2.1, *Procheiron* 39.52, *Tipukeitos* 60.2.10, *Procheiron A.* 39.153, Harmenopoulos, *Hexabiblos* 6.14.14, app3.9.1). – sheep, lamb: *probaton, arēn, arnion* (Edictum Diocl. 4.1.3,47,49, *Nomokanōn Coll.* LXXXVII cap. 405.17, *N. Mōsaïkos* 12.1, 13.1, 15, App. 3, *Basilika* 16.8.12, 44.3.60,65,81, 50.3.4, 60.3.2, 25.3–4, *N. Geōrgikos* 30, 38, 49, *Eparch. B.* 15.5,

Tipukeitos 44.3.351,374f., 50.3.24, Harmenopoulos, *Hexabiblos* 2.10.10, app3.2.2, Kekaumenos 3.88, → herd). – young bull → calf.

"Winged creatures", birds and poultry:

bird, hen: **ornis* (Edictum Diocl. 4.1.23, *Nov. Imp.* 4.30 *ornithōn agriōn ē cheiroēthōn*, *Basilika* 6.21.1, 50.1.4, 50.1.43, 50.7.5, 60.3.29, *cheiroēthēs taōn* 60.12.37, Harmenopoulos, *Hexabiblos* 2.1.7,20, App. 3.Pr. 5, *Geōponika* 14.19, 14.22). – birds, poultry: pl. **ptēna* (*Nov. Imp.* 4.30.127). – crane: *geranos* (*Basilika* 12.13.9, 20.1.15, *Nov. Imp.* 4.30). – dove, common pigeon: *peristera* (Edictum Diocl. 4.1.29, *Basilika* 42.3.8, 50.1.3–6, 50.2.2, 60.3.2, *Peira* 25.25, Harmenopoulos, *Hexabiblos* App. 3.Pr. 4). – duck: *nētta* (Edictum Diocl. 4.1.31, *Nov. Imp.* 4.30.126). – goose: *chēn* (Edictum Diocl. 4.1.21f., *Nov. Imp.* 4.30, 5.44, *Basilika* 50.1.3–6, 50.4.10, *Tipukeitos* 60.2.4, Harmenopoulos, *Hexabiblos* App. 3.Pr. 5, *Peri chēnōn*, *Geōponika* 14.22.1–16). – hen → bird. – peacock: *taōn* (Edictum Diocl. 4.1.39, *Basilika* 50.1.3–6 *cheiroēthēs taōn*, 60.12.37, *Nov. Imp.* 4.30.126, *Tipukeitos*, Harmenopoulos, *Hexabiblos* 2.1.18f., App. 3. Pr. 4, → wild animals). – pheasant: *phasianos* (Edictum Diocl. 4.1.17–20, *Nov. Imp.* 4.30.126, Physiologus 3Pin.23: *Peri phasianōn kai noumidikōn kai perdikōn kai attagōn*, *Geōponika* 14.19). – pigeon → dove.

Others, including wild animals:

animal → wild animal, → grazing animals. – bear: *arktos* (*Basilika* 19.10.40, *Eisagōgē* 52.116). – bee, wild bee: *melissa, agria melissa* (*Basilika* 16.1.9, 42.3.8, 50.1.3–6, 50.2.2, 60.3.27,49, 60.12.26, *Tipukeitos* 19.1.7, 42.3.8, 50.1.4, 60.3.27,49, 60.12.26, Harmenopoulos, *Hexabiblos* 2.1.17–19, App. 3.Pr. 3f.). – beehive, swarm of bees: **smēnos* (*Basilika* 50.1.3–6, *Tipukeitos*, Harmenopoulos, *Hexabiblos* App 3. Pr. 3, Hesychios My 721 and Sigma 1242: *smēnos, the swarm of bees, whereas the vessels smēnē*[10]). – boar → wild boar. – deer: *elaphos* (Edictum Diocl. 4.1.44, *Basilika* 50.1.3–6, 60.3.28, Leōn, *Tact.* 2.32 and 20.128: *so that not deers rule over lions, but lions over deers,*[11] *Tipukeitos* 50.1.9, 60.3.129, Harmenopoulos, *Hexabiblos* 2.1.19, App. 3. Pr. 4, Physiologus 4, → ἄγρια ζῶα). – dog: *kyōn*, hare hunting dog *lagōnikos kyōn*, shepherd dog *poimenikos kyōn*, wild dog *agrios kyōn* (N. Geōrgikos 49, 52, 54, 76, 77, 55, *Basilika* 19.10.40–42, 60.2.1,2, 60.3.2, 60.3.29, *Eisagōgē* 52.116, *Procheiron* 40.15, *Peira* 62.3.3, *Tipukeitos* 60.2.1, 60.3.1, Harmenopoulos, *Hexabiblos* 6.1.2, App 3.6.5,11). – fish 1: **ichthys* (Edictum Diocl. 5.1–12, *Basilika* 2.2.15 et passim, *Eparch. B.* 13.1 *smoked fishes (ichthyes tetaricheumenoi)*, 17.1, 17.2, 17.4 *white fishes (leukoi ichthyes)*. – fish 2: *opsarion* (*Nov. Imp.* var. 46.7). – fish 3: fishing structure, fishing ground *epochē*[12] (Leōn, *Nov.* 57.5–9, and 102–104, *Eparch. B.* 17.3, *Tipukeitos* 50.2.53f., *Procheiron A.* Parat 22.12, Harmenopoulos, *Hexabiblos* 5.11.42). – fish 4 yield, daily from fishing *agra* (*Eparch. B.*17.4). – four-footed → 1. Workanimals. – hare: *lagōs* (Edictum Diocl. 4.1.32f., Physiologus 25). – lion: *leōn* (*Basilika* 19.10.40–42, 60.3.2, *Eisagōgē* 52.116, *Peira* 42.10.30, *Tipukeitos* 19.10.35–37, Harmenopoulos, *Hexabiblos* 6.1.2, Physiologus 1). – panther: *panthēra* (*Basilika* 19.10.40–42, 60.3.2, *Eisagōgē* 52.116, *Peira* 42.10, *Tipukeitos*, Physiologus 5). – pig 1: *sys* (*Basilika* 60.3.2, *Tipukeitos* 60.3), pig 2: *hys* (N. Geōrgikos 45 and lexicographical sources [2nd-11th centuries]), pig 3: young pig, porker *choiros* (Edictum Diocl. 4.1.1,13,15,46, *Nov. Imp.* 4, 3.13, 4.30, 4.55, N. Geōrgikos 49, 52, 54, *Basilika* 44.3.65, 60.25.3–4, 60.51.15, *Eparch. B.* 15.1,6, 16, *Eisagōgē* 40.79, 52.45, *Procheiron A.* 39.151, Harmenopoulos,

Hexabiblos 6.14.15, App 3.4.4, 6.5, 9.1, Kekaumenos 3.88, → herd). – porker → pig. – wild (harmful) animal: **zōon agrion (blaptikon)*, *thērion* (*N. Geōrgikos* 42, 46, 55, 75, *Basilika* 44.12.3, 19.10.40–42, 50.1.1–5, *Eparch. B.* 21, *Peira* 38.23, *Tipukeitos* 50.1.3, 502.2, Harmenopoulos, *Hexabiblos* 2.1.10,16–21, 2.4.133, 5.11.42, App. 3. Pr. 1–6, Kekaumenos 3.88 *animals edible and for transport (zōa edōdima kai agōgima)*. – wild bee → bee. – wild boar *kapros* (*Basilika* 19.10.40–42, 50.1.53, 60.3.29, *Eisagōgē* 52.116, *Peira* 42.10, *Tipukeitos* 19.10.35–37, 50.1.53, Harmenopoulos, *Hexabiblos* 6.1.2). – wolf: *lykos* (*Basilika* 19.10.40–42, 50.1.43, *Eisagōgē A.* 52.116, *Peira* 42.10, *Tipukeitos* 50.1.7, 50.7.7, Harmenopoulos, *Hexabiblos* 6.1.2, Physiologus 3).

Terms with (additional) collective meanings:*

→ animal (*alogon, zōon*), beast for the yoke (*hypozygion*), beast of burden (*hypozygion*), beast of burden carrying on the back (*nōtophoron* sc. *zōon*), beast (*ktēnos*, pl. *ktēnē*, herd), bird (*ornis, ptēnon*), cattle (*boskēma, boukolion*), fish (*ichthys, opsarion*), four-footed, sc. animals (*tetrapoda*), grazing animals (*thremmata*), grazing livestock (*phorbas*), herd (*agelē, poimnē, poimnion*), herd of cattle (*boukolion*), herds (*ktēnē*), kicking, sc. animal (*laktistēs*), livestock (*boskēma*), poultry (*ptēna*), swarm of bees (*smēnos*), wild (harmful) animal (*thērion*).

Professions mentioned in laws

Many of the following professions are mentioned in *Basilika*, *Eparchikon Biblion* and *Hexabiblos*, and most of them are directly related to animals:

Butcher: *makelarios* (*Eparch. B.* 15). – equine surgeon: *hippoiatros* (*Basilika* 54.6.8). – fishmonger: *ichthyopratēs* (*Eparch. B.* 17). – grocer: *saldamarios*, Lat. *salgamarius* (*Eparch. B.* 13). – judge: *akroatēs*, Lat. *auditor* (*N. Geōrgikos* 7, 37, 67, *Eparch. B.* 22.2, *Procheiron* 39.75,83, *Procheiron Cal.* 10.3, 34.70, *Procheiron A.* 12.34, 17.78, 19.31, 23.37, 39.192,200 and Parat. 28.2, *Peira* 51.21, Harmenopoulos, *Hexabiblos* 3.8.41, 6.6.9, 6.12.5, 6.12.5, App 3. 3.1.18). – legatee: *lēgatarios*, Lat. *legatarius* (*Eparch. B.* 20.1–3, Harmenopoulos, *Hexabiblos* 5.11.6[13]: "In the case of bequeathing a herd, the legatee (i.e. the person to whom a legacy is bequeathed) also receives what may have been reproduced [sc. in the meantime]; but if it has been reduced, even if it has been reduced to only one bovine, the legatee receives it, even if it ceased to be a herd", and 5.11.16[14]: "If the bequeathed cattle died, the legatee does not get the meat or the skin, because after death it is no longer cattle. The bequeather cannot claim and take the cattle that is still alive but dying"). – maker of cheese: *tyropoios* (Harmenopoulos, *Hexabiblos* 2.4.22[15]: "Decree of the Prefect on producers of fish sauce: The trades of producers of fish sauce and cheese do not cause normal harm to the neighbors, because their evaporation is excessively malodorous and harmful at a great distance. Therefore, they are not allowed to settle in a town or village anywhere. But if there is a need for them for the requirements of cities and villages, they must keep a distance of three stadia [about 0.5 km]. It is necessary to know, however, that everything said before is determined for those who build new. But if there are old deeds, or even a previous commitment, the agreements that existed from the beginning in this regard must be applied"). – maker of fish sauce: *garepsos* (like maker of cheese, Harmenopoulos, *Hexabiblos* 2.4.22). – market overseer: *agoranomos* (*Basilika* 54.4.1). – pig-dealer: *choiremporos* (*Eparch. B.* 16.1,2,3,4,6), trader in sheep: *probatemporos* (*Eparch.*

B. 15.4,5). – veterinary expert: *bothros*[16] (*Eparch. B.* 21, *Tipukeitos* 19.10.201–211 (c. 27), 60.25, 60.27, 60.51, *Procheiron A.* 38.99, Harmenopoulos, *Hexabiblos* 2.4.89a, b, 6.5).

In Constantinople these (and other) tradesmen were organized in guilds (*systēmata*), mentioned in *Eparch. B.* 9.6, 13.4, 17.4, 19.4,[17] and most likely in the original text of the *Basilika* 54.20, restitutus: "On pig-dealers and grocers and the other corporate bodies".[18] *Peira* 51.7 offers an explanation for the difference between guild (*systēma*) and corporate body (*sōmateion*): "Corporate body comprises any profession practised by hand [sc. by craftsmen], ... but guild the activity not practised by hand. "[19] The guilds were led by a guild master (*prostateuōn, Eparch. B.* 14.1, 17.1,4, 21.9, or *prostatēs, Eparch. B.* 12.1, 14.2, 17.3), who was appointed on approval by the *eparchos* (latin *praefectus*) of Constantinople, *Eparch. B.* 14.2 (see also *Eparch. B.* 4.5, 5.2,3, 6.6,8,10,12,13, 7.3,4, 8.13, 9.3, 10.3, 12.2–3).

Only in non-legal sources, e.g. Digenēs Acritas (passim), the term *apelatēs*[20] has the meaning "(irregular) soldier" or "border guard". In legal sources its meaning is "cattle lifter" or "robber" (after the 14th century instead of *apelatēs* also *abisteïs* (Lat. *abigeus*, Span. *abigeo*)): *Ecloga* 17.13, *Basilika* 60.25.3–4 and 60.51.15, *Eisagōgē* 21.6, *Ecloga A* 17.10, *Procheiron* 35.bis, 38.51, *Tipukeitos* 58.9, 60.25, and *Lex.* Harmenopoulos, *Hexabiblos A.*, alpha 21, 44 and 103.[21]

Topics in legal sources

The primary issues for legal regulations are the terms and conditions of ownership of animals and their offspring, of renting animals and of leasing pastureland. *Basilika* 50.1.1–5 offer a general statement: "All animals on land, in the seas and in the air come in possession of those, who took them, and also the ones born of these at us [sc. animals in our possession]"[22] (similar *Peira* 19.41 and Harmenopoulos, *Hexabiblos* App. 3. Pr. 1–6), see also *Basilika* 15.1.5: "What is born from my brood mare and your horse, belongs to me",[23] *N. Geōrgikos* 26 and Harmenopoulos, *Hexabiblos* 2.1.10,16–21.

Legacies (*lēgata, legata,* Lat. *legataria*) of working and grazing animals are mentioned in *Basilika* 44.3–5, *Peira* 43.11 and Harmenopoulos, *Hexabiblos* 5.11.6, 16, 40, 42.

Conditions for the trade of slaves (*douloi,* also *oiketai*) and livestock (*aloga, zōa,* in detail *hippoi, onoi, kamēloi kai hosa toiauta esti*) are covered in *Basilika* 18.1.17[24]: "Whoever has charged someone with the sale and purchase of slaves or animals is liable in the proceedings (Lat. *actio*) against the principal and in the proceedings about the fine <payment> and in the appeal proceedings up to double or single", and *Eisagōgē* 23.5[25]: "In case the sold slave escapes before being handed over or is stolen by someone: In order that no action may be brought in this respect for deceitfulness or recklessness on the part of the seller, let us consider on these grounds whether the seller is to be fined, and let it be decided as follows: If the seller himself has undertaken the guarding of the sold house-slave until the handover, he owes the strictest and closest supervision, and the resulting fine shall be imposed on him. But if he has not undertaken it [the guarding] himself, he shall not be liable. The same applies to the other animals like horse, donkey, camel and what is of the kind. In any case, the seller owes the buyer legal protection and legal assistance". (This passage gives the impression that animals and slaves are considered one category in legal texts.) See also Harmenopoulos, *Hexabiblos* 23.3.58–64, and *Eparch. B.* 21.

The hiring of animals is regulated in *Basilika* 20.1.15 and 30, *Eisagōgē* 24.21/*Procheiron* 17.21 and Harmenopoulos, *Hexabiblos* 2.10.10, 3.8.14,22f., and 6.14.11, in *Peira*

37.2 and *Tipukeitos* 50.2.47f. especially for their use in pastureland (*nomadion, nomē*); *Tipukeitos* 50.2.176 has general rules relating to the allocation of winter and summer pasture (*cheimerina ē therina dasē ētoi nomadia*). – The remuneration for the hiring of draft animals (*ktēnomisthion*) is mentioned in *Basilika* 14.1.10, 49.3.20, 53.5.14. – The right of disposal of the animal's dung (*kopros*) is regulated in Harmenopoulos, *Hexabiblos* 2.10.10, 5.11.39.

Special rules for hunting (*thēra*)[26] are contained in N. *Mōsaïkos* 13 (13.177: If the hunter has stolen the animal from its owner, he shall reimburse it, but not, if he caught it in hunting), cf. similar regulations in *Basilika* 16.1.62, 19.1, 8,10, et passim, and Harmenopoulos, *Hexabiblos* 2.1.21, App. 3, Prooim. 2.

The military use of pack- and riding-animals is dealt with by Leōn, *Tact.* 1.7, 2.32, 10.13, 18.106, 20.128 et passim; → Sanctions.

Sanctions and their imposition are regulated in *Basilika* 7.17.27[27]: "Neither a horse race nor a hunting demonstration may be held on a Sunday, even if the emperor's birthday or the day of his elevation falls on a Sunday. A civil or military official who violates this shall be deposed and liable to confiscation", and *Basilika* 19.10.40–42[28]: "Whoever leads a wild dog, or boar, or wolf, or panther, or lion, or any other animal causing harm, without a leash, or not so tethered as to prevent it from causing harm, along a public way, shall, if as a result a free <man> dies, pay two hundred nomismata fine; but if he did not die, but suffered harm, he [the owner of the animal] shall be condemned at the discretion of the judge. With regard to the other facts, however, let him give double as penance", furthermore in Harmenopoulos, *Hexabiblos* 6.14.15f.[29]: "In the case of rams, cattle or horses or pigs or other herd animals meeting one another, if the one that came first was killed by the others, the owner of the one that killed shall not be responsible; but if the one that came next killed, the owner shall be responsible and either replace the animal to the owner of the one killed or compensate him"; see also *Basilika* 60.25.3–4, 60.40.1, *Eisagōgē* 52.116 and Harmenopoulos, *Hexabiblos* 6.5.13.

Special sanctions for monks (and probably also nuns) are mentioned by *Nomokanōn* Coll. XXV cap. 23 ("How the monks should form their conduct of life", e.g. that they should lead a common life and do not have their own homes, nor accumulate their own fortune …). – Special sanctions are also provided for soldiers in *Ecloga* 17.10 ("Whoever steals in an army camp or in a field camp, respectively, shall be beaten in the case of weapons, but in the case of an animal, one hand shall be cut off"). Nearly an identical decree is found in *Eisagōgē*, *Procheiron* and Harmenopoulos, *Hexabiblos* 6.5.9[30]: "For those who steal in the army camp, we command that in the case of weapons they be beaten severely, in the case of beasts of burden, however, one hand be cut off."

With regard to taxation, the legal sources contain some exemptions of dues (*ennomia*) from the emperors Theodōros I Laskaris (1205–1222) for the Venetians (*Nov. Imp.* 5.2.99), Michaēl VIII Palaiologos (1259–1282) for the *Megalē Ekklēsia* (*Nov. Imp.* var. 30.81) and Michaēl IX Palaiologos (co-emperor 1294/95–1320) and Andronikos II Palaiologos (1282–1328) for the Brontochion monastery in Mistra (*Nov. Imp.* var. 33.142, 34.132, 35.57 and 36.60). – In monastic documents the following special taxes, tithes and rents are mentioned (For all following terms see LBG, s.v.): tax on bee-keeping (*melissoennomion*), tax on pig husbandry (*choiroennomion*), tax on sheep husbandry (*probatoennomion*), tithe from pigs breeding (*choirodekatia*), tithe from sheep and pigs breeding (*probatochoirodekatia*), and rents (*pakta*) for fish ponds (*bibaropakta*). – Contributions in kind were demanded on the occasion of episcopal consecrations, among others a ram (*krios*), a sheep (*arnion*) and a varying number (up to sixty) of "birds" (*ornithes, ornithia*) (*Nov. Imp.* 4.1 and 4.27). – Exemption from taxes (*exkousseia*, Lat. *excusatio*) was granted to the Nea Monē in Chios (Kōnstantinos IX, *Nov. Imp.* 4.4, Alexios I Komnēnos, *Nov. Imp.* 4.30) on the purchase of animals: "mules, halfmules, horses, spare horses, stallions, donkey studs, she-asses, brood mares, working and in herd cattle, pigs, sheep, goats, cows, buffalos, hares, deer, dog for hare or shepherd dog, and the other four-footed animals, geese, ducks, partridges, peacocks, cranes, swans, pheasants, tame or wild birds, doves and all other winged creatures, included all the eggs from them".

Summary

With few exceptions, special laws on agriculture and animal livestock breeding did not exist in premodern times. In the Byzantine legislation these issues are dispersed over general bodies of laws, and in no Byzantine legislative text exists a systematic overview of animal-related regulations. Between the 4th and the 15th centuries more than twenty legislative corpora contain information about animals. Three of them are particularly relevant: 1) The "Farmer's Law" (*Nomos Geōrgikos*, after 741), 2) the "Imperial Laws" (*Basilika*, circa 888) and 3) the "Book of the Eparch" (*Eparchikon Biblion*, circa 900).

In general, it should be noted that most of the works of law mentioned in this chapter can only be assumed to be up to date at the time of their origin, if they are directly and con-cretely occasion-related. Otherwise, it can be assumed that they largely repeat, summarize, and sometimes also modify older law, as well as translate it from Latin into Greek (in the case of the *Basilika*).

The sources regulate legal matters, which relate to some fifty animals. They can be grouped broadly into four categories: 1) work-, pack- and riding-animals, 2) grazing animals, 3) poultry, winged creatures, birds, and 4) other, including wild animals. – Within the animal names some terms have also a collective meaning, which in part overlaps with a specific name: *agelē* (herd), *alogon* (animal), *boskēma* (cattle, livestock), *boukolion* (cattle, herd of cattle), *hypozygion* (beast for the yoke, beast of burden), *ichthys* (fish), *ktēnos* (beast, pl. herd), *laktistēs* (kicking, sc. animal), *nōtophoron* sc. *zōon* (beast of burden, carrying on the back), *opsarion* (fish), *ornis* (bird), *phorbas* (grazing livestock), *poimnē* and *poimnion* (herd), *ptēnon* (bird, pl. *ptēna*, poultry), *smēnos* (swarm of bees), *tetrapoda* (four-footed, sc. animals), *thērion* (wild and/or harmful animal), *thremmata* (grazing animals) and *zōon* (animal).

Professions are mentioned in *Basilika* (equine surgeon, market overseer), *Eparchikon Biblion* (butcher, fishmonger, grocer, judge, legatee, pig-dealer, trader in sheep, veter-inary expert), Harmenopoulos, *Hexabiblos* (legatee, maker of cheese, maker of fish sauce, judge, veterinary expert); additional evidence: "veterinary expert" also in *Tipukeitos* and *Procheiron auctum*, "judge" in *Nomos Geōrgikos*, *Procheiron*, *Procheiron Calabriae*, *Procheiron auctum* and *Peira*. – The tradesmen were organized in guilds (*Basilika*, *Eparchikon Biblion*), led by a guild master (*Eparchikon Biblion*). The difference between guilds (*systēmata*) and corporate bodies (*sōmateia*) is explained in the *Peira*.

Regulated topics are: the terms and conditions of ownership of animals, of renting animals and of leasing pastureland; legacies of working and grazing animals; conditions for the trade of slaves and livestock; hiring of animals (especially for their use in pasture-land and the remuneration for hired draft animals); the rules for hunting; the military use of pack- and riding-animals; sanctions and their imposition, including special sanctions for monks and soldiers.

Taxation is not mentioned often, it is – in the context of animals – a rather marginal issue. However, contributions in kind (ram, sheep, and birds) on the occasion of episcopal consecrations are mentioned. The main issues are exemptions (*exkousseiai*) of dues (*ennomia*), especially in monastic documents the exemption from taxes for the purchase of animals, and from special taxes (on bee-keeping, pig husbandry, and sheep husbandry), tithes (*dekatiai*: from pigs and sheep breeding) and rents (*pakta*: for fish ponds).

From the regulated issues it is evident that no systematic legislation of animal-related matters existed or was aimed for by the respective legislators. Exceptions are the amendments of laws, like the *Novellae* enacted by Justinian I and Leōn VI, and generally the *Novellae et Chrysobulla Imperatorum*. A plausible explanation could be, that non-written regional traditions of natural law have normally been sufficient for the functioning of a local agrarian society and that legislators intervened only ad hoc, if certain detail problems became evident, or if specific overriding interests of the state or the society appeared. In fact, such an interest may have been the cause for the creation of one of the few exception, the *Nomos Geōrgikos*, if this law – as I assume – can be dated to the early forties of the 8th century, on the occasion of the extremely difficult situation of Constantinople at that time.

Notes

1 Selected basic bibliography: E. S. Papagianni, "Legislation and legal practice," in *Brill's History and Culture of Byzantium, New Pauly*, ed. F. Daim and J. N. Dillon (Leiden and Boston, 2019), 190–213. Z. Chitwood, " 'The Cleansing of the Ancient Laws' under Basil I and Leo VI. Byzantine Legal Culture and the Roman Legal Tradition, 867–1056," in *Byzantine Legal Culture and the Roman Legal Tradition, 867–1056*, ed. idem (Cambridge, 2017), 16–44. Sp. N. Troianos, *Die Quellen des Byzantinischen Rechts*, transl. D. Simon and S. Neye (Berlin, 2017). W. Hartmann and K. Pennington (eds.), *The History of Byzantine and Eastern Canon Law to 1500* (Washington D.C., 2012). Sp. Troianos and A. Diarmakopoulos, "Τα ζώα ως αντικείμενο εγκληματικών πράξεων στο βυζαντινό δίκαιο," in *Animals and Environment in Byzantium, 7th–12th c.*, ed. I. Anagnostakis, T. G. Kolias, and E. Papadopoulou. *Διεθνή Συμπόσια* 21 (Athens, 2011), 435–53. H. Kroll, *Tiere im Byzantinischen Reich. Archäozoologische Forschungen im Überblick*. Monographien des Röm.-German. Zentralmuseums 87 (Mainz, 2010). A. E. Laiou, "The Agrarian Economy, Thirteenth-Fifteenth Centuries," in *The Economic History of Byzantium. From the Seventh through the Fifteenth Century*, vol. I, ed. eadem (Washington, D.C., 2002), 311–75. S. N. Troianos, "Δίκαιο και ιδεολογία στα χρόνια των Μακεδόνων," in *Byzantina* 22 (2001): 239–61. P. E. Pieler, "Lex Christiana," in *Akten des 26. Deutschen Rechtshistorikertages 1986*, ed. D. Simon. Ius commune, Sonderhefte Studien zur europäischen Rechtsgeschichte 30 (Frankfurt am Main, 1987), 485–503. N. van der Wal and J. A. H. Lokin, *Historiae iuris Graeco-Romani delineatio: les sources du droit byzantin de 300 à 1453* (Groningen, 1985). H.-G. Beck, *Nomos, Kanon und Staatsraison in Byzanz* (Vienna, 1981). P. E. Pieler, "Byzantinische Rechtsliteratur," in H. Hunger, *Die hochsprachliche profane Literatur der Byzantiner*, 2, Byz. Handbuch 2 im Rahmen des Handbuchs der Altertumswiss. 5.2 (Munich, 1978), 341–480. D. Simon, "Provinzialrecht und Volksrecht," *Fontes Minores* 1 (1976): 102–16. L. Wenger, *Die Quellen des römischen Rechts* (Vienna, 1953). E. H. Freshfield, *A Provincial Manual of Later Roman Law: The Calabrian Procheiron on Servitudes & Bye-laws Incidental to the Tenure of Real Property* (Cambridge, 1931). E. H. Freshfield, *The Procheiros Nomos* (Cambridge 1928). K. E. Zachariä von Lingenthal, *Geschichte des Griechisch-Römischen Rechtes* (Berlin, ³1892).

2 S. Lauffer (ed.), *Diokletians Preisedikt*, Texte und Kommentare. Eine altertumswiss. Reihe 5 (Berlin, 1971). *Tarif Anazarbos*: G. Dagron and D. Feissel (eds.), *Inscriptions de Cilicie*. Travaux et Mémoires, Monographies 4 (Paris, 1987), 170–85, no. 108. *Nov. et Chrysob. Imp.*: C. E. Zachariae von Lingenthal I. and P. Zepos (eds.), *Jus Graecoromanum* 1, Νεαραὶ καὶ Χρυσόβουλλα τῶν μετὰ τὸν Ἰουστινιανὸν Βυζαντινῶν Αὐτοκρατόρων. *Novellae et Chrysobulla Imperatorum*,1 (Leipzig, 1856,

repr. Athens, 1931 and Aalen, 1962), 1–258. *Nomokanōn Coll.* XXV cap.: G. E. Heimbach (ed.), "Constitutiones legum civilium ex Novellis Imperatoris Iustiniani, quae consentiunt et confirmant sanctorum patrum ecclesiasticos canones," in *Anekdota zur byzantinischen Gesetzgebung*, II (Leipzig, 1840): 145–201. *Nomokanōn Coll.* LXXXVII cap.: J. B. Pitra (ed.), *Collectio LXXXVII capitulorum sub auctore Joanne Scholastico*, Iuris ecclesiastici Graecorum historia et monumenta, II (Rome, 1868), 385–405. L. Burgmann (ed.), *Ecloga, Das Gesetzbuch Leons III. und Konstantinos' V.*, Forsch. zur byz. Rechtsgesch. 10 (Frankfurt am Main, 1983). L. Burgmann and Sp. Troianos (eds.), *Nomos Mosaïkos*, Forsch. zur byz. Rechtsgesch. 4 (Frankfurt am Main, 1979). *Ecloga Priv. Aucta*: C. E. Zachariae von Lingenthal, I., and P. Zepos (eds.), *Jus Graecoromanum*, 6 (Leipzig, 1870, repr. Athens, 1931 and Aalen, 1962). *Eisagōgē*: C. E. Zachariae von Lingenthal, and P. Zepos (eds.), *Jus Graecoromanum* 2 (Athens, 1931, Aalen, 1962), 236–368, A.D. 886. – *Basilika* (H. J. Scheltema / N. van der Wal (eds.), *Basilicorum libri LX*. Series A, vols. 1–8 (Groningen, 1955–1988). Sp. N. Troianos (ed.), *Oi neares Leontos st' tou Sophou* (Athens, 2007). Leōn, *Tact.*: G. T. Dennis (ed.), *The Tactica of Leo VI*. Corpus Fontium Historiae Byzantinae 12 (Washington, D.C., 2010). *Procheiron*: C. E. Zachariae von Lingenthal and P. Zepos (eds.), *Jus Graecoromanum* 2 (Athens, 1931, Aalen, 1962), 114–228. *Eisagōgē Aucta*: C. E. Zachariae von Lingenthal, I. and P. Zepos (eds.), *Jus Graecoromanum* 6 (Leipzig, 1870, repr. Athens, 1931 and Aalen, 1962). *Procheiron Auctum*: C. E. Zachariae von Lingenthal, I. and P. Zepos (eds.), *Jus Graecoromanum*, 7 (Leipzig, 1884, repr. Athens, 1931 and Aalen, 1962). *Procheiron Cal.*: F. Brandileone and V. Puntoni (eds.), *Prochiron Legum*. Fonti per la storia d'Italia (Rome, 1895). *Eparch. B.*: J. Koder (ed.), *Das Eparchenbuch Leons des Weisen*. CFHB 33 (Vienna, 1991). *Peira*: C. E. Zachariae von Lingenthal, I. and P. Zepos (eds.), *Jus Graecoromanum*, 4 (Leipzig, 1865, repr. Athens, 1931 and Aalen 1962). F. Dölger, C. Ferrini, S. Hörmann, G. Mercati, and E. Seidl (eds.), *M. Kritou tou Patzē Tipoukeitos sive Librum lx Basilicorum Summarium*. Studi e Testi 25, 51, 107, 179, 193 (Città del Vaticano, 1914–1957). K. G. Pitsakes (ed.), *Kōnstantinou Armenopoulou procheiron nomōn ē Exabiblos*. Byzantina kai Neohellenika Keimena, 1 (Athens, 1971). M. Th. Fögen (ed.), Harmenopoulos, *Hexabiblos aucta. Eine Kompilation der spätbyzantinischen Rechtswissenschaft*. Fontes minores 7 (Frankfurt am Main, 1986), 259–333. M. T. Fögen (ed.), *Das Lexikon zur* Harmenopoulos, *Hexabiblos aucta*. Forschungen zur byzantinischen Rechtsgeschichte 17 (Frankfurt am Main, 1990), 162–214.

3 E. Trapp (ed.), *Lexikon zur byzantinischen Gräzität*, 1–2 (Vienna, 2001–2017).

4 F. Sbordone (ed.), *Physiologus* (Rome, 1936, repr. Hildesheim, 1976). H. Beckh (ed.), *Geoponica* (Leipzig, 1895); A. Dalby, *Geōponika – Farm Work. A modern translation of the Roman and Byzantine Farming Handbook* (Totnes, 2011). M. D. Spadaro (ed.), *Raccomandazioni e consigli di un galantuomo, Stratēgikon* (Alessandria, 1998).

5 For earlier laws like e.g. the *lex Manciana*, which relate to social classes, regional or partial aspects, or even special cases, see Ch. Schubert, "Die kaiserliche Agrargesetzgebung in Nordafrika von Trajan bis Justinian, " *ZPE* 167 (2008): 251–57.

6 Theophanes the Confessor, *Chronographia*, AM 6235, ed. C. de Boor, Theophanis Chronographia, 2 vols. (Leipzig, 1883–85), vol. 1, 419f. – For the supply problems in the capital see J. Durliat, "L'approvisionnement de Constantinople", in *Constantinople and its Hinterland. Papers from the Twenty-seventh Spring Symposium of Byzantine Studies, Oxford, April 1993*, ed. C. Mango and G. Dagron (Aldershot, 1995), 9–33.

7 Cf. W. Brandes, *Finanzverwaltung in Krisenzeiten. Untersuchungen zur byzantinischen Administration im 6.-9. Jahrhundert*. Forschungen zur byzantinischen Rechtsgeschichte 25 (Frankfurt am Main, 2002), see ch. 7.2, 508–10. On the law, see now also J. Koder, *Nomos Georgikos. Das byzantinische Landwirtschaftsgesetz. Überlegungen zur inhaltlichen und zeitlichen Einordnung. Deutsche Übersetzung*, Wiener Byzantinistische Studien 32 (Vienna, 2020).

8 M. Kaplan, "L'activité pastorale dans le village byzantin du VIIe au XIIe siecle," in *Animals and Environment in Byzantium (7th – 12th c.)*, ed. Anagnostakis, Kolias and Papadopoulou, 407–20.

9 Cf. also cattle trail, *agelodromion* (See LBG, s. v.).

10 σμῆνος· τὸ μελισσῶν καὶ σφηκῶν ἄθροισμα. τὰ δὲ ἀγγεῖα σμήνη [sc. plural], Hesychios My721, ed. K. Latte, *Hesychii Alexandrini lexicon I-II* (Copenhagen, 1953), 966; Sigma 1242, ed. P. A. Hansen, *Hesychii Alexandrini lexicon III*. Sammlung griechischer und lateinischer Grammatiker 11.3 (Berlin and New York, 2005).

11 ... μὴ ἔλαφοι λεόντων ἄρχουσιν ἀλλὰ λέοντες ἐλάφων, Leōn, *Tact.* 2.32 (and 20.128). Leōn quotes his father Basileios I, who may have quoted from Maurikios, *Stratēgikon*, ed. G. T. Dennis, *Das Strategikon des Maurikios.* CFHB 17 (Vienna, 1981), 8.2.68 and 79. The saying was often repeated by Byzantine historians, *Gnomologia* attribute it to Chabrias; see K. Paidas, "Δύο παραινετικὰ κείμενα προς τον Λέοντα στ' τον Σοφό", in idem, *Κείμενα Βυζαντινῆς Λογοτεχνίας* 5 (Athens, 2009), 1.46.

12 E. Trapp, "Die gesetzlichen Bestimmungen über die Errichtung einer 'epoche'," in *Polychordia. Festschrift für Franz Dölger zum 75. Geburtstag*, ed. A. M. Hakkert and P. Wirth. *BF* 1 (Amsterdam, 1966), 329–333.

13 Ἀγέλης ληγατευομένης καὶ τὰ προστεθέντα λαμβάνει ὁ ληγατάριος· εἰ δὲ μειωθῇ, κἂν εἰς ἕνα βοῦν περιστῇ, λαμβάνει αὐτὸν ὁ ληγατάριος, εἰ καὶ ἐπαύσατο εἶναι ἀγέλη, Harmenopoulos, *Hexabiblos* 5.11.6.

14 Τελευτήσαντος τοῦ ληγατευθέντος βοὸς οὐ δίδοται τῷ ληγαταρίῳ οὔτε κρέα οὔτε βύρσα· τελευτήσας γὰρ οὐκέτι ἐστὶ βοῦς· ἔτι δὲ ζῶντα καὶ τελευτῶντα τὸν βοῦν οὐ δύναται ἀπαιτεῖν καὶ λαμβάνειν ὁ ληγατάριος, Harmenopoulos, *Hexabiblos* 5.11.16.

15 Harmenopoulos, *Hexabiblos* 2.4.22: Ἐπαρχικὸν περὶ γαρεψῶν Ἡ τέχνη τῶν γαρεψῶν καὶ ἡ τῶν τυροποιῶν οὐ τὴν τυχοῦσαν ποιεῖ τοῖς παρακειμένοις βλάβην· πολλὴ γὰρ ἡ ἀπ' αὐτῆς ἀτμὶς δυσώδης τε ἀμέτρως ἐστὶ καὶ ἐπιβλαβὴς ἐπὶ πολὺ διάστημα. Ὅθεν χρὴ τοὺς τοιούτους παντάπασι μὲν πόλιν μὴ οἰκεῖν μηδὲ κώμην. Εἰ δὲ χρεία τούτων εἴη πρὸς τὰς χρήσεις τῶν πόλεων καὶ τῶν κωμῶν, ἀφεστάναι στάδια τούτους δεῖ τρία. Εἰδέναι δὲ χρὴ ὡς ταῦτα πάντα τὰ προειρημένα ἐπὶ τῶν ἐκ νέου κατασκευαζόντων εἴρηται· εἰ δὲ χάρται εἰσὶ παλαιοὶ ἢ καὶ δουλεία τις προλαβοῦσα, κεχρῆσθαι δεῖ τοῖς ἐξ ἀρχῆς περὶ αὐτῶν συμπεφωνημένοις.

16 J. Koder, " 'Wer andern eine Grube gräbt ...' Die Bezeichnung βόθρος im 'Eparchikon Biblion'," in *Fest und Alltag in Byzanz*, ed. G. Prinzing and D. Simon (Munich, 1990), 71–76 and 194–97.

17 J. Koder, "Delikt und Strafe im Eparchenbuch: Aspekte des mittelalterlichen Korporationswesens in Konstantinopel," *JÖB* 41 (1991): 113–31.

18 Περὶ χοιρεμπόρων καὶ καπήλων καὶ λοιπῶν σωματείων, *Basilika* 54.20, restitutus.

19 ... σωματεῖον μὲν γὰρ ἐστὶ πᾶσα τέχνη, ἥτις διὰ χειρὸς ἔχει τὴν ἐργασίαν· ... σύστημα δὲ ἡ μὴ ἔχουσα διὰ χειρῶν τὴν ἐργασίαν, *Peira* 51.7.

20 I. K. Mazarakēs-Ainian, *Ellēnes stratiōtes. Apo tous Akrites-Apelates stous Armatōlous* (Athens, 2017); ODB s.v. Apelatai.

21 Ἀβίστεϊς λέγεται ὁ ἀπελάτης, *Lex.* Harmenopoulos, *Hexabiblos A.*, alpha 44, Ἀβίστεϊς· ἀγωγὴ ἡ κατὰ τῶν ἀπελατῶν, *Lex.* Harmenopoulos, *Hexabiblos A.*, alpha 103.

22 Πάντα οὖν τὰ χερσαῖα, τὰ θαλάσσια, τὰ ἐναέρια ζῶα τῶν λαμβανόντων γίνεται μετὰ καὶ ὧν τίκτουσι παρ' ἡμῖν.

23 Τὸ τικτόμενον ἐκ τῆς ἐμῆς φορβάδος καὶ τοῦ σοῦ ἵππου ἐμόν ἐστι.

24 Ὁ προστήσας τινὰ πιπράσκειν καὶ ἀγοράζειν δούλους ἢ ἄλογα ἐνάγεται πρὸς τῇ ἀγωγῇ (lat. *actio*) τῇ κατὰ τοῦ προστήσαντος καὶ τῇ περὶ χρέους ἀγωγῇ καὶ τῇ παραγγελίας ἐπὶ τῷ διπλῷ ἢ τῷ ἁπλῷ, *Basilika* 18.1.17.

25 Εἰ πρὸ παραδόσεως φύγῃ ὁ πραθεὶς δοῦλος ἢ καὶ κλαπῇ ὑπό τινος, ὡς μὴ δόλον ἢ ῥαθυμίαν ἐλέγχεσθαι περὶ ταῦτα τοῦ πράτου, σκοπήσωμεν εἰ ὁ πράτης ἐκ τούτων ὀφείλει ζημιοῦσθαι καὶ δεῖ οὕτως διαστίξαι· εἰ μὲν εἰς ἑαυτὸν ὁ πράτης ἀνεδέξατο τὴν τοῦ πεπραμένου οἰκέτου φυλακὴν ἕως τῆς παραδόσεως, χρεωστεῖ ἀκριβεστάτην (5) καὶ ὑπερβάλλουσαν παραφυλακήν, καὶ αὐτὸς ἐπιγνώσεται τὴν ἐντεῦθεν ζημίαν· εἰ δὲ μὴ ἀνεδέξατο εἰς ἑαυτόν, ἕξει τὸ ἀμέριμνον. τὸ αὐτὸ καὶ ἐπὶ τῶν λοιπῶν ζώων, οἷον ἵππου, ὄνου, καμήλου καὶ ὅσα τοιαῦτά ἐστι. πάντως δὲ χρεωστεῖ ὁ πράτης τῷ ἀγοραστῇ τὴν διαυθέντευσιν καὶ διεκδίκησιν, *Eisagōgē* 23.5.

26 Cf. A. K. Sinakos, "To kynēgi kata tē mesē byzantinē epochē (7os-12os ai.)," in *Animals and Environment in Byzantium (7th – 12th c.)*, eds. Anagnostakis, Kolias, and Papadopoulou, 71–86.

27 Μήτε δὲ ἱππικὸν ἢ κυνήγιον ἐπιτελείσθω ἐν κυριακῇ, κἂν εἰ συμβῇ τὴν γενεθλιακὴν ἡμέραν τοῦ βασιλέως ἢ καθ' ἣν ἀνηγορεύθη εἰς κυριακὴν ὑπαντῆσαι. Ὁ δὲ ταῦτα παραβαίνων ἄρχων ἢ ταξεώτης τῆς ζώνης ἐκπιπτέτω καὶ δημευέσθω, *Basilika* 7.17.27.

28 Ὁ κύνα ἄγριον ἢ κάπρον ἢ ἄρκτον ἢ λύκον ἢ πανθῆρα ἢ λέοντα ἢ ἕτερον βλαπτικὸν ζῶον λελυμένον ἢ μὴ οὕτως δεδεμένον, ὥστε μὴ δύνασθαι βλάπτειν, ἔχων ἐν δημοσίᾳ παρόδῳ, εἰ ἄνθρωπος ἐλεύθερος ἐκ τούτου τελευτήσει, διακόσια νομίσματα δίδωσιν· εἰ δὲ μὴ ἀπέθανεν, ἀλλ' ἐβλάβη, εἰς τὸ φαινόμενον τῷ δικαστῇ καταδικάζεται. Τῶν δὲ λοιπῶν πραγμάτων διπλῆν δίδωσι τὴν ζημίαν, *Basilika* 19.10.40–42.

29 Κριῶν, βοῶν ἢ ἵππων ἢ χοίρων ἢ ἄλλων τινῶν θρεμμάτων συνελθόντων, εἰ μὲν ὁ πρώτως ἐπελθὼν τοῖς ἄλλοις ἀνῃρέθη, ἀνεύθυνός ἐστιν ὁ τοῦ ἀνελόντος δεσπότης· εἰ δὲ ἀνέλῃ ὁ ἐπελθών, ὑπεύθυνος γίνεται ὁ δεσπότης καὶ ἢ ἐκδίδωσι τὸ ζῶον τῷ δεσπότῃ τοῦ ἀναιρεθέντος ἢ τὸ ἀζήμιον αὐτῷ περιποιεῖται. (16) Ὁ δόλον ποιῶν εἰς ἀσήμιον καὶ ἐξ αὐτοῦ ἐργαζόμενος καὶ πιπράσκων χειροκοπείσθω, Harmenopoulos, *Hexabiblos* 6.14.15f.

30 *Ecloga* 17.10: Ὁ κλέπτων ἐν φοσσάτῳ ἤτοι ἐν ἐκσπεδίτῳ, εἰ μὲν ὅπλα, τυπτέσθω, εἰ δὲ ἄλογον, χειροκοπείσθω. – *Eisagōgē* 40.71/*Procheiron* 39.53/ Harmenopoulos, *Hexabiblos* 6.5.9: Τοὺς ἐν φοσσάτῳ κλέπτοντας, εἰ μὲν ὅπλα, σφοδρῶς προστάττομεν τύπτεσθαι· εἰ δέ τι τῶν ὑποζυγίων, χειροκοπεῖσθαι, Harmenopoulos, *Hexabiblos* 6.5.9.

8

ANIMALS IN SATIRE

Kirsty Stewart

Animals appear in a great variety of Byzantine texts, whether as the main focus, for example in zoological material, as background details of daily life or symbolic features, as often in historiography or hagiography, as *exempla* for behaviour and incredible elements of Creation in the *Physiologus* and the hexameral literature, or even simply as decorative elements in ekphrasis. However, they also have a role to play in what we may term satire, especially in the later Byzantine period, where animals appear not simply as straightforward ways of mocking someone's appearance or habits, but as a more complex vehicle for humour and commentary.

The Mode of Byzantine Satire

Genres in Byzantine literature can be hard to define, and satire is no exception. The particular works which will be addressed here have been, and continue to be, described in scholarship variously as beast epics, mock epics, fables, animal stories and simply as entertainment literature, as well as satire. It is therefore helpful to set out some preliminary remarks on satire in Byzantium, before addressing the texts themselves and the specifically animal elements.

Even for a modern audience, satire can be hard to clearly define, since it overlaps with parody and invective, and may or may not invoke laughter in the manner of the straightforwardly comedic. Satire may comment on a moment in time, or on broader issues, arguably requiring not only knowledge of the events occurring when it was written, but also an understanding of the culture that produced it, and its intended audience. It can be accessible to all, and at the same time may require a level of understanding and background knowledge available to only a few. It may ridicule partly for fun, though in doing so pointing out to its audience behaviours to avoid, or it may more earnestly abuse its target with the serious aim of pushing for change.

In Byzantium there was no singular term for the various types of texts we may now consider satirical, and even a brief glance at those regularly referred to as such will indicate that it is not necessarily helpful to think in terms of genre, or even in terms of key literary features, but rather in terms of intent or mode. We have, for example, the begging poems of Theodore Prodromos, the Lucianic *Timarion* and *Mazaris' Journey to Hades*, the

DOI: 10.4324/9781003055877-11

liturgical *Spanos*, the poetry of Sachlikes to name a few, all of which comically criticise, but in different ways.[1]

Fundamental to all these works is a sense of humour. Humour is a core part of human society, playing a role in both reinforcing and undermining cultural norms and structures of power, in the form of individuals or institutions, depending on where the humour is directed, and acting as a mirror through which to view ourselves and our society in another way. Humour in literature can be used as a tool for "reaffirming intimacy or attempting to establish a relationship, as a means of diffusing anger or frustration, or of displaying contempt – as well as a way of negotiating complex or difficult social moments".[2] As such, humour can be an excellent tool in helping us to understand a society and its concerns, but its very specificity can also make it hard for us, at such a remove, to get the joke, or understand what it is telling us about Byzantium. As Marc Lauxtermann has put it "we laugh at what we know – and jokes made by foreigners, or worse still, by very dead foreigners, such as the Byzantines, somehow seem to lose much of their force".[3]

So, what can we say in general terms about how the Byzantines thought of satire, comedy and laughter? As always, generalisations can be dangerous, and Byzantine society was not an unchanging monolith. In any given time, in any given part of the Empire, different groups would no doubt have professed differing ideas on what was funny and why, and even on whether laughter was a good thing or not. With that caveat in place, acknowledging that the works touched on for evidence here take differing views and come from different periods, it is possible to identify some key features.

If we begin with the act of laughing, it is possible to think that the Byzantines really were a mirthless society. Basil the Great, among others, indicated that laughter was incompatible with Christianity. Indeed, he went as far as to say that Jesus never laughed.[4] Inappropriate examples of laughter appear in a number of hagiographical texts, reinforcing the concept, with physical responses including strange facial expressions, clearly frowned upon. However, some writers, even including the usually cantankerous John Chrysostom, differentiated between socially acceptable laughter and excessive mirth.[5] Chrysostom identifies laughter as being a positive reaction to seeing friends after a long time, and as a means of calming negative emotions, but is clear that it should never take over one's rationality. Laughter, Chrysostom suggests, is part of human nature, following *On the nature of man* usually attributed to Gregory of Nyssa, in which laughter is even directly connected with the positive act of repentance.[6]

Laughter, it seems, trod a fine line, and its appropriateness to a situation was key. That is not to say that we do not find transgressive humour and examples of transgressive laughter, but its dangers were acknowledged. Jokes could be misinterpreted and taken as insults, with the effect of losing one connections and support, and a fondness for laughter could equally be used to represent an individual as being weak in character. One example of this can be found in the portrayal of Emperor Michael III (r. 842–867), who reportedly enjoyed playing practical jokes and exchanging jests with his inner circle. His fondness for banter is translated into a character assassination in the *Chronicle of the Logothete*, though importantly the criticism is posthumous.[7] Here, what was arguably a clever exchange of words between the emperor and Peter the Magistros, who is specifically identified as funny, regarding Michael's new stables is presented as a haughty emperor losing his temper, and therefore being shown as thoroughly unfit to rule by a source supportive of his successor (Basil I, who usurped against Michael in 867).

In fact, it can be read as self-deprecating humour on the part of the emperor; "look at these, they'll make me famous!" exclaimed over some stables from the mouth of the most powerful man in the Empire and builder of much finer things is clearly ridiculous and said in jest. The response from Peter, which amounts to, "oh yeah?, no one even remembers who built Hagia Sophia!", is clearly a false and hyperbolic statement unlikely to have been uttered in the manner described by our source, unless Peter had a death wish.[8] Indeed, the account as presented ends with Peter being kicked out of the emperor's presence and beaten. Taking the joke out of context allows the reporting author to present Michael in a negative light. One had to be careful with jokes, and an emperor with a sense of humour was apparently less appropriate than one of a sombre nature, comedy contrasting with the role of being a ruler appointed by God, especially if he either took a joke too far, or in using humour, made himself the butt of it.

If a Byzantine author wished to be funny, he, or even she, would have some basic tools to use from their education, namely from the *progymnasmata*, which taught the essentials of rhetoric.[9] Here we find the tenets of writing comic discourse, κωμικῶς λέγειν, including parody, and specifically wordplay, incongruous content, and the cheating of audience expectations. These tools were however to be used for rhetorical purposes, that is to convince another person of an argument, not simply for entertainment. Guidance on writing *psogos* (invective), was also transmitted via these texts, essentially as the reverse of *encomium* (praise). The rules for *psogos* seem to have been interpreted more loosely, as it appears in different forms, though always to critique and ridicule. Constantine the Rhodian, a tenth-century civil servant, provides an excellent example of invective. His writings against prominent figures, often it seems at the behest of his superior Samonas, were certainly biting, are outright ridicule not friendly teasing, and could therefore be risky. Accusing someone of not only being a cuckold and corruptor of children, but also of falsifying texts and breaking treaties, had a political purpose.[10]

The devices for creating satire are not specifically outlined, and indeed it is somewhat difficult to find Byzantine terms for satire in the way we understand it. The word satire seems to have had limited use, and not necessarily to correlate with our modern understanding, though we do have a reference to the τῶν ἐμῶν σατυρικῶν (satirical writings) of the twelfth-century writer Nikephoros Basilakes, who specifically states that he destroyed them in a fit of religious zeal.[11]

Satire, if we understand it as comedic invective with a critical aim, regardless of its form, seems to have been revived in Byzantium around the eleventh century having languished since antiquity. Featuring identifiable pieces of invective, and often used for "subversion in the political domain as well as invective and personal attack in the professional sphere", it may also be seen as simply a "one-upmanship game between Byzantine males".[12] Invective, wordplay and hyperbole were key elements, and excessive rudeness, whether scatological, sexual or personal in nature, was not uncommon. The distinction, or lack thereof, between satire and invective in Byzantium adds a further level of confusion, though satire arguably tends to be more expansive. It is notably complicated to separate the two at times, and both cross into other identifiable genres or literary forms, but "[W]hether the text should be considered satire, invective, vituperation, or mocking epigram, it acquires its force because it is supposed to elicit laughter. The ultimate purpose of this laughter can of course differ greatly: liberation, insight, or humiliation – perhaps almost always a subtle combination of these".[13]

Humour and Animals

Being greedy, lustful, ugly or bad at one's profession, or even being a member of a particular profession, allowed ridicule, at least within particular groups, notably the court and literary circles. Animals appeared in Byzantine satirical works often as a core part of the insults used to mock or insult these characteristics.[14] In *Mazaris' Journey to Hades*, written c.1415, a sweet-talking character who acts like a biting adder is named Aspeitaos, ἀσπὶς being a Greek term for a snake, and Melgouzes is termed "the former milker of goats", ἀμέλγω meaning to milk, particularly in relation to sheep or goats.[15] Similar invective was used in texts and letters to insult the intelligence of other scholars. John Tzetzes, the twelfth-century scholar, described a contemporary who had criticised him as "possessed and epileptic, moonstruck son of a goat" and Theodore Prodromos refers to a colleague who wishes to "improve" on Plato as being "a pig with a brilliant jewel dangling from its snout [...] a monkey with a golden slingshot in its hands".[16] Comparison with irrational animals, driven by passion and needs rather than rationality and Christian values, is clearly negative in these contexts. The insults and humour come directly from humans and are directed at figures who are also identifiably human.

We do have texts which have the jokes come directly from the mouths of animals. Two apparently satirical texts of the twelfth century, both sometimes attributed to Theodore Prodromos, though the authorship is debated,[17] and featuring animal protagonists in the form of cats and mice are the *Katomyomachia* and the *Schede tou Myos*.[18] Both works have benefitted from recent study, and deserve some consideration here even if they are not actually satire.[19] In the *Katomyomachia* or the *War of the Cat and the Mice*, the humour is drawn largely from parody, with classical tragedy at its roots, and a socio-political element in that the mice have been described as representing the "political underworld" of the period with its "vocal demagogues".[20] The poem is written in dodecasyllables, telling the story of mice waging war on the cat that threatens them, partly told from the perspective of a Lady-mouse awaiting news of the battle's outcome.

The work utilises satirical elements, alongside the format of classical drama to create a humorous yet didactic piece, similar to the *Batrachomyomachia*.[21] While the chorus criticise excessive emotion, the general aim of this mock-epic is to entertain at the same time as show off the literary elements of the work. The *Schede tou Myos* presents a mouse who is tempted by the remains of a banquet, and before tucking in is caught and eaten by a cat, despite the mouse's protestations that he is a monk and should be spared, and this apparently mocks the gluttony of monks, using familiar fable imagery. This work also has a clear didactic purpose, introducing different rhetorical formats and words alongside the humour.

Schedography was itself a school exercise, designed to help in the understanding of syntax and grammar, as well as the use of the *progymnasmata* in longer form. While the mockery of the greed of monks is found in other satirical works, including the Ptochoprodromos poems, here it is not the focus of the work as such, rather it adds to the relationship between the cat and the mouse, with its Tom and Jerry style comedy, in creating an amusing tale that makes the school exercise more memorable and thus more effective, rather than being a targeted attack. Both these works use several common comic elements, including straightforward parody and wordplay. They are appropriate in their humour and mock concepts or groups as opposed to identifiable individuals, but this is done as part of a clear educational piece, and the tone is more one of comedic parody than satirical.[22] This didactic element in particular is a potential key in our understanding of Byzantine animal satire, as will be addressed later.

The Palaiologan Animal Satires

Let us now turn to the animal works from the so-called vernacular 'animal vegetable satires' or 'epics' of the Palaiologan period, that is the late thirteenth through fifteenth centuries. The works grouped together under this term include the *Book of Fruit*, *Book of Fish*, and the *Book of Birds* as well as the *Synaxarion of the Honourable Donkey*, and the *Entertaining Tale of Quadrupeds*.

The Book of Fruit *and the* Book of Fish

The *Porikologos* (Πωρικολόγος) or *Book of Fruit*, while it does not feature animals, is almost always addressed alongside the *Book of Fish* with which it shares a number of features.[23] It is a satirical take on Byzantine court ceremony but also of Byzantine legal practice. It may also be a tract against drunkenness or mock particular individuals represented by the fruits involved.[24] The text is written in prose and, like the other 'beast literature' it is connected with, it is anonymous. It follows the story of a central character, Grape, who is denounced for perjury before Emperor Quince, found guilty and sentenced to be beaten till all their blood is gone and drunk by man. The *Opsarologos* (Ὀψαρολόγος) or *Book of Fish* shares many features with the *Book of Fruit*, as it again mocks Byzantine legal practice and is in prose.[25] This time the denounced character is Mackerel, who is accused of conspiracy before Emperor Whale. He is also found guilty but in a less brutal conclusion is shorn of his beard.

Neither text gives us a high level of information about its characters, though court titles do appear. Certain familiar associations are drawn, physical features, like Mackerel's beard, are used for effect, and the humour, while a little violent in the case of the *Porikologos*, is essentially grounded in parody of the Byzantine legal system. Neither text is easily dateable, though they are generally placed during the last centuries of Byzantium.[26] Despite the obvious satire found in the use of technical terminology and the structure of the works, similar to *passio* or the trial of a saint in format, the wild accusations of the characters and the direct use of titles, such as Apple the Logothete, Pinecone the Cupbearer, Grand Domestikos Tuna and Protostrator Swordfish, it is hard to identify any further social, or personal, criticism within either the *Book of Fish* or the *Book of Fruit*.

The lack of complexity and short length of the works may in part relate to the choice of fish and fruit as the characters, though 75 fruits are mentioned in the *Porikologos* and a similarly wide range of fish and sea creatures in the *Opsarologos*. It is easy to anthropomorphise animals with features that are familiar and similar in some way to our own. Few fish have the most aimable of faces, and fruit obviously do not, so while both may allow for clever wordplay and straightforward parody, they limit the author or authors, and seem unlikely to have been used for more personal invective.

The Book of Birds

The *Poulologos* or *Book of Birds*[27] is a poem of around 668 lines written in unrhymed vernacular verse, and takes the form of *Rangstreitdichtung*, a debate over precedence. It is preserved in seven manuscripts, ranging from Constantinopolitanus gr. Seraglio 35 which dates itself to 1461, το χφ ιδιωτικής συλλογής Ζώρα (Z), dating from the seventeenth century.[28] In the poem, the Eagle, Emperor of the Birds, invites all the birds to his son's wedding. No pretext beyond this is given for the story which continues essentially as an exchange of insults, with one guest abusing another. The 28 birds, including the bat, appear in pairs and are roughly split between those that are virtuous and those who are not, with the virtuous generally equating to tame or domesticated birds who are pitted against birds considered useless or with traditionally poor connotations.

Eventually, in the majority of the manuscripts at least, the Eagle puts an end to the insults, reiterating that the birds are in attendance for a celebration, and threatening them with an attack from the Hawk, Peregrine, Pied Falcon and Bearded Vulture, who are not part of the wider plot. Peace is thus restored. The text is in the form of a dialogue, and the focus of the arguments centres on appearance, ideas of ancestry and concepts of nobility, using the physical aspects of the birds to reflect court costume, and linking their behaviour with human errors. The concentration on appearance and status satirises the conscious and visible depiction of status in Byzantine court ceremonial and the regalia connected with it as described in the fourteenth-century ceremonial treatise "Pseudo-Kodinos".[29]

For instance, the Partridge describes her purple stockings, stating that they befit her rank as one of noble origin.[30] However, some of the insults are very much more personal, going beyond any identifiable physical features of the birds, and more directly linking the text to the political situation of the fourteenth century. For example, we have the insults directed at the Pheasant by the παραυιαλίτης, tentatively identified as the flamingo, which stress that, based on the wearing of the *atypin* (ἀτυπίν), a term used for garments of the Persian/Turkish court which became fashionable for Byzantine courtiers, anyone would think this individual was of noble birth, whereas in fact they were nothing of the sort, but having inherited some money, had bought their position.[31] The purchase of a court rank was certainly not uncommon at the time, but clearly wasn't well thought of by those with inherited rank.

Niels Gaul has identified the Flamingo with the courtier and tax collector Alexios Apokaukos, the bird being called the son of a cobbler, the man being of unknown origin.[32] Both the man and the bird are directly connected with salt-works, the bird being accused in this text of borrowing money to invest in salt-works and subsequently failing to pay his debts, the man having been superintendent of salt-works after 1320, then becoming *megas doux* in charge of the fleet.[33] Gaul links this role with the behaviour of the Flamingo, who is further accused of running away along the coast from one stronghold to another to avoid their debts. Apokaukos had indeed reinforced various coastal defences and had become an enemy of his former patron Kantakouzenos, allying with Anna of Savoy, enjoying the support of merchants and sailors and the derision of wealthy citizens and aristocrats who supported Kantakouzenos. It would thus seem that the characters presented in this work may be identifiable in a way that is uncommon for the animal satires more generally.

It is relatively easy to see the birds' attempts at social-climbing while undermining each other as reflecting the confusion of the period generally. Human flaws are highlighted, for example in the greed for titles and extravagant clothing so that the bird characters in this work reflect human behaviour. However, like the *Book of Fish* and the *Book of Fruit*, the *Book of Birds* uses well-known practices and institutions to signpost this critique, alongside the vernacular and wordplay which adds to the satirical tone, though the work retains a more serious tenor than our next two works, staying politely on the appropriate side of comedy, perhaps as it addresses a very current argument between courtiers.

The *Synaxarion of the Honourable Donkey*

The Συναξάριον τοῦ τιμημένου Γαδάρου, or the *Synaxarion of the Honourable Donkey*, as we have it today may be found in only one manuscript, Cod. Vindobonensis theol. gr. 244, dated to the early sixteenth century.[34] The manuscript is significant as it is the single largest collection of Greek literary texts in the vernacular, and additionally contains a number of non-literary, theological and scientific works, also in the vernacular. The *Synaxarion* is written in vernacular fifteen-syllable political verse and the author is anonymous. The language of the work, as well as several stylistic features and motifs, have led scholars to date it to the late fourteenth or early fifteenth century.[35] Our clearest means of dating the work however, are the references to certain weapons, namely σκλόπους (*sclopi*, hand gun) and λουμπάρδα (cannon). These feature in other texts of the same period including the *Spanos*, the *Chronicle of Morea*, Pseudo-Kodinos, and the *Chronicle of the Turkish Sultans*, often in use *against* the Byzantines rather than by them, but which do not seem to have reached the Byzantine or Ottoman empires until c.1400, making a fifteenth-century date more likely.[36] Previous editions and studies of the *Synaxarion* have combined it with a late fifteenth- or sixteenth-century text entitled Γαδάρου, λύκου κι ἀλουποῦς διήγησις ὡραῖα.[37] This has led to some confusion in the editions of the text.[38] While they share many features, and arguably come from a shared source, the two stories feature some significant differences, so it is the definitely Byzantine *Synaxarion* that is discussed here.[39]

It tells the story of the Donkey, who is taken from his field by the Fox, or rather vixen, and the Wolf, ostensibly to be educated by them. The animals travel together on a boat to the middle of the sea, where the Fox 'prophesies' an imminent storm and their subsequent deaths. To prevent this, the animals agree to make confession and sacrifice whoever has committed the worst sin, apparently in an attempt to gain God's favour and save themselves from drowning. Inevitably, the poor Donkey is chosen despite the trifling nature of his sin. Yet the outcome is not as expected; Donkey is not eaten, as Fox had intended, but saves himself by forcefully removing the wolf from the boat and threatening Fox such that she throws herself into the sea. Wolf and Fox then declare that Donkey should no longer be referred to as a γάδαρος, from the Arabic loan-word ἀείδαρος meaning 'always-beaten', but as νικόν (=ὀνικόν), another word for donkey which plays with the word Nike, victory.

The fable elements of the work are obvious, and have recently been traced to similar Italian stories which suggest a Western influence.[40] The potential satirising of religious characters is fairly clear, whether these are simply corrupt clergy who are trying to fleece their parishioners or more specifically a heretical sect like the Bogomils.[41] However, important here is the extreme language that goes alongside the otherwise familiar behaviour and format. Pretending to repent one's sins is a common feature of animal fables, and wolves becoming monks is also fairly familiar. The fox character in this story specifically undermines the tears of contrition and repentance that were part of Byzantine religious practice;

"... I squeeze myself a bit,
And I piss on my tail, I water my eyes,
And I knit my brows together,
And it looks like tears and I have great hope
That God will hold my tears in high esteem."[42]

This is not simply an admission of greed or a pretence at repentance but a direct mockery of a significant act in the Orthodox faith, using scatological humour. It appears quite offensive, but given what we know about the Byzantine penchant for crude humour, and the importance of playing with audience expectations, this was probably very funny. Despite, or perhaps because of, the religious imagery, here too we also find innuendo contrasting it. It can be found in the euphemistic references to the 'weapon' Donkey uses to beat Wolf before kicking him from the boat. We are told that Donkey appears to have an armoury in his belly in line 337, and that the weapon is long and thick in line 326 before Fox who has jumped into the sea rather than be attacked, gives a list of military equipment which it resembles, including clubs, spears, cannon and full saddlebags. The actual act of violence against Wolf is described such that Donkey, our hero, appears to have completely lost control of himself in a manner that is obviously inappropriate, but also funny;

> "The donkey hit him, ... and beat him
> With all of his strength, and as much as he was able,
> And flogged him with anger and made a mess of him.
> At first he brayed in the middle of the open sea.
> Immediately he let out screams and clubbed him,
> wet himself frequently and farted often,
> He lifts his tail, and opens his mouth
> And brays mightily and kicks and ejects him from the world"[43]

A good chunk of the humour in this text comes from the recognisable fable elements, and the parody of the clergy, alongside the hyperbole used in describing the faux-religious behaviour of Fox and Wolf, and the 'hero's' inappropriate physical behaviour and oversized attributes. The wordplay may be clever at points, and the range of technical terms used of direct interest to historians, but it is the slapstick, if extreme, violence, humour, and the victory of the underdog Donkey that stand out in this text, in direct contrast to the apparently religious nature of the story.

The Entertaining Tale of Quadrupeds

The *Entertaining Tale of Quadrupeds*[44] is perhaps the most complex of all these works, and survives in five manuscripts, the earliest of which, the Constantinopolitanus gr. Seraglio 35, is dated to 1461 by its scribe, giving us a date *ante quem*. The four other manuscripts are Parisinus gr. 2911 (fifteenth century), Vindobonensis theol. gr. 244 (sixteenth century), Petropolitanus gr. 721 (seventeenth century, formerly Lesbiacus 92) and Petropolitanus gr. 202 (sixteenth century).

The *Tale* begins with a short prologue outlining its topic and purpose. The text continues with Emperor Lion sending an embassy on behalf of the "bloody and disgusting" beasts to the "clean and domesticated" animals in order to arrange a peaceful meeting.[45] Here they will "trade witty words and come to see the merits and the faults of each part", in a manner potentially not dissimilar to the meeting of a *theatron*, essentially a literary salon, though of differing audiences and sizes, sometimes held in the presence of the emperor, or between different schools.[46] Once an agreement is reached, the animals all come together and are encouraged by Emperor Lion, the ruler of them all, to begin to jest in turn. Arguments break out amongst the animals, broadly over how they interact with humans, but also relating to stereotypes of behaviour and appearance. These form the main body of the work, before a final battle occurs in which the clean and domestic animals are victorious over those grouped together as the bloody and disgusting, and Emperor Lion is killed.

The nanny goat is derided in sexual terms, the donkey for the violence with which he is treated by his master, and the pig for his eating habits. The author signposts this humour, so that Ox warns the audience before he discusses the size and uses of his genitals, saying;

> "Indeed, if women happen to be present
> And hear the following remark and statement
> They tend to giggle and to be amused"[47]

His claims upset Donkey, who comes farting onto the stage to upbraid him and put forth boasts on behalf of his own genitals, a scene which the illustrator of the manuscript seems to have greatly enjoyed depicting.[48]

Alongside the overt innuendo we find claims for the animals' dislike of verbosity, yet inclination towards it. The animals' tentative peace is also marked by numerous threats of violence and descriptions of unpleasant deaths. The death of the Cat as described by the Rat is perhaps the most gruesome;

> "… When people find
> that you have been committing these misdeeds,
> then you'll see sticks and clubs fall on your ribs!
> And when they beat you, then you'll get the squirts,
> and you'll start farting farts like walnuts rattling!
> They'll hit you many times around the head;
> and when you've croaked it, miserable bastard,
> they'll throw you onto the dung, poor dear,
> where pigs will eat you"[49]

The drama and very practicality of the descriptions and arguments adds to the humour, contrasting the treatment of the pet or at least cared for working animal with the extremely violent death of other creatures due to their natural behaviour, perceived as wrongdoing by humans. The humour in this work seems to be tied less to the apparent frame story of a civil war, and be more focused on the invective between animals, which is predominantly exchanged between those who are ostensibly on the same side. Again we find scatological humour and sexual innuendo, with extreme violence presented if not entirely acted upon, at least until the end of the text, and then the description is much more reserved.

The frame of the *Tale* claims a didactic purpose for the poem, and has an ending similar to the *Synaxarion* in that the victorious at the end of the work are not as expected. It has been argued that both the prologue and the epilogue to the work are therefore later additions.[50] We can say that the epilogue seems to have influenced the two later manuscripts of the *Poulologos*, in which there is a battle not included in the earlier versions, again between what are essentially the domestic and the wild creatures, though in those the Emperor Eagle and his allies are victorious. Even if the prologue and epilogue were added later, which is not certain, they are present in all our manuscript versions, so were clearly viewed as a core part of the text. The didactic purpose, and the killing of an emperor, albeit in the potentially added sections, rather than the main text, are somehow both acceptable enough in the context of an animal story to have been included unquestioningly.

Why Animals?

The summaries of these animal satires give us a number of shared characteristics, at least between the more detailed works, which may help us in considering why the authors chose to have animals as the protagonists. The works utilise fable, wordplay and parody, and note the value of education, if not also their own role as didactic material. The humour in each of these works is fairly similar, whether it is verbally aggressive or depicting violent slapstick, and all highlight contemporary issues. The authors may easily be expressing a general sense of frustration, fear and fatigue caused by the struggles of the period, lightening the mood somewhat at the same time as picking apart the roots of such conflict, including the civil wars, religious controversy and personal relationships that so impacted the period.

The Palaiologan era was after all an extended period of warfare, poverty and religious argument leading to the final collapse of an empire that had considered itself an earthly reflection of heaven, awaiting the second coming. In such an environment, how safe could humour have been? It is worth noting that noticing humour in Byzantine texts is one thing, understanding why it is funny, often quite another, and even where the humour may be reasonably obvious, there is frequently another layer, or a target, which we cannot get at without deeper knowledge. That may have been equally true for contemporaries, depending on the text and audience.

Animals as Masks

In many cultures, it has been suggested that animals were used to hide behind, the idea being that depicting people as animals either made it harder to identify the individual being targeted, or that where someone recognised themselves in the work, "the targets of such mockery will refuse to make themselves ridiculous by acknowledging that any resemblance exists".[51] Indeed Phaedrus stated that fable came into being for just this reason, allowing slaves to avoid punishment for "jesting" with stories of animals, rather than directly criticising the harsh behaviour of their masters.[52] Beast literature in its various forms has thus been seen as a means for the underdog to express dissatisfaction. However, the Byzantines are in other works perfectly happy to directly name and shame those they mock, and put their own names to such works.

In the Palaiologan satires, if we are able to identify even some of the individuals, as we are in the *Poulologus*, surely they must have been extremely obvious to the audience. We even refer regularly to the Byzantine 'talent to abuse'. The authors of these texts are also not exactly 'underdogs' in the traditional sense. All are educated, though not perhaps to the same degree, and likely to have been part of Constantinopolitan society. As Paul Magdalino has pointed out, although Byzantium was authoritarian, anonymous pamphlets ridiculing political figures were common throughout the period, and the regime itself utilised parody and satire in the form of 'shame parades', in which individuals were paraded through the streets, sat backwards on a donkey, often while wreathed in entrails, while comic versions on their transgressions were chanted by the crowd.[53] Similar activities were on occasion undertaken by the general populace to express their derision of particular individuals or events, seemingly with no direct repercussions for the participants. Indeed, it seems there were professional figures, associated with the hippodrome as songwriters, who were required to create authorised satire for such occasions, at least until the twelfth century.[54]

There is, therefore, no need for the authors of our works to be underdogs *per se*, if satire as a response to political situations, in the broadest sense, was accepted, if not exactly acceptable. The key exception is in relation to the emperor, who could only be criticised indirectly, at least until after death. It was however possible to mock members of the court, and use them as scapegoats for any criticism of a regime, or to criticise the system more generally through key figures. We may see the *Poulologos* in this category, with the birds representing different sides in the ongoing disputes at the court of Emperor John VI Kantakouzenos (r. 1347–1354). The Emperor Eagle is not directly attacked, but the many wild birds, who can be identified as those opposed to Kantakouzenos, come in for very direct and personal criticism, reinforcing Kantakouzenos' position and that of his allies. Kantakouzenos was involved in several civil wars and his position as emperor was never secure, making it a prime choice for portrayal in this kind of text. Veiled and direct attacks on one's contemporaries, whether as friendly banter or invective, were also common enough, occasionally requiring an apology if misunderstood, but equally often a regular part of the competition for power and patronage.

As such, it is no wonder that the animal satires do not stray into this territory, but are rather more focused on courtly and religious power, areas that are potentially harder to attack in a more open manner. We may question if Emperor Lion in the *Tale* is really the emperor, or if the story relates to a fictive situation, or if the 'emperor' is another ruler or person of power, but the use of the term βασιλεὺς would no doubt make a Byzantine audience think of their own ruler, and potentially feel somewhat uncomfortable at the events of the text, or the reverse, encouraging the sentiment of unrest, but providing an outlet for that. As recently argued by Ulrich Moennig, this work may not relate to a specific incidence of violence at all, but simply address civil war as a general theme.[55] If that is the case, the satire is of the period more generally, and of the ridiculousness of such conflict when compared with the day-to-day struggles of the populace noted in the references to food, religion, trade and other general activities discussed by the animals themselves.

The Carnivalesque and Satire

Mikhail Bakhtin coined the term 'Carnival laughter' or the 'carnivalesque', meaning a literary mode that subverts the status quo, in the broadest sense, through humour and chaos. This somewhat anarchic comedy, he argues, was restorative. He relates it specifically to the religious festivals that occurred throughout the Christian year, and particularly those after Lent or other periods of sacrifice.[56] The carnivalesque allowed for the expression of feelings of frustrations, and often inverted normal practice for a short period, but in doing so, it aimed to reinforce normal standards and structures thereafter. Regarding the connection between the carnivalesque and animals, and specifically the donkey, Bakhtin notes that that animal traditionally symbolises "the material bodily lower stratum, which at the same time degrades and regenerates".[57]

It is worth noting that novels in which literary scholars have identified the carnivalesque tend to have several characters with different opinions and a number of layers so that no single view is privileged, not dissimilar to our poems. The violence of this type of humour has an additional role too. The comic violence displayed presents a modern audience with a degree of unease at its sheer brutality. It is reasonable to suppose that such a response could be found amongst a Byzantine audience as well. This comic violence is entertaining, because laughter is often accompanied by an "absence of feeling" or sense of superiority, and through laughter we can remove our sense of pity long enough to find entertainment in such danger.[58] Nevertheless, the potential sense of discomfort allows an audience to question that emotion, and therefore the content of the text and its related associations. One could stretch that assumption and suggest that sympathy for the non-human characters is identifiable, particularly those with whom the audience regularly interacted, creating a connection at the same time as adding distance from the people and situation being presented through those animals.

Didactic Humour

As we have seen, rules for comic writing existed in Byzantium, and humour was understood in several ways, being used to ridicule and potentially as a rhetorical tool for convincing others to agree with the author, at the cost of an opponent, and these are core in understanding satire. However, there were other attested uses for comic writing. The tenth-century scholar Arethas of Caesarea, whose views on laughter and the comic have been discussed in detail by Aglae Pizzone, stresses that humour, if used in a didactic manner, could be an appropriate means of engaging with an audience of varied abilities, so long as it remains witty and not rude.[59] Humour as a vehicle for learning, at least for Arethas, fell into the acceptable limits of laughter. By the twelfth century attitudes appear to have relaxed still further. Eustathios of Thessalonike, in his commentaries on Homer, considers the audiences' enjoyment to be key, and returns to the physical benefits of laughter, and the psychological relief it produces.[60] For Eustathios, humour is a narrative device for entertainment, and does not require another aim or frame to justify its presence. Pizzone suggests that this gradual change in attitudes to the comic can be connected to the more positive engagement with classical texts, as well as to the performative presentations of literature in *theatron*.[61]

The fable elements of these works may add something further to this angle. It is not uncommon for us to find violence within fables either, and the moral element of these works, with their highly recognisable motifs, could also reinforce the satirical message. Obviously not all fables feature animals, but many do. Like the *schedos* featuring a cat and mouse, the animals make the message more straightforward and memorable. Fables were used in the schoolroom, but also in sermons, mirrors for princes, and a variety of other works, as they can briefly and effectively provide a lesson or guidance to their audience. Not dissimilarly, the *Physiologus*, one of the most popular works of the Middle Ages, presents certain information about animals, as well as some plants and stones, and then provides a Christian allegory of them.[62] Violence is not prevalent, but animal actions are very much highlighted as examples of good or bad human behaviour. So too in the hexaemeral literature of the Church Fathers, where again, animal behaviour is contrasted with that of humans, often to provide a lesson, Creation being readable and a means of becoming closer to God.[63] Animals, as Derrida famously stated, are good to think with, and as satire asks its audience to question, this may be another reason for animals to appear as protagonists. Not only do they act as a safety barrier between the target and the author, they encourage a degree of open-minded enquiry from the audience who is familiar with the need to look for a moral or *exempla* when animals are discussed in this manner. Add this to the vernacular language used in these texts, and we have another possible reason for the use of animals in satire, accessibility.

Animals as a Unifying Mechanism

Perhaps the jokes here are far more obvious, and less abusive than they appear. Animals in and of themselves are amusing, otherwise the prevalence of cat videos online today is hardly explicable. Living closely with a wide array of animals, though admittedly probably not so closely with elephants and lions, doesn't need to remove amusement at normal animal antics. It is also no doubt funny that the same animals that act as *exempla* via texts like the *Physiologus* can be presented as acting just as mean, supercilious and pompous as we sometimes do. The concerns of the animals in these texts, and their relationships, may as well be between family and neighbours as between courtly contemporaries, the basic level of the humour loses little from our inability to spot particular people in the depictions.

We, and the Byzantine audience, all know someone who thinks they're smarter than others, but has been tricked by one they consider lesser, or someone who shows off their wealth in an attempt to appear cultured, only to be shown up in some other way. At its heart, the simple idea of animals having human concerns, and making them appear like us rather than the usual act of humans behaving like animals, is laughable to an audience which is broadly secure in their status as God's chosen, if not in this life, then in the next. In this sense, these works act as a way of bringing together humanity in a time of strife, openly mocking earthly concerns and normal human behaviour in an inclusive manner. No one is left out, be they ecclesiastic, peasant, emperor, male or female, of Latin, Orthodox or even potentially Muslim faith. The parody in these texts, while it may be of religious learning or courtly intrigues, is also focused on humanity, and the subversion between the rational and irrational creatures of God stresses a line of difference that unites the human population of the Empire in its various forms. Fundamentally, using animals to depict conflicts between different groups makes our similarities more obvious, mitigating the conflict to an extent. The animal highlights social conditioning, and how ridiculous that often is, by removing the perceived barriers between the rational and the irrational.

Where traditionally the use of animals to critique has been seen as a means of masking the insult in order to protect the writer, here it is more of a format which allows for new extremes. Placing such words in the mouths of animals, and presenting them as ridiculing and physically harming each other, does put them at a degree of remove from their human counterparts, thus making everything marginally less personal and somehow more acceptable. It adds a further layer of licence to the use of the vernacular language itself. Animal satires in the Palaiologan world are perhaps like the 'comedy roasts' popular in America, an acceptable means of showering insults in anticipation of their acceptance with good humour. If we think that these works would have been performed in a *theatron* for example, where everyone in the audience would have been familiar with one another, even if they were not all of the same status or opinions, adding in the animal humour would have extended the works' accessibility and thus inclusiveness. Animals make the humour more universal and easier to spot, particularly if some were slower to pick up on the puns than others.

Later Reception

It is worth noting that these works have often reached us via later manuscripts. For example, the *Entertaining Tale of the Quadrupeds* comes to us in five manuscripts, the earliest of which is dated to 1461, and the latest of which comes from the seventeenth century. Equally the *Synaxarion of the Honourable Donkey* may be 150 years older than the first manuscript we have containing it. While the scribes and patrons of these manuscripts may not be at such a remove from these texts as we are, it is questionable whether they would have a much better idea about some of the specific humour than we do. Older scholarship was somewhat dismissive of these works, largely for their vernacular language and somewhat messy style, and even advocates of these works often have little praise for the skills of some, if not all, of the authors. The didactic purposes of these texts, where they are claimed, may also have been less clear by the time they were copied, and they are unlikely ever to have been considered examples of 'good Greek'. And yet they were copied. It may be suggested that this is as much for the purposes of nostalgia and a general sense of the need to preserve texts after the fall of the Empire.

The copying of vernacular texts together in particular manuscripts also indicates there was something about this use of language that encouraged their collection and grouping. Perhaps the scribes or patrons who determined the content of these manuscripts had a particular interest in the day-to-day lives of the Byzantines, but it seems just as likely that they copied these texts because someone, somewhere, found these stories funny and relatable enough to copy them out, warts and all. Indeed, the title of the *Entertaining Tale of Quadrupeds* is recorded in that format in one late fifteenth-century manuscript, while in two other corpuses dating from around the same time, it is titled "Amusing verses for one's merriment, and all its stories are most comical" or a variation thereof.[64] Clearly there is something funny in these texts that doesn't necessarily require specific knowledge, and was recognisable to Byzantines living under Western or Ottoman rule, sometimes even long after the texts are likely to have been written.

Indeed, in form and basic elements, the Palaiologan animal satires are remarkably similar to other overtly humorous texts in Byzantium, especially those which use the vernacular; they parody the familiar, contrast the religious with the sexual or violent, and polite society with the most disgusting terminology.[65] That said, the satirical destruction of powerful figures, or the carnivalesque reinforcement of the status quo may have been equally valid reasons for copying the works. The 'emperors' may not have been understood as Byzantine at all by the later audiences of these works, but the message carried within the animal stories, and the ability to laugh at human foibles they allow, can have been equally valid.

Conclusion

The uses for satirical humour in Byzantium seem hardly to be different than those which the American author and environmentalist Edward Abbey noted as being his motivations for writing in the twentieth century –

> "Not so much to please, soothe or console, as to challenge, provoke, stimulate, even to anger if necessary—whatever's required to force the reader to think, feel, react, make choices.... I write to amuse my friends, and to aggravate—exasperate—ulcerate—our enemies."[66]

The Byzantines clearly understood that satire could be educational, and that different audiences could take different things from the same text. There is no reason for the comedy to take away from the overtly stated intentions of any of these works, and the animals further the apparent aims of didacticism in their voiceless capacity as traditional vehicles for information and examples of appropriate or inappropriate behaviour. Sexual innuendo, slapstick and clever wordplay are always funny, just maybe more so when they are contrastingly presented by those who fight, eat and breed with no higher purpose. Animals in satire allow for a more violent expression of rebellion, while at the same time being capable of simply unifying those of disparate viewpoints and backgrounds. Fundamentally, animals were just as amusing in the Byzantine era as they are today.

Notes

1 See R. Bouchet, *Satires et parodies du Moyen Âge grec* (Paris, 2012); P. Marciniak and I. Nilsson (eds.), *Satire in The Middle Byzantine Period: The Golden Age Of Laughter?* Explorations In Medieval Culture 12 (Leiden, 2021).

2 See J. Haldon, "Humour and the Everyday in Byzantium," in *Humour, History and Politics in Late Antiquity and the Early Middle Ages*, ed. G. Halsall (Cambridge, 2002), 55.

3 M. D. Lauxtermann, *Byzantine Poetry from Pisides to Geometres: Texts and Contexts*. Wiener Byzantinische Studien 24/2 (Vienna, 2019), vol. 2, 122.

4 Saint Basil, *Ascetical Works*, trans. M. M. Wagner. Fathers of the Church 9 (New York, 1950), 271. Basil the Great in *Asceticon magnum sive Questiones*, ed. J.-P. Migne, PG 31 (Paris, 1857), 961:30–31.

5 S. Halliwell, *Greek Laughter: A Study of Cultural Psychology from Homer to Early Christianity* (Cambridge and New York, 2008), 509–10.

6 A. Pizzone, "Towards a Byzantine Theory of the Comic?," in *Greek Laughter and Tears: Antiquity and After*, ed. M. Alexiou and D. Cairns (Edinburgh, 2018),159.

7 *The Chronicle of the Logothete* §131.27, trans. S. Wahlgren, *The Chronicle of the Logothete*, Translated Texts for Byzantinists 7 (Liverpool, 2019), 184–85. Symeon the Logothete, *Symeonis Magistri et Logothetae Chronicon*, ed. S. Wahlgren, CFHB 44.1 (Berlin and New York, 2006), 131:26-27.

8 Lauxtermann, *Byzantine Poetry*, 119–23.

9 See for example, *Progymnasmata: Greek Textbooks of Prose Composition and Rhetoric*, trans. G. A. Kennedy (Atlanta, Ga., 2003).

10 See for example the attack on Leo Choirosphaktes, a high-ranking court official, translated in Lauxtermann, *Byzantine Poetry*, 125–26.

11 For a discussion of the use of the term satire in Byzantium see P. Marciniak, "The Art of Abuse: Satire and Invective in Byzantine Literature. A Preliminary Survey," in *Eos* 103.2 (2016): 351–53. For the destruction of Basilakes' satirical writings see P. Magdalino, "Appendix. Nikephoros Basilakes on His Own Satirical Writings" in *Satire in The Middle Byzantine Period: The Golden Age Of Laughter?* ed. P. Marciniak and I. Nilsson. Explorations In Medieval Culture 12 (Leiden, 2021), 332–34.

12 C. Messis, "The Fortune of Lucian in Byzantium" in *Satire in The Middle Byzantine Period*, ed. Marciniak and Nilsson, 38; Lauxtermann, *Byzantine Poetry*, 133.

13 F. Bernard, "Laughter, Derision, and Abuse in Byzantine Verse" in *Satire in The Middle Byzantine Period*, ed. Marciniak and Nilsson, 39.

14 P. Eliopoulos, "Τα ζώα στον προσβλητικό λόγο των Βυζαντινών," *Byzantina Symmeikta*, 31 (2021): 51–120.

15 *Mazaris' Journey to Hades: or, Interviews with Dead Men about Certain Officials of the Imperial Court*, ed. J. N. Barry, M. J. Share, A. Smithies and L. G. Westerink (Buffalo, 1975), 44:30–31 ὡς ἀσπὶς βύων τὰ ὦτα, ἐκεῖνος ὁ Ἀσπιέταος ὥσπερ ἀσπὶς δάκνυων; 44:4 ὁ τὰς αἶγας πρότερον Μελγουζῆς ἀμελγων.

16 John Tzetzes, *Commentarii in Aristophanem*, ed. L. Massa Positano, D. Holwerda et al. (Groningen and Amsterdam, 1960–62), 835:9; See G. Podesta, "Le Satire Lucianesche di Teodoro Prodromo," in *Aevum* 21 (1947): 3–12.

17 For a discussion of the attributed authorship see P. Marciniak, "A Pious Mouse and a Deadly Cat: The Schede tou Myos, attributed to Theodore Prodromos," *GRBS* 57.2 (2017): 509-12.

18 H. Hunger, *Der byzantinische Katz-Mäuse-Krieg. Theodore Prodromos, Katomyomachia* (Graz, 1968), 7; J. Fr. Boissonade, *Anecdota graeca* I (Paris, 1829), 429–35; K. Sathas, Μεσαιωνικὴ Βιβλιοθήκη VII (Venice, 1894), 114–17; K. Horna, "Analekten zur byzantinischen Literatur," in *Jahresbericht des k. k. Sophiengymnasiums* (Vienna, 1905): 12–16; J.-Th. Papademetriou, "Τὰ σχέδη τοῦ μυός: New Sources and Text," in *Classical Studies presented to Ben Edwin Perry* (Urbana, Ill., 1969), 210–22; most recently M. Papathomopoulos, "Τοῦ σοφωτάτου κυρου Θεοδώρου τοῦ Προδρόμου τὰ Σχέδη τοῦ μύος," *Parnassos* 21 (1979): 376–99.

19 P. Marciniak, "A Pious Mouse and a Deadly Cat," 507–27; and P. Marciniak and K. Warcaba, "Theodore Prodromos' *Katomyomachia* as a Byzantine Version of Mock-Epic," in *Middle and Late Byzantine Poetry. Texts and Contexts*, ed. A. Rhoby and N. Zagklas (Turnhout, 2018), 97–110.

20 A. Karpozilos and A. Kazhdan, "Mice," in *The Oxford Dictionary of Byzantium* ed. A. P. Kazhdan, A.-M. Talbot, A. Cutler, T. E. Gregory and N. P. Ševčenko (Oxford, 2005), www-oxfordrefere nce-com.ezproxy.is.ed.ac.uk/view/10.1093/acref/9780195046526.001.0001/acref-9780195046 526-e-3516?rskey=ZAIHKN&result=1. (Accessed 7 May 2023).

21 For a more detailed analysis see the chapter by K. Piotrowska in this volume.

22 C. Messis and I. Nilsson, "Parody in Byzantine Literature," in *Satire in The Middle Byzantine Period*, ed. Marciniak and Nilsson, 72–3.

23 *Porikologos: Einleitung, kritische Ausgabe aller Versionen, Übersetzung, Textvergleiche, Glossar, kurze Betrachtungen zu den fremdsprachlichen Versionen des Werks sowie zum Opsarologos*, ed. H. Winterwerb (Cologne, 1992) and M. Bartusis, "The Fruit Book: A Translation of the *Porikologos*: translated from Byzantine Greek with an Introduction" *Modern Greek Studies Yearbook* 4 (1988): 205–6.

24 It is found in the following manuscripts; Vindobonensis theol. gr. 244, Parisinus gr. 2316, Petropolitanus gr. 488, Bodleianus Seld. supra 15, Ambrosianus O 1175 supra, and Μετεώρων Βαρλαάμ 195. A few verses are also preserved in Petropolitanus 202.

25 The *Book of Fish* is found in one manuscript, χφ Escoraliensis Ψ IV 22.

26 Bartusis, "The Fruit Book," 205–6. Bartusis argues that since no court titles or functions which are exclusive to the twelfth century or earlier are found in any of the manuscripts it is likely that the text is Palaiologan.

27 Ὁ Πουλολόγος; κριτική ἔκδοση με εισαγωγή, σχόλια και λεξιλόγιο, ed. I. Tsavare, (Athens, 1987).

28 The full list of manuscripts that preserves the texts is Constantinopolitanus gr. Seraglio 35, Petropolitanus gr. 202, Vindobonensis theol. gr. 244, Petropolitanus gr. 721 (formely Lesbiacus 92), Escorialensis gr. 496, Αθηναϊκό χφ αρ. 701 της Εθνικής Βιβλιοθήκης της Αθήνας and χφ ιδιωτικής συλλογής Ζώρα (Z).

29 *Pseudo-Kodinos and the Constantinopolitan court: offices and ceremonies*, ed. R. Macrides, J. A. Munitiz and D. Angelov (Farnham, 2013).

30 See Ὁ Πουλολόγος, ed. Tsavare, verses 188–93.

31 Ibid., verses 271–81. For *atypin* see N. Gaul, "The Partridge's Purple Stockings: Observations on the Historical, Literary, and Manuscript Context of Pseudo-Kodinos' Handbook on Court Ceremonial," in *Theatron: Rhetorische Kultur in Spätantike und Mittelalter*, ed. M. Grünbart (Berlin, 2007), 93.

32 Gaul, "The Partridge's Purple Stockings," 93–4. See also G. Makris, "Zum literarischen Genus des Pulologos, " in *Origini della letteratura neogreca. Atti del Secondo Congresso Internationale. Neogreca Medii Aevi (Venezia, 7–10 Novembre 1991)* vol. 1 (Venice, 1993), 391–412.

33 R. Guilland, "Études de civilisation et de littérature byzantines, I: Alexios Apocaucos," *Revue du Lyonnais* (1921): 523–41.

34 For an in-depth discussion of the manuscript itself and its scribes, see P. Vejleskov, "Codex Vindobonensis theologicus graecus 244," in *Copyists, collectors and editors: manuscripts and editions of late Byzantine and early Modern Greek literature*, ed. H. Eideneier, A. F. Van Gemert and D. Holton (Heraklion, 2005), 179–214.

35 See U. Moennig, "Das Συναξάριον τοῦ τιμημένου Γαδάρου: Analyse, Ausgabe, Wörterverzeichnis," in *BZ* 102 (2009): 128.

36 M. C. Bartusis, *The Late Byzantine Army: Arms and Society, 1204–1453* (Philadelphia, 1997), 334–41.

37 See Moennig, "Συναξάριον," 110. See for example W. Wagner (ed.), *Carmina Graeca medii aevi* (Leipzig, 1874), 112–23.

38 See Moennig, "Συναξάριον," 131–34.

39 For more detail on both works, and the potential common source, see M. Lauxtermann and M. C. Janssen, "Asinine Tales East and West: The Ass's Confession and the Mule's Hoof," *BZ* 112.1 (2019): 105–22.

40 Lauxtermann and Janssen, "Asinine Tales East and West," here 113.

41 For the discussion on the Bogomils specifically see K. Stewart, " 'Who has the most faults?': Animal Sinners in a Late Byzantine Poem" in *Fallen Animals: Art, Religion, Literature*, ed. Z. Hadromi-Allouche (Lanham, Md., 2017), 80.

42 "Συναξάριον," ed. Moennig, ll. 203–7, "καὶ δάκρυα οὐδὲν ἔχω καὶ σφίγγομαι ὀλίγο,/καὶ τὴν οὐράν μου κατουρῶ, τὰ ὀμμάτιά μου βρέχω,/καὶ εἰς τὰ ματοφρύδια μου κρεμάζονται οἱ κόμποι,/καὶ ὁμοιάζουν δάκρυα καὶ ἔχω μέγα θάρρος,/ὅτι ὁ θεὸς τὰ δάκρυα περὶ πολλοῦ τὰ ἔχει."

43 Ibid., ll. 310–17, "Ὁ γάδαρος βολίζει τον, τσιληπουρδᾶ καὶ κροῦ τον,/ μὲ ὅλην του τὴν δύναμιν, ὅσον καὶ ἂν ἐδυνήθην,/ καὶ ἐκτύπησέ τον μὲ θυμὸν καὶ ἐχαρβάλωσέ τον./ Εἰν πρώτοις τὸν ἐρόντζεψεν στὴν μέσην τοῦ πελάγους./Εὐθὺς βάνει τὸν φώναρον καὶ ἐξεματσουκώνει,/καὶ σφυροκατουρεῖ συκνὰ καὶ συχνοπορδαλίζει,/σηκώνει τὴν οὐράκλαν του, ἀνοίγει καὶ τὸ στόμα,/καὶ οὐριάζει δυνατά καὶ τσινᾶ κ'ἐξοίκισεν τὸν κόσμον"

44 *An Entertaining Tale of Quadrupeds: Translation and Commentary*, trans. N. Nicholas and G. Baloglou (New York, 2003).

45 For a discussion of these groups, and the potential that they are a set of four not two, and on the format of *Rangstreitdichtung* see G. Prinzing, "Zur byzantinischen Rangstreitliteratur in Prosa und Dichtung," *Römische Historische Mitteilungen*, 45 (2003): 241–86.

46 *An Entertaining Tale of Quadrupeds*, ed. and trans. Nicholas and Baloglou, ll. 65–66, "λόγους νὰ συνάρωμεν τινὰς ἐκ τῶν ἀστείων/ καί ἴδωμεν τοῦ πασᾶ ἑνὸς τὸν ἔπαινον καὶ ψόγο."

47 Ibid., ll. 632–34 "πολλάκις γὰρ ἂν εὑρεθοῦν καὶ γύναια ἐκεῖσε/καὶ νὰ ἀκούσουν τὸ παρὸν ἀπόφθεγμα καὶ ῥῆμα,/ἂν τύχη νὰ γελάσουσιν καὶ νὰ ἐμνοστευθοῦσιν."

48 C Graecus Seraglio f.35v and f.54v reproduced in *An Entertaining Tale of Quadrupeds*, Nicholas and Baloglou, 231.

49 *An Entertaining Tale of Quadrupeds*, ed. and trans. Nicholas and Baloglou, ll 168–74 " ὅταν ταῦτα σὲ εὕρωσιν ποιοῦντα οἱ ἄνθρωποι,/νὰ εἶδες ῥαβδὲς καὶ ματζουκιὲς ἀπάνω στὰ πλευρά σου,/ καὶ νὰ σὲ κροῦν καὶ νὰ τζιλᾶς, νὰκλάνης καρυδᾶτα./Πολλάκις εἰς τὴν κεφαλὴν νὰ τύχη νὰ σὲ δώσουν/καὶ νὰ ψοφήσης, ἄθλιε, καὶ νὰ σὲ ῥίψουν ἔξω/ εἰς τὴν κοπρέαν, ἄτυχε, καὶ νὰ σὲ φᾶν οἱ χοῖροι." A comparable punishment for a cat caught eating meat is mentioned in Ptochoprodromos Poem IV.

50 For discussion of the prologue see R. Beaton, *The Medieval Greek Romance*, 2nd edition (London,1996), 269 and P. Vasiliou, " Κριτικές Παρατηρήσεις στην «Παιδιόφραστο Διήγηση»," in *Ελληνικά* 46 (1996), 62–68, and the same work 72–77 on the epilogue, as well as *An Entertaining Tale*, trans. Nicholas and Baloglou, 449–59.

51 J. M. Ziolkowski, *Talking Animals: Medieval Latin Beast Poetry, 750–1150* (Philadelphia, 1993), 7.

52 Babrius and Phaedrus, *Fables* ed. and trans. B. E. Perry (London, 1965), book 3, prologue, 33–37.

53 P. Magdalino, "Political Satire" in *Satire in The Middle Byzantine Period*, ed. Marciniak and Nilsson, 105.

54 Ibid., 112–13.

55 U. Moennig, "Η Διήγησις των τετραπόδων ζώων. Αφήγημα εμφυλίου πολέμου," in *The Greek World in Periods of Crisis and Recovery, 1204–2018. Proceedings of the 6th European Congress of Modern Greek Studies* (Athens, 2020), 411–26.

56 M. Bakhtin, *Rabelais and his World*, trans. H. Iswolsky (Bloomington, Ind., 1984), 99.

57 Ibid., 78.

58 H. Bergson, "Laughter" in *Comedy*, ed. W. Sypher (Garden City, N.Y., 1956), 63.

59 Pizzone, "Towards A Byzantine Theory Of The Comic?," 153.

60 Ibid., 163–5.

61 Ibid., 165.

62 See *Der Physiologus nach den Handschriften G und M*, ed. D. Offermanns (Cologne, 1966); *Physiologus*, trans. M. J. Curley (Chicago, 2009); P. Cox, "The *Physiologus*: A *Poiēsis* of Nature,"

Church History 52.4 (1983), 433–43 and M. J. Curley, "*Physiologus, Φυσιολογία* and the Rise of Christian Nature Symbolism," in *Viator,* 11 (1980): 1–10.
63 See F. E. Robbins, *The Hexaemeral Literature: A Study of the Greek and Latin Commentaries on Genesis* (Chicago, 1912); P. M. Blowers, *Drama of the Divine Economy: Creator and Creation in Early Christian Theology and Piety* (Oxford, 2012).
64 *An Entertaining Tale of Quadrupeds,* ed. Nicholas and Baloglou, 253–54.
65 For example, *Mazaris' Journey to Hades* and *Spanos,* referenced above.
66 See M. P. Branch, "Are You Serious? A Modest Proposal for Environmental Humor," in *The Oxford Handbook of Ecocriticism,* ed. G. Garrard (New York, 2014), 388.

9

MAN, BEAST AND NATURE

Descriptions of Hunting in Byzantine Literature

Charis Messis and Ingela Nilsson

Byzantine descriptions of hunting have a rich narrative potential that can be interpreted across several literary forms and which convey, in addition to practical information on hunting itself, different messages of an aesthetic, moral or ideological kind.[1] In Greco-Roman Antiquity, hunting was a subject in mythological narratives or in the particular genres of bucolic poetry and didactic literature.[2] In Byzantium, however, these two literary forms were replaced largely by the epideictic genre. Descriptions of hunting appear either as independent *ekphraseis* adopting the form of *progymnasmata*, poems or letters, or as descriptive episodes within larger narrative texts of historiography, hagiography, romance or didactic poetry. In such contexts, hunting becomes another way to speak of political power, to praise or reprimand an emperor or a general, to reveal a hidden characteristic in a person, or simply to create or stage adventure. There are also more neutral references to hunting activities in various texts and hunting descriptions can be found in *ekphraseis* of certain animals, especially in those devoted to dogs.

All these different forms and uses turn descriptions of hunting into a broader reflection on Byzantine literature and its way of conceiving and representing man, beast and nature. Here we will deal with autonomous and semi-autonomous *ekphraseis* and some other descriptions characteristic of the way in which hunting metamorphoses into a literary motif. In order to avoid a merely generic categorization, we offer in the following a discussion based on three social and ideological functions of hunting: first, the 'ordinary' hunt; second, the heroic-'imperial' hunt; and third, the sportive-'aristocratic' hunt. The first category spans the entire Byzantine period, while the other two belong to specific historical periods.

The 'Ordinary' Hunt

Hunting was originally perceived as an economic activity, either performed by ordinary people in their need for food or carried out by servants of wealthy people in order to add varied and 'exotic' meat to an already rich table.[3] The game would be hares, partridges or deer, and the hunters most often remain anonymous. This kind of activity, widespread in the countryside, does not leave many literary traces in Byzantine culture; it is considered

DOI: 10.4324/9781003055877-12

a banal activity of no relevance to literature. Exceptions may appear when the hunt is part of a setting that aims to present an exceptional fact or person, especially in hagiography; or when nature is a source of pleasure or fear, especially if hunting is described through the lens of a city man imagining the countryside and the emotions it evokes. In such descriptions, the economic aspect of hunting is often overshadowed by the glorification of human activities in nature. Let us look at a few examples, keeping in mind that these scenes play a specific narrative role in each story.

It is important to note that nature is constructed as an emotional landscape, especially in the literary register of *progymnasmata* and epistolography. The model description of an ordinary hunt, written by Libanius, dates to Late Antiquity. The author here emphasizes the beauty of the hunt and the emotions it provokes in the spectators. The hunters are people who live off or for this activity and who reside in the neighbouring town.[4] The hunt is carried out with the use of horses and dogs, but the game is not specified – Libanius speaks of a beast (θηρίον) living in the mountains. However, it becomes clear that he is not describing a specific, but a 'typical' hunt; he concludes his rhetorical exercise: "these things are a delight to hunters in view of the danger, but a crime on the part of those watching if kept silent."[5] Libanius designates the relationship that lingers throughout the Byzantine period between intellectual and hunter, between man of letters and man of action. The one is nourished by the emotions that the dangers of the hunt provide, the other is overwhelmed by the thrill of a sight worth describing. The only animals that receive any attention here are the ones who help men in hunting; the game seems to be of minor importance.

The idyllic staging of nature, which is only just touched upon in Libanius' *progymnasma*, will later become a key element in another literary context in the Byzantine period. The number of autonomous *ekphraseis* and *progymnasmata* declined considerably in Middle Byzantine literature and epistolography came to cultivate this rhetorical tradition. The tenth-century author Theodore Daphnopates, in a letter addressed to Nikephoros, *xenodochos* of Pylae, offers an idealization and idyllization of life in the countryside, where "everything happens in such a way as to rejoice the heart. Any cause of sorrow disappears, everything comes to fill the soul with joy and gaiety."[6] The hunt that Daphnopates describes lingers on the border between that of anonymous people and that of the aristocracy, to be examined below. In the framework of the bucolic autumn happiness that he depicts, hunting ranks among the most popular country activities alongside vintaging, harvesting and fishing:

It is also the season to organize hunting parties (κυνηγέσια); weapons are supplied against wild animals, the club is raised, the cutlass is sharpened, the spear is sharpened, the lance is prepared. The troop of connoisseurs is formed – servants, free men, hired servants, relatives, in a word all those who are capable of taking part in the pursuit; with you they attack the hills and jump into the wooded valleys; they explore the forests and the cover of thickets, ranged in line at a short interval and sometimes closing in a circle, they search the thick bushes to distinguish the game, they rush in pursuit. All the dogs capable of running and scouting accompany you without difficulty and guide you straight to the desired game.[7]

This image of a large-scale organized hunt gives way to the detailed description of a hare hunt.[8] The author again insists more on the beauty of the sight than on the usefulness of the game ("it is a most pleasant and exciting spectacle"[9]); his gaze is aesthetic and the reactions of the participants are limited to their sensations and not to any material profit: "So what a day for you when you caught the game! How could the word capture the intensity of your joy? It is as if the whole soul were rushing out, tense with the attention to the hunt and jubilant at its victory."[10] This literary hunt is an urban construction, an amusement and an aesthetic experience.

A different image prevails in the hunts depicted in hagiography. In the *Life of Theoktiste* (BHG 1724), written in the tenth century by Niketas Magistros, an eminent scholar at the court of Leo VI, and reproduced by Symeon Metaphrastes in his *Menologion*, the hunters organize themselves as a company (ἑταιρεία).[11] They go from the island of Euboea where they live to the desert island of Paros, "for the island [of Paros] has an abundance of game, deer, and wild goat."[12] This second type of animal offers the opportunity for a little *ekphrasis*, for these animals are "a marvel to behold and describe."[13] The purpose of this hunt, carried out by simple people and lasting for days (ch. 19), taking place in the forest (chs 12 and 18), was to stock up on meat, hunting being seen as a central activity of the countryside micro-economy. Even if the author constructs this part of the story in implicit dialogue with Dio Chrysostom's *The Euboean Discourse, or the Hunter*,[14] the image of the hunt – beyond the impression of a marvelous and idyllic nature – corresponds to what we would call an ordinary hunt.

The meeting between the hunters and Theoktiste exemplifies a topos of monastic literature: the discovery of the concealed holiness of a hermit. There are several variations on this theme and the hunters play the same role as in the *Life of Theoktiste*. In the *Life of Theodore of Cythera* (BHG 2430), the inhabitants of Monemvasia, having arrived in the desert Cythera to hunt wild donkeys and goats discover the intact corpse of the saint.[15] In the *Life of Peter the Athonite* (BHG 1505), a hunter discovers the saint in a remote place and becomes a channel of his holiness: "a hunter (θηρευτής), taking his bow and his quiver, went out towards the mountain to hunt." He came across an extraordinary deer: "Seeing that the deer was very large and very beautiful, he abandoned everything and followed it throughout the day … looking for a way to catch it."[16] The deer – the Christological animal *par excellence* – finally led him before the saint, first perceived as a terrifying vision; the saint then told his own story to the hunter, who became the means for Peter to transmit his paradigm of holiness.

The encounters between hunter and saint can also take a different turn, as in the *Life of Paul the Younger* (10th century). Here, a hunter named Theophanes takes his dogs to hunt in the nearby reeds. When the dogs sense the saint's presence they start to bark and Theophanes, thinking they are barking at an animal, urges them to attack; he takes his bow and shoots three arrows at him.[17] In the *Life of Barbaros the Younger* (BHG 220), rewritten by Constantine Akropolites, there is another version of the same theme.[18] Barbaros retires to the mountains after a military defeat and becomes a fearsome brigand; after various mischiefs and a miraculous encounter, he converts to the true faith and lives the life of wild animals. He finally meets the punishment of his former deeds by the arrows of hunters who take him for a beast. These arrows are his 'martyrdom' and, at the same time, the means of expiation for his misdeeds: "Drink, Barbaros, the cup that you have offered to others," says Barbaros to himself before dying.[19]

In hagiography of the ninth and tenth centuries, the hunter thus becomes the mediator between the civilized world (of the faithful) and the 'wild world' (of the saints). He is a liminal figure who can handle the margins and he is accordingly transformed into a powerful literary image, embodying the anonymous countryman whose main activity is to provide food.

All these texts (*progymnasmata*, letters, hagiographical narratives) show that hunting is part of a utilitarian countryside logic, subordinated to the requirements of an agricultural and pastoral economy. It is a hunt that most often takes place collectively. But these texts also share the two images that define nature in the Byzantine imagination: peaceful and idyllic, yet at the same time threatening and bloodthirsty, hospitable and inhospitable – an ambiguous space.[20] In fact, the hunting space is an uncultivated space, most often wooded and partly inaccessible to the inhabitants of neighboring towns and villages.[21] Countryside hunting is primarily a hunt for small or medium-sized animals (hares, stags) and birds (partridges), rarely for aggressive wild animals such as boars, wolves and bears. Such beasts cause problems for agricultural communities when they approach villages and monasteries or when they harm the crops. Paul the Silentary, a poet of the sixth century, for example, speaks in one of his poems of a boar, "tireless destroyer of plants laden with fruit," killed by a hunter,[22] and several *Lives* of saints mention harmful beasts that surround the community.[23]

The Heroic-'Imperial' Hunt

The hunt for large beasts (lions, bears, wild boars, large deer) is the privilege of heroes, courageous figures of legend or imagination. With this kind of hunt, we move away from food and economy; we are now dealing rather with a rite of passage that lends different forms of power to its performer.[24] It is an obligatory part of a valiant person's journey to supreme power, related to one of the oldest myths in human history and, in our case, one of the founding myths of the political ideology of the Middle Byzantine period. According to this political mythology, hunting is for man a sacrament of royalty, because power is attributed like a kind of game to the most valiant, to the best of the community. There are no *ekphraseis* proper of this kind of hunt, but the descriptions often adopt ekphrastic elements. The case of Basil I (r. 867–886) offers the most characteristic episode in historiography and will serve as our example here.

In the *Life of Basil* – an account of the first emperor of the Macedonian dynasty, written in the circle of his grandson Constantin VII – hunting offers a setting for each decisive stage in the hero's life and all dynastic issues are 'resolved' during a hunt. Basil's first exploit took place during a hare hunt organized with great pomp by Michael III. In this episode, the game does not matter – only the circumstances. During this hunt, Basil manages to ride the emperor's horse, on whose back usually only the emperor is permitted to ride. Leaving aside all the connotations that such an act could have for the reader-listener of the story, we will simply note that the success gives Basil visibility and a position in the imperial entourage as *protostrator* (responsible for the hunt).[25] The second hunting episode follows in the next chapter: it takes place at the Philopation, the imperial reserve of animals outside the walls of Constantinople.[26] This is a ritualized hunt, organized by the new *protostrator* Basil himself, carrying "the imperial club" (τὸ ῥόπαλον τὸ βασιλικόν). The appearance of a wolf creates panic in all participants and only Basil confronts it and kills it with his club (ch. 14). In the first episode, Basil rides the imperial horse; in the second, he uses the imperial club – he thus takes over all attributes with which Michael was exclusively invested. According to the author, that is why the uncle of the emperor, as well as Leo the Philosopher, have reason to see the result of this hunt as a bad omen for the fate of Michael's dynasty.

The third hunting scene offers the setting for a failed murder attempt that Michael's entourage is preparing against Basil (ch. 24). The hunt also, in an indirect manner, seals the death of the emperor: his death is the result of an illness caught by a fall during the hunt (ch. 102). Basil's contact with power is thus always defined in relation to hunting, which in the *Life* becomes almost a metonym for court and royalty. This is a choice of the biographer; in another tenth-century biography of Basil, the *Reign of the Emperors* by Genesios, Basil's hunting exploits do not play the same narrative and ideological role. In this text, the taming of the horse is disconnected from the hunting context, whereas Basil's hunting prowess is summarily presented: "Basil exceeded even the Centaurs in hunting."[27] There is also a cursory reference to his killing a deer and a wolf in the presence of the emperor and to the fact that his death was provoked by a fall while hunting.[28] In the *Chronicle* of Symeon the Logothete, the taming of the horse has no connection with hunting and there is no reference to the hunting exploits of Basil; Basil, on the other hand, is presented as defeated by a large deer and dead from the fall this caused.[29]

As we have seen, hunting in political discourse may serve as both a means of legitimization and of delegitimization. Hunting can be a sign of bravery, but also one of vanity, frivolity and lack of seriousness – signs of a tyrant. Let us return once more to the example of Basil I and the way in which he is represented in the *Life of Patriarch Euthymios* (first half of the tenth century). Whereas in the *Life of Basil*, the hunt assured the hero's ascension to the throne, here the hunt ridicules Basil – not only because it turns into a lethal situation, but because there is a complete reversal of the codes that preside over a heroic hunt:

> It was August and the emperor Basil had gone out for hunting (θηράσων) into Thrace ... when, finding a herd of deer, he gave chase, with the Senate and the huntsmen (κυνηγετῶν). They were all scattered in every direction in pursuit, when the emperor spurred after the leader of the herd, whose size and sleekness made him conspicuous. He was giving chase alone, for his companions were tired; but the stag, seeing him isolated, turned in his flight, and charged, trying to gore him; he threw his spear, but the stag's antlers were in the way, and it glanced off useless to the ground. The emperor now, finding himself helpless, took to flight; but the deer, pursuing, struck at him with its antlers, with the result that it carried him off. For the tips of the antlers having slipped under his belt, the stag lifted him from his horse and bore him away.[30]

The hero who, in the *Life of Basil*, killed wolves and stags with his bare hands, is here transformed into the unfortunate toy of a royal stag. The stag leads Basil to an almost grotesque punishment and everyone present is a witness to Basil's 'downfall', some making desperate efforts to save Basil from the horns of his adversary. And it does not end there:

> They all began running hither and thither, and just managed to catch a glimpse of the object of their search, carried aloft by the beast. They gave chase with all speed but without success; for the stag, when they were well out-distanced, stood panting and breathing hard, but when a rush brought them nearer, straightway bounded off to a good distance. So they were at a loss, till some of the Bodyguard, as it is called (τῆς καλουμένης ἑταιρείας), cut off the stag from in front before it was aware, and, scattering circle-wise in the mountains, put the stag up again by shouting. Then one of the Farganese, managing to ride alongside the deer with a naked sword in his hand, cut the horn-entangled belt through. The emperor fell to the ground unconscious.

When he returned to himself, he ordered the man who had delivered him from danger to be arrested, and ordered the cause of such insolence to be investigated. 'For', said he, 'it was to kill me, not to save, that he stretched out his sword'.[31]

Basil, in losing in his fight with the deer, also loses the generosity and justice that should characterize an emperor; he completely delegitimizes himself.

If Basil is ridiculed in order to be delegitimized, other emperors carry the signs of the tyrant. This is the case of Constantine V, who devoted himself excessively to hunting. According to Patriarch Nikephoros, such behaviour discredited his participation in theological discussions.[32] The same indignation is shared by Stephen the Deacon, author of the *Life of Stephen the Younger*, who – in the context of iconoclasm – accused the emperor of having loved paintings of hunting scenes, another sign of thoughtlessness and inability to behave like a worthy emperor.[33] The connection between the performance and painting of hunting appears in the case of Andronikos I, narrated by Niketas Choniates; we will return to this below.

In hagiographical accounts and beyond the imperial context, hunting becomes a means of punishment in texts which were written from the perspective of those who favored the veneration of icons. Thus, for example, the iconoclast bishop of Nicomedia, Eusebios: he was out hunting when he encountered a bear that knocked down his horse, he fell to the ground and was devoured by it.[34] On a more limited scale, the husband of Mary the Younger, Nikephoros, was punished for his arrogance in a hunting accident. He once saw a hare that he went after without restraints: "His horse slipped and Nikephoros fell with it, his right shoulder was displaced and his right hand has since remained inactive."[35] Ultimately, hunting imposes morality on any kind of abusive power.

The Sportive-'Aristocratic' Hunt

The third kind of hunting is playful and sportive, a hunt that, unlike the previous examples, does not attribute power, but yet illustrates it well, embellishes it and makes it explicit. This sort of hunt becomes useful in the re-elaborations of supreme power that takes place in Byzantium with the establishment, from the end of the eleventh century, of the military families of the eastern provinces of Asia Minor; it does not, however, eradicate the 'heroic' hunt in which several Komnenian emperors, such as John II (r. 1118–1143) or Manuel I (r. 1143–1180), continued to engage, as we will see later. It should be noted that this hunt becomes legitimate and escapes criticism because it is said to be a preparation for war.[36]

If the heroic hunt is carried out by the strength and skill of the hunter, the sportive and playful hunt is a spectacular hunt, carried out with the use of 'war machines'. The aristocratic hunter, always on horseback, is a director rather than a protagonist and his machines are the dogs, the leopards and the falcons – this is to say, he organizes a slaughter between animals. His game are large birds (cranes, herons) and small and medium-sized animals (hares, deer). Dogs had been hunting companions since Antiquity and falcons had appeared in literature in the late tenth century (but in practice certainly earlier); leopards or cheetahs seem new to twelfth-century Byzantium.[37] While hunting with dogs is shared by poor and rich, aristocrats and common people, hunting with a falcon or a leopard requires a considerable investment and becomes the privilege of an elite wishing to hunt with elegance. In ethnological terms, one could speak of 'passive hunting', carried out with "l'utilisation d'un objet technique <qui> introduit une distance par rapport au gibier, éloignement qui se traduit par une dilution de la responsabilité du meurtre."[38] While heroic hunting is depicted as individual encounters or as encounters of a small group of people, the playful hunt is a highly performative act that mirrors all of society. It is made up by a) specialized and auxiliary personnel (animal trainers such as falconers); b) aristocrats or the emperor, under whose auspices the hunt takes place; c) invited spectators, especially in the case of bird hunting (aristocratic women, men of the court not taking part in the actual hunt, intellectuals who immortalize the scene in their eulogies).[39]

The first trace of this kind of hunting in Byzantium appears in the tenth-century letter of Daphnopates, discussed above. The letter offers an image that falls between two representations: the anonymous and ordinary hunt of the countryside, on the one hand, and the aristocratic hunt, performed to offer leisure and amusement to Nikephoros, on the other. In another letter by the same author, addressed to Emperor Romanos II (959–963), who was a most devoted fan of hunting,[40] there is a description of this latter kind of hunt. The letter is presented as a response to a gift that the emperor had sent to the author: a wild goat and a hare. The actions of the emperor are divided into three categories according to the weapons used to catch the game: the spear that kills the goat, the dogs that attack and scare out the hares, and the falcons that fight and bring down the partridges. In order to legitimize the hunt at this rather early date, the author underlines the imperial exploits: "As for me, I saw there signs and symbols of your victorious and powerful reign against the barbarians."[41] The hunted animals represent the enemies, while the animals that assist in the hunt (dogs and falcons) represent the means available to the emperor to accomplish his exploits.

This tendency is confirmed in the eleventh-century writings of Michael Psellos. His hunting protagonists are the first emperor of the Komnenian dynasty, Isaac I (r. 1057–1059), and John Doukas, a member of the high aristocracy of the capital.[42] Psellos' description of Isaac as a hunter is seminal for the texts of the following century and worth citing in full:

Isaac was passionately devoted to hunting. No one was ever more fascinated by the difficulties of this sport. It must be admitted, moreover, that he was skilled in the art, for he rode lightly and his shouts and halloos lent wings to the dogs, besides frightening the coursing hare. On several occasions he even caught the quarry in full flight with his hand. He was, too, a dead shot with a spear. But crane-hunting attracted him more, and when the birds were flying high in the air, he still refused to give up the hunt. He would shoot them down from the sky, and truly his pleasure at this was not unmixed with wonder. The wonder was that a bird so exceptionally big, with feet and legs like lances, hiding itself behind the clouds, should, in the twinkling of an eye, be caught by an object so much smaller than itself. The pleasure he derived from the bird's fall, for the crane, as it fell, danced the dance of death, turning over and over, now on its back now on its belly.[43]

Isaac engages in an athletic hunt that demonstrates his courage and skill; he prefers the hunting of cranes and small animals to that of wild beasts. This text heralds the framework of imperial hunting in the twelfth century: the heroic hunt (of wild beasts) turning into pleasure hunting, even though both types are practiced and literarily depicted in the Komnenian period. Crane hunting becomes a specialization, at least according to the texts that have come down to us – the most noble kind of hunt for an aristocratic society.

The other hunter hero in Psellos' writings is John Doukas. The image of Doukas that he puts forward in his *Chronographia* combines the heroic and the playful. Doukas thus anticipates, like Isaac, the 'heroes' of the Komnenian dynasty:

> He indulges in all kinds of hunting, observing carefully the flight of birds and the tracks followed by wild beasts. He urges on the dogs and chases the dappled hind. He is mad about bears, too – I have often reproached him for that, but all to no purpose, for the pastime never fails to give him amusement. His life is spent in two pursuits - books and hunting: in other words, his leisure hours are devoted to the latter, and when he works, the whole world is his study, everything in its place.[44]

Psellos also uses the imagery of hunting in three letters addressed to the same person. In the first, he expresses the distance between hunter and intellectual, before presenting a romantic image of his hero as ideal hunter:

> I used to ridicule hunting and make fun of such activities; and I tried to dissuade you from them and used to advise you to instead spend time with books. But now I have changed my mind, I am not that demented. What do I prescribe for you? Ride your horses, hunt, jump through trenches, traverse rivers, gallop downhill and run up steep paths! Carry the falcon to your right, sitting unbound on your arm, and send him against geese, against partridges, against pigeons. If he captures the game in his flight, don't expect the Laconian dogs to trace the escaped animal. But if the latter has taken refuge somewhere, surround the grove, urge the dogs and don't give up until you catch it.[45]

In a second letter, Psellos presents an account of a hunt that he has heard from someone else, involving the noble reaction of John Doukas faced with the loss of his brave hunting bird.[46] In a third letter to the same addressee, Psellos imagines himself as an aristocrat hunter carrying a falcon on his hand as a model of nobility.[47] Leisure in Psellos' texts is defined in two ways: books *or* hunting, and – much more rarely – books *and* hunting. The opposition between books and hunting becomes a powerful and tenacious literary image in the writings of intellectuals until at least the fourteenth century, to which we will return below.

As we move towards the end of the eleventh century and the solid establishment of the Komnenian dynasty, hunting becomes a legitimate and essential activity of the social group that runs the empire. Alexios I and his successors cultivated their image as hunters to the extreme. Indeed, this is how their eulogists immortalize them. In his poetic eulogy of John II, for instance, Theodore Prodromos underlines the hunting exploits of the emperor.[48] Manuel I, in the writings of John Kinnamos, becomes the hero of imperial hunting; this is often combined with his bellicose character, as in the case when he goes out to hunt, but comes across enemies.[49] The 'exemplary' hunting feat, lending him the aura of an epic hero, is when he kills a panther that looks like a lion:

> It (the animal) had a double nature, taking something from both, a leopard in a lion and a lion in a leopard, a monstrous mixture of qualities, terrible in valor, courageous in frightfulness, and all the properties belonging to both in each other. Such was this

beast; most of those who attended the emperor fled when they saw it. For it was unendurable for many people to see. But when it came close, there was not one who then opposed it. But while they fled, the emperor drew the sword with which he was equipped, and rushed to strike the beast; bringing the blow down on its forehead, he drove it up to the chest. Such was the emperor in hunting.[50]

For the Komnenians, hunting guarantees their birthright to supreme power – an image that will prevail during their reign. When, after Manuel's death, the empire begins to crumble, writers rediscover old animosities towards the hunt. The picture that Niketas Choniates offers for the last years of the empire is notable: in his view, a century marked by refined hunting is now declining with hunter–emperors unworthy of power.[51] Finally, shortly before the fall, Euphrosyne, the wife of Alexios III Angelos, is turned into an indication of the complete denigration of hunting.[52] The figure of a woman-hunter – or rather a prostitute who plays with birds – personifies the decadence of an empire falling apart: long gone are the Komnenian emperors who practiced hunting as preparation for war, hunting is now pure decadence with no military purpose.

The Hunting *Ekphraseis* of the Twelfth Century

In addition to being a spectacular imperial and aristocratic practice, hunting also became an important aesthetic concern of the Komnenian period. This literary and artistic interest, expressed in *ekphraseis* and descriptions of paintings, goes beyond the narrow framework of hunting and reflects rather a new valorization of nature and the animal world.

The autonomous *ekphraseis* of hunting seem to be the result of a fruitful encounter that took place in the time of Manuel I Komnenos and his immediate successors: dazzling hunts organized under imperial auspices which gave authors the opportunity to rewrite the rhetorical tradition. These hunts allowed them to develop a metaphorical rhetoric that compared hunting with power. From this period we have two *ekphraseis* by Constantine Manasses, a teacher and rhetor in Constantinople;[53] an *ekphrasis* by Constantine Pantechnes, the bishop of Philippoupolis at the end of the century;[54] and an epistolary *ekphrasis* written by Basil Pediadites, bishop of Corfu, towards the end of the twelfth or the beginning of the thirteenth century.[55] There are also briefer hunting descriptions embedded in larger *ekphraseis*, such as in the description of a dog by Nikephoros Basilakes or the descriptions of the months in the novel by Eumathios Makrembolites.[56] In addition, there is a short *ekphrasis* embedded in the much longer historiographical narrative of Niketas Choniates that describes the paintings of hunting scenes that decorate the apartments of Andronikos Komnenos. In the following, we will offer a brief introduction to these Komnenian texts.

Two hunting *ekphraseis* by Manasses have come down to us, both on bird hunting. The first belongs to the tradition of countryside hunting as experienced by an astonished city scholar who discovers the delights of nature. This *Description of the Catching of Siskins and Chaffiniches* (ἔκφρασις ἀλώσεως σπίνων καὶ ἀκανθίδων) describes the hunting of small birds by means of glue traps.[57] The hunt is performed by a group of boys led by an old man in a place outside of Constantinople having all the characteristics of a *locus amoenus*.[58] Against this background, Manasses describes in great detail the different techniques of glue-hunting and the reactions of the participants to this sight.[59] The purpose of this hunt was to capture pretty and singing birds to put up for sale and at the same time to prepare a spontaneous meal from the game that could not be sold:

> After collecting the captured birds, they sorted out the game. All female birds were killed and thrown in a pit; they had prepared for these poor creatures a trench that one could call Hades or capacious tomb. As for the male birds, they divided them and made some prisoners, plucked the others, roasted them and devoured them whole without sparing even the bones, because they had prepared a fire there in advance.[60]

The author encapsulates in his text three small *ekphraseis* of birds: an elegant goldfinch,[61] a falcon[62] and another unspecified bird. The latter is described as follows, showing off Manasses' attention to detail:

> I then saw a bird in the hands of a lime-hunter and I admired the bounty of nature and the richness of beauty it had abundantly provided. Its beak was sharp and thin, the head black, the back was all yellowish, the lower parts were the colour of saffron and looked as if someone had woven gold on very thin linen, all of its plumage was of a natural beauty, the neck and chest were gilded, the rear parts were white as snow with black spots in a few places. It was impetuous, it was agile; you would say that he was dancing a warlike dance.[63] From his chest rose a soft song. He was so graceful to see, so pleasant to hear.[64]

The central figure in this *ekphrasis* is, however, not the birds, but the old man who leads the troop of young boys. He constitutes a grotesque figure who provokes the laughter of listeners-readers of the text with his stubbornness for perfection, his manias and his anger, his bilious character, his vanity, his rigorous discipline, and his ridiculous and hilarious falls. Ever the subject of mockery, even his baldness is revealed by chance, something that makes the participants, including the narrator, laugh.[65] We have argued elsewhere that there is a possible relationship between glue-hunting and education in the twelfth century: this motif is a recurring feature in Manasses' authorship and we suggest that the *ekphrasis* of the catching of small birds is constructed like a metonymy or a mirror game between hunting and education. The relationship between the old man and the children would then be the relationship between teacher and students in an educational system which was becoming more and more competitive.[66]

The other *ekphrasis* by Manasses takes us away from the literary-pastoral of the classroom and inscribes itself into the praise of Komnenian power.[67] It is a description of a crane hunt with the help of falcons, taking place in the presence of Manuel I Komnenos. The irony and condescension that marked the description of the glue-hunt here give way to the magnificence of the participants and the grandiosity of the spectacle. In this *ekphrasis*, which depicts an aerial battle between falcons and cranes, Manuel becomes a replica of the founder of the Komnenian dynasty, Isaac, who according to Psellos excelled at this kind of hunt.

The structure of the text follows more or less the traditional composition of an *ekphrasis*: a narrative frame containing a series of descriptions of characters and events. The imperial hunt involves several types of birds of prey and is organized as a military campaign. The hero-fighter (Manuel) engages in single combat, supported by his multiple aids (a staff responsible for the organization of the hunt) and a crowd of spectators (among whom is the narrator of the *ekphrasis*). The emperor carries a falcon, carefully depicted in much detail: it is an old noble female falcon with piercing eyes and greyish plumage. The other birds of prey fall into two categories: beginners and veterans, the latter being more valuable than the former. The hunt starts and quickly turns into a bloodthirsty war scene. The emperor does not release the female falcon, but uses another old and experienced bird for the hunt. The war goes on, a fierce battle between cranes and birds of prey:

> The cranes sensed war, and lining up and placing themselves in a phalanx, they backed away, like men who would neither dare to face the enemies in front of them nor rise up against them. They stretched out their wings, an almost gigantic thing that looked like a large shield, and after having straightened their necks like long spears and prepared the talons attached to their feet, they were ready to receive the attackers and defend themselves with beaks, talons and wings. When the old and experienced bird of prey was launched and, flying lightly into the depths of the sky, overtook the cranes and caught them in their flight, a joy mingled with fear took possession of the spectators and the part that was afraid felt joy and the part that rejoiced withdrew by fear. Such was the pleasure and at the same time the fear for the fate of that bird of prey![68]

Finally, after the intervention of other falcons, one particular crane is brought down. When the crane fell for the first time, they cut its talons and trimmed its beak before releasing it to fly and then sent young birds after it to learn to hunt without risk. The text closes with a traditional ekphrastic turn of phrase, defining the function of the description "for me as a vivid reminder of the event and for others as a clear representation of what they have not seen."[69] We will find identical phrases in all the autonomous *ekphraseis* which we will discuss in the following.

This grandiose description of an aerial hunt is composed of several small *ekphraseis* of people and animals: that of the emperor, as an ideal soldier and hunter in accordance with the rhetorical habits of Komnenian authors, that of the falcons involved in the hunt and that of the captured crane. In the following we will see several common elements in the descriptions of animals, which leads us to think that this kind of descriptions is the result of high-level rhetorical training. The text constitutes one of the most successful achievements in terms of its literary force, its finesse of description and the clarity of its ideological-political message. We read this description by Manasses as a demonstration of a new aesthetics of imperial power.[70]

The *ekphrasis* of Pantechnes is divided into two parts, of which the first is a description of a hunt with the help of "cruel hawks and mountain herons"[71] and dogs, while the second presents in great detail a hunt carried out with the help of cheetahs.[72] Pantechnes speaks of a hunt without the presence of the emperor, carried out instead by his staff under the orders of a great dignitary (ὁ μεγιστάν). This dignitary was at the same time "in charge of managing the imperial table" and the hunting party was looking in particular for "partridges and wild beasts." After having thus described the staff, the birds of prey and their equipment,[73] he offers a summarizing description of the hunt:

> They throw... the impetuous hawks for which they let go the straps. As they are used to, as soon as they are released from their bond, they take off, soar lightly into space and float from the air above in order to locate the hunted beast ... the hawk makes a

hissing sound, rushes on the animal, tears it with its talons and stops it from fleeing … The falconers then throw against the partridges the birds they have in their hands, trained for this purpose. Some flee, others attack; it is like a sort of struggle and combat between the hunter and the hunted. Most partridges finally manage to escape, but some have the unfortunate fate of being caught. The carnivorous birds dig the tips of their talons into the flesh of the partridges, tear them apart and kill them. These wretches cry out painfully and fill the air with the sound of their flapping wings. As for the proud hawk, it is perched proudly on the partridge, as if it takes pride in the spectacle, turning often to one side and the other, seeming to threaten those who would try approach at this moment.[74]

Pantechnes gives more detail than Manasses on the course of the hunt and he is more precise in the information he provides.[75] The second part of the text is a detailed depiction of a hare hunt with the help of a cheetah. This section begins with a description of cheetahs and an expression of the author's admiration, in biblical terms, for the human ability to tame wild beasts. Then follows the description of the hunt:

The pard-trainers brought them on the gelded horses which they rode, binding their necks with leashes, so that the beasts should not get out of line when it was inopportune, and jump onto what they should not. If a hare springs forth from somewhere, and the pard-trainer thinks he should send the cheetah after it, the other hunters are immediately barred from loosing the hounds or the birds; for verily, the cheetahs would attack not just the wild beasts, but them as well. And the cheetah pursues the hare alone, and reaching it with swift-turning bounds not more than two or three in number, he entangles it; and striking it with its front feet, lifts it up with swift-acting palms; and faster than speech the hare ends up under the teeth of the beast.[76]

The author then describes the skillful and dangerous way in which the pard-trainer removes game from the cheetah's mouth and the reward to which the beast is entitled. The *ekphrasis* ends with an address to the addressee, a friend (ἑταῖρε) who remains anonymous and who is not the same as the high functionary who organized the hunt.

This *ekphrasis* is rather peculiar. It is not a hunting description that turns into direct or indirect praise of the emperor or a high official despite the fact that the author follows the hunt while looking for a noble dignitary, nor is it a hunting description in a bucolic setting. It is a detailed description of the hunting techniques that became widespread during the twelfth century, namely falconry and hunting with the cheetah. What is the function of the text and who is it really addressed to? Is it a school exercise in rhetoric that demonstrates the author's meticulous observational skills? The careful language and turns of phrases could point in this direction. The small inserted *ekphraseis* of the birds of prey and the cheetah follow traditional paths of rhetoric that lingers on the bizarre and the exceptional. The falcons of Pantechnes resemble those of Manasses in several respects, while the description of the cheetah recalls zoological treatises from Late Antiquity; in addition, however, this *ekphrasis* also expresses a scholarly curiosity for the wild fauna, which also lends it an encyclopaedic and didactic character.

The letter by Basil Pediadites is probably addressed to a member of the Doukas family,[77] or – if one adopts a dating to the beginning of the thirteenth century–, to the *doux* of Corfu, Demetrios Kataphloron. The letter accompanies a gift of encaged birds captured by the author, as well as birds that serve as decoys.[78] The author describes in the first part of the letter the methods used to capture birds (glue sticks and nets), moving on to an *ekphrasis* of his garden. In the second part of the letter, he describes a hunt that he carried out in the same garden with the help of a glue stick held by himself.[79] In this *ekphrasis*, there is no precise description of the birds; their nature remains entirely undefined, beyond the title that speaks of goldfinches (ἀκανθίδες). The techniques of capture are also allusively presented and the participation of the author who, in Homeric terms, hides behind the foliage, moves on his knees and catches the bird with his spear, constitutes a very particular method of hunting. This small *ekphrasis*, which recalls in several ways Manasses' on the catching of small birds, or rather the school atmosphere which privileges the composition of such texts,[80] seems like a pure literary exercise and the gifts which accompany it could very well be imaginary birds, made of words and images.

Nikephoros Basilakes offers, among his progymnastic exercises, a praise of a dog. The dog acquired significant literary visibility in the twelfth century, especially in the writings of John Tzetzes who provides us in his *Chiliades* with moving stories of canine fidelity.[81] Basilakes confirms this tendency and offers an encomium, including all the warnings and excuses for such a choice of subject. The text describes the role of dogs in the hunt of deer and rabbit: "The dog chases down deer and rabbits and all the other wild animals upon the earth, from which he prepares a luxurious feast for the king's table."[82] The praise of the dog is linked to that of the hunt; let us not forget that the word which indicates hunting in Greek (κυνήγιον) means "to lead the dogs" and the ancient treatises on hunting are, in fact, treatises concerning the training of dogs. Also, in the case of Basilakes, hunting has both a spectacular and utilitarian character, since one of its main purposes is to bring meat to the imperial table. In addition, he proceeds to the *ekphrasis* of a hunt, as if he were describing a painting ("you also wish for me to describe for you in words, as if in a painting, the practice of hunting"[83]):

The hunter rides around on horseback, urging on the hunt, and the dogs gather around him in a circle, like an army around its general as it readies for battle. At any given time you could see one dog rolling around at the hunter's feet, whining in a fawning manner, and another exercising his legs and eagerly competing against others in a race, and still another glorying in the collar on his neck, reveling in the gems and taking pride in the golden leash ... When they arrive at the actual location of the hunt, by which I mean a plain that nourishes wild beasts or perhaps even a mountain ridge, the hunter stations the dogs all around him as they avidly watch for battle. Then, as though released from a starting gate, they all burst into the forest in a mob ... Visualize, then, the dogs attacking deer and nobly slaughtering them; one dog striking at an entire phalanx of rabbits and always killing the hindmost; another ferociously attacking a terrifying boar and with spear-like teeth devouring it from all sides; and still others tangling with various other animals ... visualize the very pleasant spectacle that happens then: no dog comes back empty-handed and unsuccessful, but each and every one comes dragging his quarry and bringing it to the master himself as if he were a tax collector.[84]

The hunt is described as a military activity, but from a specific point of view: the gaze of the author and, consequently, of the spectator, is focused on the activity of the dogs before, during and after the hunt, and the vision of the hunt is partial – it is a hunt in which the role of the protagonist is played neither by anonymous hunters, by an aristocrat, nor by noble or less noble game, but simply by dogs.

In the Komnenian novels, the presence of hunting is rather limited. Theodore Prodromos' Dosikles, like all young aristocrats of the twelfth century, engaged in hunting with his friends and likened his elopement with his beloved to a hunting party: "Yes, yes, join me, my hunting companions, in this present pursuit of the girl."[85] In the novel by Niketas Eugenianos, the protagonist Charikles describes his adolescence in the following terms:

> I had already reached adolescence, / brought up according to the norms of well-born youths; / I was happy in the company of the young men with whom I associated; / I rode, I joined in sports, as is customary for young men, / I hunted hare, I became a skilled equestrian / – for I had highly skilled companions – / but I had as yet no experience of love, / nor had down begun to shadow my chin.[86]

As the narrator indicates, Charikles follows the typical path of any young man of noble birth: sports, hunts, riding. In the novel by Eumathios Makrembolites, the hero Hysminias is not a hunter, even if his pursuit of the young girl Hysmine is once compared to "excellent hunting."[87] However, the author offers two hunting descriptions inserted into an extensive *ekphrasis* of the twelve months, a series of paintings in the garden of the heroine's father. The first of the two describes an autumn month, represented as a fowler:

> The youth that came after him was just growing his first beard, his head was not uncovered but was covered by gossamer-fine linen over both his head and his braids ... He is carrying cages for sparrows and is twisting cords, making traps for birds and keeping a close watch on their flight; he plants an entire meadow, and lets his sparrows out in the meadow but pulls them back often with a light line. The birds do not perceive the trap, they do not understand the trick; they see a pleasant meadow, with sparrows flying round on their line and others chirruping sweetly and delightfully in their cages; they come into the meadow, to the sparrows and are caught in the trap. The fowler who had set the trap catches and kills the birds and mocks their gullibility.[88]

The bird catcher is here presented as a preadolescent who uses decoy birds as his method. The game is killed, so we are dealing with an agricultural practice of hunting for food. The second description is of a spring month, presented as a typical hunter of small animals (hare) accompanied by his dogs:

> Following these there was depicted a youth with a vigorous body and a bold look, completely mad for hunting and the pursuit of game, with blood-stained hands and seeming to shout to his dogs ... A hare dangled from his left hand and with his right he was fondling the dogs, who were all rolling around at the youth's feet as though playing with him.[89]

In this scene, too, it is not the aristocratic hunter who is the emblematic figure, but a typical hunter who seeks useful game for food. The *ekphraseis* of the months are not related to the exaltation of nobility, but to the conventional and idealized representation of nature and time.[90]

The *ekphraseis* of Makrembolites constitute a link between descriptions of hunting and descriptions of paintings that contain scenes of hunting. An example of the interaction between literary and pictorial representation is the small *ekphrasis* of the palace of Andronikos, inserted in the *History* by Niketas Choniates:

> [In the palace of Andronikos], in addition to chariot races, there were scenes of the chase, with clucking birds and baying hounds; deer, hare, and wild boar hunts; and with the *zoumbros*[91] run through with a hunting spear (this animal is larger than the high-spirited bear or spotted leopard and is bred and raised by the Tauroscythians). There were also scenes of rustic life, of tent dwellers, and of common feasting on game, with Andronikos cutting up deer meat or pieces of wild boar with his own hands and carefully roasting them over the fire. Similar scenes also depicted the way of life of the man who is confident in the use of bow, sword, and swift-footed horses and who flees his country because of his own foolishness or virtue.[92]

Niketas describes paintings which we are accustomed to seeing in the Roman villas of Late Antiquity, in certain Byzantine manuscripts, and in certain houses of private individuals who wanted to exalt the hunting exploits of Manuel I Komnenos[93] – paintings which abound in varied hunting fauna, composed of familiar and exotic animals. But Choniates adds a complementary element: Andronikos is not the aristocratic hunter that the literature of the previous generation exalted: he is a cook without nobility who cuts the meat of stags and wild boars into pieces with his own knife and cooks them on the fire, thus taking on a task without prestige. Andronikos is a particular Komnenian hero (namely, a failed hero) who restores to aristocratic hunting its utilitarian character: that of nutrition.

This portrait concludes an entire period which exalted the hunting prowess of the Komnenians and generated a varied literature which made hunting a subject of primary importance in the manifestation of literary culture and in the praise of power. These cultural, ideological and political conditions (the spectacular quality of oratory and patronage) are almost completely gone after 1204, even though the imperial hunting as a practice continued.

Digenis Akritis: A Case of Heroic Pleasure Hunting

Digenis Akritis is a warrior poem with romance flavour of uncertain date; the oldest manuscript version dates to the fourteenth century and the story ties in with the novels and romances of the previous centuries.[94] It is accordingly a text that lingers between the Middle and the Late Byzantine period and, more importantly, hunting completely invests the protagonist and hero Digenis. According to version G, the heroic fate of Digenis is first sealed when his father bestows on him the claws and teeth of a lion he has killed.[95] Then after his studies, Digenis desires to hunt and participates with his father and maternal uncles in his very first hunt – a kind of initiation rite where he confronts and successively kills a bear, a stag and a lion. To kill the first beast he uses a club, for the second his own hands, and for the third a sword.

In this episode there is scaling not only in relation to the game, but also in relation to the tools used (*Digenis* G, 4.102–86). The bear asks for the force of nature, that one transforms himself into a woodsman; the stag imposes the agility of feet and hands, while the lion – the noblest and least 'natural' of the three beasts – requires the sword. Defeating and killing a lion implicitly refers to David and Samson and is "un rite de passage qui consacre les héros et les saints."[96] These three animals are killed to demonstrate heroic valour, not to provide food. It should be noted that the vapours and liquids of these beasts provoke miasma (μίασμα): "you will change what you are wearing, for it is stained with foam from the wild beasts and the lion's blood" (*Digenis* G, 4.206–7). After this hunt, Digenis becomes an adult and begins his own career among the brigands.

The other great episode in the career of Digenis is his encounter with an emperor who visits the region. What is most striking is that the hunting episodes that accompany this scene are drawn from the *Life of Basil* (ch. 13 and 14).[97] Digenis first tames an indomitable imperial horse (*Digenis* G, 4.1054–65) and immediately kills a lion which appeared and frightened everyone (in the case of Basil I, it was a wolf) and offers it to the emperor: " 'Accept,' he said, 'your servant's prey, lord, hunted for you.' " (*Digenis* G, 4.1066–75).[98]

In the sixth song, the hunt and the logic behind it become metaphorical language to indicate the abduction of a woman, in this case the attempted abduction of the wife of Digenis by Philopappous and his allies (*Digenis* G, 6.430–33 and 462). This is not an initiation hunt for wild animals and women, reserved for young men and found in the Escorial version of *Digenis* and in demotic songs, but rather a provocative hunt for pleasure, carried out by marginal beings who defy their fellow men. Once Digenis has decided to settle down and live as a lord, hunting takes on a similar character of amusement. The hunt is here no longer an initiation, nor an investiture, but a pastime of an idle social group in times of peace:

> He spent time with them for several days and achieved many outstanding deeds of valour, going out each day with them on the chase ... for nothing had yet been found that could outstrip him, but everything that existed fell into his hands, whether it was a lion or deer or any other wild beast. He did not have dogs with him or fleet-footed leopards, he did not sit upon a horse, he did not use swords but his hands alone and feet were everything to him.
>
> (*Digenis* G, 8.20–22 and 25–30)

However, this description is far from a praise of the aristocratic hunt as it developed and legitimized itself from the eleventh century onwards, examined in some detail above. Digenis hesitates here to be related to the aristocrats, to be one among the others, refuses even to use the horse and the sword; he insists on remaining an eternal seeker of legitimization by pursuing a 'wild' and archetypal hunt – man against beast. Finally, in the hero's epitaph, where the great deeds of his life are summarized, hunting is among his exploits: Digenis was "fearsome to lions and all wild beasts ... for if ever the marvelous young man went out to hunt, all the wild beasts ran for cover in the marsh" (*Digenis* G, 8.254 and 262–63). To sum up, we would say that *Digenis* G, the oldest version at least in relation to its manuscript, even if it polishes certain wild aspects of hunting, underlines the importance of hunting as a marker of social and sexual prerogatives of the hero.

In the Escorial version (E), hunting plays the same role: as an initiation and, once the trials have been passed, amusement and recreation in the life of the hero. Most episodes that deal with hunting are identical in the two versions, even if sometimes another light shines on certain episodes in version E. Among the guerrillas, hunting and its trophies are much more emphasized than in version G. Philopappous is resting on his hunting trophies: "There he found Philopappous reclining on his couch, with many animal skins all around him: he had a lion and a boar for his pillow" (*Digenis* E, 646–48), while the trials the hero must perform consist in starving, killing and skinning a lion and kidnapping a young bride (*Digenis* E, 658–68). This last element underlines, better than in version G, the perception of women as game of choice. In version E, it is the mother of Digenis who addresses her son in this way, alluding to his obligation to marry in the following terms: "Welcome, my child, if you have brought me game from hunting," and receiving an answer from Digenis in the same register: "my marvellous game will come and you will see it" (*Digenis* E, 807–809).[99] The hunt becomes a metaphor of the amorous conquest, an imagery often employed in love poetry, novels and romances.

Digenis finally makes the distinction between countryside hunting and the heroic hunt clear, addressing his father: "How long shall I be hunting hares and partridges? Hunting partridges is what peasants do, but young lords and the sons of the high-born hunt lions and bears and other fierce beasts" (*Digenis* E, 744–47). Hunting edible game (rabbits and partridges) is utilitarian hunting and is carried out to put food on the table in agropastoral societies, whereas wild animal hunting is free and rewarding, with no economic value. Digenis, as a hero, consumes values, for values are the product of his hunting; the small game that he devours in his moment of rest comes from the sweat of his subordinates, the common people, the men who admire him.

The Literary Presence of Hunting in the Late Byzantine Period

In the Palaiologan period, despite the fact that hunting became widespread as an aristocratic and imperial practice, imperial hunting is treated with suspicion, as a symbol of decadence, and its literary presence is rather limited. In order to find any positive reference to hunting, we must turn to texts that praise military leaders of the countryside, such as Charles I Toccos at the end of the fourteenth century, who "with falcons, sparrowhawks, swift stone herons, / he hunts cranes, partridges and doves...".[100] Or we turn to the Palaiologan romance, where aristocratic hunting still lingers in the stories. In *Livistros and Rhodamne*, there is an image that draws on Manasses' depiction of Manuel I Komnenos in his *ekphrasis* of a crane hunt. The hero Livistros, "rode a war-horse, he held a hawk, behind him followed a dog on a leash," but unlike the emperor, Livistros lived from hunting in his wanderings: "living by my hawk and nourished by my hound."[101] Hunting with a falcon, his usual game was partridges; on one occasion, the heroine Rhodamne even has to go hunting with a falcon to meet the hero.[102]

In *Velthandros and Chrysanza*, there are also several hunting scenes and here, too, the hero hunts with a falcon:

> Velthandros and his squires mounted and joined in the hunt behind the prince. When they found a hare in a lofty place on a mountain, they released a falcon. A swift eagle appeared flying in the sky, swooped down and seized the falcon. The prince saw what had happened and was grieved and distressed but Velthandros quickly drew his bow. Taking good aim, he hit the eagle's wing. In its pain it released the unscathed falcon— it was completely without harm or injury. The prince marvelled at Velthandros's bowmanship and praised his great skill effusively.[103]

Apart from the romances, the literary occasions to exalt hunting are rare in this period. Demetrios Kydones makes the connection between hunting and literary pursuits, in the same vein as Michael Psellos some three centuries earlier, in a letter to a certain Andreas Asanes.[104] The letter was written in 1373/1374 from Lesbos, where Kydones lived at the court of Francesco I Gattilusio, and it offers a short but humorous description of the hunting activities of local aristocrats:

> Will you not believe that I have neglected literary pursuits and my customary activities and am spending my time in hunting? Will you not believe that I set out with the hounds after the prey before daylight and, until the setting of the sun's rays, I wasted my time tracking it? I rode my horse along the cliffs, just about joining the falcon in their flight and filling the air with the cries of a madman. There was a time when I criticised you for these things; it seemed that you wanted to live with wild animals rather than with human beings. When others tell you these things about me, you will think it incredible. Yet, I am afraid that I might let everything lapse into oblivion and make hunting the sole purpose of my life.[105]

Kydones goes on to exalt the pleasure of hunting and the abundance of both hunting companions, horses, dogs, falcons, and game: "Here, the partridges are more numerous than the owls in Athens. As I told you, I am afraid to confuse education (παιδεία) with play (παιδιά) and become a laugh to you because of this exchange." The reason, he explains, is the lack of learning and city pleasures in Mitylene.[106] Kydones thus playfully laments his fate, which forces him to hunt instead of indulging in intellectual activities. The hunt he describes, carried out with the help of falcons, targets the partridges that apparently abound. Kydones underlines both the pleasure that hunting provides and its utilitarian aspect, as the game feeds a small society: the falcon-hunter to whom part of the game belongs by right, the participants in the hunt, and the neighboring farmers.

Just like the letter of Kydones recalls Middle Byzantine epistolography, the praise for a dog by Theodore of Gaza, written in the fifteenth century, recalls the praise composed by Basilakes in the twelfth century. Like his predecessor, Theodore devotes considerable attention to hunting. He first underlines the etymological connection between hunting and dogs (κυνηγεσία and κύων) and then proceeds to a sort of history of the hunt through examples that refer to the gods and heroes of Antiquity – to the Homeric warriors, to the Spartans, to the Macedonians and the Persians – to demonstrate the contribution of hunting to the formation of an efficient and strong soldier. Theodore then cites Plato who, in *Laws* (6.763b 1–6) prescribes hunting with dogs so that young men learn well the territory of their city, and closes by underlining the contribution of dogs to hunting.[107] His conclusion paraphrases the *Oration on Artemis* by Libanius (5.13): "Without hunting there would be wild beasts all over, harming men and besieging them by keeping them locked up in the cities; we delivered ourselves from all these beasts thanks to dogs and the hunt."[108] Ultimately, hunting only assures the balance of nature, it is an activity synonymous with the civilizing intervention of man on nature.

Hunting is an almost universal economic and social practice which, in certain periods, in certain societies, and under certain cultural conditions, takes on the characteristics of a literary subject which signifies, conveys and figuratively expresses conceptions about political, economic, social, and literary values. Here we have followed some general lines of the Byzantine ideas on hunting and through them, we have touched upon the constants or developments that concern both the reality and the imagination. A full history of hunting in Byzantium remains to be written.

Funding

The writing of this article has been undertaken within the frame of the research programme Retracing Connections (https://retracingconnections.org/), financed by Riksbankens Jubileumsfond (M19-0430:1).

Notes

1　On hunting in Byzantium, see F. Koukoulès, *Βυζαντινῶν Βίος καὶ Πολιτισμός*, 6 vols (Athens, 1948–56), vol. 5, 387–423 (published previously as "Κυνηγετικὰ ἐκ τῆς ἐποχῆς τῶν Κομνηνῶν καὶ τῶν Παλαιολόγων," *Epeteris Etaireias Byzantinon Spoudon* 9 [1932]: 3–33). See also E. Patlagean, "De la chasse et du souverain," *DOP* 46 (1992): 257–63; L. Delobette, "L'empereur et la chasse à Byzance du XIe au XIIe siècle," in *La forêt dans tous ses états: de la Préhistoire à nos jours*, ed. J.-P. Chabin (Besançon, 2005), 283–96; A. Sinakos, "Το κυνήγι κατά τη μέση βυζαντινή εποχή (7ος–12ος αι.)," in *Ζώα και περιβάλλον στο Βυζάντιο (7ος–12ος αι.)*, ed. I. Anagnostakis, T. Kolias and E. Papadopoulou (Athens, 2011), 71–86; C. Messis and I. Nilsson, "The *Description of a Crane Hunt* by Constantine Manasses: Introduction, Text and Translation," *SJBMGS* 5 (2019): 9–89, esp. 9–41.
2　On this literature, see in general. MacKinnon, "Hunting," in *The Oxford Handbook of Animals in Classical Thought and Life*, ed. G. L. Campbell (Oxford 2014), 203–15.
3　On the consumption of game in Antiquity, see C. Chandezon, "Le gibier dans le monde grec. Rôles alimentaire, économique et social," in *Chasses Antiques. Pratiques et représentations dans le monde gréco-romain (IIIe siècle av. – IVe siècle apr. J.-C.)*, ed. J. Trinquier and C. Vendries (Rennes, 2009), 75–95; for Byzantium, see H. Kroll, *Tiere im Byzantinischen Reich. Archäozoologische Forschungen im Überblick* (Mainz, 2010). See also the chapter on the consumption of animals and animal products by Kokoszko and Rzeźnicka in this volume.
4　Libanius, *Hunt*, ed. R. Foerster, *Libanii Opera. Vol. VIII. Progymnasmata – Argumenta orationum demosthenicarum* (Leipzig, 1963), 487:13.
5　Libanius, *Hunt*, ed. Foerster, 489:5–6; trans. C. Gibson, *Libanius's Progymnasmata, Model Exercises in Greek Prose Composition and Rhetoric* (Atlanta, 2008), 451.
6　Daphnopates, *Letters*, no. 37:2–4, ed. J. Darrouzès and L. Westerink, *Daphnopatès Théodore, Correspondance* (Paris, 1978) 207. The attribution of this letter to Daphnopates is not certain; see ibid., 24–25 and cf. D. Chernoglazov, "Beobachtungen zu den Briefen des Theodoros Daphnopates. Neue Tendenzen in der byzantinischen Literatur des zehnten Jahrhunderts," *BZ* 106 (2013): 623–44.
7　Daphnopates, *Letters*, no. 37:52–62, ed. Darrouzès and Westerink, 211.
8　Daphnopates, *Letters*, no. 37:64–88, ed. Darrouzès and Westerink, 211–13.
9　Daphnopates, no. 37:69, ed. Darrouzès and Westerink, 211.
10　Daphnopates, no 37:86–88. ed. Darrouzès and Westerink, 213.
11　*Life of Theoktiste, AASS* Nov. IV, col. 224–33, trans. A. Hero, "Life of St. Theoktiste of Lesbos," in *Holy Women of Byzantium: Ten Saints' Lives in English Translation*, ed. A.-M. Talbot (Washington, D.C., 1996), 95–116. For ἑταιρεία, see chs 12 (τὴν ἑταιρείαν), 14 (τῆς ἑταιρείας), 17 (τοὺς ἑταίρους) and 19 (τοὺς ἑταίρους). This *Life* has been thoroughly studied; most recently, see C. Messis, "Fiction and/or Novelisation in Byzantine Hagiography," in *The Ashgate Research Companion to Byzantine Hagiography. Volume II: Genres and Contexts*, ed. S. Efthymiadis (Farnham, Surrey, 2014), 313–41, here 329–32; C. Høgel, "Beauty, Knowledge, and Gain in the *Life of Theoktiste*," *Byz* 88 (2018), 219–36.
12　*Life of Theoktiste*, ch. 11, trans. Hero, 108.
13　*Life of Theoktiste*, ch. 17, 112.
14　P. Desideri, "City and Country in Dio," in *Dio Chrysostom: Politics, Letters, and Philosophy*, ed. S. Swain (Oxford, 2002), 93–107. In the *Life of Mary of Egypt*, Zosimas is presented

metaphorically as a hunter; see A. Kazhdan, "Hagiographical Notes. 9. The Hunter or the Harlot?," *BZ* 78 (1985) 49–50.

15 *Life of Theodore of Cythera*, ed. N. Oikonomidis, "Ὁ Βίος του αγίου Θεοδώρου Κυθήρων (10ος αι.) (12 Μαϊου – BHG, αρ. 2430)," in *Τρίτον Πανιόνιον Συνέδριον, Πρακτικά* (Athens, 1967), 264–91, here v. 243–46 (289). On the relation between this life and that of Theoktiste, see S. Efthymiadis, "Hagiography from the 'Dark Age' to the Age of Symeon Metaphrastes," in *The Ashgate Research Companion to Byzantine Hagiography. Volume I: Periods and Places*, ed. S. Efthymiadis (Farnham, Surrey, 2011), 95–142, here 125.

16 *Life of Peter the Athonite*, ed. K. Lake, *The Early Days of Monasticism on Mount Athos* (Oxford, 1909), 18–39, here 31. The presence of an extraordinary deer as an indication of the holiness of a person is a recurring motif in hagiography; for further references, see Messis and Nilsson, "The *Description of a Crane Hunt*," 13–14.

17 H. Delehaye (ed.), "Vita S. Pauli Junoris," ch. 25, in *Der Latmos*, ed. T. Wiegand (Berlin, 1913), 105–57, here 119–20.

18 Note the distinction between the *Passion* of Barbaros, martyred under Julian (BHG 219), ed. H. Delehaye, "Les Actes de S. Barbarus," AB 29 (1910), 276–301, and the *Life* of Barbaros who lived in the time of Michael II, (re)written by Constantine Akropolites (BHG 220), ed. A. Papadopoulos-Kerameus, Ἀνάλεκτα Ἱεροσολυμιτικῆς Σταχυολογίας, vol. I (St Petersburg, 1891), 405–20.

19 *Life of Barbaros*, ed. Papadopoulos-Kerameus, 416:24.

20 The same ambiguous representation of nature is clear in pictorial hunting scenes, e.g., in the illuminated Oppian; see I. Spatharakis, *The Illustrations of the Cynegetica in Venice. Codex Marcianus Graecus Z 139* (Leiden, 2004); see also E. Dautermann Maguire and H. Maguire, *Other Icons: Art and Power in Byzantine Secular Culture* (Princeton, NJ, 2007), 47–51 and 82–89.

21 See J. Lefort and J.-M. Martin, "L'organisation de l'espace rural: Macédoine et Italie du Sud (Xᵉ-XIIIᵉ siècle)," in *Hommes et richesses dans l'Empire byzantin*, 2 vols, ed. V. Kravari, J. Lefort and C. Morrisson (Paris, 1991), vol. 2, 11–26, here 24.

22 Paul the Silentiary, VI.168, ed. P. Waltz, *Anthologie grecque. I. Anthologie Palatine. Tome III, livre VI* (Paris, 1931), 93.

23 On protection against wild beasts, see I. Anagnostakis, "Ὁ φράκτης, ο αγριόχοιρος και η άρκτος," in Anagnostakis, Kolias and Papadopoulou, *Ζῷα*, 195–233.

24 Patlagean, "De la chasse," 257.

25 *Life of Basil*, ch. 13, ed. I. Ševčenko, *Chronographiae quae Theophanis Continuati nomine fertur liber quo vita Basilii imperatoris amplectitur* (Berlin, 2011). On the significance of hunting in this text, see Patlagean, "De la chasse," 258–59. See also C. Messis, "Est-elle possible une lecture subversive de la *Vie de Basile*? Stratégies narratives et objectifs politiques dans la cour de Constantin VII Porphyrogénète," in *Storytelling in Byzantium: Narratological Approaches to Byzantine Texts and Images*, ed. C. Messis, M. Mullett and I. Nilsson (Uppsala, 2018), 201–22.

26 On the Philopation, see N. Ševčenko, "Wild Animals in the Byzantine Park," in *Byzantine Garden Culture*, ed. A. Littlewood, H. Maguire and J. Wolschke-Bulmahn (Washington, D.C., 2002), 69–86; Delobette, "L'empereur," 286; H. Maguire, "The Philopation as a Setting for Imperial Ceremonial and Display," BF 30 (2011): 71–82.

27 Genesios, *History*, IV. 26 and IV. 39, ed. A. Lesmueller-Werner and I. Thurn, *Iosephi Genesii regum libri quattuor* (Berlin, 1978), 78 and 89.

28 Genesios, *History*, IV.40 and 42, ed. Lesmueller-Werner and Thurn, 89–90 and 91.

29 Symeon the Logothete, *History*, ch. 131:60–63 and 132:174–82, ed. S. Wahlgren, *Symeonis Magistri et Logothetae Chronicon* (Berlin, 2006), pp. 235 and 270.

30 *Life of Euthymios*, ed.-trans. P. Karlin-Hayter, "Vita Euthymii," *Byz* 25–27 (1955–57): 1–172, ch. 1, 8:2–17.

31 *Life of Euthymios*, ch. 1, 8:20, ed. Karlin-Hayter.

32 Nikephoros, *Against the Iconoclasts*, ed. J.-P. Migne, PG 82, 229cd.

33 *Life of Stephen the Younger*, ch. 26, 121 and 215, ed. M.-F. Auzépy, *La Vie d'Etienne le Jeune par Etienne le Diacre* (Aldershot, 1997). On representations of hunting scenes in which emperors take part, see A. Grabar, *L'empereur dans l'art byzantin* (Paris, 1936), 57–62 and 134–44.

34 *Life of Kosmas and John Damascus*, ed. A. Papadopoulos-Kerameus, Ἀνάλεκτα Ἱεροσολυμιτικῆς σταχυολογίας, v. IV (St Petersburg, 1897), 271–302, here 228.

35 *Life of Mary the Younger, AASS* Nov. IV, col. 703b.

36 On this aspect, see Messis and Nilsson, "The *Description of a Crane Hunt*," 12–17.

37 The terminology is unclear, to say the least: T. Buquet, "Le guépard médiéval ou comment reconnaître un animal sans nom," *Reinardus* 23 (2011): 12–47, has argued for the difference between leopards and panthers, while N. Nicholas, "A Conundrum of Cats: Pards and Their Relatives in Byzantium," GRBS 40/3 (1999): 253–98, proposes that the Greek word πάρδαλις corresponds to the cheetah. On the hunting with leopards/cheetahs, see A. Papagiannaki, "Experiencing the Exotic: Cheetahs in Medieval Byzantium," in *Discipuli dona ferentes: Glimpses of Byzantium in Honour of Marlia Mundell Mango*, ed. T. Papacostas and M. Parani (Turnhout, 2017), 223–57.

38 B. Hell, *Sang noir. Chasse, forêt et mythe de l'homme sauvage en Europe* (Paris, 2012), 30.

39 Unfortunately, the presence of women in the audience is reported only when linked to other significant events. We learn, for example, from Nikephoros Gregoras in the fourteenth century that Empress Irene was rendered sterile by a terrible fall from her horse, "when she went out with her husband and emperor to watch a hunting party;" Gregoras, *History*, ed. L. Schopen, *Nicephori Gregorae Byzantina Historia* (Bonn 1829–1830), 44:7–10.

40 On Romanos II and hunting, see Patlagean, "De la chasse," 259.

41 Daphnopates, *Letters*, no 14:7–38, ed. Darrouzès and Westerink, 151. Our translation. On this letter, see also Messis and Nilsson, "The *Description of a Crane Hunt*," 14–15.

42 See also the letter addressed to Sagmatas, *synkellos* and *protonotarios* of the drome: Psellos, *Letters*, no 251, ed. S. Papaioannou, *Michael Psellus Epistulae* (Berlin, 2019), 623–26, on which see G. Dennis, "Some Notes on Hunting in Byzantium," in *ANAΘHMATA EOPTIKA: Studies in Honor of Thomas F. Mathews*, ed. J. Alchermes, H. Evans and T. Thomas (Mainz, 2009), 131–34, here 133–34.

43 Psellos, *Chonographia*, VII.72, ed. D. R. Reinsch, *Michaelis Pselli, Chronographia* (Berlin, 2014), 243; trans. E. R. A. Sewter, *The* Chronographia *of Michael Psellus* (London, 1953), 244. On this hunt, see also Patlagean, "De la chasse," 259, and Delobette, "L'empereur," 288. Cf. also the references to the same emperor in Psellos, *Letters*, no 142:56–64, ed. Papaioannou.

44 Psellos, *Chonographia*, VII.180–81, ed. Reinsch, 294–95; trans. Sewter, *The* Chronographia, 298.

45 Psellos, *Letters*, no. 54:3–14, ed. Papaioannou.

46 Psellos, *Letters*, no. 67, ed. Papaioannou.

47 Psellos, Letters, no. 76:36–39, ed. Papaioannou.

48 Theodore Prodromos, *Poems*, no 25:18–20, ed. W. Hörandner, *Theodoros Prodromos: historische Gedichte* (Vienna, 1974), 336. On the death of John II, caused by a wound he suffered at a wild boar hunt, see John Kinnamos, *History*, ed. A. Meineke, *Ioannis Cinnami Epitome rerum ab Ioanne et Alexio Comnenis gestarum* (Bonn, 1836), 24:10–23; Choniates, ed. A. Van Dieten, *Nicetae Choniatae Historia* (Berlin, 1972), 40.

49 Kinnamos, *History*, ed. Meineke, 189:2–23; see also 126–27.

50 Kinnamos, *History*, ed. Meineke, 266:22–267:13; trans. C. Brand, *Deeds of John and Manuel Comnenus* (New York, 1976), 200.

51 For Alexios III, see Choniates, *History*, ed. Van Dieten, 223; for Isaac Angelos, ibid., 399. Cf. a different and very positive image of the hunting exploits of Alexios IV, the son of Isaac Angelos, in Nikephoros Chrysoberges, *Discourse to Alexios IV*, ed. M. Treu, *Nicephori Chrysobergae ad Angelos orations tres* (Breslau, 1892), 30:20–31:11.

52 Choniates, *History*, ed. Van Dieten, 520; on Euphrosyne, see L. Garland, *Byzantine Empresses: Women and Power in Byzantium AD 527–1204* (London, 1999), 210–24; Messis and Nilsson, "The *Description of a Crane Hunt*," 36.

53 On this author see I. Nilsson, *Writer and Occasion in Twelfth-Century Byzantium: The Authorial Voice of Constantine Manasses* (Cambridge, 2021).

54 Details are sparse and partly contradictory; see E. Miller, "Description d'une chasse à l'once par un écrivain byzantin au XIIe siècle de notre ère," *Annuaire de l'Association pour l'encouragement des études grecques* 6 (1872): 28–52, here 31, and K. Horna, "Die Epigramme des Theodoros Balsamon," *Wiener Studien* 25/2 (1903): 165–217, here 209.

55 Pediadites, *Letter*, ed.A. Karpozilos, "Βασιλείου Πεδιαδίτη Ἔκφρασις ἁλώσεως ἀκανθίδων," *Ἠπειρωτικά Χρονικά* 23 (1981), 284–98. See also K. Manafis and I. Polemis, "Βασιλείου Πεδιαδίτου Ἀνέκδοτα ἔργα," *Ἐπετηρὶς Ἑταιρείας Βυζαντινῶν Σπουδῶν* 49 (1994–1998), 1–62, here 2–12.

56 For Basilakes, see A. Pignani, *Niceforo Basilace Progymnasmi et Monodie* (Naples, 1983), 133–38 (Greek texts) and 309–12 (Italian trans.); J. Beneker and C. Gibson, *The Rhetorical Exercises of Nikephoros Basilakes: Progymnasmata from Twelfth-Century Byzantium* (Cambridge, MA, 2016),130–41 (Greek text and English trans.). For Makrembolites, see M. Marcovich, *Eustathius Macrembolites de Hysmines et Hysminiae amoribus libri XI* (Munich, 2001). On ekphraseis in the literature of the twelfth century, see also I. Taxidis, *The Ekphraseis in the Byzantine Literature of the 12th Century* (Alessandria, 2021).

57 On glue-hunting in antiquity and Byzantium, see C. Vendries, "L'auceps, les gluaux et l'appeau. A propos de la ruse et de l'habileté du chasseur d'oiseaux," in Trinquier and Vendries, *Chasses Antiques*, 119–40.

58 *Description of the catching of siskins and chaffinches*, ed. K. Horna, *Analekten zur byzantinischen Literatur* (Vienna, 1905), 6–12, here lines 32–40. For a new edition with English translation, see C. Messis and I. Nilsson, "The *Description of the Catching of Siskins and Chaffinches* by Constantine Manasses: Introduction, Text and Translation," *SJBMGS* 8 (2022): 9–66.

59 On these techniques, see C. Messis and I. Nilsson, "L'ixeutique à Byzance: pratique et représentation littéraire," SJBMGS 7 (2021): 81-107.

60 *Description of the Catching of Siskins and Chaffinches*, 79–85, ed. Horna. To our knowledge, this treatment of male and female birds is not attested in any other text, but Manasses does not seem to make it up because this practice is common in modern bird hunting.

61 *Description of the Catching of Siskins and Chaffinches*, 55–61, ed. Horna.

62 *Description of the Catching of Siskins and Chaffinches*, 105–12, ed. Horna. This is a description filled with irony: "he who not long ago rose above the clouds was now touched by the hands of small children."

63 πυρρίχην: on this military dance, attested since Antiquity, see J.-Cl. Poursat, "Les représentations de danse armée dans la céramique attique," *Bulletin de correspondance hellénique* 92 (1968): 550–61.

64 *Description of the catching of siskins and chaffinches* 190–98, ed. Horna.

65 On this figure, see also K. Chrysogelos, "Κωμικὴ Λογοτεχνία και γέλιο τον 12ο αι. Η περίπτωση του Κωνσταντίνου Μανασσή," *Byzantina Symmeikta* 26 (2016): 141–61, here 149–51.

66 Messis and Nilsson, "The *Description of the catching of siskins and chaffinches*."

67 Edited, translated and commented in Messis and Nilsson, "The *Description of a Crane Hunt*."

68 Messis and Nilsson, "The *Description of a Crane Hunt*," 211–22; trans. 75.

69 Messis and Nilsson, "The *Description of a Crane Hunt*," 329–31; trans. 79.

70 For another interpretation, see H. Maguire, "'Signs and Symbols of your always victorious reign': The Political Ideology and Meaning of Falconry in Byzantium," in *Images of the Byzantine World: Visions, Messages and Meanings. Studies Presented to Leslie Brubaker*, ed. A. Lymberopoulou (Farnham, 2011), 135–45.

71 Ed. M. Miller, "Description d'une chasse à l'once par un écrivain byzantin au XIIe siècle de notre ère," *Annuaire de l'Association pour l'encouragement des études grecques* 6 (1872): 28–52, here 47: τοὺς ἐπιβούλους ἱέρακας καὶ τοὺς πετραίους ἐρωδιούς.

72 On the terminology, see above n. 37.

73 Pantechenes, *Description of the Hunt*, ed. Miller, 47–48.

74 Pantechenes, *Description of the Hunt*, ed. Miller, 49; trans. Messis and Nilsson, "The Desciption of a Crane Hunt," 32.

75 Pantechnes speaks, for example, about the reward for the bird of prey after a successful flight (*Description of the Hunt*, ed. Miller, 49), and of a staff specialized in the care of falcons.

76 Pantechnes, *Description of the Hunt*, ed. Miller, 50–51; trans. Nicholas, "A Conundrum of Cats," 290–91.

77 Pediadites, *Letter*, 296:3, ed. Karpozilos: δουκικὸν κάρα τρισόλβιον.

78 Pediadites, *Letter*, 296:7–8, ed. Karpozilos: στρουθοὺς ἀνδραποδιστάς.

79 Pediadites, *Letter*, 298:66–68, ed. Karpozilos: ὃν ἐν ταῖς χερσὶ δόνακα, ἔχοντα προσαρμοστὸν λύγον ἀκροτελεύτιον, εὖ μάλα καλυμμένον ἰξῷ, τῆς τοῦ στρουθοῦ κατέφερον πτέρυγος.

80 Karpozilos, "Βασιλείου Πεδιαδίτη," 295, thinks that this *ekphrasis* was influenced by Manasses.

81 John Tzetzes, *Histories*, IV.160–290, ed. P. Leone, *Ioannis Tzetzae Historiae* (Naples, 1968), (six stories). On dogs in Byzantium, see A. Rhoby, "Hunde in Byzanz," in *Lebenswelten zwischen Archäologie und Geschichte. Festschrift für Falko Daim zu seinem 56. Geburtstag*, ed. J. Drauschke et al. (Mainz, 2018), 807–20. For dogs as diplomatic gifts, see the chapter by Drocourt in the present volume.

82 Basilakes, *Praise of the Dog*, 18–20, ed. Pignani, *Niceforo Basilace*, 133; trans. Beneker and Gibson, *The Rhetorical Exercises*, 131. See also C. Gibson, "In Praise of Dogs: An Encomium Theme from Classical to Renaissance Italy," in *Our Dogs, Our Selves: Dogs in Medieval and Early Modern Art, Literature, and Society*, ed. L. Gelfand (Leiden, 2016), 19–40, esp. 25–31 on Basilakes. See also Taxidis, *The Ekphraseis*, 56–57.

83 Basilakes, *Praise of the Dog*, 34–35, ed. Pignani, 134; trans. Beneker and Gibson, *The Rhetorical Exercises*, 133.

84 Basilakes, *Praise of the Dog*, 35–65, ed. Pignani, 134–35; trans. Beneker and Gibson, *The Rhetorical Exercises*, 135.

85 Prodromos, *Rhodanthe and Dosikles*, bk. 2:415–16, ed. M. Marcovich, *Theodori Prodromi de Rhodanthes et Dosiclis amoribus libri IX* (Stuttgart, 1992); trans. E. Jeffreys, *Four Byzantine Novels* (Liverpool, 2012), 48. In this novel, there is also a tale of a "foolish and senseless" hunter: bk. 5:341–52.

86 Eugeneianos, *Drosilla and Charikles*, bk. 3:53–60, ed. F. Conca, *Nicetas Eugenianus de Drosillae et Chariclis amoribus* (Amsterdam, 1990); trans. Jeffreys, *Four Byzantine Novels*, 375.

87 Makrembolites, *Hysmine and Hysminias* 4.4.2, ed. Marcovich; trans. Jeffreys, *Four Byzantine Novels*, 201.

88 Makrembolites, *Hysmine and Hysminias* 4.12, ed. Marcovich; trans. Jeffreys, *Four Byzantine Novels*, 204.

89 Makrembolites, *Hysmine and Hysminias* 4.15, ed. Marcovich; trans. Jeffreys, *Four Byzantine Novels*, 205.

90 In a poem on the twelve months, attributed to Kallikles (ed. R. Romano, *Nicola Callicle Carmi* (Naples 1980), no 37), October is represented as a fowler (v. 43–45), while December is a hunter (v. 55–56). In this case, the game is presented as nourishment for the rich.

91 On this animal, to be identified with the European bison, see A. Pontani, *Niceta Coniata Grandezza e catastrofe di Bisanzio*, Vol. II: *Libri IX-XIV* (Milan, 1999), 672–73, n. 81; for animals in Choniates, see L. Bossina, "La bestia e l'enigma. Tradizione classica e cristiana in Niceta Coniata," *Medioevo Greco* 0 (2000): 35-68.

92 Choniates, *History*, ed. Van Dieten, 333; trans. H. Magoulias, O *City of Byzantium, Annals of Niketas Choniates* (Detroit, 1984), 184.

93 See Kinnamos, *History*, ed. Meineke, 266:6–9 on the habit of high functionaries to have representations in their houses of the military and hunting exploits of the emperor.

94 On the dating of Digenis, see A. Goldwyn and I. Nilsson, "Troy in Byzantine Romances: Homeric Reception in Digenis Akritis, the Tale of Achilles and the Tale of Troy," in *Reading the Late Byzantine Romance: A Handbook*, ed. A. Goldwyn and I. Nilsson (Cambridge, 2019), 188–210, here 191–92.

95 *Digenis Akrites* G, 3.102–6, ed. and trans. E. Jeffreys, *Digenis Akritis: The Grotaferrata and Escorial Versions* (Cambridge, 1998), 50. On hunting in *Digenis*, see D. Ricks, "The Pleasures of the Chase: a Motif in Digenes Akrites," *BMGS* 13 (1989): 290-95 and A. Goldwyn, *Byzantine Ecocritism: Women, Nature, and Power in the Medieval Greek Romance* (Cham, 2018), 48–81.

96 M. Pastoureau, *Une histoire symbolique du Moyen Âge Occidental* (Paris, 2004), 62.

97 For a slightly different interpretation of this episode, see G. Prinzing, "Historiography, Epic and the Textual Transmission of Imperial Values: Liudprand's Antapodosis and Digenes Akrites," in *Reading in the Byzantine Empire and Beyond*, ed. T. Shawcross and I. Toth (Cambridge, 2018), 336–50.

98 On the affinities with the *Life of Basil*, see A. Markopoulos, "Ο Διγενής Ακρίτης και η βυζαντινή χρονογραφία. Μια πρώτη προσέγγιση," *Ariadni* 5 (1989): 165–71, here 170.

99 On women as game in the anthropology of hunting, see Hell, *Sang noir*, 288–89.

100 *Chronicle of Tocco*, 3466–67, ed. G. Schiro, *Ignoti auctoris Chronaca Tocchorum Cephalleniensium* (Rome, 1975).

101 *Livistros and Rhodamne*, 40 and 104, ed. P. A. Agapitos, Ἀφήγησις Λιβίστρου καὶ Ροδάμνης (Athens, 2006); trans. P. A. Agapitos, *The Tale of Livistros and Rodamne: A Byzantine Love Romance of the 13th Century* (Liverpool, 2021), 56 and 58.
102 *Livistros and Rhodamne*, 3377–78 and 2245–46, ed. Agapitos.
103 *Belthandros et Chrysantza*, ed. E. Legrand, *Bibliothèque grecque vulgaire I* (Paris, 1880), 770–79; trans. G. Betts, *Three Medieval Greek Romances* (New York, 1993), 19.
104 *PLP 143.*
105 Trans. G. Dennis, "Some Notes on Hunting in Byzantium," *Anathemata heortika: Studies in Honor of Thomas F. Mathews* (Mainz, 2009), 131–34, here 133.
106 Kydones, *Letters*, no 135, ed. R. Loenertz, *Démétrius Cydonès Correspondance* (Vatican, 1960), vol. II, 4–5. Cf. also *Letter* no 229:14–19, to the same addressee.
107 Cf. John Eugenikos on hunting and dogs in his description of a miniature illumination (?); ed. F. Boissonade, *Anecdota nova* (Paris, 1844), 331–35, here 232–33.
108 Theodore of Gaza, *Praise of the Dog*, ed. J.-P. Migne, PG 161, 985–98, here ch. 4, 992; I. Tziffa and K. Staikos, Θεόδωρος Γάζης Κυνός ἐγκώμιον (Athens, 2017), 56. On the author and the texts, see Tziffa and Staikos, 15–46 and 81–100; J. Kindstrand, "Notes on Theodorus Gaza's Canis Laudatio," *Eranos* 91 (1993): 93–105; Gibson, "In Praise of Dogs," 31–35. For the text of Libanius, see R. Foerster, *Libanii opera, v. I, fasc. I: Orationes I–V* (Leipzig, 1903), 308–9.

10
ANIMALS IN BYZANTINE MOSAICS

Henry Maguire

Introduction: metonymy, metaphor, and talisman

In addition to their role as literal illustrations of creatures from terrestrial nature of land, sea, and air, the animals portrayed in Byzantine mosaics carried overtones of meaning that were multifarious and often overlapping. These meanings can be classified broadly under three headings, namely metonymy, metaphor, and talisman. For the purposes of this chapter a metonymical image can be defined as one in which the part stands for a larger whole, as when a cross represents the crucifixion, or a hand portrayed above a cross signifies the Father in heaven. A metaphor can be described as a motif that conveys an idea about something or someone, but has no literal connection with it, such as when a vine represents Christ or his church, or a spring of water represents the Virgin Mary. A talismanic image is one that, in addition to its literal or metaphorical meanings, carries a power within itself to affect the world around it for good or for evil, as in the case of a magical sign. In Byzantium, representations of animals may fall into any of these categories, or into more than one at the same time. Nevertheless, the three categories form a useful framework for analysing animal imagery in Byzantium, and its changing reception by viewers over the centuries.

Animal mosaics in Early Byzantine churches

A clear example of metonymy in Byzantine floor mosaics can be found in the north transept of the basilica of Dumetios at Nikopolis, in Greece, which dates to the second quarter of the sixth century.[1] This mosaic, which is square in format, has a central field that displays a line of nine trees of different species. Eight birds are portrayed flying around their upper branches, while three more birds of varying sizes peck at the ground below. No other creatures are to be seen in the central panel, but it is surrounded by an inner border made up of small circular medallions each containing a small bird with a ribbon around its neck. Around this border there is a broader band, which frames the whole mosaic. It contains lines representing waves, in which diverse water creatures swim, including large and small

DOI: 10.4324/9781003055877-13

fishes, octopuses, shellfish, and water birds. An inscription in a panel beneath the central panel identifies the subject of the mosaic for the viewer:

Here you see the famous and boundless ocean,
containing in its midst the earth,
bearing round about in the skilful images of art everything
that breathes and creeps,
the foundation of Dumetios, the great-hearted archpriest.

Thus, the floor illustrates a geographic concept that was well-known in Antiquity, in which the inhabited earth, here represented by the central panel, was conceived as being completely surrounded by the ocean, here shown by the outer border.[2] But, in apparent contradiction of the inscription, the central field does not contain "images of everything that breathes and creeps", but only depicts birds. Therefore, the birds, by metonymy, stand in for the full variety of terrestrial creatures, including all species of mammals and reptiles, as well as birds.

A similar kind of metonymy can be seen on a smaller scale in the motifs of the nave pavement of a church at Delphi, which dates to the late fifth or early sixth century (Fig. 10.1).[3] The mosaic is made up of repeated octagons, each one of which is framed by four hexagons. The octagons and the hexagons are linked by crosses filled with a swastika meander. Each of the octagons contains a different land animal, either a quadruped, such as a camel, a dog, or a boar, or else a bird, such as a cock. The hexagons that frame the octagons contain various sea creatures, so that every unit of an octagon and four hexagons represents on a miniaturized scale the geographical schema of the Earth surrounded by the ocean. The crosses linking the units may convey ideas expressed in Early Christian exegesis, according to which the cross signified the cardinal directions of the world;[4] a commentary on Mark's Gospel attributed to St. Jerome, for example, called the cross "the squared form of the world".[5]

In the case of the mosaic at Delphi, the creatures evoking Earth and Ocean, in addition to their role as metonymies of terrestrial creation, may have borne talismanic overtones, for similar designs were found on contemporary amulets that were worn on the body to bring good fortune. They also appeared woven into people's clothing. For example, a Late Antique gold phylactery, which is now in Berlin, bears paired images of dolphins, birds, snakes, and ivy leaves depicted in relief on the body of the cylinder (Fig. 10.2).[6] On the other side of the amulet there are letters reading, in reversed mirror-writing, "ΕΠ ΑΓΑΘΩ", or, "For a Good Cause". A similar combination of a snake, fishes, and birds together with ivy leaves can be found on the bands from a tunic now divided between three museums, which dates to the sixth century.[7] Such parallels from the domestic sphere suggests that the creatures of land, sea, and air depicted on the floors of churches were not only metonymical signs for the whole of nature, but also expressions of a wish for prosperity, whether of the community or the individual.

Another pavement that portrays the earth surrounded by the ocean occupies the narthex of the large basilica at Heraklea Lynkestis, in Macedonia; it is dated to the turn of the fifth and the sixth centuries (Fig. 10.3).[8] The mosaic is more complex than the preceding examples, because in this case the basic metonymical scheme is overlaid with additional layers of both metaphorical and talismanic significance. The whole of the floor is framed by

Figure 10.1 Delphi, Christian basilica, nave mosaic with animals representing earth and ocean.

Source: Author

Figure 10.2 Gold Phylactery with animals of land and sea. Berlin, Staatliche Museen zu Berlin, Museum für Spätantike und Byzantinische Kunst. Copyright: Staatliche Museen zu Berlin, Skulpturensammlung und Museum für Byzantinische Kunst / Antje Voigt Public Domain Mark 1.0.

a wide border containing a series of thirty-six octagons, each containing one or more sea creatures or water birds. Within the border a long rectangular field displays a line of ten trees of different species, including cedar, cypress, apple, fig, and pomegranate. One of the trees is depicted bare of leaves, as in winter. Birds of different varieties fly among the upper branches of the trees, while below them are depicted beasts, some of which are in combat with each other; they include, from left to right: a goat, a lion, and a bull charging at each other, a dog, and a cheetah disembowelling a newly killed hind (Fig. 10.3). These plants and the animals must illustrate terrestrial nature rather than Paradise, for there were no seasons in Eden,[9] and no death and no violence.[10] Eden was a place of harmony, where, according to a sermon attributed to Basil the Great, even the vultures ate grass rather than flesh.[11]

At the centre of the mosaic the line of trees and animals is disrupted by the insertion of an oval frame created by the curving stems of an acanthus plant, which enclose vines growing from a large handled vase (Fig. 10.4). Two beasts, a stag and a doe, flank the vase in symmetrical poses; above them stand two peacocks, also posed symmetrically. These animals can be understood on two levels. On one level, they can be seen as part of earthly nature, like the

Figure 10.3　Heraklea Lynkestis, Large Basilica, floor mosaic in the narthex, detail of a cheetah disembowelling a hind.

Source: Author

other beasts and birds depicted on the floor. But, on a second level, they can be interpreted as metaphors. The deer, for example, may refer to the first verse of Psalm 41, "As the hart longs for the water fountains, so longs my soul for thee, O God", which was associated by Early Christian writers both with baptism and with the Eucharist.[12] In a well-known floor mosaic found in the antechamber of the baptistery at Salona this verse was inscribed above a depiction of two stags drinking from a vase.[13] As for the peacocks, they were not only praised as among the most beautiful creatures of terrestrial creation,[14] but also they were symbols of eternal life.[15] For this reason they were frequently represented in funerary art, in tomb paintings,[16] in carvings on sarcophagi,[17] and in mosaics that covered graves.[18]

Thus, the Heraklea mosaic creates a distinction between two types of animal representation according to their interpretation. Outside the central frame created by the acanthus, the animals were intended to be read on only one level, as literal illustrations selected from the natural world of conflict and seasonal change, that is, as metonyms. Inside the acanthus, however, the animals had the potential to be read on two levels, both as literal illustrations and as symbols of spiritual concepts, so that they acted both as metonyms and as metaphors at the same time. The distinction was embodied in the manner in which the artists portrayed the animals. Outside the acanthus frame, the animals are rendered in active, naturalistic poses, whether in flight in the air, or in combat on the ground. Within the frame, however, the animals stand in symmetrical poses on either side of the vase, in a motionless harmony outside of earthly time.

Figure 10.4 Heraklea Lynkestis, Large Basilica, floor mosaic in the narthex, detail of the central motif with deer and peacocks flanking a vase.

Source: Author

It is possible that some of the beasts portrayed in the entrance porch of the large basilica at Heraklea embodied even a third level of significance, which was talismanic. Portrayals of fierce creatures and of animal combat performed an apotropaic function in Greek and Roman art, and this role continued into the Christian era. The ferocity of the animals was believed to protect the thresholds of buildings from the entry of harm. For example, a sixth-century panel depicting a tiger seizing a doe was set in the mosaic border immediately inside the southwest entrance of the Cathedral of Apamea in Syria.[19] At Heraklea the powerful image of a cheetah disembowelling a doe was set directly in front of the door to the south aisle of the church (Fig. 10.3); in this location it may have performed a protective function in addition to its metonymical significance as a part of terrestrial creation.

Animals appear with less frequency in Early Byzantine wall and vault mosaics than they do in pavements, in part because animals, as representing the Earth, were considered a more suitable subject for floors that were walked upon than for the upper parts of a church. However, some notable depictions of animals on church walls and vaults do survive, among which can be highlighted the sixth-century mosaic in the apse of the church of Panagia Angeloktistos at Kiti, in Cyprus (Fig. 10.5).[20] Here we find a central image of the Virgin standing and holding her child, which is framed by a rich border depicting three slender vases on either side, each rising from a bed of fleshy acanthus leaves and flanked by a pair

Figure 10.5 Kiti, Panagia Angeloktistos, apse mosaic with border containing stags, parrots, and ducks flanking vases.

Source: Author

of animals in symmetrical poses. At the top we see a pair of stags, in the middle two green parrots, and at the bottom two long-necked ducks.

At the literal level, these animals once again stand as metonyms for the creatures of land, air, and water. In this sense, when seen in conjunction with the portrayal of the Virgin with her child, they expressed the idea that the created world had been sanctified through the incarnation, and was thus "changed into a world of beauty", as the sixth-century patriarch Anastasios of Antioch wrote in a sermon on the Annunciation.[21] On another level, however, the animals and the vases depicted in the border of the mosaic at Kiti can be read metaphorically. As at Heraklea, the vases flanked by pairs of stags can be associated with the hart and the water fountains of Psalm 41, and with the symbolism of Baptism and the Eucharist. Additionally, the Virgin frequently was praised with metaphors drawn from nature in Early Byzantine hymns and homilies, as in a fifth-century sermon by Hesychios of Jerusalem, where among other images she becomes a dove and a turtle dove.[22] The water evoked by the vases in the mosaic echo the language of a fifth-century sermon attributed to Proklos, which describes the Virgin of the Annunciation as a fountain and a river.[23] In the *Akathistos* hymn she is the river of many streams and the rock giving water.[24]

In summary, depictions of animals played an important role in the imagery of the floors, walls, and vaults of Early Byzantine churches. They were not merely decorative, but carried a panoply of meanings, which could be literal, symbolic, or apotropaic. Until the eighth century animals played an important role in the visual setting of Christian cult, a space in which, in later centuries, they were to become less welcome.

Animal mosaics in Early Byzantine imperial art

The excavations of the Great Palace of the Byzantine emperors in Constantinople have revealed a magnificent sixth-century mosaic pavement which once decorated a peristyle courtyard in front of an apsidal hall, situated to the east of the hippodrome.[25] The mosaics, which have survived on three sides of the court, preserve a large number of figures and vignettes scattered against an imbricated white background. Many of the scenes involve animals. Some creatures are portrayed on their own; others fight each other, such as an elephant engaged with a lion, or an eagle with a serpent (Fig. 10.6); or else they prey upon each other, such as a bear attacking a kid or a fawn. Yet other beasts are being fought or hunted by humans, such as a leopard attacked by a man brandishing a sword (Fig. 10.7). In addition, several of the animals evoke peaceful scenes of rural daily life, such as goats shown alongside their herdsmen (Fig. 11.6).

At the Great Palace, then, we find a variety of individual animals together with displays of animal violence, hunts, and pastoral settings. This range of subject matter can be matched in many other Late Antique mosaic floors from wealthy villas;[26] it expresses, in general terms, ideas about the fecundity of the natural world, its exploitation, and its control by the master of the domain. In this particular setting, in the imperial palace, it expresses through metonymy the universal dominion of the emperor over the whole terrestrial world that is evoked by the individual motifs.[27] In addition, the fierce beasts depicted in the mosaics could have conveyed a metaphorical content. Their symbolic message can be compared to that expressed by the well-known Barberini ivory in the Louvre, which also dates to the sixth century (Fig. 10.8).[28] At the centre of the ivory we see the emperor, perhaps Justinian, on horseback, triumphing over his enemies, while one of his generals, on the left, brings him a victory. At the bottom, Persians and Indians bring tribute from east and west, accompanied

Figure 10.6 Constantinople, Great Palace, floor mosaic, detail showing an eagle killing a snake beside a group of goats.

Source: Author

by lions, tigers, and elephants. A personification of the earth, cradling fruits in her lap, supports the emperor's right foot.

Here, as in the mosaics of the Great Palace, the general meaning appears to be one of imperial dominance over the resources of the Earth, including its agricultural bounty, while the wild beasts express an additional metaphorical content, namely the wild and savage natures of the empire's barbarian enemies. The Persians and Indians seen at the bottom of the reliefs in their exotic costumes both lead the wild animals to the emperor and are themselves the wild animals, whom the emperor subjugates and tames. The poet Corippus, in his account of the accession of Justinian's successor, Justin II, compared the Avar ambassadors entering the throne room to savage tigers entering the hippodrome in Constantinople, which circled the arena in front of the spectators, before lying down gently before the throne of the emperor, subdued.[29]

Further to their roles as metonyms and metaphors, some of the animals portrayed in the Great Palace floor may have carried a talismanic significance. The motif of an eagle killing a serpent seen at the left of Figure 10.6, for example, was both a symbol of the fight of good against evil and a protective device in its own right.[30] As we know from Roman sculpture, raptors clasping their prey were portrayed upon the lappets of armour, in addition to other powerful devices such as the heads of Gorgons.[31] A medieval text, the collection of

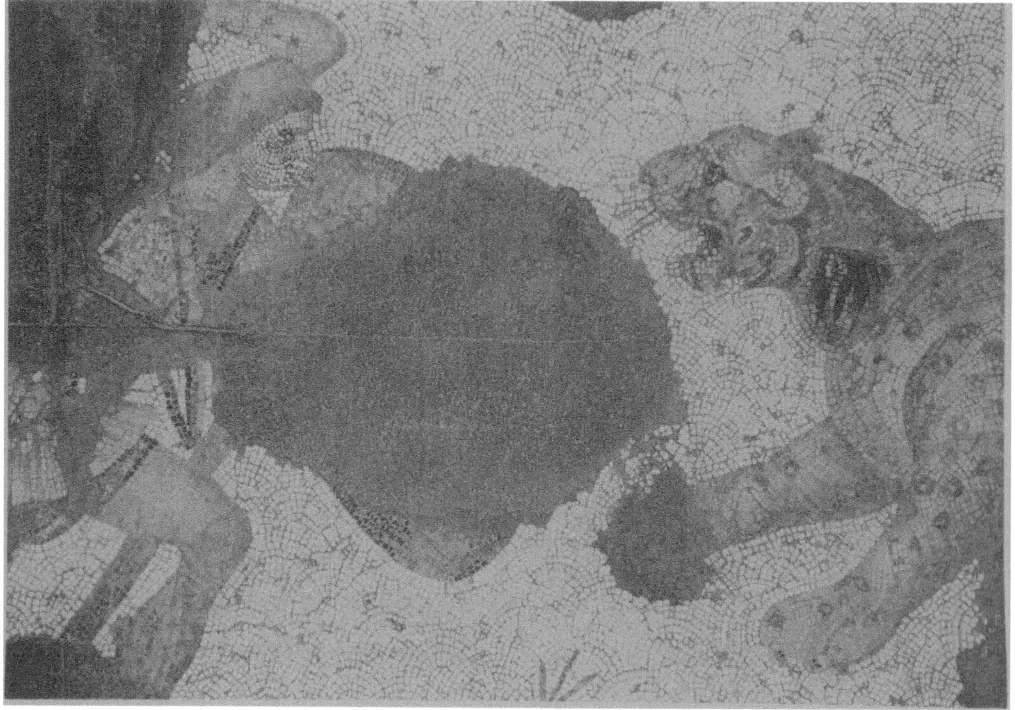

Figure 10.7   Constantinople, Great Palace, floor mosaic, detail of hunter attacking a leopard.
Source: Author

chronicles known as Theophanes Continuatus, informs us that the tenth-century emperor Constantine VII built a guardhouse of porphyry in front of his chamber in the palace in Constantinople, in which there was a silver sculpture of an eagle stifling a serpent that was coiled around its feet.[32] Presumably, since this group was placed in the emperor's guardhouse, it was intended to be protective.

In summary, the meanings conveyed by the animals portrayed in the floor mosaics of the Great Palace can be classified under the same three headings as those of contemporary churches. Together they acted as metonyms, in that they stood for the whole of earthly life that ideally was subject to imperial rule; in addition, some of them were metaphors of defeated barbarian enemies or of evil in general; and some functioned as talismans, intended to protect the emperor and his domain.

The iconophobic reaction

The animals that appeared so frequently on the floors of Early Byzantine churches were not to everybody's liking.[33] There was a strong current of opposition to such imagery, which is exemplified by a letter written early in the fifth century by Saint Neilos of Sinai to the prefect Olympiodoros who was about to build a church. The prefect had proposed decorating the nave of the new building with "the pictures of different birds, and beasts, reptiles, and plants". To this, the saint replied: "may I say that it would be childish and infantile

Figure 10.8 Barberini ivory, mounted emperor in triumph over the earth and barbarians accompanied by wild beasts. Paris, Musée du Louvre. Copyright: Bridgeman Images XIR159131.

to distract the eyes of the faithful with the aforementioned [trivialities]". Instead, Neilos recommended that the prefect fill the nave of his church with many crosses. "Whatever is unnecessary", he concluded, "ought to be left out".[34]

Neilos of Sinai composed his letter during a period at the turn of the fourth and the fifth centuries when the floors of churches underwent an aniconic phase, in which they tended to be made up of geometrical patterns rather than animal or human imagery.[35] Examples include the cruciform church at Qaousiyé, at Antioch in Syria, which is dated by its inscription to 387.[36] Here each of the four naves is filled with an exclusively geometric pavement. Another example is the basilica of Ayia Trias, near Yialousa, on Cyprus, which has been dated by coin evidence to the early fifth century. Its mosaic floors are entirely abstract, except for three small panels in the north aisle, two depicting pairs of sandals, and the third a branch bearing pomegranates.[37] Like Saint Neilos, whoever composed floor mosaics of this kind clearly did not feel that animals, symbolic or otherwise, were necessary to the decoration of a church.

In later centuries, some Christians came to see the portrayals of animals in a church not just as childish, but rather as threatening. The change in attitude came about as a result of the rise of the cult of sacred portrait icons, which in the sixth and later centuries increasingly became the objects of veneration. The worship of sacred images brought the danger that people could venerate animal images too, just as the pagans had done. In other words, the fear was that people would no longer view animal images merely as metonymies or metaphors, but instead they would see them in an unchristian way as idols, that is as divine powers in their own right. The issue eventually came to a head with the advent of the iconoclastic dispute in the eighth and ninth centuries, when the legitimacy of religious images came under fierce debate. In their arguments, the defenders of icons were forced to make a distinction between sacred portraits of Christ and his saints, which they believed were allowable, and portrayals of animals such as had been worshipped by pagans, which were not.

The key text in this debate was Saint Paul's Epistle to the Romans, chapter 1, verses 23–25, where the apostle first says: "And they exchanged the glory of God, who is incorruptible, with the likeness of an image of man who is corruptible", but then goes on to add the following qualifying phrase, "or of birds, quadrupeds, or reptiles, and they paid respect to and worshipped the creature rather than the Creator". The defenders of icons argued that this passage did not condemn Christian portrait icons, but only the nature-derived imagery associated with pagan cults.[38] The logical consequence of this stance would be that the supporters of images, in order to keep their veneration of Christian portrait icons, had to exclude animals and other nature-derived subjects from their churches. This reasoning was particularly evident in the churches of Palestine, which had been outside the Byzantine empire since the Arab conquests of the seventh century. Here Christians appear to have departed from the official Byzantine iconoclastic policy which was introduced early in the eighth century, in that they continued to accept figural icons, while at the same time they rejected animal imagery.

In Palestine the pavements of churches were, until the beginning of the eighth century, especially lavish in their displays of nature, including both wild and domestic animals. But at some time between 718 and 756, Christians living in this region actually defaced or destroyed animals that previously had been portrayed on the floors of their churches, even while, as will be seen below, they appear to have maintained the veneration of portrait icons which were displayed above.[39] In this respect the Christians of Palestine appear

Figure 10.9 Umm al-Rasas, Church of St. Stephen, floor mosaic in nave, detail showing iconoclastic interventions.

Source: Author.

to have differed from those within the borders of the Byzantine empire, where during the iconoclastic dispute there is some evidence for the destruction of church mosaics portraying human figures, but little for the erasure of animals.

Many examples of the erasure of animals survive in the church floors of Palestine, including those that took place in the nave of St. Stephen at Umm al-Rasas in Jordan.[40] Here a rich pavement was laid in 718, in which, according to the familiar schema, hunting and farming scenes metonymically representing the Earth were surrounded by a border containing aquatic motifs evoking the water. Shortly after the creation of this mosaic, its images were partially erased, as for example the fishes seen in Figure 10.9. In this case, as in several other Palestinian floors that suffered iconoclasm, the carefully carried out restoration of the damaged areas demonstrates that the congregation continued to use the floor after the destruction had taken place.

Animals in mosaics after iconoclasm

The iconoclastic episode in the mosaics of Palestine ended by the middle of the eighth century; after this time, church floors laid in this region became primarily aniconic in their composition.[41] Sometimes the pavements contained crosses, but they displayed little if any animal imagery, in accordance with the advice given centuries earlier by St. Neilos of Sinai. In the nave of the Church of the Virgin at Madaba in Jordan, for example, a sixth-century floor mosaic, which may have been figural, was replaced in 767 by a purely geometrical composition, in which there are no animals, but only a few stylized plant forms.[42] At the centre of the new mosaic is a medallion containing an inscription, which calls upon the viewer to contemplate an icon of the Virgin Mary, demonstrating that here the portrait icon was venerated even while images of animals were excluded from the pavement of the church.

A contemporary mosaic survives in a church at Shunat Nimrin in Jordan, where the floors in all three aisles of the building are almost entirely aniconic, being filled with geometrical designs based on repeated squares and octagons.[43] Into these patterns many crosses were discreetly inserted, as can be seen in the eastern panel of the nave, which is composed of four units each composed of two interlaced squares. At the centre of the panel there is a circle with a small black cross at its centre. Crosses reappear in the western panel of the nave, but in a disguised way. Here there is an overall design of repeating squares framed by hexagons, which in their turn frame poised squares. The points of the poised squares are joined by lines which can be read as reiterated crosses linking the elements of the composition. The artist has encouraged the viewers to read these crosses as meaningful signs, rather than mere accidents of ornament, by making the bars that connect the poised squares thicker than those linking the regular squares, thus emphasizing the crosses. Only in the central panel of the nave do a few animals appear, in the form of ducks surrounding the dedicatory inscription. Thus, in accordance with the precepts of St. Neilos, the nave has been filled with a multitude of crosses, while the animal imagery has been reduced to a minimum.

A somewhat larger selection of animals can be found in the floor mosaics of the Euthymius Monastery at Khan el-Ahmar on the West Bank, which have been dated to the middle or the second half of the eighth century.[44] But here, too, the pavements are primarily aniconic. The central panel in the nave of the church is paved with slabs of stone in the *opus sectile* technique, creating a design that is completely without figures. The surviving tessellated mosaics in the south aisle also are for the most part aniconic, presenting panels filled with interlace and with complex geometrical compositions incorporating propitious designs such as interlaced squares. However, the first panel in the south aisle, immediately in front of the western entrance, contains representations of animals. The composition contains six large poised squares. The square at the centre of the bottom row frames a knotted snake emerging from a vase. Another vase appears in the square above. The four squares on either side contain four more creatures, which are somewhat crudely rendered: at the top left we can recognize a beribboned parrot, and at the bottom left what may be an ostrich. At the bottom right is some kind of cat, and above it another bird.

Thus, the pavement at the Euthymios monastery, like that at Shunat Nimrin, is primarily aniconic, but it also incorporates a few animals into its design. It is possible that the range of animals, a reptile, a quadruped, and birds, may have been intended as an evocation of terrestrial creation, as in earlier floor mosaics. The knotted snake directly in front of the threshold, however, must also have been a talisman, since knots frequently acted as apotropaic devices in medieval visual culture, including floor mosaics.[45] Inserted into the geometrical design at the western end of the nave mosaic at Shunat Nimrin there is a large octagon enclosing a double Solomon's knot, which is framed by the inscription: ΜΕΘ ΗΜΩΝ Ω ΘC ("God with us").[46] This device, like the knotted snake at Khan el-Ahmar, must have been intended to protect the entrance to the building; the knotting of the snake served to increase its apotropaic powers.

In churches constructed outside of Palestine in the Byzantine empire after the final defeat of iconoclasm in 843 there was a transition from tessellated floors with their rich nature-derived imagery to more abstract compositions executed in opus sectile composed of plaques or strips of marble and coloured stone.[47] Here, as in the post-iconoclastic pavements of Palestine, geometrical compositions were preferred. Animals only occurred occasionally in these floors, usually in the form of small tessellated inserts into the surrounding plaques of marble. An example is the opus sectile pavement that covers the nave and aisles of the Katholikon of the Sagmata Monastery in Greece, which may date to the twelfth century.[48] This floor was made up of repeated square and rectangular panels filling the nave and aisles of the basilican church, of which over twenty survive.

The panels are filled with interlaced designs, for the most part based upon squares and circles. Only four of them contain animal motifs: a knotted serpent and a group of six birds were placed in front of the sanctuary; a raptor fighting a snake was positioned in the centre of the nave; and two confronted quadrupeds were located to the left of the entrance to the building. Thus, the animals appear as isolated motifs within largely abstract compositions of multicoloured pieces of stone. The relative prominence given to the snakes, at the entrance to the bema and in the centre of the nave, indicates that, as at Khan el-Ahmar, their function was primarily talismanic, to protect the church and its congregation from evil. Another example of an opus sectile floor containing an inserted animal can be found in the nave of the church of the Blachernai Monastery at Arta, which dates to the thirteenth century.[49] This pavement has a quincunx design composed of five interlaced circles created out of bands of white marble and arranged within a square, of which the central circle frames a frontal eagle set in tesserae.

The major exception to the relative absence of animals on the post-iconoclastic floors of Byzantine churches is the splendid pavement of the south church of the Pantokrator Monastery (the Zeyrek Camii) in Constantinople, which dates to the 1130s.[50] Here, created out of opus sectile, we find almost the full range of animal motifs that had appeared on early Byzantine church floors, including various kinds of birds, fishes and beasts, as well as fragments of hunting scenes (Fig. 10.10). It appears that the medieval marble floor set in the church of the Stoudios monastery in Constantinople contained a similar variety of animals, since it is described in the sixteenth-century diary by Stephan Gerlach as being "adorned with figures of birds and other animals",[51] but it is so poorly preserved today that the individual subjects can no longer be deciphered, although it is evident that animals were portrayed.[52]

As in church pavements, depictions of animals are rarely found in Byzantine wall mosaics during the centuries after iconoclasm, except where their presence was required by the illustration of a biblical story. The major exception comes in the last phase of Byzantine art, in the well-known mosaics in the church of the Chora Monastery (the Kariye Camii) in Constantinople, which date to the early fourteenth century.[53] In the inner narthex of the church we find several prominent depictions of birds. One is portrayed in the scene of the Annunciation to St. Anne, which includes a view of her garden, complete with an ornamental fountain and a grove of trees. In the upper branches of the trees, a bird is shown flying down to her young in their nest.[54] The motif of the bird and her nestlings was a traditional metaphor of regeneration associated with praises of the Annunciation, both in the art and the literature of the Byzantine church.[55] But at the Chora Monastery we find four other birds playing a prominent role in the mosaics devoted to the early life of the Virgin. The scene of Mary being caressed as an infant by her parents, Joachim and Anne, is flanked

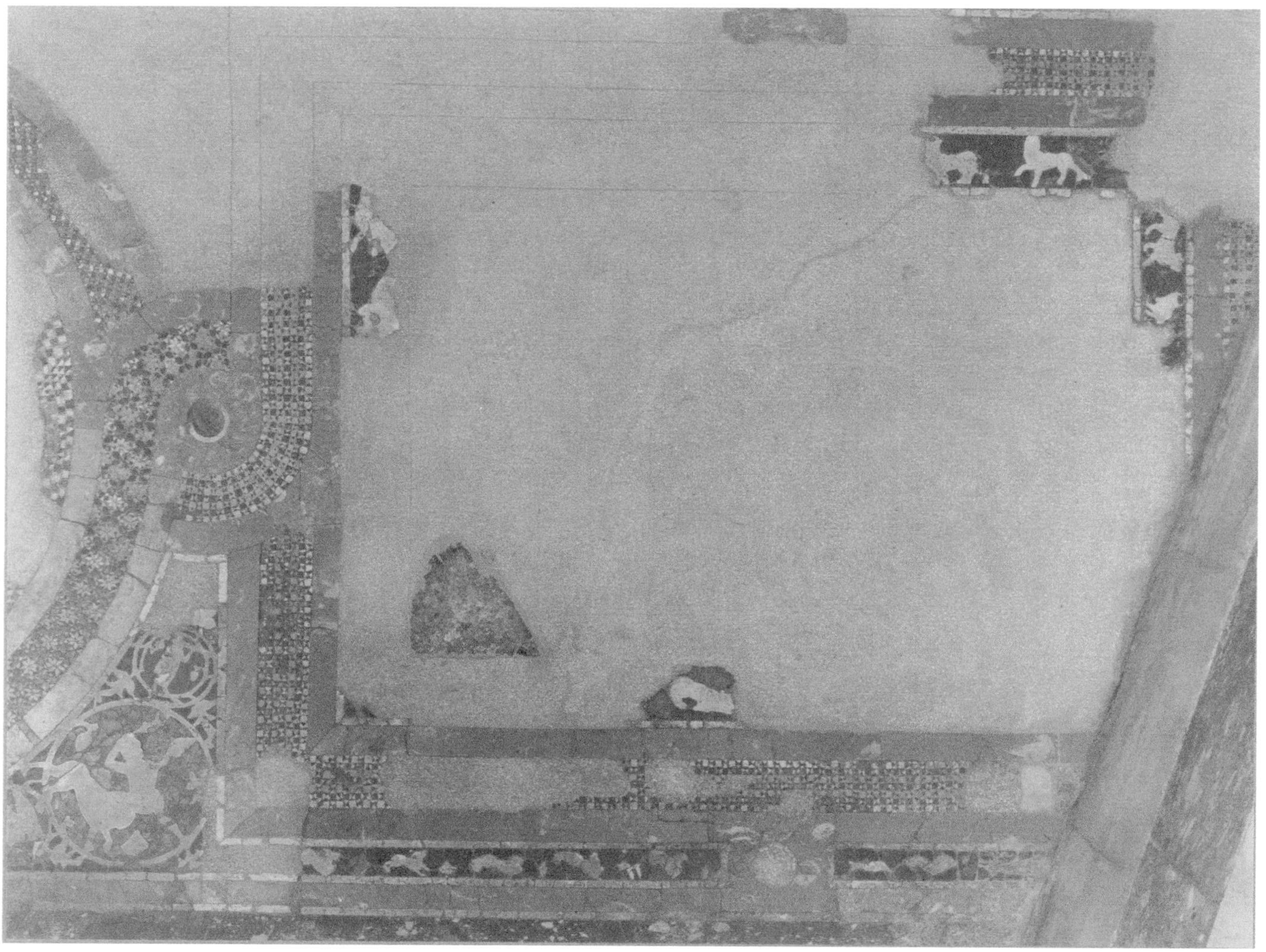

Figure 10.10 Constantinople, Pantokrator Monastery (Zeyrek Camii), opus sectile pavement with animals. Copyright: Creative Commons CC0 1.0 Universal license; Dumbarton Oaks Research Library and Collection, Image Collections and Fieldwork Archives, Washington, D.C.

by a pair of large and resplendent peacocks (Fig. 10.11),[56] while the Presentation of the Virgin in the temple is bracketed by a handsome male and female pheasant.[57] Yet another bird is portrayed in the outer narthex in the mosaic depicting John the Baptist bearing witness to Christ prior to His Baptism. Here, in the water beneath Christ, the artist depicted a long-necked bird that pecks at a snake (Fig. 10.12).[58]

As in the Early Byzantine church floors, these animals can be interpreted as metonymies, as metaphors, and as talismans. As we have seen, Early Byzantine writers such as Gregory of Nazianzos praised the peacock as a supremely beautiful ornament of all creation.[59] Thus, in the context of the infancy of the Virgin it is possible to see the bird as a metonymy for the beauty of all of creation, which was sanctified by the incarnation. In addition, the peacock may still have carried its traditional metaphorical meaning as a symbol of eternal life. As for the water bird attacking a snake, this was a well-known apotropaic motif found in Early Byzantine mosaics and amulets,[60] as well as in the sgraffito designs of medieval Byzantine pottery.[61] In the context of the mosaics in the Chora monastery, the bird and snake motif can be associated with the customary exorcism of the waters before the ceremony of Baptism.[62]

Very little survives of the mosaic decoration of the walls and vaults of medieval palaces in Byzantium, nor do surviving texts explicitly describe mosaics depicting animals in these locations, although animal mosaics preserved in palaces outside the empire may provide more or less distant echoes of lost works in its capital.[63] However, a few medieval floors survive in the Great Palace of Constantinople, and we have extensive literary descriptions of the medieval pavements laid there, a few of which appear to have contained animal imagery. Notwithstanding these exceptions, it appears that generally speaking in the palace, as in contemporary churches, the multi-figured tessellated compositions characteristic of earlier centuries were replaced by geometrical pavements executed in larger pieces of stone. The peristyle court in the Great Palace provides a striking illustration of this change, for its sixth-century mosaics (Figs. 10.6–10.7), many of which were in good condition, were overlaid by a later floor composed of plain slabs of Prokonnesian marble, which concealed the multifarious animals of the older work.[64]

We know from Theophanes Continuatus and from other sources that many splendid marble pavements were put down in the Great Palace during the ninth century, some composed of slabs of Prokonnesian marble like the one covering the peristyle mosaic, others of opus sectile.[65] Only two of the mosaics described by the medieval texts were said to have contained images of animals. One of them adorned the church of the Virgin of the Pharos, located near the audience hall known as the Chrysotriklinos in the Great Palace, which will be discussed below. The other pavement that included animal imagery was made in a bedchamber constructed by the ninth-century emperor Basil I, according to his biography, the *Vita Basilii*. It appears from the description in the text, which is somewhat difficult to interpret, that the design of this floor was a quincunx composed of five circles of coloured marble arranged in a square. The circles enclosed tessellated inserts depicting birds, four eagles "made of fine variegated tesserae" in the four outer circles, and a peacock, "all of gleaming tesserae", in the central medallion.[66]

In the medieval churches an important motivation for the change from figured tessellated floors, with their abundance of animal imagery, to aniconic pavements made of stone slabs or opus sectile must have been a lingering fear of accusations of idolatry, such as earlier had caused the iconoclasm in the pavements of Palestine and the subsequent aniconic turn that had followed there. But the mosaics in the Great Palace, where religious iconophobia was

Figure 10.11 Constantinople, Monastery of the Chora (Kariye Camii), inner narthex, mosaic showing peacocks flanking the Virgin being caressed by her parents Copyright: Creative Commons CC0 1.0 Universal license; Dumbarton Oaks Research Library and Collection, Image Collections and Fieldwork Archives, Washington, D.C.

Figure 10.12 Constantinople, Monastery of the Chora (Kariye Camii), outer narthex, mosaic showing a water bird attacking a snake beneath John the Baptist bearing Witness to Christ. Copyright: Creative Commons CC0 1.0 Universal license; Dumbarton Oaks Research Library and Collection, Image Collections and Fieldwork Archives, Washington, D.C.

not such an important issue, suggest that there must also have been other reasons for the switch from tessellated figural floors to abstract compositions in plaques of stone, reasons that could have applied both to imperial and to ecclesiastical buildings. One of these factors may have been a fear of sensory overload caused by the relatively small size of the medieval structures compared to their early Byzantine predecessors. In a small church or palace chamber with richly decorated walls and vaults, a highly figured tessellated pavement may have provided too much competition with the decoration displayed above, whether sacred images in a church, or, in the case of Basil's bedchamber in the palace, golden portraits of the emperor, the empress and their children surrounding a gleaming cross.[67]

The floor of the church of the Virgin of the Pharos in the Great Palace, mentioned above, seems to have provoked a concern of this kind in the mind of at least one observer, the patriarch Photios, who wrote an ekphrasis of the building after its restoration by the Emperor Michael III, probably in 864. Photios describes the rich interior decoration of the church, including its gilded capitals and cornices, its golden chains, its multi-coloured marble revetments, and the gold mosaics on its walls depicting Christ, the Virgin, and an accompanying retinue of angels, apostles, martyrs, patriarchs, and prophets. He also gives an account of the "minute work" of the tessellated floor, which depicted a variety of creatures, saying: "The pavement, which has been fashioned into the forms of animals and other shapes by means of variegated tesserae, exhibits the marvellous skill of the craftsman... .". It is uncertain whether this floor dated to the time of the restoration by Michael III, or whether it was taken over from an earlier building. In either case, the overall effect of the interior of the Pharos was sumptuous, and possibly too much so, for, immediately after his description of the building, Photios makes a critical comment:

In one respect only do I consider the architect of the church to have erred, namely that having gathered into one and the same spot all kinds of beauty, he does not allow the spectator to enjoy the sight in its purity, since the latter [the spectator] is carried and pulled away from one thing by another, and is unable to satiate himself with the spectacle as much as he may desire.[68]

This desire for a "purity" of viewing without distraction may have reinforced concerns about the pagan connotations of animal imagery, leading to a medieval preference for marble pavements over the highly figured tessellated floors of earlier times. With relatively few exceptions, it was no longer as appropriate for animals, at best unnecessary and at worst idolatrous, to be depicted in the mosaics of floors and walls, where they could compete either visually or conceptually with the sacred images.

Notes

1 E. Kitzinger, "Studies on Late Antique and Early Byzantine Floor Mosaics, I: Mosaics at Nikopolis," *DOP* 6 (1951): 83–122, figs. 20–22; H. Maguire, *Earth and Ocean: The Terrestrial World in Early Byzantine Art*. College Art Association Monograph Series 43 (University Park, 1987), 21–24, figs. 10–12.

2 For the illustration of this concept in textiles, see H. Maguire, "The Mantle of Earth," *Illinois Classical Studies* 12 (1987): 221–28.

3 P. Asemakopoulou-Atzaka, *Syntagma ton palaiochristianikon psephidoton dapedon tes Hellados*, 2, *Peleponnesos-Sterea Hellada* (Thessaloniki, 1987), 194–97, pls. 337b, 338a.

4 Maguire, *Earth and Ocean*, 29.

5 *Evangelium secundum Marcum*, 15, ed. J.-P. Migne, PL 30 (Paris, 1846), col. 638.

6 V. H. Elbern, "Per speculum in aenigmate. Die 'imago creationis' an einem frühchristlichen Phylakterion," in *Studien zur spätantiken und byzantinischen Kunst Friedrich Wilhelm Deichmann gewidmet*, ed. O. Feld and U. Peschlow, vol. 3 (Bonn, 1986), 67–73, pl. 17.

7 Metropolitan Museum of Art, inv. no. 90.5.154; N. Kajitani, "Coptic Fragments," [in Japanese], *Textile Art* 13 (1981): 49, fig. 64a; H. Maguire, "Garments Pleasing to God: the Significance of Domestic Textile Designs in the Early Byzantine Period," *DOP* 44 (1990): 217, fig. 9; Museum of Fine Arts, Boston, inv. no. 35.87; F. D. Friedman (ed.), *Beyond the Pharaohs: Egypt and the Copts in the 2nd to 7th Centuries A.D.*, exhibition catalogue, Rhode Island School of Design (Providence, 1989), 158, no. 67; and the Victoria and Albert Museum, London, inv. nos. 334-1887 and 335-1887; A. F. Kendrick, *Catalogue of Textiles from Burying-Grounds in Egypt*, vol. 1, *Graeco-Roman Period* (London, 1920), 66, no. 62, pl. 14; Kajitani, "Coptic Fragments," 49, figs. 64b and c.

8 G. Cvetković-Tomašević, *Heraclea*, vol. 3, *Mosaic Pavement in the Narthex of the Large Basilica at Heraclea Lyncestis* (Bitola, 1967); eadem, "Mosaïques paléochrétiennes récemment découvertes à Héracléa Lynkestis," in *La mosaïque gréco-romaine*, vol. 2 (Paris, 1975), 385–99, figs. 183–92; R. E. Kolarik, "The Floor Mosaics of Eastern Illyricum," in *Eisegeseis tou Dekatou Diethnous Synedriou Christianikes Archaiologias*. Hellenika 26 (Thessaloniki, 1980), 173–203, esp. 193; Maguire, *Earth and Ocean*, 36–40.

9 H. Maguire, "Paradise Withdrawn," in *Byzantine Garden Culture*, ed. A. Littlewood, H. Maguire, and J Wolschke-Bulmahn (Washington, DC, 2002), 23–35, esp. 24.

10 H. Maguire, "Adam and the Animals: Allegory and the Literal Sense in Early Christian Art," in *Studies on Art and Archaeology in Honor of Ernst Kitzinger on his Seventy-Fifth Birthday*, ed. W. Tronzo and I. Lavin, *DOP* 41 (1987): 363–73, esp. 365–66.

11 *In scripturae verba, "Faciamus hominem ad imaginem et similitudinem nostram," Oratio II*, 6, ed. A. Smets and M. Van Esbroeck, *Basile de Césarée, Sur l'origine de l'homme*, Sources Chrétiennes 160 (Paris, 1970), 242.

12 P. A. Underwood, "The Fountain of Life in Manuscripts of the Gospels," *DOP* 5 (1950): 41–138, esp. 51–52; Maguire, *Earth and Ocean*, 38; B. Shilling, "Fountains of Paradise in Early Byzantine Art, Homilies and Hymns," in *Fountains and Water Culture in Byzantium*, ed. B. Shilling and P. Stephenson (Cambridge, 2016), 208–28, esp. 212.

13 E. Dyggve, *History of Salonitan Christianity* (Oslo, 1951), 31–33, figs. II, 25-30.

14 Maguire, *Earth and Ocean*, 39; H. Maguire, *Nectar and Illusion: Nature in Byzantine Art and Literature* (New York, 2012), 52–53, 58.

15 On the symbolism of the peacock, see H. Lother, *Der Pfau in der altchristlichen Kunst: eine Studie über das Verhältnis von Ornament und Symbol*. Studien über christliche Denkmäler, NS 18 (Leipzig, 1929), esp. 56–83; F. Cabrol and H. Leclercq, *Dictionnaire d'archéologie chrétienne et de liturgie*, vol. 13, 1 (Paris, 1937), cols. 1075–97; Maguire, *Earth and Ocean*, 39–40.

16 For example, in a finely painted fourth-century tomb at Nicaea: B. Brenk, *Spätantike und frühes Christentum*. Propyläen Kunstgeschichte, Supplementband 1 (Frankfurt, 1977), 169, pl. 135.

17 As on fifth- and sixth-century sarcophagi at Ravenna: G. Bovini, G. Valenti Zucchini, and M. Bucci, *"Corpus" della scultura paleocristiana bizantina ed altomedioevale di Ravenna*, vol. 2, *I sarcofagi a figure e a carattere simbolico* (Rome, 1968), nos. 28a (sarcophagus in S. Apollinare in Classe), 35a and b (S. Francesco), and 45 (Museo Nazionale).

18 For example, a mosaic covering a double tomb in a church at Kelibia: M. Yacoub, *Le Musée du Bardo* (Tunis, 1996), 59, fig. 52.

19 J. Balty, *Mosaïques d'Apamée* (Brussels, 1986), 12, pl. 25; P. Donceel-Voûte, *Les pavements des églises byzantines de Syrie et du Liban: décor, archéologie et liturgie*, vol. 1 (Louvain-la-Neuve, 1988), 205, 212, figs. 186, 192.

20 Maguire, *Nectar and Illusion*, 79–81, pl. 7; Shilling, "Fountains of Paradise," 214–21, figs.11.3–11.4.

21 *In Annuntiationem*, ed. J.-P. Migne, PG 89 (Paris, 1865), col. 1384.

22 Hesychios of Jerusalem, *Homilies* 5.1–3, ed. M. Aubineau, *Les homélies festales d'Hésychius de Jérusalem*, vol. I, SubsHag 59 (Brussels, 1978), 158–64.

23 *In Sanctissimae Deiparae Annuntiationem* 4, ed. J.-P. Migne, PG 85 (Paris, 1864), col. 436A.

24 L. Peltomaa, *The Image of the Virgin Mary in the Akathistos Hymn* (Leiden, 2001), 8–11, 16–17.

25 G. Brett in *The Great Palace of the Byzantine Emperors, First Report* (Oxford, 1947), 64–97; D. Talbot-Rice, ed., *The Great Palace of the Byzantine Emperors, Second Report* (Edinburgh, 1958); J. Trilling, "The Soul of the Empire: Style and Meaning in the Mosaic Pavement of the Byzantine Imperial Palace in Constantinople," *DOP* 43 (1989): 27–72; W. Jobst, H. Vetters, eds, *Mosaikenforschung im Kaiserpalast von Konstantinopel*, Österreichische Akademie der Wissenschaften, phil.-hist. Klasse, Denkschriften 228 (Vienna, 1992).

26 L. Schneider, *Die Domäne als Weltbild: Wirkungsstrukturen der spätantiken Bildersprache* (Wiesbaden, 1983).

27 K. M. D. Dunbabin, *Mosaics of the Greek and Roman World* (Cambridge, 1999), 235.

28 *Byzance: L'art byzantin dans les collections publiques françaises* (exhibition catalogue, Musée du Louvre, Paris, 1992), 63–66, no. 20.

29 Corippus, *In Praise of Justin II*, 3. 231–54, ed. and trans. A. Cameron, *Flavius Cresconius Corippus: In Laudem Iustini Augusti Minoris. Libri IV* (London, 1976).

30 Trilling, "The Soul of the Empire," 59, fig. 40.

31 R. A. Gergel, "The Eagle Vanquishing a Hare. A Flavian Victory Motif," *American Journal of Archaeology* 91 (1987): 303; J. M. Padgett, ed., *Roman Sculptures in the Art Museum, Princeton University* (Princeton, 2001), 27–33, no. 7.

32 Theophanes Continuatus, *Chronographia*, ed. I. Bekker. CSHB (Bonn, 1838), 451; trans C. Mango, *The Art of the Byzantine Empire 321-1453, Sources and Documents* (Englewood Cliffs, 1972), 208.

33 On this topic, see, in general, H. Maguire, "Christians, Pagans, and the Representation of Nature," *Riggisberger Berichte* 1 (1993): 131–60, esp. 147–53; idem, *Nectar and Illusion*, esp. 11–47.

34 *Epistulae*, 4.61, ed. J.-P. Migne, PG 79 (Paris, 1865), cols. 577–80; trans. Mango, *Art of the Byzantine Empire*, 33. See also H. G. Thümmel, "Neilos von Ankyra über die Bilder," *BZ* 71 (1978): 10–21.

35 On this aniconic phase, see E. Kitzinger, "Stylistic Developments in Pavement Mosaics in the Greek East from the Age of Constantine to the Age of Justinian," in *La mosaïque gréco-romaine, Colloques internationaux du Centre National de la Recherche Scientifique* (Paris, 1965), 341-52, esp. 343–44 (reprinted in idem, *The Art of Byzantium and the Medieval West: Selected Studies* (Bloomington, 1976), 64–88); Maguire, "Christians, Pagans, and the Representation of Nature," 132–39.

36 D. Levi, *Antioch Mosaic Pavements* (Princeton, 1947), 283–85, figs. 112–13, pls. 113–15, 139.

37 D. Michaelides, *Cypriot Mosaics* (Nicosia, 1987), 38–40, pls. 27–28.

38 Especially in the refutation of the iconoclast council of 754 by the Orthodox council of 787: G. D. Mansi, *Sacrorum conciliorum nova et amplissima collectio*, vol. 13 (Paris, 1902), col. 285, trans. D. Sahas, *Icon and Logos. Sources in Eighth-Century Iconoclasm* (Toronto, 1986), 111–12.

39 R. Talgam, *Mosaics of Faith: Floors of Pagans, Jews, Samaritans, Christians, and Muslims in the Holy Land* (Jerusalem and University Park, 2014), 425–30; D. Reynolds, "Rethinking Palestinian Iconoclasm," *DOP* 71 (2017): 1–64.

40 M. Piccirillo and E. Alliata, *Umm al-Rasas, Mayfa'ah*, vol. 1, *Gli scavi del complesso di Santo Stefano* (Jerusalem, 1994); S. Ognibene, *Umm al-Rasas: La chiesa di Santo Stefano ed il 'problema iconofobico'* (Rome, 2002); Talgam, *Mosaics of Faith*, 387–91.

41 Talgam, *Mosaics of Faith*, 395–404.

42 M. Piccirillo, *Madaba, le chiese e i mosaici* (Milan, 1989), 41–66; idem, *The Mosaics of Jordan* (Amman, 1993), 50; Maguire, *Nectar and Illusion*, 38–41, figs. 1.18–1.19; Talgam, *Mosaics of Faith*, 396–98, fig. 478.

43 M. Piccirillo, "A Church at Shunat Nimrin," *Annual of the Department of Antiquities of Jordan* 26 (1982): 355-43; idem, *The Mosaics of Jordan*, 322-23; Talgam, *Mosaics of Faith*, 401–2, fig. 482.

44 P. Donceel-Voûte, "Les pavements omeyyades: traditions, recherches et innovations (fin du VIIe-VIIIe siècle)," in *La mosaïque gréco-romaine*, vol. 7, *Actes du VIIème colloque international pour l'étude de la mosaïque antique*, ed. M. Ennaïfer and A. Rebourg, 2 vols. (Tunis, 1999), I:151–70, pl. 70,1; Talgam, *Mosaics of Faith*, 402–4, figs. 483–87.

45 K.-H. Clasen, "Die Überwindung des Bösen: ein Beitrag zur Ikonographie des frühen Mittelalters," in *Neue Beiträge deutscher Forschung: Wilhelm Worringer zum 60. Geburtstag*, ed. E. Fidder (Königsberg, 1943), 13–36; E. Kitzinger, "The Threshold of the Holy Shrine: Observations on Floor Mosaics at Antioch and Bethlehem," in *Kyriakon, Festschrift Johannes Quasten*, ed. P. Granfield and J. A. Jungmann, vol. 2 (Münster, 1970), 639–47; U. Zischka, *Zur sakralen und profanen Anwendung des Knotenmotivs als magisches Mittel, Symbol oder Dekor* (Munich, 1977); E. Kitzinger, "Interlace and Icons: Form and Function in Early Insular Art," in *The Age of Migrating Ideas: Early Medieval Art in Northern Britain and Ireland*, ed. R. M. Spearman and J. Higgitt (Stroud, 1993), 3–15.

46 Talgam, *Mosaics of Faith*, 401, fig. 482.

47 On medieval opus sectile floors in Byzantium, see U. Peschlow, "Zum byzantinischen opus sectile-Boden," in *Beiträge zur Altertumskunde Kleinasiens. Festschrift für Kurt Bittel*, ed. R. M. Boehmer and H. Hauptmann (Mainz, 1983), 435–47; A. Guiglia Guidobaldi, "L'opus sectile pavimentale in area bizantina," in *Atti del I Colloquio dell'Associazione Italiana per lo Studio e la Conservazione del Mosaico* (Ravenna, 1993), 643–63; H. Maguire, "The Medieval Floors of the Great Palace," in *Byzantine Constantinople: Monuments, Topography and Everyday Life*, ed. N. Necipoğlu (Leiden, 2001), 153–74; Y. Demiriz, *Interlaced Byzantine Mosaic Pavements* (Istanbul, 2002).

48 A. K. Orlandos, "E en Boiotia Mone tou Sagmata," *Archeion ton Byzantinon Mnemeion tes Hellados* 7 (1951): 98–109, esp. 103, fig. 28 and pl. A.

49 A. K. Orlandos, "E para ten Artan Mone ton Blachernon," *Archeion ton Byzantinon Mnemeion tes Hellados* 2 (1936): 1-56, esp. 30, fig. 25; H. Kier, *Der mittelalterliche Schmuckfussboden* (Düsseldorf, 1970), 27, fig. 320.

50 A. H. S. Megaw, "Notes on Recent Work of the Byzantine Institute in Istanbul," *DOP* 17 (1963): 333–71, esp. 337-38; R. Ousterhout, "Architecture, Art, and Komnenian Ideology at the Pantokrator Monastery," in *Byzantine Constantinople*, ed. Necipoğlu, 133–50.

51 S. Gerlach, *Stephan Gerlachs des Aeltern Tage-Buch* (Frankfurt, 1674), 217.

52 Megaw, "Notes on Recent Work of the Byzantine Institute," 339; E. Kudde, N. Melvani, and T. Okçuoğlu, *Stoudios Monastery in Istanbul: History, Architecture and Art* (Istanbul, 2021), 80-94, figs. 65–70, drawings 10–12.

53 P. A. Underwood, *The Kariye Djami*, 4 vols. (London, 1967–75).

54 Ibid., 2: pls. 93–95.

55 H. Maguire, *Art and Eloquence in Byzantium* (Princeton, 1981), 42–52.

56 Underwood, *The Kariye Djami*, 2: pls. 114, 116–18.

57 Ibid., 2: pls. 120, 125.

58 Ibid., 2: pls. 115-16.

59 Gregory of Nazianzos, *Homilies*, no. 28.24, ed. P. Gallay and M. Jourjon, *Grégoire de Nazianze, Discours 27-31 (Discours Théologiques)* (Paris, 1978), 152.

60 For an example in mosaic, see the fifth-century pavement from the transept of the Ilissos basilica in Athens, now preserved in the Byzantine and Christian Museum, Athens; *The World of the Byzantine Museum* (Athens, 2004), 90, figs. 74, 76. For amulets, see C. Bonner, *Studies in Magical Amulets* (Ann Arbor, 1950), 303, nos. 304-305, pl. 15; J. Spier, "An Antique Magical Book Used for Making Sixth-Century Byzantine Amulets?" in *Les savoirs magiques et leur transmission de l'Antiquité à la Renaissance*, ed. V. Dasen and J.-M. Spieser (Florence, 2014), 43–66, esp. 46, fig. 2.

61 E. D. Maguire and H. Maguire, *Other Icons: Art and Power in Byzantine Secular culture* (Princeton, 2007), 81–82, fig. 78; H. Maguire, "Magic in Byzantine Pottery: the Other Within," in *Identity and the Other in Byzantium: Papers from the Fourth International Sevgi Gönül Byzantine Studies Symposium*, ed. K. Durak and I. Jevtić (Istanbul, 2016), 205–20, esp. 207–8, fig. 3.

62 R. Greenfield, *Traditions of Belief in Late Byzantine Demonology* (Amsterdam, 1988), 146–47.

63 The prime example is the Norman Palace in Palermo, on which see: E. Kitzinger, "Some Preliminary Observations on the Mosaics in the Torre Pisana in Palermo," in *Mosaïque, Hommages à Henri Stern* (Paris, 1983), 239–43; E. Kitzinger, *I mosaici del periodo normanno in Sicilia*, vol. 6, *La Cattedrale di Cefalù, la Cattedrale di Palermo e il Museo Diocesano, mosaici profani* (Palermo, 2000); D. Knipp, *The Mosaics of the Norman Stanza in Palermo: a Study of Byzantine and Medieval Islamic Palace Decoration* (Leuven, 2017).

64 Brett, *The Great Palace*, 8–16; Talbot Rice, ed., *The Great Palace*, 10-23.

65 Theophanes Continuatus, *Chronographia*, ed. Bekker, 140–46; trans. Mango, *The Art of the Byzantine Empire*, 161-65; Maguire, "Medieval Floors of the Great Palace." An opus sectile pavement still survives in the Boukoleon palace; N. Asgari, "Istanbul Temel Kazılarından Haberler – 1983," *Araştırma Sonuçları Toplantısı* 2 (1984): 43–62, figs. 13–15.

66 *Vita Basilii*, 89, ed. I. Ševčenko, *Chronographiae quae Theophanis continuati nomine fertur liber quo Vita Basilii imperatoris amplectitur* (Berlin, 2011), 146; trans. Mango, *Art of the Byzantine Empire*, 197. For a reconstruction of the floor based on the text and on other surviving pavements, see Maguire, "Medieval Floors of the Great Palace." 4–7.

67 *Vita Basilii*, 89; ed. I. Ševčenko, 147; trans. Mango, *Art of the Byzantine Empire*, 198.

68 Photios, *Homilies*, no. 10.5, ed. B. Laourdas, *Photiou Homiliai*, Hellenika, suppl. 12 (Thessaloniki, 1959), 101; trans. C. Mango, *The Homilies of Photius Patriarch of Constantinople* (Cambridge, Mass., 1958), 187.

11

ANIMALS IN BYZANTINE MANUSCRIPTS

Nancy P. Ševčenko

There are few illuminated Byzantine manuscripts that do not include representations of animals in one way or another, and, although they may look the same, the meaning of any animal we see represented may vary from manuscript to manuscript according to the text it accompanies. Or the meaning may elude us entirely. Given the vast amount of visual evidence, this chapter will be very limited and deal only with the representations of land animals. Birds, reptiles and marine creatures, and most fantasy creatures, sadly, will be left by the wayside.[1] In the chapter that follows, I will explore the following subjects: "group portraits" of animals; animals in handbooks; animals intrinsic to the content of a narrative but not its focus; the exegetical use of animals, and animals in illuminated ornament.

"Group portraits"

Some Byzantine miniatures take the form of a sort of visual catalog of miscellaneous animals, piled up randomly on the page. Such compositions are found primarily in Old Testament manuscripts of the Octateuch, to illustrate God's creation of terrestrial creatures on the Fifth Day (Genesis 1:24–25), to show Adam being given dominion over all living creatures (Genesis 1:26–29; cf. Psalm 8:6–8), and Adam giving them names (Genesis 2:19–20).[2] Animals are shown entering Noah's ark (Genesis 6 and 7) and leaving it when the ark reaches dry land (Genesis 8:17–19).

What then are these animals? The Creation passage in Genesis speaks generically of quadrupeds, reptiles, wild beasts and livestock (τὰ κτήνη). Given the vagueness of this text and the later passages, artists of the Octateuch manuscripts, which range in date from the eleventh to the fourteenth century, had relatively free rein to include whatever animals they liked. The choices vary from manuscript to manuscript, but in most of the Old Testament miniatures are of the barnyard animals (sheep, goats, dogs, horses and donkeys, oxen, sometimes camels, or even squirrel intruders), their predators (wolf and fox), the regular game animals (deer, boar, hare, wild ox) as well as the powerful wilderness creatures (lion, bear, leopard), and the occasional exotic animal that would not have been familiar from either the barnyard or the adjoining forest (elephant, giraffe). Even some fabulous animals are included (unicorn, griffin, chimaera) indicating that these composite animals were

DOI: 10.4324/9781003055877-14

accepted as real, also created by God, even if they live somewhere very far away and are not often seen.

In the Creation scenes, the animals are depicted with little relation to each other in space or in scale.[3] In the twelfth-century Seraglio Octateuch, for example, the gazelle is larger than the elephant, which is about the size of the fox.[4] These miniatures are an assembly of individual portraits, apparently pulled out of model books; each animal assumes the long-standing portrait form of itself familiar from Early Christian or even ancient Roman monuments: the gazelle bites its raised rear left leg; the bear appears to be falling forward down a cliff; the sheep is drinking from a stream, the bull is resting, the fox is leaping, the hare is hiding.[5]

The setting is different in these Octateuch manuscripts once Adam enters the narrative and is presented as sovereign over the animals: now the setting is Paradise, with flowery hillsides and trees laden with fruit, and now the animals are limited to the usual domestic ones, plus wilder creatures such as deer, and fiercer ones like the boar, the bear or the lion: no exotic or imaginary animal is included here. Adam is shown seated and appears to be addressing them; they face him, lined up, side by side, listening and deferential, most of them appearing in pairs (see Figure 11.1).[6] This Genesis passage is echoed in Psalm 8:6–8 and illustrated in a number of marginal Psalter manuscripts.[7]

When it comes to Adam naming the animals, the landscape features fade. The animals are still face to face with the seated Adam; a mouse and a leopard have been added, and a unicorn and a griffin restored, to the overall list.[8] In the fourteenth-century Florence Octateuch, the animals inhabit different hillsides, all sparsely vegetated, with the domestic animals (which include the giraffe) and birds clustered closest to Adam, the game and other wild animals next, and the less savory animals farthest away from him. Nonetheless, despite the landscape setting and the late date of the Florence Octateuch, many of the animals retain their traditional "portrait" pose.

The domestic animals entering and leaving the box-like Noah's ark are sometimes in pairs, but the odder ones – in Vatican gr. 746 (twelfth century), an elephant, plus a unicorn and a griffin – enter alone, by a different door.[9] The fantasy animals do not seem to have survived the voyage and are not seen leaving the ark.

When it comes to the distinction between the clean and unclean animals (the edible and non-edible, Leviticus 11), the Bible text is more precise, and, given the everyday relevance of the distinctions that are being set forth, we would presume that the animals depicted would faithfully adhere to the text. This they don't: many animals shown are not included among the unclean in Leviticus. To be sure, there are no fantasy creatures, but in Vatican gr. 747 (eleventh century) there is a giraffe (without markings), and in Athos Vatopedi 602 (thirteenth century) an elephant and a hedgehog along with the usual suspects. The composition is similar to the Genesis ones, but the creatures of the air are clearly separated from the quadrupeds, and the standing figure of Moses, now guided by the hand of God from heaven, replaces Adam.[10]

Similar assemblages of animal portraits exist in manuscripts other than the Octateuchs as well. A set illustrates, chapter 39 of the Book of Job, where God proclaims his knowledge of, and power to provide for the animals he has created.[11] Although the images are usually strung out over several folios in Job manuscripts, they are piled up one on top of the other in the right margin of the page in the ninth-century codex of the *Sacra Parallela* where the Job text is cited.[12]

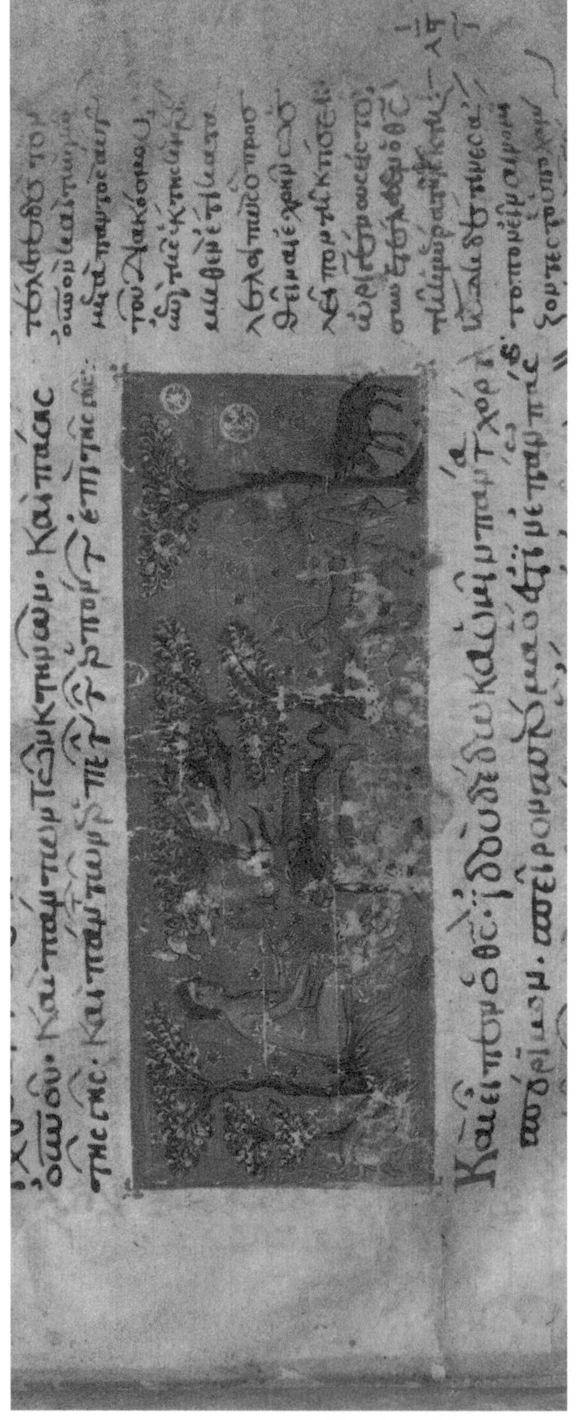

Figure 11.1 Adam sovereign over the animals. Octateuch. Biblioteca Apostolica Vaticana, Vat. gr. 746, fol. 31r. Copyright: Biblioteca Apostolica Vaticana.

And in a purely secular manuscript, the *Cynegetica* of Pseudo-Oppian in Venice, a hunting manual illustrated in the eleventh century (see below), the two opening miniatures present comparable collections of beast images.[13] On fol. 2r the miniature represents "the deadly wild beasts on the hills" that are hunted. Since none is referred to by name, the artist was presumably free to choose which to paint. Eight animals, one a rhinoceros, are lined up in two rows facing the huntress Artemis, and facing Oppian (fox and wolf), as though before Adam himself.[14] Man's dominion over the animal kingdom, though not alluded to in the text, is conveyed in these *Cynegetica* miniatures by the use of this familiar composition.

Animals in handbooks

When it is important that an animal be recognizable, as is the case in the illustrations of more scientific texts, the animal is identified by caption, and a certain effort is made to have it look somewhat more like its real self than its traditional "portrait."

The *De Materia Medica* of Dioscurides (first century A.D.), for example, is a manual for finding the medicinal properties of all living things.[15] The text, which focused mostly on plants, was copied and illustrated in most periods of Byzantine art.[16] A tenth-century copy in New York (Morgan Library M 652) illustrates Book II, the section on remedies offered by animal parts or dung; the animals reflect much of the basic repertory found in the Octateuchs, but it includes some animals whose parts would not perhaps have been that easy to come by, such as the beaver, wild goat and the wild boar.[17] The animals are beautifully observed; each has its own isolated position in the center of the page, above the relevant text. Although this manuscript may have been an imperial commission for a hospital, simpler versions could well have served as the basis for many a later animal portrait.

The splendid *Cynegetica* manuscript in Venice, already mentioned, illustrates a long poem written in the third century A.D. by the author commonly referred to as Pseudo-Oppian.[18] The text deals with the hunting of wild animals, probably in the Syrian hinterlands, and includes chapters on the breeds of horses and hunting dogs deemed suitable for each type of hunt.[19] Of the many different wild animals depicted, some (bear, lions, hyenas, leopards, etc.) were hunted down out of necessity, being threats to people and livestock, but others (deer, wild goats, boar) were hunted for sport or for food.[20] The remarkable aspect of these miniatures is their attention not only to the accurate physical appearance of the animals but also to their peculiar habits and habitats. Lions lurk in the tall reeds, bear in caves, and boar in the woods, and each must be hunted in a different manner. The landscape setting changes accordingly. The text is illustrated in narrative strips that are, with the exception of the two folios discussed earlier, full of action and entirely focused on the animal being hunted (See Figure 11.2a and b). The dominion over the animal kingdom enjoyed by Adam is here something hard-earned, and, despite God's assurances, the contest did not always end happily for the hunter himself.[21]

Interest in the habits of animals, though not a proper manual, also characterizes a section of another work, Book XI:1–12 of the *Christian Topography* by Cosmas Indicopleustes, a sixth-century merchant who describes animals of India and Ethiopia, most of which he claims to have actually seen.[22] The text of Book XI was illustrated in Cosmas manuscripts of the eleventh and twelfth centuries, but the author's and the artists' unfamiliarity with the actual animals (when they existed at all) has resulted in some peculiar images. The rhino, for example, is depicted as a horse with two horns, and the musk deer, called a moschos, was visualized as an ordinary fox (also moschos), in the Sinai manuscript 1186.[23] Sections

Figure 11.2   Boar hunting and lion hunting. Cynegetica. Biblioteca Nazionale Marciana, Gr. Z. 479 (= 881), fols. 56v–57r. Su concessione del Ministero della Cultura – Biblioteca Nazionale Marciana. Divieto di riproduzione.

of the text of Cosmas' Book XI were included in the Smyrna manuscript of the *Physiologus* and illustrated accordingly (see below).

Animals as intrinsic to a narrative, and for the setting of a narrative

In this group are included animals that are germane to, embedded in, even essential to the text, but are not its focus. Among them, by far the most frequently used in storytelling are horses, sheep, and lions. In extended narrative cycles, such as those in the Octateuchs, in manuscripts of *Barlaam and Joasaph* (a popular romance involving the Indian prince Joasaph who was converted to Christianity by the monk Barlaam), the historical chronicle of Skylitzes (twelfth century), or the *Alexander Romance*, horses show movement from place to place: whether it is an individual riding away to deliver a message or a whole troop of riders galloping forward in battle, their spirited images help propel the narrative along and bind one episode to the next.[24] The trappings differ, and the horses vary in color, but all are on the move, trotting with heads held high, or rearing to take off.[25] Then there are the simple beasts of burden who transport goods or gifts; these are primarily mules or donkeys with drooping heads; one carries the Virgin during the Flight into Egypt and another bears Christ into Jerusalem. Camels, which, of course, tend to be associated with Egypt, figure prominently in illustrations to the Biblical story of Joseph, but as much to establish the geographical setting of the narrative as to show a traditional means of transport.

Animals often serve to define the setting in which the action takes place. Flocks of sheep and goats in a landscape identify the setting as a pastoral one, and also make clear the occupation of the hero, as with the scenes of David among his flocks found in Psalter illustration (see Figure 11.3).[26] All that is needed is a pair of sheep, a pair of goats and a dog to mind them. Depending on the artistic level of the illuminator, the representations of these animals can range from sizable goats with long hair and splendid spiral horns to sheep that are little more than a dot of white on the parchment. Sheep may also accompany images of the shepherd saints Mamas and Sozon.[27] Job himself, no shepherd of course, but a rich landowner, owned vast flocks, and the illustrated Job manuscripts have images of the animals that constituted his possessions and displayed his wealth. These include not only sheep and goats, but cattle, donkeys and camels; the precise numbers of each are given in the Bible text.[28] Horses, it should be noted, have no place in this pastoral imagery: if we believe what we see, horses are possessed by soldiers, or by big game hunters, but not by the local farmer.[29]

Farm animals also convey the seasonal round of agriculture, as exemplified by the oft-repeated image of a man plowing a field with a team of oxen. It is used to illustrate the ekphrasis of spring in Gregory of Nazianzus' homily for the first Sunday after Easter, or the bucolic life left behind by the aspiring monk in a manuscript of the Heavenly Ladder of John Climacus, and elsewhere.[30] In the Old Testament, farm animals were also subject to ritual slaughter: images abound in manuscripts of the Octateuchs, the Book of Job, the Book of Kings and other Old Testament manuscripts. Sometimes the sacrificial animals, usually sheep, but also goats and bulls, are shown having their throats slit, or lying atop an altar enveloped in flames.[31]

Animals of the wilderness have their own place in narrative illustration, although the animal as existential foe, such a familiar theme in ancient art and in Byzantine sculpture and textiles, was by comparison a rather minor theme in Byzantine manuscript illumination. To be sure, David was celebrated for his fearless defense of his flocks, and in illustrated Psalters

Figure 11.3 David and Melodia. Psalter. Paris, BnF grec 139, fol. 1v. Copyright: Bibliothèque nationale de France.

he is shown singlehandedly wrestling and killing a marauding bear and a lion.[32] And to be sure, wild animals abound in scenes of Christian martyrdom, when the agent of death for a saint was a wild beast – a lion, bear or leopard – released into the arena by the authorities to tear him or her apart.[33] Yet in this hagiographical context these beasts were not actually read as foes, but welcomed as agents of God who in this way enable the saint to achieve his dream of joining the elect through martyrdom.[34] In some cases, the animals yield to his charisma and fail to attack at all. This is the case with Daniel in the Bible (Daniel 6:16–22) and with Sts. Thecla, Euphemia and Panteleimon, who are depicted as flanked by subservient, and often quite foolish-looking, lions.[35]

Even though more prosaic miracles involving wild animals are commonplace in hagiographical texts, very few are illustrated in manuscripts.[36] The images that we do have do not involve any outright confrontation with the devil but present the miracle as the saint's involvement with the ordinary perils of rural life. Two such miracles involve Saints Cosmas and Damian: in an opening miniature in a Menologion of 1063 the saints first drive out the snake a man had inadvertently swallowed in his sleep, then heal the injured leg of a camel.[37]

Animals as metaphor, allegory or symbol: the exegetical use of animal imagery

The prime Christian text in this category is the *Physiologus*, a collection of Christian interpretations of animal lore involving the habits of a number of different animals.[38] Composed somewhere between the second and the fourth century, the text was hugely influential in the Middle Ages, especially as the basis for the Western Bestiary, and it certainly contributed to the reputations, both good and bad, of the animals it described.[39]

The most important illustrated version of the *Physiologus* is an eleventh-century manuscript formerly in Smyrna, that was unfortunately destroyed by fire in 1922.[40] Now all we have are grainy black and white photos of some of the miniatures, taken before the fire. There were at one time sixteen images of animals, real and otherwise, in this manuscript, not including the birds or reptiles. It is quite a mix, from the lion and other familiar animals to the unicorn, the siren and centaur, and the ant-lion.[41]

More than their appearance, it is the habits of each animal that are the basis for a Christian interpretation. Some animals in the *Physiologus* were consistently identified as evil, e.g., the viper, hedgehog, fox, siren and centaur, wild ass, monkey, etc., while others become symbols of Christ, His sacrifice and resurrection: e.g., the father lion who brings his cubs to life, the panther lying three days in a cave (see Figure 11.4).[42] The Smyrna manuscript includes the text of Book XI of Cosmas' *Topographia Cristiana* (see above), so images of ten of more animals were added.[43] The artist in this manuscript valiantly endeavored to represent the animals described in both these texts, but, given that many of them were unfamiliar or had to be distorted to fit their Christian interpretations, the animals we see are often barely recognizable.[44]

But the moral values assigned to the animals in *Physiologus* do not seem to have had a decisive effect on their images elsewhere: though widely circulated as a text, it was not often illustrated. In fact, few of the animals appear in Byzantine illustrated manuscripts elsewhere. In short, the Christian message of a particular *Physiologus* animal does not necessarily accompany it in every context in which we see its image.[45]

The illustrated marginal Psalters are rich in animal imagery, much of it negative, animals being used throughout as metaphors for the enemies of the faithful.[46] This characterization

Figure 11.4   Physiologus, olim Library of the Smyrna Greek Evangelical School B. 8, p. 56 (fol. 28v). Copyright: Bernabò et al., *Il Fisiologo ̌di Smirne. Le miniature del perduto codice B. 8 della Biblioteca della Scuola Evangelica di Smirne* (Florence, 1998), fig. 31.

echoes the use of animal imagery in verbal invective, especially in the period of Iconoclasm.[47] But alongside the negative images, e.g. of lions, such as is found in these Psalters, is the contemporary belief that the lion is a powerful symbol of strength.[48] The meaning here is not fixed, but follows context.

Other texts that may have affected the animal imagery that we find in Byzantine manuscripts are the verbal sketches of particular animals found in commentaries on the first six days of Creation (the *Hexaemeron*). The homily on the Fifth Day of Creation by Basil the Great (fourth century) includes pithy one-line characterizations of a number of different animals, both as evidence for God's creation and care, and as exempla of human virtues or vices.[49] Although Basil's homilies themselves were not illustrated with anything other than an author portrait, a work based on his homilies does have miniatures (the Chronicle of Michael Glykas).[50] Basil's more nuanced, less allegorical approach provides a better parallel to the rather balanced approach to most animals that we find in Byzantine miniatures across the board, than does the polemical approach of Psalter illustration or the strictly exegetical approach of the *Physiologus*.[51]

Symbols of the Evangelists, and the zodiac

There are two sets of animals that do not fit neatly into any of these categories: the four symbols of the Evangelists, and the signs of the zodiac. Each Evangelist was assigned a particular animal based on the vision of Ezekiel (Ezek. 1:1–10; Rev. 4:6–8). The assignation was a bit arbitrary and changed over time. The four creatures were the eagle, the bull, the lion, and man himself, and they are commonly shown grasping the book of the relevant Gospel in their claws, hoofs, talons, or hands. They may be shown full length or just as busts, but either way, the images of the four beasts are fairly decorative and heraldic.[52]

Astronomical illustration contains its own set of creatures. The zodiac in an early ninth-century Ptolemy manuscript is displayed as a wheel divided into sections, one for each month; the sections contain an easily identifiable, if conventional, image of the animal that forms the relevant constellation.[53] The animals Ares, Taurus, and Leo are stretched out as though running, to fit the long narrow space that each is allotted. This is of course a fixed series, though the representations themselves, as with the evangelist symbols, change stylistically over time.[54]

Animals in ornament

The place to find the greatest number and variety of animals is in manuscript ornament: in the canon tables of Gospel books, and everywhere in headpieces, and in initials. Yet the reasons for their presence in these contexts is often difficult to understand. Even tracking the role of a specific animal is a challenge: some species – lions and foxes, rabbits and deer, say – appear over and over, and are linked to many different types of texts. Establishing an overall pattern of use or meaning of any one animal, or even of the assemblages of animals that we find in Byzantine ornament, is unlikely. Yet the subject is well worth further exploration.

Canon tables and headpieces

Many illuminated Gospel books are prefaced by a series of ten canon tables, lists of the passages that are shared by each of the four Gospels, by three or two Gospels, or occur in just one.[55] Each canon table is given an architectural frame: these range from simple ink-drawn arches to magnificent displays of complex vegetal designs.

Animals had, of course, constituted part of the ornamental apparatus of manuscripts for ages.[56] But the peak period was the eleventh and twelfth centuries, when different sections of a manuscript began to receive special attention in the form of ever more elaborate ornament.[57] Two luxury manuscripts, both from the third quarter of the eleventh century, will serve as examples.

Like many a simpler manuscript, the Gospel book in Parma has pairs of birds flanking a central fountain atop its nine canon tables.[58] But here there are also pairs of jackals, of lions, foxes, cheetahs, and hares, and even a siren and a leopard-centaur, all presumably eager for the living waters offered by the fountain (see Figure 11.5).[59] The headpieces also have animals on top of them: griffins over the Letter of Eusebius, deer over the page containing the seated Christ in Majesty, and monkeys over a miniature that precedes the portrait of Matthew and the beginning of his Gospel (see Figure 11.6).[60] The animals throughout are lively, treading in grassy areas strewn with flowers; they provide a fine contrast to the tight abstract designs of the ornament below.

The animals have no evident connection to the content of the canon tables on which they are poised. Yet the sequence of images in the Parma Gospels is worth a look. The first miniature, a headpiece for the Letter of Eusebius (fol. 3r), is topped by a pair of guardian griffins;[61] then comes the Christ in Majesty page (fol. 5r) with the usual deer on top, then the first canon table (fol. 8r) where there are two other hybrids (the siren and the leopard-centaur, both playing instruments),[62] then come wild animals (jackals opposite lions, foxes (wolves?) flanking peacocks, fols. 8v–9v), followed by pairs of more domesticated creatures (guinea fowl, cheetahs opposite hare, ducks opposite parrots, fols. 10r–12r) and finally pheasants opposite monkeys atop the two pages with narrative scenes (fols. 12v–13r). Below the pair of monkeys is a Nativity scene and standing together below it, the holy pair of saints Constantine and Helen.[63]

The sequence here may be entirely random. But perhaps not as random as it first appears. It may be helpful to look at the canon table decoration of another Gospel manuscript of this same period (Paris BnF grec 64).[64] Atop the first canon table here is a pair of griffins, and on the recto facing them are wild horses (fols. 3v–4r). The second pair of tables features domesticated animals with their black-skinned, nude minders carrying whips: there is a camel and an elephant opposite a pair of oxen and an excited horse (both tame: note the cowbell on the ox and the saddle on the horse) (fols. 4v–5r). The third set of canon tables has hunting scenes: on the left folio is a man with a falcon which has brought home a rabbit and is now attacking a family of partridges, while opposite a hunter is about to release a cheetah on some fleeing deer (fols. 5v–6r). The fourth pair and the final, single, canon table all have very carefully observed prized birds: herons and peacocks opposite fancy ducks and guinea fowl (fols. 6v–7r), and then some exquisite pheasants opposite herons (?) and partridges (fol. 7v–8r). Each pair of opposing canon tables is thus devoted to a particular theme: griffins face wild horses; somewhat foreign beasts of burden (the camel and the elephant) with their dark-skinned handler face traditional domesticated animals (oxen and the horse, also with human tenders); then come two modes of aristocratic hunting practice, and, finally, pairs of exotic and highly valued birds. At the risk of over-interpretation, one can suggest that from the wild and untamed we are being led step by step into an aristocratic garden, perhaps represented by the floral headpiece with its birds and their nests, over the beginning of the Gospel of Matthew (fol. 10r).

Figure 11.5 Canon table. Gospel book. Parma, Biblioteca Palatina Palat. 5, fol. 11r Su concessione del Ministero della Cultura – Complesso monumentale della Pilotta, Biblioteca Palatina.

Similar creatures can be found also *inside* the spectacular so-called "carpet" headpieces that preface various sections of a Gospel book or Lectionary, such as the Letter of Eusebius, or the opening page of each of the four Gospels, or the start of the Synaxarion (calendar).[65] Lavish headpieces can also be found in some Psalters (though not in other Old Testament manuscripts), and in collections of homilies, saints' lives, and liturgical rolls. At first, it would seem that only the birds are allowed in, to peck at fruit and nest among the vines.

Figure 11.6 Nativity, Sts. Constantine and Helen. Gospel book. Parma, Biblioteca Palatina Palat. 5, fol. 13r Su concessione del Ministero della Cultura – Complesso monumentale della Pilotta, Biblioteca Palatina.

Then (presented here in order of complexity without implying a clear chronological path from one to another) we encounter a simple and popular formula, the symmetrical pairing of two griffins and two lions inside the tight coils of headpiece vegetation; this starts in the late eleventh century but can be found at least into the late twelfth.[66] The creatures are placed within the coils of the design, but given their own space: they do not engage with or merge with the dominant vegetation. These two animals, being paradigms of strength and dominance, were apparently placed in an opening headpiece to protect what follows.[67] In the twelfth century the lions and griffins tend to be joined or replaced by rabbits and deer, even leopards and monkeys, animals who can have had little to do with protection and much to do with the multiple joys of Paradise.

The interweaving of ornament and animals becomes ever more complex and reaches a peak before the mid-twelfth century in the remarkable headpieces of the so-called Kokkinobaphos school which was to influence the ornament of several later generations.[68] Animals no longer cavort atop these headpieces but are instead nestled inside them. In the center may be a multi-level element evoking a candelabra or fountain, made up of acanthus leaves (e.g. Vatican gr. 1162, fols. 3r, 55r). In the way it attracts birds and animals, it apparently takes the place of the fountains that used to surmount the headpieces, and may carry the same revivifying message (see Figure 11.7).[69] Hiding nearby are tiny lions, griffins and hare and many kinds of birds, all so well integrated into the now leafy, less coiled, vegetal design that it becomes a challenge for the viewer to spot them at all: enjoyment of these panels could almost be called a form of armchair hunting.

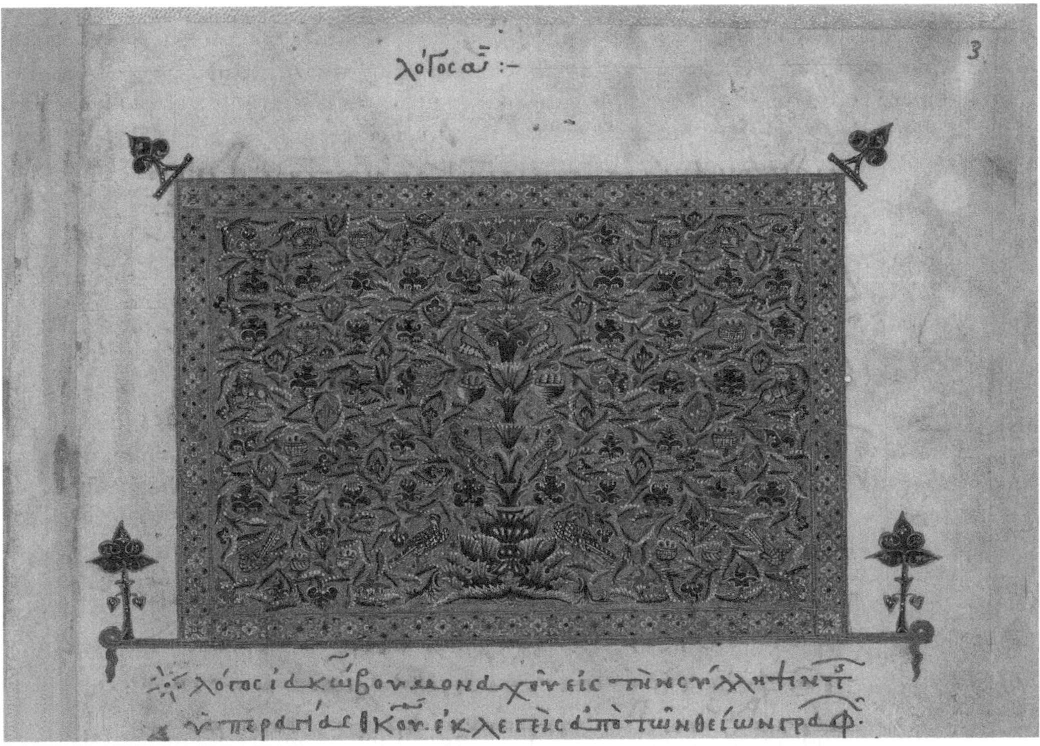

Figure 11.7   Headpiece. Homilies of Kokkinobaphos. Biblioteca Apostolica Vaticana, Vat. gr. 1162, fol. 3r. Copyright: Biblioteca Apostolica Vaticana.

In the second half of the twelfth century and into the early thirteenth, the animals are again inserted into more abstract coils, but are out of scale with the setting; their integration into the ornament is disrupted and they become a dramatic counterpoint to, or even dominate, the vegetal ornament. A good example is a twelfth-century manuscript of the homilies of Gregory of Nazianzus, in Paris.[70] The range of animals remains pretty well the same, but their exaggerated swagger and tendency toward the fantastic is a far cry from the secretive creatures hiding in the earlier headpieces.[71] Most of the later, Palaiologan, headpieces no longer continue the tradition: Palaiologan ornament, although closely dependent on Middle Byzantine forerunners, is relatively devoid of animal life.[72]

It is hard to determine just how much Christian baggage these animals carry with them into this ornamental context. The serpents and their ilk that so dominate the initial letters (see below), are entirely absent. To be sure, the lion may share the headpiece with its timid and vulnerable prey, the deer and the hare. But the animals are rarely involved in outright aggression, as they are in the initials; within the confines of the headpiece, predator and prey are kept well apart.[73] It would seem that here, at least, predation itself is halted.

Initials

The aggression absent in the garden is fully present in the initials. As far back as the ninth century, Byzantine initials began to attach, and ultimately incorporate, the figures of living creatures, such as snakes, dragons, birds or fish.[74] Starting as flourishes of the pen of the scribe or artist, these creatures gradually take over more and more of the letter and begin to eat at each other: the snakes and dragons and fish bite each other's heads or tails, or devour other creatures, even the occasional human, while still managing to keep the initial letter recognizable. Given their physical malleability, snakes form the greatest number of such letters. Animals too, but their natural movements – standing on their hind legs or fleeing – are generally retained and used to form the upright, diagonal, or horizontal bars of a letter. They are not deformed, but simply exploited to create the letter shape. Rarely do they constitute the entire letter: there is usually a bar or base of vegetation or a sort of platform to provide structure.

The main content of these zoological initials is the timeworn one of predator and prey, involving many raptors and their small-bird prey, but also lions and leopards, dogs and foxes, whose prey is mainly hare and deer, or (for the fox) poultry.

There are certain very popular, oft-repeated initial types illustrating such a combat: e.g. the oval letter E formed by two cheetahs springing from top and bottom at a doe which forms the central bar of the letter (see Figure 11.8), or the letter A with a standing fox as its left bar attacking a cock or something on the right bar.[75] Sometimes when the fight is between two equals, it is not entirely clear which animal is actually winning: the two aggressive forces seem to be neutralizing each other within the confines of the letter.

The purpose of these initials is varied. They help the reader locate the beginning of a particular text. They animate the page. They may reflect age-old animal tales.[76] But what they convey most of all is the violence that the animal kingdom perpetrates on itself, violence that is deliberately distanced from the enlightening and salvific text which the initial introduces.[77]

Occasionally, however, a closer connection between initial and text can be discerned. There is a particular initial B, for example, that appears in two eleventh-century manuscripts of the lives of saints (Menologia). The letter has the figure of a standing lion as its vertical bar,

Figure 11.8   Initial E. Homilies of Kokkinobaphos. Paris, BnF grec 1208, fol. 80r. Copyright: Bibliothèque nationale de France.

with a snake's coils forming the rounded parts of the letter. The snake is devouring the lion's front paws which are held out before it to form the horizontal. In both these Menologia, this lion-initial B prefaces a Vita that starts with the words βασιλεύοντος ("ruling") followed by the name of the emperor who ruled at the time the saint was martyred.[78] Thus the initial is drawn into the text: through the word it introduces, it becomes associated with the evil ruler. The lion, then, can perhaps be read as that emperor being devoured by a serpent as his just reward. In another contemporary Menologion, the initial Λ (L) is formed by a leopard on its hind legs as the left diagonal bar; it is grabbing a standing doe by the throat, the doe forming the right bar. The initial starts the first word of the life of St. Severian, Likinnios (= Licinius), the emperor under whom Severian was martyred; one could argue that the fatal attack on the saint for which Likinnios is responsible is being played out in the initial.[79] These initials are reminiscent of the so-called "revenge" initials found in a number of Menologia where, when the first line of the Vita text refers to the evil emperor, the opening initial shows this very emperor, labeled, crowned and in imperial garb, being devoured by a serpent.[80]

But such a tight connection is not found everywhere: other Vitae, even those with the emperor's name in the first line, begin with peaceful initials. And the lion-initial B is frequently used to introduce the opening words of the Gospel of Matthew itself, Biblos genesios.[81] So, however tempting it is to posit these connections, we cannot assume that a particular animal initial will carry the same meaning in relation to any text it introduces, or even that its content will be relevant to the text at all.

Two special cases: the giraffe and the zebra

There are two cases where specific historical circumstances, rather than any narrative or exegetical significance, might explain the presence of an animal in a Byzantine miniature. I refer here to two African animals, the giraffe and the zebra.

A giraffe is seen for the first time in Byzantine manuscripts in the eleventh century: in the Turin Homilies of Gregory of Nazianzus, its long front legs forming the right bar of the initial A, and in the scene of Adam naming the animals in the Octateuch Vatican gr. 747, thought to date around 1070–1080.[82] It appears again in this same Vatican Octateuch: perhaps as one of the animals entering Noah's ark (without a mate) and, more surely, as one of the animals deemed unclean and not to be eaten (see Figure 11.9).[83] It reappears in the twelfth-century Seraglio Octateuch as one of the unclean animals, but not in any of the three Genesis scenes in this manuscript.[84] After that, with one exception, the giraffe is no longer seen in Byzantine manuscripts.[85]

It is conceivable that the appearance of a giraffe in manuscripts of this period and shortly thereafter may be related to the presence of a particular giraffe in Constantinople in the mid-eleventh century. As is well known, the emperor Constantine IX Monomachos received two unusual presents from the Fatimid caliph, an elephant and a giraffe, in 1053.[86] Although the former was a reasonably familiar animal (and frequently depicted), the giraffe was not, and its arrival was reported with enthusiasm by several prominent figures of the time.[87] To be sure, giraffes figured in late antique mosaic floors, and it can be argued that the eleventh-century manuscript images derived from a model of this kind.[88] But it is tempting to view the incorporation of the giraffe into illustrated manuscripts of precisely this period as more than sheer coincidence.

Another animal makes a surprise appearance in a Job manuscript of the fourteenth century in Jerusalem.[89] Here, in place of the wild ass that traditionally illustrates Job 11:12, we encounter the well-executed image of a zebra (see Figure 11.10).[90] Why the switch? There is one account, preserved only in Arabic sources, that may be relevant here. In 1263, the Mamluk sultan Baybars sent off a shipload of animals as gifts to the Mongol khan Berke. To reach Berke, the ship had to pass through the Bosporus, but the Byzantine Emperor Michael VIII Palaiologos, for political reasons, detained the shipment in Constantinople for more than two years. The animals were kept on the boat the whole time, and many of them died. According to Behrens-Abuseif, the Arabic sources state that one of the animals on that shipment was a zebra.[91]

To be sure, images of zebras were known from times past, as was the case for the giraffe. However, the markings on this zebra are so precise as to suggest an origin in a drawing from life rather than a borrowing from antiquity.[92] So, it is tempting once again to associate the image of the zebra that appears in the Jerusalem Job with a specific historical event – despite the considerable time lapse between the two.[93]

In neither case did the image enter the mainstream: whatever their origin, these two lack the long visual tradition enjoyed by most other animals represented in Byzantine manuscript illumination. But they are a reminder that, given the importance of animals and birds in trade and diplomatic exchanges in the Byzantine period, we should not overlook

Figure 11.9 The unclean animals. Octateuch. Biblioteca Apostolica Vaticana, Vat. gr. 747, fol. 133v. Copyright: Biblioteca Apostolica Vaticana.

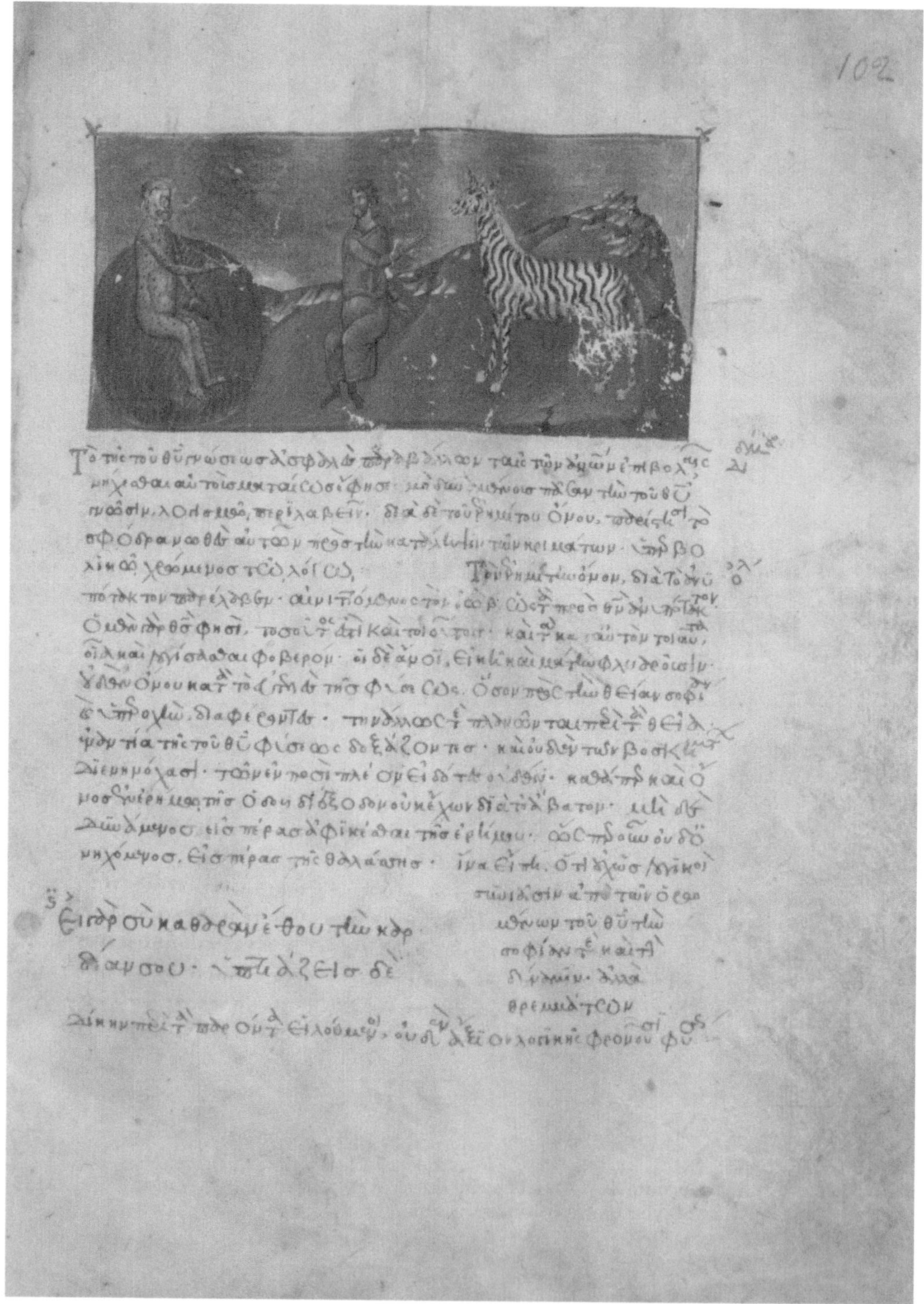

Figure 11.10   Job and the wild ass. Book of Job. Jerusalem, Greek Patriarchate, Taphou 5, fol. 102r. Copyright: Courtesy of the Panagiotes Vokotopoulos Archive at the Academy of Athens.

the influence that geographical and economic expansion or contraction may have had on Byzantine animal imagery.

General considerations

As has become clear, the animals in Byzantine manuscripts appear in a variety of contexts: they illustrate the composition of the natural world offered by Genesis, accompany handbooks for the purpose of precise identifications, illustrate narratives that need codes, serve as exemplars for ethical or religious tenets, and enrich the vocabulary of Byzantine ornament.

To approach these images solely from the point of view of their Christian or ethical interpretation, or their zoological accuracy, is to underestimate what they may be able to tell us. Animal imagery in Byzantine manuscripts needs to be approached from a much wider perspective than was possible here. What follows are but a few of many possible future lines of inquiry.

1. The theme of man's dominion over the animal world. Man's gift from God of an elevated position over the animal world was conveyed mostly through the image of a group of subservient creatures ranged before Adam, or through the image of a single beast defeated by a hero such as David, or a hunter, in one-on-one combat. But in Byzantine illumination the beasts, even at their most threatening, are not depicted as demonic: the animals are never caricatured.

The accepted fact of man's dominion also underlies the more placid images of the ploughman and his team of oxen, of the shepherds and their flocks. But we need to ask whether the theme of dominion is always the primary message. The images of the oxen, the sheep and the dog make clear their subservience to man, but they also serve to stress their vital economic importance to him, whether on the farm, the field, or in the hunting of a wild animal as a source of food. And there are other, more subtle, messages, such as a patron's level of aristocratic prestige, that may be conveyed by the animals inhabiting the more exotic ornament of the Komnenian period. And more can surely be found.[94]

2. The imagery itself. Animal imagery across the board can be viewed in terms of sets of contrasts. The existential contrast between good and evil, though an underlying theme throughout, is rarely spelled out explicitly, but there is the constant contrast between wild and domestic, between tamed and untamed, the controlled versus the uncontrolled, between the symmetry and asymmetry of pairs. There is the contrast between clean and unclean, the animals that can be eaten and those that must not be eaten. And always there is the contrast between the human hunter and the hunted, between the animal predator and its prey, between the victor and the victim, the strong and the weak. Other contrasts that we ourselves might be likely to set up, such as the familiar versus the exotic, the real versus the fantastic, the beautiful versus the ugly, were not apparently of particular concern to the Byzantines.

3. The artistic conventions. The animal assemblies described in the first section above smack of the pattern book, as we have seen. Animals in narrative settings do not appear to derive from such models, yet they too make use of timeworn formulae, if of a different kind: each animal in a narrative is shown busy doing whatever it is that it is best known for: horses prance, donkeys hang their heads or bray with open mouths, sheep drink at a stream, dogs sit and watch, oxen draw a plow, and stags bound up hillsides.[95]

The animals above and within the panels of ornament have their own sets of conventions: many appear to be frozen in place, heraldic in their symmetry. They don't look real, but like statues of themselves embedded in the thriving vegetation. We know that statues of animals such as griffins, lions and bears abound as guardians to entrances, or as parts of fountain structures in the densely planted aristocratic gardens described in the literary Romances. "The bounding of lions, the leaping of leopards, and the swaying of bears, as well as the images of other wild animals that the craftsman had excellently carved, were so close to moving that the beholder wished he could withdraw somewhere far away, lest the beasts suddenly leap on him and tear him to pieces."[96] In another such text, the hero encounters a fountain statue of a griffin spouting water; after a bit the griffin gets up and walks away.[97] In a manuscript of the Homilies of Gregory of Nazianzus in the Vatican, dated 1062, griffins sit on pedestals alongside the headpiece, while elsewhere within ornamental panels, there are heads or busts of bears, monkeys, etc., protomes that suggest fountain spouts.[98] There are certainly more connections between these two imaginary worlds, that of the garden headpiece and the Romance, still to be found.

4. The element of whimsy. Whimsy, humor – and perhaps even puns we don't recognize – pervade the world of the illuminated initial in Byzantine manuscripts, and should not be discounted. Will we ever get those inside jokes? Unlikely, but perhaps literary historians and further study of Byzantine satire will help us one day to de-code this elusive visual language.[99]

Notes

1 Important studies on animal imagery in Byzantium are the works of Z. Kádár, *Survivals of Greek Zoological Illuminations in Byzantine Manuscripts* (Budapest, 1978); H. Maguire, *Earth and Ocean. The Terrestrial World in Early Byzantine Art* (University Park and London, 1987); E. D. Maguire and H. Maguire, *Other Icons. Art and Power in Byzantine Secular Culture* (Princeton and Oxford, 2007); S. Lazaris, *Le Physiologu grec*, 2 vols. (Florence, 2016, 2021). For a view "on the ground" see H. Kroll, "Animals in the Byzantine Empire. An Overview of the Archaeozoological Evidence," *Archeologia medievale* 39 (2012): 93–121. For a very helpful recent treatment of Byzantine zoological knowledge and its contemporary interpretations, see A. Zucker, "Zoology," in *A Companion to Byzantine Science*, ed. S. Lazaris (Leiden and Boston, 2020), 261–301

2 The five illustrated Octateuch manuscripts are Vatican gr. 747 (eleventh century; images online); Smyrna, Greek Evangelical School A.1 (twelfth century, destroyed); Istanbul, Topkapi Sarayi (Seraglio), gr. I.8 (twelfth century); Vatican gr. 746 (twelfth century; images online); Athos Vatopedi 602 (thirteenth century), and Florence, Laur. Plut. 5.38 (now redated to the fourteenth century; it has only six miniatures, from Genesis; images online). All Octateuch miniatures have been published in K. Weitzmann, M. Bernabò, in collaboration with R. Tarasconi, *The Byzantine Octateuchs*, 2 vols. (Princeton, 1999). See also J. Lowden, *The Octateuchs. A Study in Byzantine Manuscript Illumination* (University Park, PA, 1992), and J. Anderson, *The Creation of the Illustrated Octateuch* (Wiesbaden, 2022).

3 Weitzmann and Bernabo, *Octateuchs*, 22–23; figs. 3b, 40–42. See Maguire, *Earth and Ocean*, 56–59.

4 Seraglio, fol. 32v: Weitzmann and Bernabò, *Octateuchs*, 22–23; fig. 41. There was a similar composition in the Smyrna Octateuch fol. 7r: ibid., p. 23; fig. 42.

5 It has been suggested that these "vignettes" or animal "portraits" reveal an origin in illustrated zoological treatises, but a derivation from mosaic floors where animals are often firmly separated by geometrical designs, is more likely (see H. Maguire in this volume).

6 Vatican gr. 746, fol. 31r. See also Vatican gr. 747, fol. 19v; Seraglio, fol. 37v, Smyrna fol. 9v. Weitzmann and Bernabò, *Octateuchs*, pp. 25-26; figs. 55-58.

7 Chludov Psalter, Moscow State Historical Museum gr. 129, fol. 7r: M.V. Shchepkina, *Miniatjury Khludovskoi Psaltiri* (Moscow, 1977); Athos Pantokrator 61, fol. 21r: S. Dufrenne, *L'Illustration*

des psautiers grecs du moyen age, I. Pantokrator 61, Paris grec 20, British Museum 40731 (Paris, 1966), pl. 2; Bristol Psalter, London BL Add. 40731, fol. 16r: ibid., pl. 48; Theodore Psalter, London BL 19352, fol. 6v: S. Der Nersessian, *L'Illustration des psautiers grecs du moyen age. II. London, Add. 19352* (Paris, 1970), fig. 13. Adam stands apart in these miniatures, which have little or no landscape, and the animals are just strewn around above or before him.

8 Vatican gr. 747, fol. 22r; Seraglio, fol. 42v; Florence, Laur. Plut. 5.38, fol. 6r: Weitzmann and Bernabò, *Octateuchs*, color fig. 1, and figs. 5a,79–82. In the naming scene in the Genesis frontispiece in the tenth-century Leo Bible (Vatican Reg. gr. 1, fol. IIr; images online), the only unusual animals are a camel and an elephant. *La Bible du Patrice Léon. Codex Reginensis Graecus 1*, ed. P. Canart (Vatican City, 2011), 108–13 (by S. Dufrenne), pl. VIII. See also H. Maguire, "Adam and the Animals: Allegory and the Literal Sense in Early Christian Art," *DOP* 41 (1987): 363–73.

9 Entering the ark: Vatican gr. 747, fol. 29r; Smyrna, fol. 20r; Vatican gr. 746, fol. 53v. Leaving the ark: Smyrna fol. 21r; Vatican gr. 746, fol. 55v. Weitzmann and Bernabò, *Octateuchs*, 51–52, figs. 131–34; 54–55, figs. 141–42.

10 Vatican gr. 747, fol. 133v; Seraglio 8, fol. 282r; Athos Vatopedi 602, fol. 28v: ibid., 188–89; figs. 840–44.

11 Job 12:7–9; 38:39–39:1–30: S. Papadaki-Oekland, *Byzantine Illuminated Manuscripts of the Book of Job* (Athens, 2009), 228–29, figs. 260–61; 197–202, figs. 215–16, 224–25.

12 Paris BnF grec 923, fol. 198r: ibid., fig. 217 (images online). Once again, the choice of animals is not based on the Biblical text, but apparently dependent on models at hand. A similar range as in these Old Testament miniatures reappear as bases for the columns of canon tables in some twelfth-century Gospel manuscripts: in a Gospel in Venice, for example, we encounter elephants, foxes, leopards, dogs, bull, lion, goat, elephant, unicorn, camel, monkey, a lion eating a hare, and a bear: Marciana Z 540, fols. 2v–9v: I. Furlan, *Codici greci illustrati della biblioteca Marciana*, 6 vols. (Milan, 1978–1997), vol. 2 (1981), 13–17; pls. 2–3, figs. 3–7.

13 Text in *Oppian, Colluthus, Tryphiodorus*, trans. A. W. Mair (Cambridge, MA, 1928, rp. 1987), 6–10. Venice Marciana gr. Z 479 (881): Facsimile: *Cynegetica: tratado de caza y pesca, Cod. Gr. Z. 479 (= 881)* (Valencia, 1999); Commentary: *Tratado de caza: Oppiano Cynegetica, Biblioteca nazionale Marciana de Venecia cod. gr. Z.479 (=881)*, ed. P. Eleuteri (Valencia, 2002). See also I. Spatharakis, *The Illustrations of the Cynegetica in Venice* (Leiden, 2004).

14 The text on these two folios (fols. 2v–3r) compares the ease of fishing and fowling with the dangers of hunting these wild beasts, the imbalance illustrated on fol. 3r with contrasting pairs of creatures. Here they follow exactly the sequence of the text (I:45–71), which enables us to identify the lion-like animal on both folios 2r and 3r with its single sharp, upward-pointing, nostril, as a rhinoceros. Artemis had commanded Oppian, presumably the figure depicted at the right on fol. 2r, to "Sing the battles of wild beasts and hunting men" (I:35).

15 H. Gerstinger, *Dioscurides: Codex Vindobonensis med. gr. 1 der Österreichischen Nationalbibliothek*, 2 vols. (Graz, 1970).

16 The most important of the illustrated ones are the sixth-century Dioscurides in Vienna (ÖNB med. gr. 1; some images digitized) (see above) and the late ninth-early tenth-century version in New York City (Morgan Library M 652 (images online)): facsimile *Pedanii Dioscuridiis Anazarbaei de Materia Medica*, 2 vols. (Paris, 1935). N. Kavrus-Hoffmann, "Catalogue of Greek Medieval and Renaissance Manuscripts in the Collections of the United States of America. Part IV, 2: The Morgan Library and Museum," *Manuscripta* 52/2 (2008): 208, 212–30; figs. 1–4. The miniatures in these two manuscripts form the basis of the book by Zoltan Kádár, a Byzantinist as well as a zoologist: Z. Kádár, *Survivals*, pls. 32–45, 64–85, 96–101. Another work, the *Hippiatrika*, a veterinary manual for the treatment of equine ailments, is illustrated in a fourteenth-century manuscript in Paris (BnF grec 2244), but the horses are almost like diagrams, the artist being concerned with matters other than the usefulness, beauty or strength of the equine patients. A. McCabe, *A Byzantine Encyclopaedia of Horse Medicine: the Sources, Compilation and Transmission of the Hippiatrica* (Oxford, 2007); S. Lazaris, *Art et science vétérinaire à Byzance. Formes et fonctions de l'image hippiatrique* (Turnhout, 2010)

17 For the distribution of these animals, based on the evidence of their bones, see Kroll, "Animals," esp. 100–101. Bones of the boar, for example, are found mostly in the region of the Lower Danube, and only occasionally elsewhere.

18 Venice, Marciana gr. Z 479. See note 13 above.

19 For example, some horses and dogs were bred specifically for lion-hunting, others for leopard-hunting, others for bears. Fine horses and hunting dogs were highly prized gifts in this period: N. Drocourt, "Les animaux comme cadeaux d'ambassade entre Byzance et ses voisins (VIIe-XIIe siècle)," in *Hommage à Alain Ducellier*, eds. B. Doumerc and C. Picard (Toulouse 2004), 67–93 and his contribution in this volume.

20 Other wild animals included here are lynx, wolf, jackals, wild horses, wild asses, bulls and the elephant. The text includes the hunting of tigers, though they were no longer a threat by the eleventh century. For the threats posed by these animals, especially lions, see the memoirs, *An Arab-Syrian Gentleman and Warrior in the Period of the Crusades*, tr. P. Khuri Hitti (New York, 2000). Usamah, as a skilled lion hunter in twelfth-century Syria, was often called in by villagers to deal with lions that were threatening their livestock or the villagers themselves. The parallels with the *Cynegetica* accounts are striking. On game animals in the Byzantine period, see Kroll, "Animals," 99–101 and Messis and Nilsson in this volume.

21 Venice, Biblioteca Marciana gr. Z. 479, fols. 56v–57r. See N. P. Ševčenko, "Eaten Alive: Animal Attacks in the Venice *Cynegetica*," in *Zoa kai periballon sto Byzantio (7os–12os ai.)/Animals and Environment in Byzantium (7th-12th c.)*, eds. I. Anagnostakis, T. G. Kolias and E. Papadopoulou (Athens, 2011), 115–35.

22 W. Wolska-Conus, *Topographie chrétienne*, 3 vols. (Paris, 1968–73), vol. 3, 314–41.

23 Florence Laur. Plut. 9.28, fols. 267r–268v (images online), and Sinai 1186, fol. 202r: K. Weitzmann and G. Galavaris, *The Monastery of Saint Catherine at Mount Sinai. The Illuminated Greek manuscripts* (Princeton, NJ, 1990), fig. 180. The animals are the rhinoceros, bull-deer (ταυρέλαφος, a buffalo), giraffe (καμηλοπάρδαλις), wild bull (ἀγριόβους, a yak), musk (deer) (μόσχος), unicorn, pig-deer (χοιρέλαφος, a boar) and hippopotamus. Cosmas admits to having seen the unicorn only in statues and the hippo not at all.

24 Octateuchs: see above, note 2; S. Der Nersessian, *L'Illustration du roman de Barlaam et Joasaph*, 2 vols. (Paris, 1937); V. Tsamakda, *The Illustrated Chronicle of Ioannes Skylitzes in Madrid* (Leiden, 2002); N. S. Trahoulia, *The Greek Alexander Romance* (Athens, 1997) (facsimile of Venice, Hellenic Institute gr. 5, fourteenth century). On these, see the individual articles in *A Companion to Byzantine Illustrated Manuscripts*, ed. V. Tsamakda (Leiden and Boston, 2017).

25 The ideal hunting horses are described and illustrated in the Venice *Cynegetica*, while on fol. 8v the image of Bucephalus must certainly represent the contemporary ideal (see note 19 above)

26 Paris, BnF grec 139, fol. 1v. David is also shown out among his flocks, e.g. Theodore Psalter, fols. 28r, 188v, 189v–190r: Der Nersessian, *L'illustration*, 99, figs. 49, 295–97, and seated among them, with Melodia: A. Cutler, *The Aristocratic Psalters in Byzantium* (Paris, 1984), *passim*.

27 E.g. in the eleventh-century Lectionary Vatican gr. 1156, fols. 243r, 246r (images online). St. Mamas is milking a doe, surrounded by his flocks, in several manuscripts of the Homilies of Gregory of Nazianzus; Galavaris, *Gregory*, 99–103.

28 Job 1:3, 42:12. Papadaki-Oekland, *Job*, 58–65, 93–108; figs. 24–30, 72–91. Theodore Metochites and emperor John VI Kantakouzenos, both living in the fourteenth century, list the animals they own or owned, the latter providing exact numbers for each species: John Kantakouzenos, *History*, ed. L. Schopen, *Historiarum Libri IV*, 3 vols. (Bonn, 1828–32), 2:185.3–8 (cattle, oxen, mares, camels, mules, donkeys, pigs and sheep, most numbering in the thousands). These are much the same ones that Job owned, also in large quantities. Like Job himself, both of these men were deprived of their animal possessions by their adversaries.

29 But see Kroll, "Animals," 99.

30 For the Gregory manuscripts, see Galavaris, *Gregory*, 149–64. For the Klimakos manuscripts, see Vatican gr. 354, fol. 12v (images online): J. R. Martin, *The Illustrations of the Heavenly Ladder of John Climacus* (Princeton, NJ, 1954), 53–56, fig. 73. Both authors connect the image with the labor of the month of November. See A. Boonen, "Étude iconographique des scènes bucoliques illustrant le Discours 44 (chap. 10–11) de Grégoire de Nazianze," in *Studia Nazianzenica* II (Turnhout, 2010), 1–41, esp. 14–17; S. Germanidou, "The Plough in the Byzantine Material Culture Compared to its Western Medieval Counterpart," in *Art and Archaeology in Byzantium and Beyond: Essays in honour of Sophia Kalopissi-Verti and Maria Panayotidi-Kesisoglou*, ed. D. Mourelatos (Oxford, 2021), 157–63.

31 For the numerous Octateuch images, see Weitzmann and Bernabò, *Octateuchs*, Index s.v. "animals, slaying of". On animal sacrifice in the ancient world, see I. S. Gilhus, *Animals, Gods and Humans*.

Changing Attitudes to Animals in Greek, Roman and Early Christian Ideas (London/New York, 2006), esp. 114–121.

32 Venice Psalter, Marciana. gr. Z 17, fol. IV v: *Oriente Cristiano e Santità*, ed. S. Gentile (exh. cat., Milan, 1998), no. 7, p. 157, pl. p. 158; Theodore Psalter, fol. 188v; Der Nersessian, *L'Illustration*, fig. 295.

33 There are splendid examples in the Menologion of Basil II, Vatican gr. 1613, pp. 258, 290, 376, etc. (images online). Facsimile: *El "Menologio" de Basilio II Emperador de Bizancio (Vat. gr. 1613)* (Citta del Vaticano - Madrid, 2005). See also A. Cutler and N. P. Ševčenko, "A Recently Discovered Ivory of St. Ignatios and the Lions," in *The Material and the Ideal. Essays in Medieval Art and Archaeology in Honour of Jean-Michel Spieser*, ed. A. Cutler and A. Papaconstantinou (Leiden/Boston, 2007), 113–27. Several saints (Orestes, Zotikos, Eleutherios, Nikon, Hippolytos) were individually dragged to their death by horses. On animals and martyrs see also Papavarnavas in this volume.

34 Ševčenko, "Eaten Alive," 125–26, figs. 10–11.

35 Daniel and Panteleimon together: Athos Pantokrator 61 (Psalter), fol. 182r (lions, and a leopard): Dufrenne, *Psautiers*, pl. 27; Theodore Psalter fol. 169r: Der Nersessian, *Psautiers*, fig. 270; Thekla: London Add. 11870, fol. 174v: N. P. Ševčenko, *Illustrated Manuscripts of the Metaphrastian Menologion* (Chicago, 1990), 122; Euphemia: Venice, Marciana gr. Z 586, fol. 124r (where two of the lions have submitted, but one is still gnawing at her thigh, drawing blood): ibid., p. 177; Theodore Psalter, fol. 163v: Der Nersessian, *Psautiers*, fig. 263. On Pantaleon/Pantalemon see Papavarnavas in this volume.

36 Hagiographical manuscripts in Byzantium have little illustrated narrative in general.

37 Of course the evil attached to a snake lends a greater resonance to the story: the saint is both healing a man and counteracting a force of evil. Sinai gr. 500, fol. 5r: I. Hutter, "Le copiste du Metaphraste. On a Center for Manuscript Production in Eleventh-Century Constantinople," in *I manoscritti greci tra riflessione e dibattito. Atti del V colloquio internazionale di paleografia greca*, ed. G. Prato, 3 vols. (Florence, 2000), vol. 3, pl. 10. In a related Menologion, St. Panteleimon heals a boy bitten by a snake, Moscow, State Historical Museum 9 (Vlad. 382), fol. 101r: Ševčenko *Menologion*, 68, fiche 2A10.

38 F. Sbordone, *Physiologus* (Milan 1936, rp. Hildesheim/New York, 1991); Lazaris, *Physiologus grec*, and his essay in this volume.

39 E. Morrison, with the assistance of L. Grollemond, *Book of Beasts. The Bestiary in the Medieval World* (Los Angeles, 2019) (based on an exhibition at the J. Paul Getty Museum, Los Angeles, in 2019).

40 Smyrna Greek Evangelical School B. 8: M. Bernabò, in collaboration with G. Peers and R. Tarasconi, *Il Fisiologo di Smirne* (Tavernuzze/Florence, 1998). Present scholarship ascribes its miniatures to the eleventh century, overpainted in the fourteenth, ibid., 101–106.

41 No photographs have survived of five of the animals, but Strygowski's descriptions remain.

42 *Physiologus*, olim Library of the Smyrna Greek Evangelical School B. 8, p. 56 (fol. 28v).

43 See note 40 above. Wolska-Conus, *Topographie*, vol. 1, 94–107, esp. 106. Bernabò et al., *Fisiologo*, 70–75. The Cosmas animals included in the Smyrna manuscript are the beaver, the unicorn, the pig-elephant, hippopotamus, plus others of which no photos survive, namely the rhinoceros, elephant-bull, giraffe, and the wild bull. The beaver and the unicorn appear in both the *Physiologus* and the Cosmas texts and are depicted twice in the Smyrna manuscript. They barely resemble each other.

44 The beaver looks like a fox, the crocodile like a huge dog, the hippopotamus a winged blue horse with tusks, the ichneumon or mongoose a sort of little bear (pp. 25 and 77, 81, 36, 95). In Byzantine manuscripts in general, it can be a genuine challenge to tell the difference between a canine and a feline, even, or an antelope and a wild goat. It thus pays to be cautious when assigning meaning to a particular animal image: if we see a wolf where the artist meant to depict a fox or a dog, for example, we can easily be led astray in our interpretation.

45 But see the siren and centaur images in the Lectionary in the Gospel book in Parma 5, fol. 8r (see p. 247 in this chapter).

46 See, for example, the illustrations for Psalm 7:3 in both the Chludov and Theodore Psalters (on fol. 5v in both cases) and the illustration for Psalm 9:30 in the Theodore Psalter fol. 9v. For a list of the psalters see note 7 above.

47 See the comprehensive analyses of T. Schmidt, *Politische Tierbildlichkeit in Byzanz. Spätes 11. bis frühes 13. Jahrhundert* (Wiesbaden, 2020) and his article in this volume; N. Koutrakou, " 'Animal Farm' in Byzantium? The Terminology of Animal Imagery in Middle Byzantine Politics and the Eight 'Deadly Sins'," in *Zoa kai periballon*, ed. Anagnostakis, Kolias and Papadopoulou, 319–77

48 Schmidt, *Tierbildlichkeit*, esp. 75–122. If in Psalm 7:2–3, the lion is David's enemy, in the Canticle of Moses the lion executes the wishes of Moses (Deut. 32:24): roughly the same image is used is for both: e.g. Theodore Psalter, fols. 5v and 195v, Der Nersessian, *L'Illustration*, figs. 9 and 308.

49 Basil of Caesarea, *Homilies in Genesis*, ed. and trans. S. Giet, *Basile de Césarée, Homélies sur l'Hexaéméron* (2nd edn, Paris, 1968), esp. Homily 9, 478–523; *The Nicene and Post-Nicene Fathers*, vol. 8: *Basil Letters and Select Works*, ed. P. Schaff (Edinburgh, 1895). On the influence of these homilies, see Giet, 70–73, 548. On the "literal" and "partially allegorical" strands in *Hexaemera* literature, and on the relevance of these commentaries to works of art, see Maguire, "Adam and the Animals"; idem, *Earth and Ocean*, 31–40, 57–59; idem, *Nectar and Illusion. Nature in Byzantine Art and Literature* (Oxford, 2012), 51–53, 58. On *hexaemera* literature in general, see F. E. Robbins, *The Hexaemeral Literature. A Study of the Greek and Latin Commentaries on Genesis* (Chicago, 1912). See also Zucker, "Zoology," 276–80, and Lazaris in this volume.

50 The section of the *Annals* by Michael Glykas (twelfth century) that deals with the animals of Creation (Michael Glykas, *Annals*, ed. I. Bekker, *Michaelis Glycae Annales* (Bonn, 1836), esp. 92–126) was illustrated in a manuscript in Venice (Marciana gr. Z 402, fols. 37v, 38r, 43v, 44r, 46v: thirteenth century) with small marginal images of the animals of the text: Furlan, *Marciana*, vol. 3, 49–57, pl. 11 and figs. 47–53. The Venice manuscript is being studied by Martin Hinterberger, whom I wish to thank warmly for his helpful emails

51 Basil's antipathy to allegory is expressed at the beginning of Homily 9 (Basil of Caesarea, *Homilies* no 9, ed. Giet, 478.11–480.10). The allegorical strain in *Hexaemera* literature is best represented by Anastasius of Sinai.

52 R. S. Nelson, *The Iconography of Preface and Miniature in the Byzantine Gospel Book* (New York, 1980) and G. Galavaris, *The Illustrations of the Prefaces in Byzantine Gospels* (Vienna, 1979).

53 Vatican gr. 1291, fols. 9r; 22r–37v (images online). I. Spatharakis, *Corpus of Dated Illuminated Greek Manuscripts to the Year 1453*, 2 vols. (Leiden, 1981), no. 3 p. 6; figs. 8–9

54 Although the zodiac images in a manuscript dated 1322, now in the Vatican (Vatican gr. 175, fol. 39r) do not differ markedly from those in the ninth-century Vatican Ptolemy, those in a virtually contemporary manuscript dated 1346 (Athos, Vatopedi 1199) pair the signs of the zodiac with the far more naturalistically painted labors of the months that extend over a sequence of folios, instead of around a wheel. Spatharakis, *Corpus*, no. 235, p. 59, fig. 425; P. Chrestou et al., *Hoi Thesauroi tou Hagiou Orous. A. Eikonographemena Cheirographa*, vol. 4 (Athens, 1991), figs. 313–24 and pp. 320–22.

55 K. Nordenfalk, *Die spätantike Kanontafeln*, 2 vols. (Göteborg, 1938). On Armenian canon tables and two Armenian literary sources that interpret their iconography, see T. Mathews, *Armenian Gospel Iconography. The Tradition of the Glajor Gospel* (Washington DC, 1991), esp. 166–76, 206–11.

56 On early Byzantine ornament see K. Weitzmann, *Byzantinische Buchmalerei des 9. und 10. Jahrhunderts* (Berlin 1935, rpr. Vienna, 1996); I. Hutter, "The Decoration," in *La Bible du Patrice Léon. Codex Reginensis Graecus 1*, ed. P. Canart (Città del Vaticano, 2011) (commentary volume to the facsimile), 195–272.

57 Hutter, "Le copiste," vol. 2, 535–86; vol. 3, pls 1–39; eadem, "Decorative systems in Byzantine manuscripts, and the scribe as artist: evidence from manuscripts in Oxford," *Word & Image 12* (1996): 422; R. S. Nelson, "Theoktistos and Associates in Twelfth-Century Constantinople: An Illustrated New Testament of A.D. 1133," *The J. Paul Getty Museum Journal* 15 (1987): 53–78, esp. 59–66.

58 Biblioteca Palatina, Palat. 5 (images online): P. Eleuteri, *I manoscritti greci della Biblioteca Palatina di Parma* (Milan, 1993), 3–13; pls. 1–4; Nelson, *Preface*, 57–60, 119–21; Galavaris, *Prefaces*, 74–93; R. S. Nelson and J. L. Bona, "Relative Size and Comparative Value in Byzantine Illuminated Manuscripts: Some Quantitative Perspectives," *Paleografia e codicologia greca. Atti del II colloquio internazionale*, ed. D. Harlfinger, G. Prato et al. (Alessandria, 1991), 339–53, figs. 1–4, esp. 341–42, fig. 1. P. Degni, "Il Tetravangelo della Biblioteca Palatina di Parma, Pal. 5 (con

un Riconsiderazione del Par. Gr. 64)," *Scripta* 14 (2001), 79–94. The manuscript has attracted interest mainly because of its Gospel scenes; not all of its canon tables have been published.

59 Parma, Biblioteca Palatina Palat. 5, fol. 11r. See also fols. 8v, 9r, 9v, 10v and 8r, respectively. The fountain is generally interpreted as a reference to the waters of life of Psalm 42:1–2. See P. Underwood, "The Fountain of Life in Manuscripts of the Gospels," *DOP* 5 (1950): 41–138. See also the many useful articles in *Fountains and Water Culture in Byzantium*, eds. B. Shilling and P. Stephenson (Cambridge, 2016). The birds and some of the animals often have wide open mouths, perhaps to emphasize their thirst. The fountain structures in Parma 5 are wonderfully varied. For more fountain sculpture imagery, see p. 257 in this chapter.

60 Parma, Biblioteca Palatina Palat. 5, fol. 13r. See also fols. 3r and 5r. On the two facing pages with narrative scenes (fols. 12v and 13r), the fountain is replaced by a fruit tree for the pheasants and a date tree for the monkeys. The related but more modest version of the Parma Gospels (Oxford, Bodleian E. D. Clarke 10; some images online) lacks the animals. Nelson, "Relative size," 341–42. From the online photo, the griffin on fol. 3r of the Parma manuscript appears to have been pounced from the verso, perhaps to be copied into some other manuscript.

61 On the protective role of such creatures, see Maguire, *Other Icons*, esp. 9–11, 86–87, 96; A. McClanan, "Illustrious Monsters: Representations of Griffins on Byzantine Textiles," in *Animals in Text and Textile. Storytelling in the Medieval World*, eds. E. Wetter and K. Starkey (Riggisberg, 2019), 133–45.

62 The siren plays the harp, the leopard-centaur a pair of cymbals. Note a siren and a regular horse-centaur, both with instruments, are paired in *Physiologus*, where, however they have a bad reputation, as do most hybrid creatures except the griffin: e.g., Smyrna, Greek Evangelical School B. 8, p. 20: Bernabò et al., *Fisiologo*, 24, fig. 10.

63 The monkeys have red straps around their haunches. The left-hand monkey has white whiskers, what looks like a string of beads around its chest, and a red rear end reminiscent of that of a baboon (but both monkeys have long tails). The tight red collars on the cheetahs, and the red straps on the monkeys indicate their domesticated status. One wonders if the monkeys are meant to relate to the beginning chapters of Matthew's Gospel, which has a long genealogy of Christ. Two of the angels in the Nativity scene look directly up at the monkeys, rather than at the arc of heaven. The presence of Constantine and Helen below the Nativity remains unexplained. There is a pair of peacocks over the portrait of Matthew on fol. 13v, and then the sequence stops: there are no animals over the headpiece to Matthew or later.

64 Images online. *Byzance. L'art byzantin dans les collections publiques*, exh. Cat. (Paris, 1992), no. 268; A. Saminsky, "The Origin and Purpose of the Illuminated Four Gospels in the Bibliothèque Nationale, Paris, cod. grec 64," (in Russian), *Moschovia* 1 (2001): 405–18. See also Degni, "Tetravangelo."

65 Hutter, "Decorative systems".

66 Hutter, "Le copiste," 565, note 140 cites the late-eleventh-century Lectionaries Athos Chilandari 105, fol. 1r; Athens, National Library gr. 2363, fol. 1r (images online). A slightly later example is London, BL Harley 5785, fol. 1r (eleventh-twelfth century; images online), and Athens National Library gr. 163, fol. 5r (late twelfth century; images online): A. Marava-Chatzinikolaou and C. Toufexi-Paschou, *Catalogue of the Illuminated Byzantine Manuscripts of the National Library of Greece*, 3 vols. (Athens, 1978–97), vol. 1, no. 46, figs. 481–511. Other similar Lectionaries have only griffins: Athens National Library gr. 2804, fol. 2r (images online), Athens, Benaki Museum TA 316, fol. 1r (images online), and Athos Iviron 1404, fol. 1r. A pair of griffins appears already in the Barberini Psalter of around 1060: Vatican Barb. gr. 372, fol. 6r (images online): *Biblioteca Apostolica Vaticana. Liturgie und Andacht im Mittelalter*, exh. cat. Köln (Zurich, 1992), no. 21, 125; Hutter, "Copiste," 557, pl. 16:2. Note this particular grouping appears only in the opening headpieces.

67 On lions and griffins guarding entrances, see now G. Vaxevanis, " 'Lions Frighten wild beasts...'. An Inscribed Marble Arch of the Middle Byzantine Period from the Chalkis Region, Euboea, Greece," in *Art and Architecture in Byzantium and Beyond. Essays in Honour of Sophia Kalopissi-Verti and Maria Panayotidi-Kesisoglu*, ed. D. Mourelatos (Oxford, 2022), 113–34, esp. 125 and note 91 with bibliography.

68 The apogee is found in two homily manuscripts, Paris BnF grec 1208 (images online), and Vatican gr. 1162 (images online). I. Hutter and P. Canart, *Das Marienhomiliar des Mönches Jakobos von*

Kokkinobaphos. Codex Vaticanus Graecus 1162 (commentary to facsimile) (Zurich, 1991); J. C. Anderson, "The Seraglio Octateuch and the Kokkinobaphos Master," *DOP* 36 (1982): 83–114.

69 Vatican gr. 1162, fol. 3r. See also fol. 55r. Some animals are no more than protomes of themselves, spitting out vines as though they were water. See p. 257 in this chapter.

70 Paris BnF grec 550, fols. 49r, 59v, 72r (images online): Galavaris, *Gregory*, 242–45, esp. figs. 411–13

71 H. Buchthal, "Studies in Byzantine Illumination of the Thirteenth Century," *Jahrbuch der Berliner Museen*, 25 (1983): 27–102 discusses the quite evident Armenian influence exhibited by these bold creatures.

72 R. S. Nelson, "Palaeologan illuminated ornament and the Arabesque," *Wiener Jahrbuch für Kunstgeschichte* 41 (1988): 7–22; figs. 1–30. There are exceptions: the headpiece on fol. 153r of the Gospels Patmos 81 (dated 1335), has a rather familiar pair of lions in the lower coils, but no other headpiece in the manuscript has animals: ibid., fig. 6.

73 There are exceptions: in the Gospel book Paris BnF grec 71, fol. 71r (twelfth-century; images online), a cheetah pursues a deer inside the headpiece to Mark. But the cheetah occupies a circle of stems on the left, the deer a circle on the right: they are unlikely ever to meet. The canon table birds and animals act as predators only upon the command of a falconer or a hunter who releases the raptor or the cheetah onto a desired prey. For another exception, see Paris BnF grec 550, fol. 166v, where twice within the headpiece, the predator and the prey occupy the same coil. Galavaris, *Gregory*, fig. 423.

74 C. Franc-Sgourdeou, "Les initials historiées dans les manuscrits byzantins aux XIe-XIIe s.," *Byzantinoslavica* 28 (1967): 336–54; L. Brubaker, "The Introduction of Painted Initials in Byzantium," *Scriptorium* 45 (1991): 22–46; Hutter "Decorations." See also Weitzmann, *Byzantinische Buchmalerei, passim.*

75 Initial E: Paris, BnF grec 1208, fol. 80r. Cf. Vatican gr. 1162, fol. 59v (where a fox and a cheetah chase a central deer); both are Kokkinobaphos manuscripts of the twelfth century. Cf. also the Psalter Jerusalem, Greek Patriarchate, Taphou 51, fol. 109r (a fox and an eagle) (second half of the thirteenth century). Initial A: Athos Dionysiou 4, fol. 113r; Athos Iviron 55, fol. 70r (both thirteenth century). Such initials do run in families, often hallmarks of specific groups of manuscripts, such as the Kokkinobaphos group, or their later successors.

76 On Aesop, R. Ousterhout and A. Sitz, "Reading Aesop in Cappadocia," in *After the Text. Byzantine Enquiries in Honour of Margaret Mullett*, eds. L. James, O. Nicholson and R. Scott (London/ New York, 2022), 259–72. On the illustrated Fables manuscript Morgan Library New York M 397 (late tenth-early eleventh century), see Kavrus-Hoffmann, "Catalogue," 101–12.

77 See Maguires, *Other Icons*, esp. 58–67.

78 In Athens National Library gr. 2522, fol. 112v (images online), the ruler is Maxentios, in Venice Marciana gr. Z 585, fol. 54v it is Likinnios. Marava-Chatinikolaou and Toufexi-Paschou, *Catalogue*, vol. 3, no. 45, fig. 272; *Oriente Cristiano*, 170 no. 16.

79 Venice, Marciana gr. Z 586, fol. 78v: Ševčenko, *Menologion*, 176, fiche 5C5. The life of St. Euphemia in this same manuscript (fol. 124r) starts with the name of the emperor Diocletian: the opening initial Δ (D) is formed by a dog attacking a bear, both on their hind legs: eadem, 177, fiche 5D5.

80 Ševčenko, *Menologion, passim.* The tone is very much that of iconoclast invective of the ninth century, see Koutrakou, "Animal Farm."

81 It is a particularly favored initial for the Matthew Gospel among the Decorative style Gospels and their derivatives but is found as early as the Paris Gospels BnF grec 64 (fol. 12r; images online) (eleventh century).

82 Turin Biblioteca Nationale Universitaria C.1.6, fol. 73v: V. Kepetzis, "Scenes of Performers in Byzantine Art, Iconography, Social and Cultural Milieu: The Case of Acrobats," in *Medieval and Early Modern Performance in the Eastern Mediterranean*, ed. A. Öztürkmen and E. B. Vitz (Turnhout, 2014), 260, fig. 22.4b; Vatican gr. 747, fol. 22r: Weitzmann and Bernabò, *Octateuchs*, 30–31, fig. 79.

83 Vatican gr. 747, fols. 29r and 133v: ibid., 51–52, 188-1-89, figs. 131 and 840.

84 Istanbul Seraglio 8, fol. 82r: ibid., 189, fig. 841.

85 In the Florence Octateuch, the giraffe is a tiny creature in the Naming scene, barely visible: Weitzmann and Bernabò, *Octateuchs*, 30, color fig. 5a.

86 See articles by Drocourt and Schmidt in this volume.

87 N. P. Ševčenko, "Wild Animals in the Byzantine Park," in *Byzantine Garden Culture*, eds. A. Littlewood, H. Maguire and J. Wolschke-Bulmann (Washington DC, 2002), 77–78; eadem, "The giraffe that came to Constantinople," in *The Eloquence of Art. Essays in Honour of Henry Maguire*, eds. A. O. Lam and R. Schroeder (Abingdon/New York, 2020), 336–49. Its arrival in Constantinople was noted by Michael Psellos and Michael Attaleiates. The exchange of exotic animals across the Mediterranean has been thoroughly studied in recent years, above all by Nicholas Drocourt and by Thierry Buquet, esp. his "La belle captive. La giraffe dans les ménageries princières au moyen âge," in *La bête captive au Moyen Âge et à l'époque moderne*, eds. C. Beck and F. Guizard (Amiens, 2012), 65–90.

88 The giraffe images that illustrate the text in Cosmas (and *Physiologus*) manuscripts look nothing like a real giraffe: Wolska-Conus, *Topographie*, vol. 3, 320–21, fig. 2, and Bernabò et al., *Fisiologo*, 66

89 Jerusalem, Greek Patriarchate Taphou 5, fol. 102r: Papadaki–Oekland, *Job*, 228, fig. 259. The manuscript has been dated to the second quarter of the fourteenth century by Erich Lamberz, who has identified its scribe as scribe E of Athos Vatopedi 390. Scribe E is known to have worked for both Theodore Metochites and for John VI Kantakouzenos: E. Lamberz, "Joannes Kantakuzenos und die Produktion von Luxushandschriften in Konstantinopel in der frühen Palaiologenzeit," *Actes du VIe colloque international de paléographie grecque*, ed. B. Atsalos and N. Tsironi, 3 vols. (Athens, 2008), vol. 1, 133–57, esp. 146–47, 153 note 90.

90 Jerusalem, Taphou 5, fol. 102. There is one other manuscript in which the wild ass of the Greek text is illustrated with the representation of a zebra: the Job manuscript in Paris dated 1362 (BnF grec 135, fol. 100r; images online): Papadaki-Oekland, *Job*, 228; image online). But in the Paris manuscript the representation is almost laughable: the usual ass is just painted all over with parallel rows of black lines. The Job passage has been variously translated, and its meaning remains obscure. It speaks of the ὄνος ἐρημίτη, the ass of the desert (otherwise known as the ὄνος ἄγριος).

91 D. Behrens–Abuseif, *Practicing Diplomacy in the Mamluk Sultanate. Gifts and Material Culture in the Medieval Islamic World* (London, 2014), 28, 61–63. But the French translators of the Lives of Baybars say the gift contained wild asses and do not mention zebras: S. Sublet and A. Miguel, *Les trois vies du sultan Baybars* (Paris, 1992), 95–98. One of the problems in tracking the zebra in Greek is that the Byzantines had no word for it other than wild ass; unlike the Arabs, they never refer to the stripes. See T. Buquet, "Nommer les animaux exotiques de Baybars, d'Orient en Occident," in *Les non-dits du nom. Onomastique et documents en terres d'Islam. Mélanges offerts à Jacqueline Sublet*, ed. C. Müller and M. Roiland-Rouabah (Beirut, 2013), 375–402.

92 The pattern of the stripes differs according to the species of zebra. On the cultural history of the zebra, see C. Plumb and S. Shaw, *Zebra* (London, 2018).

93 There were other embassies in the Palaiologan period: M. T. Mansouri, *Recherche sur les relations entre Byzance et l'Egypte (1259–1453) d'après les sources arabes* (Tunis, 1992), 117. He has a list of them, although the gifts are not always named.

94 See Schmidt on animal semiotics in this volume.

95 According to Basil the Great, "Each animal is distinguished by peculiar qualities. The ox is steady, the ass is lazy, the horse has strong passions, the wolf cannot be tamed, the fox is deceitful, the stag timid, the ant industrious, the dog grateful and faithful in his friendships," Basil of Casearea, *Homilies in Genesis*, ed. Giet, 489.

96 Ekphrasis (fourteenth century) on the garden of St. Anne, tr. by M-L. Dolezal and M. Mavroudi, "Theodore Hytakenos' Description of the Garden of St. Anna," in *Byzantine Garden Culture*, 105–58, esp. 126–28, 144–45. The poet goes on to describe the statues of birds around the rim of the fountain that "seemed to dip their beaks and drink from the water… ."

97 *Belthandros and Chrysantza* (fourteenth century), tr. G. Betts, *Three Medieval Greek Romances* (New York/London, 1995), 10–11; McClanan, "Griffins," 142–43; Dolezal and Mavroudi, as above. Some of the animal statues in the Romances appear to be automata: Dolezal and Mavroudi, "Hyrtakenos," 130–32.

98 Vatican gr. 463, fol. 4r (images online): Spatharakis, *Corpus*, fig. 135. One of the fountains in the Parma Gospels (see note 58 above) shows a pair of quadrupeds (lions? bears?) seated back to back as supports for the water basin; water is pouring from their jaws (fol. 9v). At least two eleventh-century manuscripts have a pair of lions clinging to the sides of the water basin, tails

hanging down, over their opening headpieces; the lions are either extremely thirsty or part of the fountain structure: Sinai 503, fol. 1r (Menologion): Hutter, "Copiste," pl. 13, and Athos, Iviron 1404, fol. 1r: (Lectionary): eadem, pl. 22. A similar pair forms the initial M, the lions using the fountain as a shower, in the Vatican Psalter gr. 752, fol. 3r (images online): F. D'Aiuto, "Il *Vat. Gr. 752: fattura materiale, scrittura, mise en page* (con qualche osservazione sul Salterio Hierosol. S. Sepulchri 53)," in *A Book of Psalms from Eleventh-Century Byzantium: The Complex of Texts and Images in Vat. Gr. 752*, eds. B. Crostini and G. Peers (Città del Vaticano, 2016), 43–156, at 121–22, fig. 39; S. Gillingham, "David and Christ sing the Psalms: The Psalter as Prophecy and Liturgy," in ibid., 241–59, at 247, fig. 4. The headpiece of the liturgical roll in Athens (National Library 2759, second half of the twelfth century) contains probably the most animals and birds of any headpiece of the period, including animal protomes: Spatharakis, *Corpus*, fig. 565. Cf. also Paris, BnF grec 71, fol. 25r (images online)

99 See the articles in *Satire in the Middle Byzantine Period: the Golden Age of Laughter?*, eds. P. Marciniak and I. Nilsson (Leiden–Boston, 2021), including that of H. Maguire, "Parody in Byzantine Art," 127–51. See also the chapter by Stewart in the present volume.

PART III

The Materiality of Animals in Byzantium

12

ANIMALS AS A SOURCE OF FOOD DURING THE BYZANTINE PERIOD

Dietetic Advice and Dietary Reality

Maciej Kokoszko and Zofia Rzeźnicka

Scope of the Study

The main purpose of this study is to provide the reader with a general overview of the consumption of meat and animal products in the Byzantine period. In order to accomplish this objective, we have tried to present the issue in the broadest possible perspective. Therefore, we do not focus solely on listing the types of meat and animal products consumed by the Byzantines, but we also discuss the ways in which the foodstuffs were prepared. At the same time, we must stress that it is impossible to scrutinise each way all the different types of food obtained from animals were used within the present chapter. In consequence, we have been forced to select judiciously from the collected information. Upon an in-depth analysis, we decided to exclude topics concerning fish, seafood and honey. As regards the former two, we opine that the question of their presence in the Byzantine menu deserves a separate and comprehensive study.[1] As for honey, the ample material by medical and agronomic authors convinced us that the topic is equally too complex to be discussed only in brief.[2] Consequently, we chose to focus primarily on animals that were reared on the majority of farms. In addition, we also included information concerning certain avian species whose keeping was not widespread but still common or considered luxurious.

Sources

In the Byzantine period, no single treatise was written that contained all the data we need. For instance, with the exception of the Latin compilation of recipes attributed to Apicius, known as *On the Subject of Cooking* (fourth century),[3] and the Latin *opusculum* by Anthimus entitled *On the Observance of Foods* (sixth century)[4] we know of no cookery books from that period. Moreover, the latter treatise does not strictly represent the genre of culinary literature, since it is, in fact, a dietetic work. Therefore, it points us towards the most extensive corpus of sources covering the subject of food, i.e., medical literature, and especially those extant works which discuss dietetics and pharmacology. In order to paint

DOI: 10.4324/9781003055877-16

a holistic image of the consumption of meat, offal, dairy and eggs in the Byzantine diet, one has to analyse not only works from the Byzantine period, but also treatises composed as early as in antiquity, since these were later used extensively by generations of Byzantine physicians.

When it comes to ancient doctors, the greatest impact on the development of dietetics and pharmacology in Byzantium came from Galen of Pergamon (ca. 129–215), who refined the concept of the therapeutic function of food, dating back to Hippocrates (fifth/fourth century BC). Dietetics defined food as nourishment,[5] which, due to its processing and absorption,[6] ultimately was turned into human flesh. The subject was of such importance that Galen devoted a separate treatise (known as *On the Capacities of Foodstuffs*) to the nature of key edibles. Pharmacology, in turn, treated foods as medicines,[7] because physicians believed that the vast majority of foods affected the characteristics of the body.[8] As a result, Galen also analysed the properties of food in his most important work on pharmacology, i.e., *On the Capacities of Simple Drugs*. Following his theory,[9] Byzantine doctors such as Oribasios (ca. 350–400), Aetios of Amida (ca. 500–550), Paul of Aegina (ca. 630–670), the anonymous author of the dietetic treatise *On Foods* (seventh century),[10] or Symeon Seth (eleventh/twelfth century) also took into account the dual role of food, thereby preserving particularly exhaustive information on the subject.

Attention should also be paid to the manner in which medieval physicians used the achievements of their predecessors. Namely, recent research shows that they included in their writings the information they considered useful to their own medical practice, at the same time correcting and supplementing[11] the preserved material in line with their current knowledge and needs.[12] This means that each of the aforementioned authors documented the realities of his own times, taking into account the food consumption patterns typical for his lifetime. Here, it should also be noted that, as far as the significance of foods is concerned, ancient and Byzantine physicians did not limit themselves in their reflections to the field of *ars coquinaria*, but treated the issue holistically, also discussing subjects that were of interest to agronomy, which we know thanks to the Byzantine agricultural encyclopaedia *Geoponika* (tenth century). The information it provides allows us to supplement the data from medical works. Obviously, the present text is additionally based on a number of other sources whose content is intended to make the picture more comprehensive.

The Most Popular Food Animals

Let us begin with the quadrupeds which are covered in detail in Byzantine medical literature. Not all information is straightforward to interpret. For instance, in their chapters devoted to the said creatures, Oribasios and Aetios of Amida, after presenting the dietetic characteristics of the meat of farm animals, i.e. pork, beef, goat and mutton, proceed to the discussion on hares, deer, wild donkeys, horses, camels, bears, lions, leopards, panthers, dogs and foxes.[13] Although the physicians described how the consumption of each of these meats impacts the human body, it does not mean that all of them were a regular sight on Byzantine tables.[14] In all probability, the inclusion into the treatises of such a comprehensive list was only intended to show that, in a world where a constant shortage of food was an everyday phenomenon and diet depended on locally available products, people were keen not to miss any opportunity to acquire food. Therefore, depending on which part of the Empire they lived in and what species of animals were available to them, humans would eat any type of meat.

As a result, in order to determine the species that were eaten most commonly, we should instead turn to another extract from Oribasios' *Medical Collections*, namely to a fragment borrowed by the medical writer from Athenaeus of Attalia (first century). In the passage, Athenaeus describes changes in the seasonal quality of meat, though exclusively focussing on farm animals, i.e. pigs, goats, sheep and oxen.[15] Interestingly, the meat of exactly the same animals (and in the same order) is also analysed by Paul of Aegina in the seventh century, who, however, expands the list with hares and deer.[16] As the last two were wild animals, and thereby their presence in the Byzantine diet was rather limited (to those who had time to hunt), we can conclude that inhabitants of the Mediterranean most frequently consumed pork, goat meat, mutton and beef (only sporadically enriching their diet with venison).[17]

As far as beef is concerned, Johannes Koder[18] argues that it was consumed in the Byzantine period as a last resort (because cows and oxen were used mainly for labour, and consequently their meat was not appreciated).[19] As a result, he, interpreting Chapter 15 of *Book of the Eparch* (ninth/tenth century), hypothesizes that there was no organised supply of the capital in beef and that the Constantinopolitan butchers would have traded in lamb and mutton only.[20] Nevertheless, his final conclusions are hard to accept. Firstly, beef and young beef (or veal) are mentioned by Anthimus in entries 3 and 11 of his treatise, which makes us believe that it was still appreciated in the sixth century (at least in the West). Secondly, cows and calves are put on the list of supplies taken on military campaigns for the indulgence of the imperial table.[21]

Thirdly but most importantly, cattle breeding and consumption in the Byzantine period has been confirmed by archaeological research.[22] The latest Constantinopolitan evidence[23] shows that, although cattle were used as beasts of burden and draught animals in the city,[24] a high number of their bones, which, on the one hand, have no pathological changes due to prolonged physical stress but, on the other, display butchering marks, leads to the conclusion that the cattle were slaughtered for meat. As the majority of cattle bones (71.69 per cent) have been proved to come from creatures which were between four and seven years old, one can also conclude that the Constantinopolitans did not refrain from slaughtering animals capable of providing other benefits, i.e. milk and labour. Since the rest of the bones come from young animals, one should also surmise that veal was not absent from Constantinopolitan tables either. What is more, the consumption of beef must have been relatively high because the osteal material recovered in the Port of Theodosius includes as much as 25.68 per cent of the bones belonging to cows, oxen and buffalos; these finds score the second place after those of ovicaprids, which are predominant in the capital of the Empire.

The poultry hierarchy can be reconstructed, in turn, on the basis of the dietary characteristics of the birds compiled by Oribasios, Aetios of Amida and Paul of Aegina. The three list partridges, hazel grouse, domestic pigeons, hens, pheasants, thrushes, blackbirds, sparrows, turtledoves, common wood pigeons, ducks, peafowl, geese, ostriches and cranes, seemingly without indicating any priorities.[25] The catalogue is, however, quite straightforward to interpret. First, the vast majority of the enumerated avian species are either wildfowl (some of which were kept in captivity for meat[26]) or semi-domesticated birds (e.g., domestic pigeons),[27] which implies that the birds were reared only periodically (in places with specialised infrastructure), and that their meat was consumed rather sporadically. Additionally, as some of the domesticated birds (i.e. pheasants[28] and peafowl[29]) were considered luxuries,[30] they were consumed by few Byzantines. Last but not least, the habitat of ostriches was limited to Northern Africa and the Middle East,[31] which means that their meat was a proportion of the daily diet therein exclusively. All in all, out of all the species, we may deduce that it was only hens and geese that were truly domestic and ubiquitous in the Mediterranean husbandry.

In order to establish which of these two avian species was more and which less popular, we will once again refer to yet another part of the above analysed extract from Athenaeus of Attalia's output, notably, the one in which the author focuses on fowl that constitutes the most nutritious food in autumn and winter. It is striking that he includes in this group almost exclusively wild birds, while only mentioning hens out of domestic species, adding as an aside that they do not thrive in winter.[32] When one attempts to understand what he intended to say, one can arrive at the conclusion that, in all probability, he wanted to suggest that, unlike domestic poultry, wild fowl did not depend on the fodder which was provided by men, since such birds obtained their food in the wild. Hence, at the time when peasants began to run low on food stocks and on animal feed, wild birds still remained a substitute food source. Since hens are mentioned by Athenaeus as the only birds which lose weight in winter,[33] the said remark suggests that during that season hens had limited access to fodder, thus proving that these birds were exclusively farm animals.

One can additionally surmise that hens were the only farm birds mentioned in the passage because they were the quintessential domestic fowl (i.e. they far outnumbered other poultry kept on farms).[34] The supposition is confirmed by Oribasios, Aetios of Amida and Paul of Aegina, who provide data that hen eggs were regarded as the best for consumption, followed by pheasant eggs, and then those of geese and ostriches.[35] Since eggs were always obtained primarily from domestic poultry, and having in mind the fact that pheasants were a luxury and the habitat of ostriches was very limited, we should surmise that, while hens were the most productive in terms of eggs, geese were the second most widespread domestic fowl[36] (also appreciated for providing eggs but considerably less than hens). Interestingly, support for our conclusion can be also found in a tenth-century military treatise. Notably, its author lists hens and geese among the provisions taken on military campaigns, writing that the average number of chickens in the supplies was 200, while the number of geese amounted to 100.[37] It is likely that the said ratio reflects, at least roughly, the scale of keeping the birds.

Meat Consumption Patterns

As mentioned before, one should also note that the Byzantines ate the meat of wild animals, and although game never became a significant element of their everyday diet,[38] its dietetic properties were diligently described by the physicians, who generally regarded it as dry (therefore tough) and hard to digest.[39] Returning to the main topic of our discussion on the consumption of meat from agricultural production, it is noteworthy to mention that the Mediterranean is not a favourable location for keeping farm animals. One of the reasons behind this is the limited access to fresh fodder caused by the short vegetation period. As a result, meat was generally in short supply in the region. For the present study, it is particularly important that most of the mentioned quadrupeds and fowl brought many other benefits, both inside and outside the household. Oxen were primarily used for cultivating the soil, they were draught animals or beasts of burden.[40] Cows, goats and sheep were bred for milk. In turn, hens and geese laid eggs, which were eaten as such or were utilised as an ingredient of complex dishes. All farm animals provided manure for soil.[41] Moreover, the majority of the creatures produced valuable raw materials. For instance, geese were a source of feathers,[42] while sheep of fleece.[43] As a result, it seems reasonable to surmise that most of the said animals were sent to slaughter only as a last resort.[44]

An exception to this rule were pigs, which, besides their meat (and manure), did not provide people with any significant additional benefits.[45] As they did not require any special infrastructure, and could easily survive on household leftovers, many of them were kept close to human abodes or even shared the same space with people, becoming less and less vulnerable to the seasonal changes of nature which other farm animals were subject to.[46] The last phenomenon is particularly visible when we once more turn to Athenaeus of Attalia's writings. Notably, he maintains that pork displays its best properties in the period from the descent of the Pleiades[47] until spring,[48] which would not be possible, unless pigs relied on fodder provided by their owners. On the other hand, goat meat was (according to Athenaeus) best from early spring to the setting of Arcturus,[49] mutton from the spring equinox[50] to the summer solstice,[51] and beef from late spring until the end of summer,[52] i.e. all of them were said to be top quality only when pasturelands were full of fresh vegetation.[53]

There are many reasons why we believe that pork was the favourite and thereby very popular meat in Byzantium.[54] First, apart from the fact that the animals are not demanding in terms of their food and living conditions, pigs are fairly prolific, and on average have many more young than any other farm quadrupeds. A sow can have over a dozen piglets in a single litter, which can be slaughtered as early as the suckling stage or as they grow.[55] This does not mean, however, that pork was uniformly present on the tables of all peasants and city-dwellers across the Mediterranean, since both small-scale breeders and the owners of large pig herds profited most from selling the finest animals (i.e. those that were at least one-year-old[56]) mainly in cities. Pigs were sold and purchased in the capital (in *Forum Tauri*).[57] We also have evidence testifying that keeping pigs in urbanised areas of the Empire existed, even in the capital itself.[58] Such a case is described, for example, by John Tzetzes (twelfth century).[59]

Secondly, the popularity of pork is implied in medical literature, which analyses the qualities of the meat so systematically and in such great detail that we can surmise that it was fairly common in the diet. Moreover, one can infer that it was not only popular but also held in high regard, as medical sources attribute to the food high nutritiousness,[60] maintain that it stimulated the human body to the production of good juices, and classify pork as easy to digest.[61] Remarks on the dietetic properties of pork allow us to assume that the Byzantines preferred the meat of mature though not old animals.[62] The conclusion is drawn from the fact that both the meat of old pigs and piglets was considered a less valuable nourishment: the former was characterised as sinewy and dry, hence, difficult to digest,[63] while the latter was said to contain excessive amounts of moisture, because of which it did not provide the body with enough nourishment.[64] Be that as it may, there is substantial evidence that in Byzantium the meat of suckling pigs was considered a delicacy.[65]

The dietary characteristics of beef, goat and sheep meat was considered to be governed by a similar rule: the meat of younger animals (due to higher moisture content) was said to have better properties than that of the old ones,[66] as they tended to grow increasingly dry (and thereby tough) with age to become difficult to digest.[67] The medical authors unanimously attributed the greatest dryness to beef (which is usually indicated by the remarks that its consumption leads to diseases caused by the accumulation in the human body of the dry and cold humour called black bile).[68] Because of the fact that medical texts maintain that meat from old animals is tough, and generally of inferior quality,[69] the rules of dietetics they were based on make us think that the vast majority of meat (including pork) accessible to an average or less affluent purchaser was obtained from animals which had already ceased to be useful for the household.[70] The conclusion is based on the premise that such meat, in terms of cuisine, hard to process, and, in terms of medicine, hard to absorb by the body, must have been in lesser demand and thus cheaper on the Byzantine market. Accordingly, the poor would consume aged boars, billy goats and rams that no longer played a reproductive role, sows which had become infertile with age, cows, nanny-goats and ewes that had stopped giving milk, and also oxen that could no longer be used in farm work.

Such meat must have been relatively affordable, and likely to be consumed even in the countryside. Young animals, on the other hand, must have been more expensive, because their slaughter, especially if they belonged to a smaller herd, would have deprived their owners of the future benefits the creatures were expected to bring. Therefore, any decision to kill a younger animal for food would have followed some measures taken to minimise the resulting losses, and concerned mainly those creatures which were regarded as unnecessary in the herd, i.e. young billy goats, male lambs and young bulls.[71] Consequently, prime quality meat from quadrupeds, i.e. that from young and adult but not aged animals, must have been relatively expensive (and therefore it was more often offered in cities, where buyers tended to have more means to purchase it),[72] with young beef being on the top of the list of delicacies.[73] This phenomenon is reflected in Byzantine epistolography, which teaches us that both young animals or their fresh meat were considered a treat, as we learn, for instance, from another story told by John Tzetzes. Notably, the scholar received a lamb, some chickens and a partridge, which Tzetzes was not happy with, and would have preferred some cooked or cured meat instead.[74]

We may conclude then that meat was a luxury foodstuff in Byzantine times, served on festive days or during public feasts given by the Emperor or the ruling elite. Probably the most vivid picture of the Byzantine craving for meat was painted by Ptochoprodromos (twelfth century). In one of his poems he refuses to eat vegetables (a normal diet for the poor) and instead calls for μονόκυθρον, i.e. a one-pot dish, thick with ingredients such as meat, or παστομαγειρεία, prepared with a lot of lamb fat.[75] As for public feasts, one can get a picture of how they looked like, for instance, on the basis of a speech delivered by Eustathios of Thessalonike on the occasion of the wedding of Alexios, Manuel I Komnenos' son, and Agnes, a daughter of Louis VII, king of France, which took place in 1180.[76] Among the celebrations held in Constantinople was a feast in the hippodrome, for which no expense was spared by the Emperor, and, though we do not know much about the dishes served to the guests, we are informed that the menu included large quantities of slaughtered quadrupeds and poultry.[77]

While in the country people both reared and slaughtered the animals themselves, in cities of the Empire meat was generally only available from traders. According to *Book of the Eparch*, in Constantinople in the ninth and the tenth centuries meat could be bought at the grocer's, directly from the capital's slaughterhouse or at the *Forum Tauri*. The document reveals that grocers ran their retail shops or stalls in the main streets and squares of the city.[78] They (allegedly) traded in cured meat exclusively.[79] In turn, the inhabitants of Constantinople bought their fresh lamb, mutton and possibly beef from the butchers working in the capital's slaughterhouse,[80] while they would go to the *Forum Tauri* to purchase their fresh pork. The same text allows us to hypothesize that meat was bought by the end-user in relatively small pieces, as, before the sale of their meat, the animals were eviscerated by the butchers (thereby providing valuable offal), their heads and legs were cut off (thus becoming coveted ἀκροκώλια). These parts were sold separately from the rest of the carcass, which, in turn, was cut by the butcher into smaller pieces.[81]

It has been already implied that pork was the favourite kind of meat, both in the countryside and in the city, which is probably why the Byzantine physician Anthimus, who spent a substantial part of his life in the West, included in his treatise a noticeably higher number of recipes for pork and pork offal than for the meat (and offal) of other quadrupeds.[82] As a medical doctor he valued young pork the most,[83] as a gourmet versed in the contemporary cuisine, he advised eating it boiled or roasted,[84] while he preferred piglets to be served stewed[85] with a sauce made from honey and wine vinegar.[86] Anthimus also knew bacon, called by him *laridum*.[87] His testimony is interesting for our study, as it enables us to make a supposition about how the product was usually consumed in the East. Namely, Anthimus (who wrote his treatise as a gift to the Frankish ruler Theuderic) stressed that the Franks ate *laridum* raw.[88] The fact that he was surprised by the practice, and that he recommended eating *laridum* after it was carefully cooked and cooled,[89] makes us suppose that it was exactly the manner which was practised in the East. Apart from bacon, the Byzantines also used pork lard and/or tallow (λαρδή), which was, for example, on the list of military imperial provisions.[90]

Anthimus' work also testifies to the wide availability of pork offal on the market. Among these products one could distinguish both delicacies and less expensive items. Given the ancient Greco-Roman culinary tradition, we may conclude that the former included, *inter alia*, uteri and udders.[91] Another delicacy known since antiquity was the liver of fig-fattened pigs,[92] called *ficatum*.[93] On the other hand, an example of a lower quality offal are kidneys, the outer parts of which were exclusively considered to be acceptable as a foodstuff by Anthimus.[94] Last but not least, Oribasios, Aetios of Amida and Symeon Seth included in their treatises the dietetic properties of the already mentioned ἀκροκώλια, i.e., pig's ears, snouts and trotters, which implies that those by-products of animal slaughter were also appreciated during the Byzantine period.[95]

We must also bear in mind that the slaughter of animals results in a pressing need to process large amounts of meat, which would otherwise quickly go off in the hot Mediterranean climate.[96] One of the techniques of its preservation was salt-curing.[97] And since the method is often mentioned in the Byzantine medical[98] as well as agronomic[99] literature, we may assume that salted meat was commonly used by the inhabitants of the Empire. It was also regularly taken on military campaigns by soldiers.[100] On the basis of Oribasios' writings (who borrowed the information from Galen), we can suppose that among the meats of quadrupeds, the process was most effective with the meat of mature pigs.[101]

As for the culinary use of fresh beef, mutton and goat meat, the relatively few remarks on the subject in Anthimus' *On the Observance of Foods* suggest that such foodstuffs were not as favourite part of the diet as pork. On the other hand, the analysed literary sources and archaeological evidence teach us that beef, mutton and goat meat were consumed in large quantities in the Byzantine Mediterranean. In all probability, young mutton and goat meat[102] were eaten more often than beef. When writing on the former, Anthimus argues that it could be eaten stewed or roasted.[103] Despite the fact that even relatively regular consumption of mutton had no negative impact on health according to Anthimus, he considered lamb a better food,[104] being on a par with the meat of young goats.[105]

In the case of dry and tough meats like beef, the physician preferred those obtained from younger animals. His predilection can be confirmed by his remark made in the eleventh chapter of his work, where he recommends beef that he calls *tenerior*, that is: 'tender' or 'soft', thus implying that it should come from a younger creature.[106] Due to the limited succulence of the meat, Anthimus advised boiling or steaming as the most beneficial method of its preparation, although he also allowed for roasting as long as the food was not prepared directly over a flame.[107] An exhaustive recipe from the third chapter of *On the Observance of Foods* suggests that, in Anthimus' opinion, the best method to prepare mature beef was stewing. He writes that, first, the meat ought to be soaked in water in order to remove its unpleasant odour. Next, it should be pre-cooked for a short time in water and then stewed in a sauce made from vinegar,[108] leek, penny-royal, celery and fennel. Later, the sauce was to be enriched with honey and ground spices (pepper, costus, spikenard and cloves).[109]

Milk and Dairy Products

Cows were generally not only a source of meat but also of milk,[110] which Oribasios describes as the thickest and fattest among that given by other quadrupeds.[111] Owing to the high fat content in cow's milk, it was often used for making butter,[112] which in the Mediterranean was primarily used as a pharmaceutical agent, and not a foodstuff.[113] From Anthimus we learn that cow's milk was not always processed, but was also consumed fresh,[114] although it is worth noting that, when writing about milk, he more often refers to that obtained from goats.[115] The lower popularity of cow's milk most probably stemmed from the fact that cows in the uplands of the Mediterranean region had no access to suitable pasturelands at high altitudes, which were easily accessible to and sheep. In consequence, cows could only graze in limited areas, which implies that they were greatly outnumbered on the majority of farms by other milking livestock, i.e., goats and sheep.[116] As a result, we may presume that significantly larger amounts of goat's and sheep's milk were available for either drinking or processing in Byzantium.[117]

The above conclusions are confirmed by the data from *Geoponika*. In the treatise we read that farmers had methods to secure the continuous availability of sheep's milk, which they achieved by means of staggering the periods when sheep were lambing in order to ensure the animals gave birth at different times of the year.[118] It goes without saying that such methods would not have been used had the farmers not been interested in all-year-round milk production. The importance of milk from goats is further stressed by the fact that the author of the work, when discussing the benefits of keeping these creatures, lists milk and cheese first,[119] which allows us to presume that the species was primarily kept for milk.[120] What is more, the compiler of *Geoponika* placed the main *corpus* of advice on cheesemaking (i.e., information on testing milk quality,[121] its coagulation,[122] storage of ὀξύγαλα,[123] making cheese[124] and obtaining μέλκη[125]) in the book devoted to sheep and goats, and not to cows and oxen. In all likelihood, he would not have done so if the former two had not been the greatest provider of milk for cheesemaking. Finally, the indispensability of goat's and sheep's milk in cheesemaking is often mentioned by Byzantine physicians. They characterise the former as of medium thickness[126] (thus, perfect for drinking as well as for cheesemaking), whereas the latter they describe as thicker,[127] which implies that it was even more useful in terms of cheese production than that of goats.

One should stress the fact that, in general, medical treatises contain a cornucopia of data on milk, as it was not only an important foodstuff but also an indispensable medicament.[128] From the writings we can assume that its quality depended *inter alia* on its thickness and freshness.[129] The former was conditioned by the season of the year when milk was obtained. Being thinnest in spring, it would gradually become thicker and, thus, more nutritious over the following months and reach its best properties in mid-summer.[130] As far as milk's freshness is concerned, it was short-lived, because in the hot Mediterranean climate milk tended to quickly go sour.[131]

In order to prevent fresh milk from going off quickly, it was boiled. It seems that, to this end, the Byzantines, on top of boiling milk in the pot, were still using the method invented in antiquity based on immersing hot stones or iron discs into a vessel with the milk.[132] From medical treatises we also learn about a practice of serving milk with the addition of some salt and honey,[133] which were supposed to facilitate digestion, and, as we can conclude from Anthimus' narration, also to slow down milk's natural fermentation.[134] What is more, medical writings inform us about the practice of boiling milk with cereal products in order to prepare nutritious soups.[135] Since it was the peasants who had a regular access to both milk and cereals, one might assume that such dishes were most popular among the Byzantine villagers.[136] On the other hand, the analysed source texts equally suggest that by modifying the list of ingredients, one could turn a common gruel into a luxurious dish. For example, the rich would have these dishes prepared with rice,[137] i.e., a cereal which only a few could afford at that time.[138] We also read about the tradition of preparing ὀξύγαλα. It can be concluded that this term was used *inter alia* in reference to soured milk.[139] However, ancient agronomic literature proves that the word ὀξύγαλα also meant cheese curd,[140] and this meaning is corroborated in *Geoponika*, whose author describes the practice of dipping ὀξύγαλα in olive oil or wrapping it in terebinth leaves.[141]

As for cheese, we may suppose that it was an important constituent of the Byzantine diet, since cheesemaking was the most effective method of utilising surplus milk.[142] Although it seems that the majority of it was eaten by the peasantry,[143] it was a foodstuff consumed by all social strata because it is also mentioned in the list of provisions taken by the emperor on his military campaigns.[144] And since it was also documented that sheep with their lambs and cows together with their calves were part of the supply train, one can surmise that the ruler could partake not only of mature, but also of freshly produced cheese,[145] as could his courtiers.[146] One of such, namely Michael Psellos, knew much about factors making cheese palatable and healthy. These are described in his letter 105, where he first delves into the way milk is produced in the body[147] before concerning himself with the manner in which cheese is obtained. From the text we also learn that he especially liked the variety of mature cheese (which was full of holes) produced in Paphlagonia.[148] When elaborating on the properties of cheese,[149] the authors of medical treatises list both the already mentioned curd cheeses as well as hard, matured ones. Depending on the level of maturity, cheeses were categorised as soft and hard, thick or thin-textured, sticky or prone to crumbling. As for flavours, they were described as piquant and salty, or with a hint of sweetness.[150]

Poultry and Eggs

Having discussed the foodstuffs obtained from quadrupeds, let us now focus on fowl. The fact that Oribasios, Aetios of Amida and Paul of Aegina described the dietetic properties of meat from birds in detail, proves sufficiently that the product was part of the diet in early Byzantium. As we learn from the said physicians, poultry was generally considered easy to digest, though less nourishing than the meat of four-legged livestock.[151] The above information suggests that it was not valued as much as the meat of quadrupeds, and, thus, it seems that it was also less frequently consumed.[152] Hence, in all likelihood, birds (and especially hens) were not primarily reared for their meat itself, but for eggs. Moreover, the comment on the digestibility of poultry meat suggests that birds were often slaughtered when there was a need to provide light food for those suffering from health problems. Such patients were advised to consume the meat of fully-fledged but not fully mature animals, as the physicians argue that this kind of meat is better than that from the very young, fully mature or old creatures.[153] It seems that the slaughtering of laying hens was not economically viable, and thus, the first to be killed were presumably those birds that were naturally incapable of producing eggs, namely young roosters.[154] And as for older birds, in all probability they were cocks that were past their reproductive role in the flock and hens which were past their egg laying period.[155] Such aged, i.e. three-year-old hens are mentioned, for example, by the author of the satire *Timarion* (twelfth century),[156] where slaughtered and eviscerated birds are pictured as awaiting their buyers, with their fat removed from the belly and displayed squarely on each side of the carcass.[157]

Hens were the most popular poultry species during the Byzantine period, with doctors considering their meat to be light.[158] In the eleventh century, Symeon Seth was even more specific, and distinguished between hens and young chickens, arguing that the former were better for those who were physically active, while the latter were recommended for those with sedentary lifestyle.[159] In terms of digestibility and nutritiousness, hen meat was said to be on a par with that of pheasants, although the latter was generally regarded as tastier.[160] On the other hand, completely different properties were ascribed to goose and peafowl meat, as both were considered to be stodgy,[161] tough,[162] sinewy,[163] and, thus, not fully digestible.[164]

The above characteristic suggests that their preparation was more time-consuming, since both required a longer thermal processing. Moreover, in order to facilitate the digestion of geese and peafowl, one should use ingredients stimulating the process. These were, mostly, expensive spices, which many could not afford. Accordingly, geese would have been eaten mainly in well-off circles, which is probably why they were kept in Constantinople itself, as is testified to in Prokopios of Caesarea's story about Theodora's alleged public indecent behaviour.[165] No wonder then that geese were also taken on military campaigns to be prepared for the imperial table and feasts the ruler treated his army to.[166] One should add that geese were also kept for eggs. However, due to a short period of egg laying, and thus limited egg yield,[167] they were significantly less productive than hens. What is more, not all farms could provide breeding conditions which were appropriate for the species, such as access to water.[168]

Another species of fowl, namely ducks, were regarded as game[169] for most of the Byzantine period.[170] And although their domestication had already begun in antiquity,[171] on the basis of *Geoponika* we might conclude that, at the time when the treatise was published, the process was still in progress.[172] Symeon Seth's mention of the meat of domesticated ducks[173] allows us to assume that there was a gradual increase in its percentage share in the diet of the Byzantines at his time. Duck meat was also mentioned in the poem *On Medicine* by Michael Psellos, where it was characterised as tough, hard to digest and sinewy, which makes us believe that the author classified ducks as belonging to wildfowl.[174] The fact that the majority of ducks still lived in the wild and that even those domesticated required access to bodies of water (which could not be provided by every farm),[175] suggests that duck meat was not eaten commonly.

The above conclusion can be reinforced by the fact that physicians emphasise the toughness of duck meat in their treatises,[176] which indicates that they were discussing meat obtained from wild birds, i.e. hunted ones. The meat of wild ducks was also mentioned by Anthimus (who by recommending only the consumption of the tender breast meat from ducks suggests that it must have come from a young creature). This conclusion is based on the fact that the physician approved of its consumption only occasionally.[177] One should add that Anthimus never mentions the use of duck eggs in his culinary advice, which implies that they were not a significant element of the Byzantine diet.[178]

As far as dishes from chicken are concerned, Byzantine literature provides us with a number of examples. For instance, Eustathios of Thessalonike, the learned twelfth-century Archbishop, writes that once he was treated to a stuffed chicken. The bird was deboned (with the exception of its wings and legs) and floated on the surface of a wine-based, nectar-like sauce.[179] Another time, Eustathios feasted on a bird stuffed with almonds (and a host of other ingredients) and then baked in pastry. The pastry itself constituted the pan in which the bird was seated and a dome-shaped pastry cover secured the meat against excessive temperature. Sealed inside, the bird was cooked perfectly, remaining juicy and acquiring an exquisite aroma. The chicken was served on a platter together with the pastry pan and cover.[180]

Geese were also a valuable source of food, and the most desired products were obtained from birds which were fattened.[181] For instance, Oribasios mentions the gizzards[182] and livers[183] of fattened geese, with the latter also being noted by Aetios of Amida.[184] Despite neither of the medical authors mentioning how the said offal was to be prepared, some clues on the subject can be found in *On the Observance of Foods*. According to the data, we may assume that the best method to prepare gizzards was boiling,[185] while livers presumably should be prepared in the same way as *ficatum*.[186] The fact that the physicians considered such offal nutritious, light and tasty[187] suggests that they were consumed even by affluent Byzantine Epicureans. One should however note that the practice of fattening geese would also have been considered as having a negative impact on the quality of the meat itself. Otherwise Anthimus would not have warned against consuming meat from the rump of geese, which, as we may infer from his narrative, was excessively fatty and thus difficult to digest,[188] recommending their breasts instead.[189] Whereas we understand from Oribasios and Aetios of Amida that goose wings were considered as having beneficial nutritional properties.[190]

The data on fattening pheasants provided by the author of *Geoponika* indicate that these birds were classed among domestic poultry in the tenth century.[191] We can infer from the analysed medical treatises, however, that the keeping of these birds had not been a long-standing tradition in the Mediterranean region. Hence, we can deduce that the importance of their meat must have increased gradually, alongside the growing popularity of the species. In order to explore this issue, let us analyse passages on the application of pheasant fat in Byzantine therapeutics. First of all, it should be emphasised that the substance is absent from the medicaments recommended by Oribasios, although it was mentioned by Philagrios of Epirus (ca. 300–340),[192] who was active in the first part of the same century as Oribasios. Thus, we may assume that the tradition of using pheasant fat in medicine in the fourth century was still not fully established, which may indicate that the product was not widely accessible. Nevertheless, since pheasant fat is recommended on several occasions by Aetios of Amida,[193] we may suppose that, over time, it must have gained in popularity. The existence of an available resource of pheasant fat is confirmed by his contemporary – Anthimus – advising the reader not to consume the rump of pheasants due to its fattiness.[194] Therefore, it seems that breeding pheasants in the sixth century was already so widespread that, firstly, physicians had little trouble obtaining pheasant fat for medical purposes and, secondly, the meat of fattened birds was a more frequent sight on Byzantine tables. The latter inference is supported by Alexander of Tralles' (ca. 550–605) writings, who makes us think that pheasant meat was so widely available in the second half of the sixth century that, when writing on healthy diets including the discussed food, the physician had to emphasise that it was the lean meat of the birds that should be consumed.[195] As for scanty culinary information on pheasants, Aetios of Amida preserved a recipe for a therapeutic pheasant (or partridge) broth. The soup was prepared from water, olive oil, some *garum*, a pinch of salt, and dill, as well as pulverised polypody. The dish (due to the latter) was supposed to help in removing excessive amounts of black bile from the body.[196]

Byzantine physicians convey notably less information on peafowl meat, and they only emphasise its extraordinary toughness. Possibly, such brevity was, on the one hand, caused by the unfavourable dietary properties of the meat itself, and, on the other, by its rarity value. The latter is, for instance, confirmed by an extract from *Geoponika*, stating that peafowl lay eggs as rarely as twice a year,[197] which implies that both the eggs as well as peafowl meat were not widely available.[198] Another clue to assessing the importance of the discussed product in the Byzantine diet is the fact that the said treatise lacks any data on fattening the birds,[199] which allows us to presume that their meat was consumed rather rarely. Therefore, we may conclude that peafowl were bred primarily for their aesthetic qualities, i.e., colourful plumage, and to showcase the owner's high economic standing.[200] Be that as it may, Anthimus did include a peafowl dish recipe in his work.[201] It was prepared from tenderised meat,[202] which was stewed and then served in its own gravy, with the addition of honey and pepper.[203]

As we have already mentioned, birds provided people not only with meat but primarily with eggs. The latter, especially those obtained from hens[204] and geese,[205] would have been a supplement to the diet of both the rich and the poor.[206] The affluent, however, would also consume coveted pheasant eggs. As for peafowl eggs, a complete lack of any remarks on their dietary properties allows us to surmise that they were an extreme culinary rarity in the early Middle Ages. Nevertheless, their consumption in the Mediterranean is testified to in *Learned Banqueters* by Athenaeus of Naucratis (second/third century). From the work we learn that ancient authors of culinary books, Epaenetos (first century BCE) and Herakleides of Syracuse (fourth century BCE), considered peafowl eggs to be tastier than the eggs of both Egyptian geese and hens.[207] And although we are unable to pinpoint how popular they were in that period, since peafowls were regarded as luxurious birds, we may assume that their eggs were consumed mostly by the wealthiest.

As medical treatises recommend that eggs be eaten fresh, it is of little surprise that the most sought after were those which had been freshly laid.[208] The analysis of the source material proves that Byzantine physicians were perfectly aware that acquiring high-quality eggs did not guarantee that they would become a beneficial foodstuff. Hence, in their works, they devoted a lot of attention to the question of the thermal processing of eggs, giving us an insight into the various culinary techniques applied during the preparation of the product.[209] Oribasios, Aetios of Amida and Paul of Aegina recommended the consumption of completely runny eggs (τρομητά), which, in all likelihood, were just warmed up in water. All authors classified soft-boiled eggs (ῥοφετά) as less nutritious (but still a valuable element of diet),[210] whereas Oribasios and Aetios of Amida list both hard-boiled eggs (ἐφθά), and eggs baked in hot ashes (ὀπτά) as not being a beneficial food.[211] It does, however, seem that they were considered less harmful than eggs fried in the pan (ταγηνιστά), which were thought to be the worst type of foodstuff.[212] Even such eggs, but with the addition of special ingredients, were recommended by the anonymous author of *Selection of Medicines* to patients suffering from intestinal problems.[213]

Last but not least, Symeon Seth mentions ἐξεφθά eggs in his treatise, which are believed to be the equivalent of today's poached eggs.[214] The Byzantines were also aware of an egg-dish called ᾠὰ πνικτά/πηκτά. When writing on its preparation, medical authors explain that after cracking them, the eggs should be thickened in a double-bottom pot known as δίπλωμα. Most likely, the delicacy was the equivalent of today's coddled eggs. And since the recipe for the dish was preserved by Oribasios, Aetios of Amida and Paul of Aegina, we may assume that it was popular among the inhabitants of the Empire. In the analysed treatises, we read that the eggs, seasoned with olive oil, fish sauce and some wine, were heated until they gained an appropriate thickness. One could also crack eggs onto the surface of a casserole dish, in a way that is still popular today. Seemingly, the key was to choose the right moment to move the pot away from the fireplace (before the eggs coagulated) so that the food would not lose its unique dietetic benefits by gaining the properties of typical hard-boiled or baked eggs instead.[215]

More data on using eggs in Byzantine *ars coquinaria* is given by Anthimus. For instance, the physician explains how to prepare *ova sorbilia*,[216] i.e., most probably eggs that were consumed completely runny. Namely, we learn that the eggs were put into tepid (or even better into cold) water, then heated on a low fire, while being stirred constantly,[217] so that they could gradually and evenly reach the appropriate temperature. Such eggs should be served with a pinch of salt. Anthimus also reveals a way of preparing the dish called *afratus*,[218] which was based on whipped egg whites mixed with chicken meat,[219] fish,[220] or scallops.[221] From the data contained in the recipe, we learn that the blend of egg foam and meat was shaped into a dome then put into a shallow vessel containing a mixture of meat stock and diluted fish sauce (*egrogarium*). Next, the dish was put over smouldering charcoals until, due to evaporating liquids, the foam congealed. When ready, it was sprinkled with wine and some honey.

Conclusions

To sum up, our deliberations prove that although meat and animal products never became a staple of the Byzantine diet, they were still an important supplement. The conclusion can be drawn from both literary (including medical) sources as well as archaeological evidence. Other research methods do not disprove the statement either.[222] Byzantine literary heritage, however, gives us a lot of valuable detail which allows to paint a more vivid picture of everyday and festive diet, and better understand results of other research methods.

As for detail, the available evidence demonstrates that, while the most commonly eaten by the Byzantines were supposedly cheese and eggs, meat only appeared regularly on the tables of the wealthy inhabitants of the Empire. And it was only the wealthy who could afford to buy its prime cuts. Less valuable ones and most offal were more commonly consumed by less wealthy Byzantines. Any excess meat remaining after slaughter was preserved with salt. In all probability, this situation with regard to the rich and poor also applied to the most valuable cured products, e.g., hams, which were more available for the financially privileged.[223]

It was commonly known that the quality of meat obtained from goat, sheep and cow depended on the season. Therefore, we can assume that the menu of rich Byzantines included kid meat and lamb from spring till late summer, while the wealthiest connoisseurs would even indulge their taste for young beef. Moreover, medical texts suggest that the meat from old animals was accessible throughout the year. Its availability on the market was not determined by seasonality, as it became available when the animals were no longer profitable to the farm. It is likely that such meat was less expensive, and, thus, more accessible to the less affluent inhabitants of the Empire. Pork was also an all-year-round foodstuff, as the slaughtering of pigs did not bring any major losses for the farm economy. It was considered to be a high-quality foodstuff, since any surplus could be effectively preserved with no significant deterioration in the meat's properties. Hence, it comes as no surprise that pork was the favourite type of meat in Byzantium.

As for poultry, the most commonly consumed was hen meat. However, one must bear in mind that since the birds were primarily kept for eggs in Byzantium, they were killed rather sporadically, which means that their meat was consumed less frequently than goat, mutton, beef and pork. The meat of geese and ducks was eaten even less often than hen meat, which was prompted by the fact that these birds were not so commonly bred as hens were. Moreover, the preparation of the two required longer thermal processing as well as the use of expensive ingredients, which may imply that they were mainly consumed by wealthy inhabitants of the Empire, who could also afford to include such delicacies as pheasant or peafowl meat on their menus.

Owing to their popularity and relatively high egg yield, hen eggs were consumed far more frequently than the eggs of other birds. We may suppose that the poorer ate them as a stand-alone dish, while eggs may also have been one of the ingredients of more sophisticated dishes (such as *afratus*) for the wealthy. Texts also confirm the consumption of goose and pheasant eggs, however it seems that they were eaten relatively rarely. The lack of information on the dietetic properties of duck and peafowl eggs allows us to assume that their presence in the Byzantine diet was very limited.

The analysed treatises prove that the most common milk animals were sheep and goats. However, there is no indication that milk was an everyday drink for the Byzantines.[224] Instead, it was mainly used to produce cheese, which was available even to impecunious peasants, whose consumption of cheese exceeded that of meat. Those who could afford meat were mostly the rich residents of cities, hence, we can conclude that the share of cheese in their diet was lower than in the case of poorer people living far away from the urban areas. Peasants would primarily eat fresh cottage cheese, while cheese which was matured or preserved in seawater or brine[225] (i.e. which could be easily transported over long distances with no risk of rapid deterioration in its quality[226]) was mainly eaten by city dwellers.

Finally, let us add that the amount of meat and animal products consumed by the inhabitants of the Empire throughout the year was greatly impacted by the fast days imposed by the Church.[227] In total, almost half of the calendar year was filled with days when Byzantines abstained from consuming meat and animal products on religious grounds.[228] Even more strict were the rules which monks had to obey.[229] In the light of our analyses, it seems that even though the rules of fasting were identical for the inhabitants of rural and urban areas, they impacted the menu of peasants and city dwellers in a different way. In all probability, the ban on eating meat mainly influenced the daily diet of the latter, while it was much less severe for the former as they simply could not afford regular consumption of meat.

Notes

1 On fish and seafood in Byzantium, see for instance D. Mylona, *Fish-Eating in Greece from the Fifth Century b. c. to the Seventh Century a.d.: A Story of Impoverished Fishermen or Luxurious Fish Banquets?* (Oxford, 2008); M. Chrone-Vakalopoulos and A. Vakalopoulos, "Fishes and Other Aquatic Species in Byzantine Literature: Classification, Terminology and Scientific Names," *Byzantina Symmeikta* 18 (2009): 123–57; H. Kroll, *Tiere im Byzantinischen Reich: archäozoologische Forschungen im Überblick* (Mainz, 2010), 200–26; E. Ragia, "The Circulation, Distribution and Consumption of Marine Products in Byzantium: Some Considerations," *Journal of Maritime Archaeology. Special Issue: The Bountiful Sea: Fish Processing and Consumption in Mediterranean Antiquity. Proceedings of the International Conference Held at Oxford, 6-8 September 2017* 13.3 (2018): 449–66; H. Baron and N. Marković, "Fish Consumption and Trade in Early Byzantine Caričin Grad (Justiniana Prima)," in *Animal Husbandry and Hunting in the Central and Western Balkans Through Time*, ed. N. Marković and J. Bulatović (Oxford, 2020), 154–66; V. Onar, "Animals in Food Consumption During the Byzantine Period in Light of the Yenikapı Metro and Marmaray Excavations, Istanbul," in *Multidisciplinary Approaches to Food and Foodways in the Medieval Eastern Mediterranean*, ed. S. Y. Waksman (Lyon, 2020), 340.

2 On honey in Byzantium, see, for instance, P. Androudis, "Το μέλι και το κερί στη μοναστηριακή ζωή των βυζαντινών χρόνων," in *Η μέλισσα και τα προϊόντα της: Πρακτικά του Έκτου Συμποσίου της ΕΤΒΑ με θέμα Η ΜΕΛΙΣΣΑ ΚΑΙ ΤΟ ΚΕΡΙ (Νικήτη Χαλκιδικής, 12–15 Σεπτεμβρίου 1996)* (Nikiti, 1998/ Athens, 2000), 211–20; I. Anagnostakis, "Byzantine Delicacies," in *Flavours and Delights: Tastes and Pleasures of Ancient and Byzantine Cuisine*, ed. I. Anagnostakis (Athens, 2013), 87– 92; S. Germanidou, *Βυζαντινός μελίρρυτος πολιτισμός: Πηγές, τέχνη, ευρήματα* (Athens, 2016); I. Anagnostakis, "Wild and Domestic Honey in Middle Byzantine Hagiography: Some Issues Relating to its Production, Collection and Consumption," in *Beekeeping in the Mediterranean from Antiquity to the Present*, ed. F. Hatjina, G. Mavrofridis, and R. Jones (Nea Moudania, 2017), 105–18; S. Germanidou, "Honey Culture in Byzantium: An Outline of Textual, Iconographic and Archaeological Evidence," in *Beekeeping in the Mediterranean from Antiquity to the Present*, ed. F. Hatjina, G. Mavrofridis, and R. Jones (Nea Moudania, 2017), 93–104; eadem, "Μελισσοκομεία, Νερόμυλοι, Περιστεριώνες," in *Όψεις του καθημερινού βίου στο Βυζάντιο*, ed. P. Androudis (Thessaloniki, 2020), 499–532.

3 Apicius, *On the Subject of Cooking*, ed., trans., and intr. Ch. Grocock and S. Grainger, *Apicius: A Critical Edition with an Introduction and an English Translation of the Latin Recipe Text Apicius* (Blackawton, 2006).

4 Anthimus, *On the Observance of Foods*, ed. and trans. E. Liechtenhan, *Anthimi De observatione ciborum ad Theodoricum regem Francorum epistula* (Berlin, 1963), CML 8.1.

5 In his works, Galen defines τροφή on a number of occasions. For instance, in his treatise *On the Capacities of Foodstuffs*, he accounts that food is a substance received from the outside which does not change the constitutive features of the body, but is transformed into tissue, replacing that which is naturally lost to the human body (and thus contributes to its preservation in an unchanged state), see Galen, *On the Capacities of Foodstuffs*, 1.1.26, ed. G. Helmreich, *Galeni De alimentorum facultatibus libri III* (Leipzig, 1923), CMG 5.4.2:210.2–14 = Kühn (*Claudii Galeni Opera Omnia*, vol. 1-20, Leipzig 1821–1833), 6:468–69.

6 A brief definition of the nutrition process, see Galen, *On the Natural Capacities* 1.8, ed. G. Helmreich, "Galeni De facultatibus naturalibus," in *Claudii Galeni Pergameni Scripta minora*, ed. J. Marquardt, I. von Müller, and G. Helmreich, 3 vols. (Leipzig, 1884–93), 3:114.6–17 = Kühn 2:18–19. It is complex and consists of three major stages, i.e., application of the nutrient (πρόσθεσις), its adhesion (πρόσφυσις), and assimilation (ὁμοίωσις), see Galen, *On the Natural Capacities*, ed. Helmreich, 1.11.118.3–119.24 = Kühn, 2:24–26; 3.1.204.8–206.3 = Kühn, 2:143–45.

7 Galen provided a number of definitions of the term φάρμακον. The one in the treatise *On the Capacities of Foodstuffs* states that it is a substance acting on the human body, which, as a result

of coming into contact with the body, begins to change its normal characteristics by increasing or decreasing its temperature, or by desiccation or increasing its moisture, see Galen, *On the Capacities of Foodstuffs*, 1.1.25, ed. Helmreich, CMG 5.4.2:209.23–25 = Kühn, 6:468.

8 Galen, *On the Capacities of Foodstuffs*, 1.1.26, ed. Helmreich, CMG 5.4.2:210.1–2 = Kühn, 6:468. The distinction between the nutritional and pharmacological functions of food is discussed, for instance, by John Wilkins, "Good Food and Bad: Nutritional and Pleasurable Eating in Ancient Greece," *Journal of Ethnopharmacology* 167 (2015): 7–10.

9 For Byzantine doctors, Galen's output was definitely a medical classic, see V. Nutton, "From Galen to Alexander: Aspects of Medicine and Medical Practice in Late Antiquity," *DOP* 38 (1984): 2; idem, "Galen in Byzantium," in *Material Culture and Well-Being in Byzantium (400–1453): Proceedings of the International Conference (Cambridge, 8–10 September 2001)*, ed. M. Grünbart, E. Kisslinger, A. Muthesius et al. (Vienna, 2007), 171–76; P. Bouras-Vallianatos, "Galen in Late Antique Medical Handbooks," in *Brill's Companion to the Reception of Galen*, ed. P. Bouras-Vallianatos and B. Zipser (Leiden, 2019), 38–61; idem, "Galen in Byzantine Medical Literature," in *Brill's Companion*, 86–110; P. Degni, "Textual Transmission of Galen in Byzantium," in *Brill's Companion*, ed. P. Bouras-Vallianatos and B. Zipser, 124–39.

10 The treatise was dedicated to Emperor Constantine IV Pogonatos (r. 668–685). See F. Z. Emerins, "L. S.," in *Anecdota medica Graeca* (Leiden, 1963), XI–XII; A. Dalby, *Tastes of Byzantium: The Cuisine of a Legendary Empire* (London, 2010), 18; M. Kokoszko, K. Jagusiak, and Z. Rzeźnicka, *Cereals of Antiquity and Early Byzantine Times: Wheat and Barley in Medical Sources (Second to Seventh Centuries AD)*, trans. K. Wodarczyk, M. Zakrzewski, and M. Zytka (Łódź, 2014), 24.

11 A good example of the process is Symeon Seth's *Treatise on the Capacities of Foodstuffs* (*Simeonis Sethi Syntagma de alimentorum facultatibus*, ed. B. Langkavel (Leipzig, 1868)), expanded by the author with descriptions of new substances, such as jujube (ζ:40.9–18), hashish (κ:60.22–61.7), ambergris (α:26.1–14), and the preparation called julep (ζ:41.5–13). See Bouras-Vallianatos, "Galen's Reception," 438–39.

12 On this subject, see Ph. Van der Eijk, M. Geller, L. Lehmhaus et al., "Canons, Authorities and Medical Practice in the Greek Medical Encyclopaedias of Late Antiquity and in the Talmud," in *Wissen in Bewegung: Institution – Iteration – Transfer*, ed. E. Cancik-Kirschbaum and A. Traninger (Wiesbaden, 2015), 195–221; E. Gowling, "Aëtius' Extraction of Galenic Essence: A Comparison Between Book 1 of Aetius' Libri Medicinales and Galen's On Simple Medicines," in *Collecting Recipes: Byzantine and Jewish Pharmacology in Dialogue*, ed. L. Lehmhaus and M. Martelli (Berlin, 2017), 83–101; A. Touwaide, "Medicine and Pharmacy," in *A Companion to Byzantine Science*, ed. S. Lazaris (Leiden, 2020), 364–67.

13 Oribasios, *Medical Collections* 2.28.1–16, ed. J. Raeder, *Oribasii Collectionum medicarum reliquiae: Libri I–VIII; IX–XVI; XXIV–XXV; XLIII–XLVIII; XLIX–L, Libri incerti, Eclogae medicamentorum*, 4 vols. (Leipzig, 1928–33), CMG 6.1.1:36.5–37.14; Aetios of Amida, *Tetrabiblon* 2.121, ed. A. Olivieri, *Aetii Amideni Libri medicinales I–IV; V–VIII*, 2 vols. (Leipzig, 1935–50), CMG 8.1:196.30–198.8. Cf. Galen, *On the Capacities of Foodstuffs*, 3.1.1–19, ed. Helmreich, CMG 5.4.2:332.8–337.10 = Kühn, 6:660–68.

14 Another argument for this supposition is the fact that the above list contains animals which were said by the Byzantines to be unclean, as their consumption could lead to undesired behaviour (especially as far as one's sexual life is concerned) or were associated with the dietary habits of the barbarians. Therefore, eating hares, horses, bears, dogs, etc. was either avoided or banned. See B. Caseau, "Dogs, Vultures, Horses and Black Pudding: Unclean Meats in the Eyes of the Byzantines," in *Multidisciplinary Approaches*, ed. S. Y. Waksman, 231–32, 234–35. It does seem, however, that the restrictions were not strictly followed, and the Byzantines were happy to consume hares (Onar, "Animals," 335), horses on top of other equids (Kroll, *Tiere*, 169–72; V. Onar, H. Alpak, G. Pazvant et al., "A Bridge from Byzantium to Modern Day Istanbul: An Overview of Animal Skeleton Remains Found During Metro and Marmaray Excavations," *Journal of the Faculty of Veterinary Medicine of Istanbul University*, 39.1 (2013): 4, 6–7; V. Onar, G. Pazvan, H. Alpak et al., "Animal Skeletal Remains of the Theodosius Harbor: General Overview," *Turkish Journal of Veterinary and Animal Sciences*, 37 (2013): 82, 83 (figs. 1 and 2); Onar, "Animals," 333, 335) and other species galore. They did it, however, sporadically, and definitely not on a regular basis.

15 Oribasios, *Medical Collections* 1.3.1–7, ed. Raeder, CMG 6.1.1:8.15–9.7.

16 Paul of Aegina, *Epitome of Medicine* 1.84, ed. J. L. Heiberg, *Paulus Aegineta: Libri I–IV; V–VII*, 2 vols. (Leipzig, 1921–24), CMG 9.1:60.26–61.9.

17 General research confirms that the most common farm quadrupeds were sheep, goats, pigs and cattle, for instance, see Ch. Bourbou, "Fasting or Feasting? Consumption of Meat, Dairy Products and Fish in Byzantine Greece: Evidence from Chemical Analysis," in *Ζώα και περιβάλλον στο Βυζάντιο (7ος–12ος αι.) / Animals and Environment in Byzantium (7th–12th c.)*, ed. I. Anagnostakis, T. G. Kolias, and E. Papadopoulou (Athens, 2011), 99. Purely archaeological research, for instance, by Henriette Kroll (*Tiere*, 157–68, esp. 157; eadem, "Animals in the Byzantine Empire: An Overview of the Archaeozoological Evidence," *Archeologia Medievale: Cultura materiale. Indediamenti. Territorio* 39 (2012): 97; H. Baron, "An Approach to Byzantine Environmental History: Human-Animal Interactions," in *A Most Pleasant Scene and an Inexhaustible Resource: Steps Towards a Byzantine Environmental History*, ed. H. Baron and F. Daim (Mainz, 2017), 174; see also her chapter in the present volume) and by Onar ("Animals," 333) suggests that goat and sheep meat was consumed by the Byzantines in larger quantities than that of pigs and cattle. Out of the first two it was the meat of sheep that appears to have been eaten in larger quantities than that of goats, for instance, see D. Makowiecki and Z. Schramm, "Preliminary Results of Studies on Archaeozoological Material from Excavations in Novae (Season 1992)," in *NOVAE: Studies and Materials I*, ed. A. B. Biernacki (Poznań, 1995), 80; Kroll, *Tiere*, 157; G. Pazvant, V. Onar, H. Alpak et al., "Osteometric Examination of Metapodial Bones in Sheep (*Ovis aries* L.) and Goat (*Capra hircus* L.) Unearthed from the Yenikapı Metro and Marmaray Excavations in İstanbul," *Kafkas Üniversitesi Veteriner Fakültesi Dergisi* 21.2 (2015): 147–53, esp. 153; Onar, "Animals," 334 (fig. d). Regarding the hunt, see also the chapter by I. Nilsson and Ch. Messis in the present volume.

18 Refuting Oliver Schmitt's conjectures (O. Schmitt, "Zur Fleischversorgung Konstantinopels," *JÖB* 54 (2004): 135–57).

19 J. Koder, "Über die Liebe de Byzantiner für Rindfleisch," *BZ* 102.1 (2009): 103–108, esp. 108; idem, "Παρατηρήσεις για τη χρήση βοοειδών στο Βυζάντιο," in *Ζώα και περιβάλλον στο Βυζάντιο*, ed. Anagnostakis, Kolias, and Papadopoulou, 23–38, esp. 26–28, 36–38. Also cf. idem, "Natural Environment and Climate, Diet, Food, and Drink," in *Heaven & Earth: Art of Byzantium from Greek Collections*, ed. A. Drandaki, D. Papanikola-Bakirtzi, and A. Tourta (Athens, 2013), 215, etc. The view is shared by other scholars, for instance, cf. T. Kolias, "Die Versorgung des byzantinischen Marktes mit Tieren und Tierprodukten," in *Handelsgüter und Verkehrswege. Aspekte der Warenversorgung im östlichen Mittelmeerraum (4. bis 15. Jahrhundert). Akten des Internationalen Symposions Wien, 19.–22. Oktober 2005*, ed. E. Kislinger, J. Koder, and A. Külzer (Vienna, 2010), 183; Caseau, "Dogs," 233.

20 J. Koder, "Stew and Salted Meat," in *Eat, Drink and Be Merry (Luke 12:19): Food and Wine in Byzantium*, ed. L. Brubaker and K. Linardou (Aldershot, 2007), 64–65; idem, "Über die Liebe," 103–104.

21 Constantine Porphyrogenitus, *Three Treatises on Imperial Military Expeditions: Introduction, Translation, and Commentary, Text C*, 147, ed. J. F. Haldon (Vienna, 1990), 102. Accordingly, they are present in the context of sophisticated diet of the imperial entourage (which also implies their priciness). The argument, however, is weakened due to the fact that another passage of the same treatise suggests that cows were not taken with the troops to supply the Byzantine soldiers but as provisions for their foreign allies, see Constantine Porphyrogenitus, *Three Treatises, Text C*, 596, ed. Haldon, 132. Cf. Koder, "Über die Liebe," 104–105.

22 For instance, see Kroll, *Tiere*, 161–65, esp. 162; Baron, "An Approach," 174, etc.

23 Onar, "Animals," 331–42, esp. 333–34.

24 Cf. V. Onar, K. O. Kahvecioğlu, D. Kostov et al., "Osteological Evidences of Byzantine Draught Cattle from Theodosius Harbour at Yenikapı, Istanbul," *Mediterranean Archaeology and Archaeometry* 15 (2015): 71–80.

25 Oribasios, *Medical Collections* 2.42.1–5, ed. Raeder, CMG 6.1.1:40.20–41.5; Aetios of Amida, *Tetrabiblon* 2.130–132, ed. Olivieri, CMG 8.1:200.6–202.2; Paul of Aegina, *Epitome of Medicine* 1.82, ed. Heiberg, CMG 9.1:60.1–15. Cf. Galen, *On the Capacities of Foodstuffs* 3.18.1–19.2, ed. Helmreich, CMG 5.4.2:356.1–358.7 = Kühn, 6:700–703. Within the mentioned list, cranes were considered unclean because they sometimes fed on carrion. Nevertheless, they were eagerly hunted by the Byzantine aristocracy, see n. 38. Also see B. Caseau, "Quelques réflexions sur les interdits

alimentaires dans le christianisme byzantin," in *Religions et interdits alimentaires*, ed. B. Caseau and H. Monchot (Leuven, 2022), 102.

26 The practice is confirmed in *Geoponika* for partridges (*Geoponika* 14.19.1–4, ed. H. Beckh (Stuttgart, 1994), 423.13–424.10), ducks (*Geoponika* 14.23.1–5, 428.5–17), thrushes (*Geoponika* 14.24.5–7, 429.16–430.5), etc.

27 Cf. *Geoponika* 14.1.1–7, ed. Beckh, 405.7–406.10; 14.2.1–5, 406.11–407.9; 14.3.1–3, 407.10–408.2; Symeon Seth, *Treatise on the Capacities of Foodstuffs* π:86.11–13, ed. Langkavel.

28 On breeding pheasants, see further parts of this paper.

29 On breeding peafowl, see below.

30 Ever since antiquity, both species had been considered exotic birds, though regularly bred and kept by eastern rulers in private menageries, see M. Leontsini, "Οικόσιτα, ωδικά και εξωτικά πτηνά: Αισθητική πρόσληψη και χρηστικές όψεις (7ος–11ος αι.)," in *Ζώα και περιβάλλον στο Βυζάντιο*, ed. Anagnostakis, Kolias, and Papadopoulou, 293. For instance, they were kept in the palace complex of Khosrow II, see Theophanes the Confessor, *Chronographia* 1:322.9–14, ed. C. de Boor, *Theophanis Chronographia*, 2 vols. (Leipzig, 1883–85). We can suppose that they were also kept in Constantinople (cf. M. Leontsini, "Hens, Cockerels and other Choice Fowl: Everyday Food and Gastronomic Pretensions in Byzantium," in *Flavours and Delights*, ed. I. Anagnostakis, 115).

31 For instance, see Kroll, *Tiere*, 97, 135–36, 144–45, 191; eadem, "Animals," 105.

32 Oribasios, *Medical Collections* 1.3.4–5, ed. Raeder, CMG 6.1.1:8.34–9.3.

33 Oribasios, *Medical Collections* 1.3.5, ed. Raeder, CMG 6.1.1:9.1–2.

34 Also archaeological research proves that in the Byzantine period chickens were the most popular domestic fowl, for instance, see N. Benecke, "Archäozoologische Untersuchungen an Tierresten aus dem Kastell Iatrus," in *Iatrus-Krivina: Spätantike Befestigung und frühmittelalterliche Siedlung an der unteren Donau. Band 6: Ergebnisse der Ausgrabungen 1992–2000*, ed. G. von Bülow, B. Böttger, S. Conradet et al. (Frankfurt, 2007), 397; Z. Boev and M. J. Beech, "The Bird Bones," in *Nicopolis ad Istrum: A Late Roman and Early Byzantine City. The Finds and the Biological Remains*, ed. A. G. Poulter (London, 2007), 248–49; Kroll, *Tiere*, 177–79; eadem, "Animals," 102; Onar, "Animals," 340. The exact extent of breeding chickens in Byzantium is impossible to determine, for instance, see Baron, "An Approach," 176–77. For the time being, given imperfect excavation methods, it is safe to conclude that hens, similarly to other domestic fowl, played a secondary role in the Byzantine diet.

35 Oribasios, *Medical Collections* 2.45.1, ed. Raeder, CMG 6.1.1:42.2–3; Aetios of Amida, *Tetrabiblon* 2.134, ed. Olivieri, CMG 8.1:201.19–20; Paul of Aegina, *Epitome of Medicine* 1.83, ed. Heiberg, CMG 9.1:60.17–18. Cf. Anthimus, *On the Observance of Foods* 38, ed. Liechtenhan, CML 8.1:18.3. Analogous data, cf. Galen, *On the Capacities of Foodstuffs* 3.21.1, ed. Helmreich, CMG 5.4.2:359.19–21 = Kühn, 6:706.

36 The second position of geese among domestic fowl in most Mediterranean and neighbouring territories appears to have been confirmed by archaeological material, for instance, see Benecke, "Archäozoologische Untersuchungen," 397; Boev and Beech, "The Bird Bones," 248; Onar, "Animals," 340. Relative popularity of geese is also suggested by Kroll, *Tiere*, 179–80; eadem, "Animals," 102.

37 Constantine Porphyrogenitus, *Three Treatises, Text C*, 535, ed. Haldon, 128.

38 We may presume that game constituted a relatively high percentage of meat consumed by the rich since hunting was a popular aristocratic pastime. According to Michael Psellos, emperor Constantine VIII (r. 1025–1028) was renowned for taking delight in the activity, and the same text allows us to surmise that, in his youth, his brother Basil II (r. 976–1025) did not refrain from hunting either, see Psellos, *Chronography* (Basil II) 1.22.16–21 (*Chronographie ou histoire d'un siècle de Byzance (976–1077)*, ed., trans. É. Renauld, 2 vols. (Paris, 1926–28), 1:14; (Constantine VIII) 2.8.10–12, 1:30). Isaac Komnenos (r. 1057–1059) shared Constantine's and Basil's love for hunting, and he would kill hares and cranes galore, see Psellos, *Chronography* (Isaac I) 7.72.1–15, 2:128. On hare preparation, see M. Kokoszko, "Anthimus the Dietician," *Almanach Historyczny* 23 (2021): 11–37; Z. Rzeźnicka, "Hare in Sauce According to Anthimus' Recipe: Meat," *Studia Ceranea*, 12 (2022): 758–77. In the fourteenth century cranes were considered to be a luxury foodstuff; their legs were eaten cooked in wine, see *Pulologus*, 83–85 (Ὁ Πουλολόγος: κριτική έκδοση με εισαγωγή, σχόλια και λεξιλόγιο), ed., intr., comm., I. Tsavari (Athens, 1987), 254; Leontsini, "Hens," 129; Psellos classifies them as hard to digest but wholesome, see Psellos, *Poem* 9.200–201

(*Michaelis Pselli Poemata*, ed. L. G. Westerink (Stuttgart, 1992), 197). Michael VII Doukas (r. 1071–1078) was also keen on hunting hares, deer and bears, see Psellos, *Chronography*, (Michael VII) 7.6.3–16, 2:175–76; 7.17.1–8, 2:181–82. Although Psellos does not mention deer in his medical poem, the animals are described in the work by his contemporary Symeon Seth, who writes in detail about methods of hunting deer, and classifies the meat as having bad juices, being hard to digest, and melancholic, see Symeon Seth, *Treatise on the Capacities of Foodstuffs* ε:35.22 – 36.10, ed. Langkavel. On the other hand, the poor must have come into possession of such food when they chanced upon it or when they were forced to resort to it for lack of other edibles, cf. Anthimus, *On the Observance of Foods* 25, ed. Liechtenhan, CML 8.1:14.3–6. On hunting by the rich and the poor, see A. K. Sinakos, "Το Κυνηγι κατα τη Μεση Βυζαντινη Εποχη (7ος–12ος αι.)," in *Ζώα και περιβάλλον στο Βυζάντιο*, ed. Anagnostakis, Kolias, and Papadopoulou, 71–86. Latest archaeological finds, cf. Onar, "Animals," 335. See also the chapter by I. Nilsson and Ch. Messis in the present volume.

39 On the differences between the meat of domesticated and wild animals, see Oribasios, *Medical Collections* 2.41.1–2, ed. Raeder, CMG 6.1.1:40.13–19; Aetios of Amida, *Tetrabiblon* 2.252, ed. Olivieri, CMG 8.1:244.15–16; Paul of Aegina, *Epitome of Medicine* 1.85.1, ed. Heiberg, CMG 9.1:62.2–3. Cf. Galen, *On the Capacities of Foodstuffs* 3.13.1–2, ed. Helmreich, CMG 5.4.2:344.24–345.7 = Kühn, 6:680–81.

40 For instance, see Kroll, *Tiere*, 163; M. Kaplan, "L'activité pastorale dans le village byzantin du VIIe au XIIe siècle," in *Ζώα και περιβάλλον στο Βυζάντιο*, ed. Anagnostakis, Kolias, and Papadopoulou, 411–12; Koder, "Παρατηρήσεις," 30–36; Kroll, "Animals," 98; J. Koder, "Everyday Food in the Middle Byzantine Period," in *Flavours and Delights*, ed. I. Anagnostakis, 146; Onar, Kahvecioğlu, Kostov et al., "Osteological Evidences," 71–80; A. G. Yangaki, "On Clay Milking Vessels in Byzantium: Evidence and Preliminary Observations," in *Latte e Latticini: Aspetti della produzione e del consumonelle società mediterranee dell'Antichità e del Medioevo. Atti del Convegno Internazionale di Studiopromosso dall'IBAM – CNR e dall'IRS – FNER nell'ambito del Progetto MenSALeAtene, 2–3 Ottobre 2015*, ed. I. Anagnostakis and A. Pellettieri (Lagonegro, 2016), 243; Onar, "Animals," 333.

41 Cf. *Geoponika* 2.21.4–9, ed. Beckh, 60.17–61.20.

42 *The Edict on Maximum Prices*, ed. M. Giacchero, *Edictum Diocletiani et collegarum de pretiis rerum venalium* 18.1a, 2 vols. (Genova, 1974), 1:172–73; Kroll, *Tiere*, 179 (although she generally refers to the Roman period).

43 For instance, see Kroll, *Tiere*, 159.

44 See the chapter by H. Baron, p.300 in this volume, confirming that sheep were killed relatively late and used for wool and milk before. For beef-dedicated cattle, see ibid., pp. 301–302.

45 Of course, pigs provided manure which would have been gathered mostly in winter when they were kept in the pigsty, cf. *Geoponika* 19.6.4, ed. Beckh, 506.13–17. However, the product was not highly valued when it came to fertilising sowing crops, see *Geoponika* 2.21.8, 61.12–17.

46 A good illustration of this phenomenon are pig remains from early Byzantine Sagalassos. Namely, they show that there was a practice of slaughtering animals, which although being at the same age (between fourteen and eighteen months), were born in different seasons of the year (i.e., at the turn of winter and spring or at the turn of summer and autumn). Therefore, the city dwellers were regularly provided with high quality pork throughout the year. Cf. D. Frémondeau, B. De Cupere, A. Evin et al., "Diversity in Pig Husbandry from the Classical-Hellenistic to the Byzantine Periods: An Integrated Dental Analysis of Düzen Tepe and Sagalassos Assemblages (Turkey)," *Journal of Archaeological Science: Reports* 11 (2017): 47.

47 Early November.

48 Oribasios, *Medical Collections* 1.3.1, ed. Raeder, CMG 6.1.1:8.28–29.

49 I.e., till mid-September.

50 I.e., from the 20th or 21st of March.

51 I.e., till the 21st of June. Cf. Koder, "Everyday Food," 146.

52 Oribasios, *Medical Collections* 1.3.2–3, ed. Raeder, CMG 6.1.1:8.29–33.

53 Cf. Oribasios, *Medical Collections* 2.28.16, ed. Raeder, CMG 6.1.1:37.6–14.

54 For a similar conclusion, see Kolias, "Die Versorgung," 182–83; V. N. Vlysidou, "Ο χοίρος ως σύμβολο ευδαιμονίας του βυζαντινού ανθρώπου," in *Ζώα και περιβάλλον στο Βυζάντιο*, ed. Anagnostakis, Kolias, and Papadopoulou, 46; J. Koder, *Die Byzantiner: Kultur und Alltag im*

Mittelalter (Vienna, 2016), 218. On the other hand, Kroll's research (*Tiere*, 164–68, esp. 164) proves that pigs were no longer the most frequently consumed species of domestic meat animals after the sixth century. Her opinion appears to be confirmed by the latest Constantinopolitan evidence: pig (both wild and domestic) bones excavated in the Port of Theodosius amount to mere 6.75% of the total, see Onar, "Animals," 333–34. However, the figures published by Onar can be explained by the fact that the area of the Port of Theodosius was a specific place on the map of the capital, the activities of which required a large number of working animals, i.e., cattle and equids. Ultimately, one can also have in mind the fact that the published evidence does not reflect the consumption of cured meat in the area because a large fraction of the salted pork was deboned.

55 For instance, see F. Frost, "Sausage and Meat Preservation in Antiquity," GRBS 40 (1999): 243–44.

56 Symeon Seth, *Treatise on the Capacities of Foodstuffs* χ:119.17, ed. Langkavel.

57 *Book of the Eparch* 16.2, ed., intr., trans., indexes J. Koder, *Das Eparchenbuch Leons des Weisen* (Vienna, 1991), 124.635–636. Cf. Kroll, *Tiere*, 167–68.

58 On keeping pigs in Byzantine cities, see M. J. Beech, "The Large Mammal and Reptile Bones," in *Nicopolis ad Istrum*, 165, 186; Kroll, *Tiere*, 168; eadem, "Animals," 98; Baron, "An Approach," 178.

59 John Tzetzes, *Letter* 18, ed. P. L. M. Leone, *Ioannis Tzetzae Epistulae* (Leipzig, 1972), 32.22–33.19. Cf. M. Günbart, "Store in a Cool and Dry Place: Perishable Goods and their Preservation in Byzantium," in *Eat, Drink and Be Merry*, ed. L. Brubaker and K. Linardou, 47.

60 Oribasios, *Medical Collections* 2.28.2, ed. Raeder, CMG 6.1.1:36.8–9; Aetios of Amida, *Tetrabiblon* 2.121, ed. Olivieri, CMG 8.1:196.29–30; Paul of Aegina, *Epitome of Medicine* 1.84, ed. Heiberg, CMG 9.1:60.27–28; Psellos, *Poem* 9.202–203, ed. Westerink, 197.

61 Oribasios, *Medical Collections* 2.28.1, ed. Raeder, CMG 6.1.1:36.5–7; Aetios of Amida, *Tetrabiblon* 2.121, ed. Olivieri, CMG 8.1:196.29 – 197.1; Symeon Seth, *Treatise on the Capacities of Foodstuffs* χ:119.16–120.16, ed. Langkavel. Cf. Paul of Aegina, *Epitome of Medicine* 1.84, ed. Heiberg, CMG 9.1:60.28–61.2.

62 From the treatise *On Foods* we learn that pigs were best for slaughter at the age between one and two years, see *On Foods* 5 ("De cibis," in *Anecdota medica Graeca*, ed. F. Z. Ermerins, (Leiden, 1840), 243). Archaeological evidence confirms that the meat coming from adult animals was preferred in the Byzantine period, for instance, see Makowiecki and Schramm, "Preliminary Results," 77 (fig. 7); D. Makowiecki, "Animal Economy in the Microregion of Novae in the Light of its Archaeozoological Data," in *Der Limes an der Unteren Donau von Diokletian bis Heraklios*, ed. G. von Bülow and A. Milčeva (Sofia, 1999), 138; Beech, "The Large Mammal," 167–68 (fig. 10.14); Benecke, "Archäozoologische Untersuchungen," 388, 391 (fig. 7), 392 (fig. 8); Kroll, *Tiere*, 16, 32, 49–50, 73, 92; D. Frémondeau, B. De Cupere, A. Evin et al., "Diversity in Pig Husbandry from the Classical-Hellenistic to the Byzantine Periods: An Integrated Dental Analysis of Düzen Tepe and Sagalassos Assemblages (Turkey)," *Journal of Archaeological Science: Reports* 11 (2017): 42 (fig. 3); Onar, "Animals," 334, etc.

63 Oribasios, *Medical Collections* 2.28.9, ed. Raeder, CMG 6.1.1:36.26–28; Aetios of Amida, *Tetrabiblon* 2.121, ed. Olivieri, CMG 8.1:197.18–20. Symeon Seth (*Treatise on the Capacities of Foodstuffs* χ:121.8–11, ed. Langkavel) also dissuades his readers from consuming the meat from old pigs.

64 Oribasios, *Medical Collections* 2.28.6, ed. Raeder, CMG 6.1.1:36.18–20; Symeon Seth, *Treatise on the Capacities of Foodstuffs* χ:121.8–11, ed. Langkavel.

65 Kroll, *Tiere*, 73, 134, 168, 229; eadem, "Animals," 97; Ch. Bourbou and S. Garvie-Lok, "Bread, Oil, Wine, and Milk: Feeding Infants and Adults in Byzantine Greece," *HespSuppl* (*Archaeodiet in the Greek World: Dietary Reconstruction from Stable Isotope Analysis*) 49 (2015): 174; Onar, "Animals," 334. Cf. the price of pork (*The Edict on Maximum Prices* 4.1a, ed. Giacchero, 1:142–43) versus the price of piglet meat (*The Edict on Maximum Prices* 4.46, 1:144–45).

66 Oribasios, *Medical Collections* 2.28.5, ed. Raeder, CMG 6.1.1:36.15–18; 2.28.8, CMG 6.1.1:36.20–24; Aetios of Amida, *Tetrabiblon* 2.121, ed. Olivieri, CMG 8.1:197.9–16; Paul of Aegina *Epitome of Medicine* 1.84, ed. Heiberg, CMG 9.1:61.7–9.

67 The conclusion is still present in Symeon Seth's treatise (*Treatise on the Capacities of Foodstuffs* α:20.10–15, ed. Langkavel) and it can also be seen in Psellos' medical work (*Poem* 9.204–207, ed. Westerink, 197), as his general conclusions considering the meat of adult animals are similar to those given by Symeon.

68 Characteristics of the humour, for instance, see Galen, *On Hippocrates' On the Nature of Man* 1.41, ed. J. Mewaldt, *Galeni In Hippocratis de natura hominis commentaria III*, CMG 5.9.1:51.31–32 (Leipzig, 1914) = Kühn, 15:98. Beef as a melancholic foodstuff, cf. Oribasios, *Medical Collections* 2.28.3, ed. Raeder, CMG 6.1.1:36.9–13; Aetios of Amida, *Tetrabiblon* 2.121, ed. Olivieri, CMG 8.1:197.1–9; Paul of Aegina, *Epitome of Medicine* 1.84, ed. Heiberg, CMG 9.1:61.5. The opinion is in force in the eleventh century, cf. Symeon Seth, *Treatise on the Capacities of Foodstuffs* β:26.15–19, ed. Langkavel. On diseases caused by an excess of black bile, for instance, cf. K. A. Stewart, *Galen's Theory of Black Bile: Hippocratic Tradition, Manipulation, Innovation* (Leiden, 2019), 129–48; M. Kokoszko, "Anthimus and His Work, or on Aromatics and Wildfowl in De observatione ciborum," *Symbolae Philologorum Posnaniensium Graecae et Latinae* 31.2 (2021): 106–107 n. 128.

69 General remarks on the dietary properties of the meat of old animals, cf. Oribasios, *Medical Collections* 2.28. 9, ed. Raeder, CMG 6.1.1:36.24–26; Aetios of Amida, *Tetrabiblon* 2.121, ed. Olivieri, CMG 8.1:197.17–18; Paul of Aegina, *Epitome of Medicine* 1.84, ed. Heiberg, CMG 9.1:61.7–9.

70 Accordingly, younger creatures were slaughtered for special occasions and served to demanding consumers. A good example of the pattern are the results of the excavations carried out by Polish archaeologists in Novae in the south-western part of the early-Christian basilica complex, where they found a kitchen and rich osteoarchaeological material (bones, in the first place, of pigs, with a lower number of those of cattle, sheep and goats), interpreted as remains after feasts attended by the local elite members, see D. Makowiecki, "Wyniki badań zwierzęcych szczątków kostnych z okolic zespołu pomieszczeń pomocniczych bazyliki biskupiej w Novae (Bułgaria)," in *Biskupstwo w Novae (Moesia Secunda) IV–VI w.: Historia, architektura, życie codzienne. Tom 2: Życie codzienne*, ed. A. B. Biernacki, E. Ju. Klenina, and E. Genčeva (Poznań, 2013), 295–304; on the high social position of the residents of the Episcopal Basilica complex and the distribution of all osteal finds over the area of the excavations also cf. Makowiecki, "Animal Economy," 133, 138–39. The Polish researchers found osteal remains of as many as 183 pigs. Eighty-four animals could be identified in terms of their age, out of which only 2 were over 42 months old, while the vast majority of finds belonged to adult pigs between 12 to 38 months of age (ibid. 304 (fig. 7)). They were butchered, divided, and cooked in the kitchen situated nearby. The majority of cattle (i.e. 17 out of 28 animals) served during the feasts were between 42 and 60 months old, with only one over 120 months of age (ibid. 303 (fig. 6)). As far as sheep and goat meat is concerned, out of 33 animals, the majority of the creatures (i.e. 18) were killed at the age between 24 and 48 months (ibid. 304 (fig. 8)). All in all, one can come to the conclusion that prime quality meat (since such must have been served to the prominent members of the local society) was obtained from adult but not aged animals, which concurs with our conclusions drawn from medical sources perfectly.

71 The conclusions appear to be confirmed by at least some excavations in present Bulgaria. For instance, Makowiecki and Schramm's ("Preliminary Results," 78) research at Novae showed the predominance of consumed male pigs over female ones. The same results were obtained by Mark J. Beech ("The Large Mammal," 171 (fig. 10.19)) at Nicopolis ad Istrum and Norbert Benecke ("Archäozoologische Untersuchungen," 388, 390) at Iatrus-Krivina. Similarly, the material from the villa of Bela Voda shows that bulls were slaughtered ten times more often for food than cows; the latter were also much older than the former (because they must have been bred for milk as well as labour), see Kroll, *Tiere*, 49.

72 See Kroll, *Tiere*, 159. Also cf. the price of kid meat and lamb (*The Edict on Maximum Prices* 4.47, ed. Giacchero, 1:144–45; 4.48, 1:144–45) versus the price of the meat of mature animals of the same species (*The Edict on Maximum Prices* 4.3, 1:142–43). In the document we also read that an Italic pound of beef cost 8 denarii, see *The Edict on Maximum Prices* 4.2, 1:142–43. Of course, veal must have been more expensive. Young beef as an expensive foodstuff, cf. Kroll, *Tiere*, 163. Cf. J.-C. Cheynet, "La valeur marchande des produits alimentaires dans l'Empire byzantin," in *Βυζαντινών διατροφή και μαγειρείαι: Πρακτικά Ημερίδας «Περί της διατροφής στο Βυζάντιο». Θεσσαλονίκη Μουσείο Βυζαντινού Πολιτισμού 4 Νοεμβρίου 2001 (Food and Cooking in Byzantium. Proceedings of the Symposium "On Food in Byzantium." Thessaloniki Museum of Byzantine Culture, 4 Novembre 2001)*, ed. D. Papanikola-Bakirtzi (Athens, 2005), 40 n. 48.

73 The decision to slaughter a calf was hard to take because an adult cow was usually very expensive. In the fourth century a top-quality cow cost two thousand denars, while a quality sheep was

worth four hundred, see *The Edict on Maximum Prices* 30.16, ed. Giacchero, 1:208–209; 30.18, 1:210–11. In turn, in the period between the twelfth and the thirteenth centuries a cow was over twenty times more expensive than a sheep, see C. Morrison and J. C. Cheynet, "Prices and Wages in the Byzantine World," in *The Economic History of Byzantium: From the Seventh through the Fifteenth Century* (Washington, D.C., 2002) 2:839–40 (fig. 11).

74 John Tzetzes, *Letter* 93, ed. Leone, 134.11–136.4, esp. 135.17–136.3. See A. A. Demosthenous, "The Scholar and the Partridge: Attitudes Relating to Nutritional Goods in the Twelfth Century from the Letters of the Scholar John Tzetzes," in *Feast, Fast or Famine: Food and Drink in Byzantium*, ed. W. Mayer and S. Trzcionka (Leiden, 2007), 25–31, esp. 25; Günbart, "Store," 46. Keeping chickens in Byzantine cities, cf. Kroll, *Tiere*, 65, 106, 179; eadem, "Animals," 102.

75 Ptochoprodromos, *Poem* 2.101–106, ed., intr., and trans. H. Eideneier, *Ptochoprodromos: Einführung, kritische Ausgabe, deutsche Übersetzung, Glossar* (Cologne, 1991), 115. On the dish, see Ph. Koukoules, *Βυζαντινῶν Βίος καί Πολιτισμός*, vol. 5, *Αἱ Τροφαὶ καὶ τὰ Ποτά...* (Athens, 1952), 34, 78; A.-M. Talbot, "Mealtime in Monasteries: the Culture of Refectory," in *Eat*, 119; A. Dalby, *Cheese: A Global History* (London, 2009), 100–101; idem, *Tastes*, 176; Anagnostakis, "Byzantine Delicacies," 87, 101; Koder, "Everyday Food," 144–45; idem, "Natural Environment," 215; idem, "Cuisine and Dining in Byzantium," in *Byzantine Culture: Papers from the Conference 'Byzantine Days of Istanbul' held on the occasion of Istanbul being European Cultural Capital 2010 Istanbul, May 21–23 2010*, ed. D. Sakel (Ankara, 2014), 230; idem, *Die Byzantiner*, 220; M. Leontsini and G. Merianos, "From Culinary to Alchemical Recipes: Various Uses of Milk and Cheese in Byzantium," in *Latte e Latticini: Aspetti*, ed. I. Anagnostakis and A. Pellettieri, 212–13.

76 A. F. Stone, "Eustathios and the Wedding Banquet for Alexios Porphyrogennetos," in *Feast, Fast or Famine*, ed. W. Mayer and S. Trzcionka, 32–42, esp. 35, 38.

77 Eustathios of Thessalonike, *Oration* 10, ed. P. Wirth, "Eustathii Thessalonicensis orations," in *Eustathii Thessalonicensis opera minora: Magnam partem inedita*, ed. idem (Berlin, 2000), 171.47–52.

78 *Book of the Eparch* 13.1, ed. Koder, 118.560–62.

79 We refer to Koder's ("Das Eparchenbuch," in *Das Eparchenbuch*, ed. Koder, 119; idem, "Stew," 64; idem, "Everyday Food," 146; idem, "Natural Environment," 216; idem, *Die Byzantiner*, 219) view, who argues that the grocers would sell cured meat exclusively. He does not, however, give any reasons for his idiosyncratic interpretation of the term κρέας appearing in *Book of the Eparch* (which is a neuter noun, used in accusative singular, and cannot be denoted by the epithet τεταριχευμένους, being employed in the form applicable to masculine plural nouns), see *Book of the Eparch* 13.1, ed. Koder, 118.563. Neither does he explain who was responsible for the production of cured meat products available at the grocer's. Robert I. Curtis (*Garum and Salsamenta: Production and Commerce in Materia Medica* (Leiden, 1991), 209) suggests that such preserves were manufactures in the same facilities in which salted fish were produced. As far as salted meat for private consumption is concerned, it is considered to have been produced at home, see Grünbart, "Store," 47; Kroll, *Tiere*, 168.

80 *Book of the Eparch* 15.1–6, ed. Koder, 122.607–615, 124.616–631.The butchers were not allowed to buy and store pork, see *Book of the Eparch* 15.1, 122.608. We are informed that they would buy sheep and lambs from sheep traders, see *Book of the Eparch* 15.4–5, 124.621–628.

81 *Book of the Eparch* 15.2, ed. Koder, 122.611–615. The practice of consuming offal and ἀκροκώλια is confirmed by the latest Constantinopolitan evidence, see Onar, "Animals," 334. For butchering methods, see Beech, "The Large Mammal," 172–73.

82 In total, Anthimus devotes seven chapters to pork and pork offal, cf. *On the Observance of Foods* 9, ed. Liechtenhan, CML 8.1:7.1–7; 10, CML 8.1:7.8–13; 14, CML 8.1:9.8–10.5; 16, CML 8.1:10.12–14; 18, CML 8.1:11.1; 19, CML 8.1:11.2; 21, CML 8.1:11.5–9.

83 Taking into account the Byzantine physicians' views on the properties of the meat of young and old animals, we believe that the term *recentior* (*On the Observance of Foods* 9, ed. Liechtenhan, CML 8.1:7.1–2) used by the author does not refer to pork's freshness, as assumed by Edward Liechtenhan and Mark Grant, but it indicates that pork should come from a young, and therefore not fully mature, animal. On both translators' interpretation of the term, see E. Liechtenhan, "Brief des Anthimus des erlauchten Comes und Gesandten und den ruhmreichen Theoderich, der König der Franken, über Speisediät" 9, in *On the Observance of Foods*, CML 8.1:36.17;

M. Grant, "On the Observance of Foods," 9, in Anthimus, *On the Observance of Foods: De observatione ciborum*, ed., trans., M. Grant (Blackawton, 2007), 53. The evidence that it cannot have been a very young piglet lies in the fact that such specimens are described by Anthimus (*On the Observance of Foods* 10, CML 8.1:7.8) as *lactantes*.

84 Ibid., 9, CML 8.1:7.1; 10, CML 8.1:7.8–9.

85 Ibid., 10, CML 8.1:7.8.

86 Ibid., 10, CML 8.1:7.8–13.

87 Ibid., 14, CML 8.1:8.9–10.5. For the Greek equivalent of the term i.e. λάρδος, see Michael Italikos, *Letter* 42, ed. P. Gautier, Michel Italikos, *Lettres et Discours* (Paris, 1972), 237.2. We share Andrew Dalby's (*Food in the Ancient World from A to Z* (London, 2003), 269) opinion that the foodstuff described by Anthimus should be identified as bacon. Also cf. C. Deroux, "The Franks and Bacon according to Doctor Anthimus (*De obs. cib.* 14)," in *Studies in Latin Literature and Roman History*, ed. C. Deroux (Bruxelles, 2008), 14:518–28; Z. Rzeźnicka, M. Kokoszko, and K. Jagusiak, "Cured Meats in Ancient and Byzantine Sources: Ham, Bacon and Tuccetum," *Studia Ceranea* 4 (2014): 253–55. For less precise definition of the term, see Ph. Koukoules, "Βυζαντινών τροφαί και ποτά," Ἐπετηρὶς Ἑταιρείας Βυζαντινῶν Σπουδῶν 17 (1941): 37–38; Vlysidou, "Ο χοίρος," 47; M. Leontsini, "Butter and Lard Instead of Olive Oil? Fatty Byzantine Meals," in *Identità euro-mediterranea e paesaggi culturali del vino e dell'olio: Atti del Convegno Internazionale di Studio promosso dall'IBAM–CNR nell'ambito del Progetto MenSALe Potenza, 8–10 Novembre 2013*, ed. A. Pellettieri (Foggia, 2014), 220; eadem, "Διατροφικές συνήθειες και υγεία: Παρατηρήσεις για τη διατροφή με ζωικά λίπη στις βυζαντινές διαιτητικές πραγματείες (7ος–12ος αι.)," in Ἰατρικὴ θεραπεία ἔστι μέν που καὶ σώματος, ἔστι δ' ἄρα καὶ ψυχῆς: Ὄψεις της Ιατρικής στο Βυζάντιο (14 Δεκεμβρίου 2018, Ιστορικό Αρχείο του Πανεπιστημίου Αθηνών) ΠΡΑΚΤΙΚΑ, ed. K. Nikolaou and K. Gardika (Athens, 2021), 58–59.

88 Anthimus, *On the Observance of Foods* 14, ed. Liechtenhan, CML 8.1:9.8–9.

89 Ibid., CML 8.1:8.13 – 9.2.

90 Constantine Porphyrogenitus, *Three Treatises, Text C*, 146, ed. Haldon, 102. The issue was discussed by Maria Leontsini ("Butter," 220, 228–31; eadem, "Διατροφικές συνήθειες," 58–59).

91 Cf. *The Edict on Maximum Prices* 4.4, ed. Giacchero, 1:142–43; 4.5, 1:142–43.

92 Symeon Seth (*Treatise on the Capacities of Foodstuffs* χ:120.13–15, ed. Langkavel) confirms the fact that pigs were still being fed on figs in the eleventh century.

93 Cf. *The Edict on Maximum Prices* 4.6, ed. Giacchero, 1:142–43.

94 Anthimus, *On the Observance of Foods* 16, ed. Liechtenhan, CML 8.1:10.12–14.

95 Oribasios, *Medical Collections* 2.30.3, ed. Raeder, CMG 6.1.1:38.1–2; 3.15.1–22, CMG 6.1.1:76.14–78.6 (3.15.14, CMG 6.1.1:77.20–21 – legs, snouts, and ears); Aetios of Amida, *Tetrabiblon* 2.252, ed. Olivieri, CMG 8.1:244.1–24 (2.252, CMG 8.1:244.14–15 – legs, snouts, and ears); Symeon Seth, *Treatise on the Capacities of Foodstuffs* χ:120.23 – 121.2, ed. Langkavel. Cf. Galen, *On the Capacities of Foodstuffs* 3.3.5, ed. Helmreich, CMG 5.4.2:338.28 – 339.1 = Kühn, 6:671.

96 Such decomposed meat was sent as a gift to John Tzetzes, *Letter* 4, ed. Leone, 7.20–25. Of course, the scholar was not pleased.

97 See Frost, "Sausage," 244–46, 250–52; D. L. Thurmond, *A Handbook of Food Processing in Classical Rome: For her Bounty no Winter* (Leiden, 2006), 209–19; Z. Rzeźnicka, "Rola mięsa w okresie pomiędzy II a VII w. w świetle źródeł medycznych," in *Dietetyka i sztuka kulinarna antyku i wczesnego Bizancjum (II–VII w.)*, część II, *Pokarm dla ciała i ducha*, ed. M. Kokoszko (Łódź, 2014), 240–43; Rzeźnicka, Kokoszko, and Jagusiak, "Cured Meats," 247–50. There is no exhaustive archaeological material on the salt-curing meat of quadrupeds, which may stem from the fact that it was deboned prior to salt-curing, see *Geoponika* 19.9.3, ed. Beckh, 510.8–9. Cf. Grünbart, "Store," 48; Kroll, *Tiere*, 166–67.

98 Oribasios, *Medical Collections* 4.1.35–40, ed. Raeder, CMG 6.1.1:96.10–97.1; Aetios of Amida, *Tetrabiblon* 2.149, ed. Olivieri, CMG 8.1:207.9–22. Cf. Galen, *On the Capacities of Foodstuffs* 3.40.2–6, ed. Helmreich, CMG 5.4.2:383.22–385.10 = Kühn, 6:745–47.

99 *Geoponika* 19.9.1–6, ed. Beckh, 510.1–20.

100 On its military use, for instance, see Prokopios of Caesarea, *Wars* 1.8.15 ("Procopii Caesariensis De bellis," in *Procopii Caesariensis Opera omnia*, ed. J. Haury and G. Wirth, 4 vols. (Munich, 2001), 1:39.9–11); Constantine Porphyrogenitus, *Three Treatises, Text C*, 146, ed. Haldon, 102.

See also Kroll, "Animals," 97; Koder, *Die Byzantiner*, 219; J. Haldon, "Feeding the Army: Food and Transport in Byzantium, ca 600–1100," in *Feast, Fast or Famine*, ed. W. Mayer and S. Trzcionka, 86–88. Cured meat provisions could be harmful if mishandled, cf. Prokopios of Caesarea, *Wars* 6.25.16–17, ed. Haury and Wirth, 2:263.21–264.3.

101 Oribasios, *Medical Collections* 4.1.36, ed. Raeder, CMG 6.1.1:96.15–19. Cf. Galen, *On the Capacities of Foodstuffs* 3.40.4, ed. Helmreich, CMG 5.4.2:384.11–16 = Kühn, 6:746.

102 The majority of sheep and goats were slaughtered at the age between 24 and 48 months. For instance, see Makowiecki and Schramm, "Preliminary Results," 78 (fig. 8); Makowiecki, "Animal Economy," 137–38; Beech, "The Large Mammal," 169 (fig. 10.15); Benecke, "Archäozoologische Untersuchungen," 390–91, 395 (fig. 11), 396 (fig. 12); Onar, "Animals," 333–34 (fig. d).

103 Both culinary technologies required significant amounts of quality fuel (i.e. charcoal or firewood, and not brushwood because it produces a lot of smoke and lower temperature), being an expensive commodity, for instance, cf. Koder, *Die Byzantiner*, 211–12; idem, "Historical Geography," in *The Archaeology of Byzantine Anatolia: From the End of Late Antiquity until the Coming of the Turks*, ed. Ph. Niewöhner (Oxford, 2017), 19. As a result, long-time thermal processing could be afforded mostly by the better-off. On the other hand, the poor used for this purpose less effective fuel such as brushwood, cones, bark, nut shells, or (usually in the country) a combination of sun-dried animal dung and straw, for instance, see Koder, *Die Byzantiner*, 129, 212; idem, "Historical Geography," 19. Also cf. Kaplan, "L'activité pastorale," 414.

104 The conclusion held strong in the eleventh century, cf. Symeon Seth on lamb (*Treatise on the Capacities of Foodstuffs* α:20.7–15, ed. Langkavel) versus on mutton (*Treatise on the Capacities of Foodstuffs* π:88.8–12).

105 Symeon Seth, *Treatise on the Capacities of Foodstuffs* ε:36.27 – 37.5, ed. Langkavel. Archaeological sources confirm that sheep and goats were usually slaughtered for meat when they were young. For instance, see Kroll, *Tiere*, 15, 32–33, 46, 49–50, 73, 117, 132–33, 140; Onar, "Animals," 333.

106 The interpretation can be confirmed by the contents of Symeon Seth's reflections on beef, cf. *Treatise on the Capacities of Foodstuffs* β:26.15–27.6, ed. Langkavel. Archaeological material confirms that good quality beef was obtained, first and foremost, from adult but young animals, for instance, see Makowiecki and Schramm, "Preliminary Results," 77 (fig. 6); Makowiecki, "Animal Economy," 137–38; Beech, "The Large Mammal," 167–68 (fig. 10.13); Benecke, "Archäozoologische Untersuchungen," 386–88 (fig. 3), 389 (fig. 4); Kroll, *Tiere*, 16, 33–34, 50, 73, 162; Onar, "Animals," 333.

107 Anthimus, *On the Observance of Foods* 11, ed. Liechtenhan, CML 8.1:7.14–16.

108 Anthimus frequently recommends the use of vinegar in preparing various dishes, whereas, in *On the Observance of Foods* one can find only two examples on using wine in food preparation, which leads to the conclusion that the Franks did not frequently make use of wine in cooking. Nevertheless, one should remember that both ingredients were commonly used in Greco-Roman and Byzantine cuisine. For instance, in the recipes for stewed beef preserved in *On the Subject of Cooking* we find two recipes requiring vinegar (Apicius, *On the Subject of Cooking* 8.5.3, ed. Grocock and Grainger, 268; 8.5.4, 268) and one calling for both vinegar and wine (Apicius, *On the Subject of Cooking* 8.5.1, 268). Moreover, from the same work we learn that mutton and goat meat, which (just like beef) were described as tough and dry, could be stewed in a sauce based only on wine, i.e., with no vinegar. Thus, one cannot exclude that in the central and eastern area of the Mediterranean the analysed recipe may have called for wine rather than vinegar. On the other hand, Symeon Seth (*Treatise on the Capacities of Foodstuffs* β:27.1–4, ed. Langkavel) gives vinegar as a typical ingredient of meat stocks and sauces as late as in the eleventh century.

109 Anthimus, *On the Observance of Foods* 3, ed. Liechtenhan, CML 8.1:4.16–5.15. Exotic spices were held in high esteem (even among the barbarians). Notably, Theophylakt Simokattes regales us with a story that took place during emperor Maurice's stormy reign (582–602) which mentions a gift, handed over to the Avar khagan, consisting of pepper, tejpat, cinnamon cassia and costus, see Theophylakt Simokattes, *History* 7.13.5–6 (*Theophylacti Simocattae Historiae*, ed. C. de Boor (Leipzig, 1887), 267.20–268.2). Such καρυκεῖαι were used at the court of the emperor. For instance, Constantine VIII was a gourmet, and, with the use of an abundance of condiments, was able to prepare himself very complex dishes, see Michael Psellos, *Chronography* (Constantine VIII) 2.7.1–9, ed. Renauld, 1:29. Outside the capitol, the imperial baggage train

was also provided with a lot of exotic additives, including mastic, sugar, saffron and cinnamon and cinnamon cassia, cf. Constantine Porphyrogenitus, *Three Treatises, Text C*, 219–222, ed. Haldon, 108.

110 See Kroll, *Tiere*, 14, 46, 48, 60; Onar, "Animals," 333.

111 Oribasios, *Medical Collections* 2.59.1, ed. Raeder, CMG 6.1.1:57.15–18.

112 Ibid., 2.59.3, CMG 6.1.1:57.23–26; Aetios of Amida, *Tetrabiblon* 2.104, ed. Olivieri, CMG 8.1:189.23–24; Paul of Aegina, *Epitome of Medicine* 7.3, ed. Heiberg, CMG 9.2:202.14–15. On the other hand, Koder ("Stew," 64) and Leontsini ("Butter," 222) argue that butter was mainly produced from sheep's and goat's milk (probably because both species were the main milk animals in Byzantium). Indeed, butter obtained from goat's milk was mentioned, for instance, in *Geoponika* (20.22.1, ed. Beckh, 522.13). Moreover, Dioscorides' writings (*Pedanii Dioscuridis Anazarbei De materia medica libri V* 72.1, ed. M. Wellmann, 3 vols. (Berlin, 1907–14), 1:146.13–16) suggest that the tradition of obtaining butter from sheep's and goat's milk dates back to antiquity. One must however note that Dioscorides' information was denied by Galen who explained that the only butter he knew was produced from cow's milk, since it is the fattest, and that is why the final product is called in Greek βούτυρον, i.e. cow's cheese, see Galen, *On the Capacities of Simple Drugs* 10.10, ed. C. G. Kühn, *Galeni De simplicium medicamentorum temperamentis ac facultatibus*, in *Claudii Galeni Opera omnia*, 20 vols. (Leipzig, 1821–33), 12:272. Galen's opinion was shared by Oribasios, Aetios of Amida and Paul of Aegina. Given the above evidence, we argue that butter was mainly produced from cow's milk in Byzantium.

113 For instance, see Anthimus, *On the Observance of Foods* 77, ed. Liechtenhan, CML 8.1:28.12–29.3. The same application of butter is proved by Symeon Seth (*Treatise on the Capacities of Foodstuffs* β:27.7–19, ed. Langkavel). *Book of the Eparch* (13.1, ed. Koder, 118.564) reveals that butter in Constantinople could be bought at the grocer's but it is unclear whether it was treated by the author of the document as a foodstuff or not, see Koder, *Die Byzantiner*, 220. On the other hand, Leontsini ("Butter," 222–23, 228) argues that butter was fairly popular in the Byzantine period, being an expensive and prestigious foodstuff consumed by the upper classes. On butter in late antiquity and early Byzantium, see Z. Rzeźnicka and M. Kokoszko, *Milk and Dairy Products in the Medicine and Culinary Art of Antiquity and Early Byzantium (1st–7th Centuries AD)* (Łódź, 2020), 129–39.

114 Anthimus, *On the Observance of Foods* 76, ed. Liechtenhan, CML 8.1:28.9–10.

115 Anthimus, *On the Observance of Foods* 70, ed. Liechtenhan, CML 8.1:26.4; 71, CML 8.1:26.10; 76, CML 8.1:28.9.

116 See Kroll, *Tiere*, 152, 157, 161.

117 See Kroll, *Tiere*, 163; eadem, "Animals," 99; Koder, *Die Byzantiner*, 219; Yangaki, "On Clay Milking Vessels," 243.

118 *Geoponika* 18.3.4, ed. Beckh, 488.1–4.

119 Ibid., 18.9.2, ed. Beckh, 490.18–19.

120 The special status of goat's milk is confirmed by the fact that it was used to feed infants, see Koder, "Everyday Food," 145; Ch. Bourbou, *Health and Disease in Byzantine Crete (7th–12th centuries AD)* (Farnham, 2010), 165; eadem, "Breasts and Bottles: The Contribution of Bioarchaeology to the Study of Infant Feeding Practices in Byzantine Greece (6th–15th c. AD)," in *Latte e Latticini: Aspetti*, ed. I. Anagnostakis and A. Pellettieri, 97–100; Koder, *Die Byzantiner*, 220.

121 *Geoponika* 18.20, ed. Beckh, 499.1–6.

122 Ibid., 18.12.2, ed. Beckh, 492.9–11.

123 Ibid.,18.12.3, ed. Beckh, 492.11–12.

124 Ibid., 18.19.1–9, ed. Beckh, 497.6–498.21.

125 Ibid.,18.21.1–2, ed. Beckh, 499.7–19.

126 Oribasios, *Medical Collections* 2.59.1, ed. Raeder, CMG 6.1.1:57.17–18; Aetios of Amida, *Tetrabiblon* 2.87, ed. Olivieri, CMG 8.1:180.21.

127 Oribasios, *Medical Collections* 2.59.1, ed. Raeder, CMG 6.1.1:57.18; Aetios of Amida, *Tetrabiblon* 2.87, ed. Olivieri, CMG 8.1:180.22–24.

128 For instance, see Oribasios, *Medical Collections* 2.59.1–10, ed. Raeder, CMG 6.1.1:57.14–58.19; 2.61.1–10, CMG 6.1.1:59.6 – 60.9; Aetios of Amida, *Tetrabiblon* 2.86, ed. Olivieri, CMG 8.1:180.5–12; 2.87, CMG 8.1:180.13–181.3; 2.88, CMG 8.1:181.4–9; 2.89, CMG 8.1:181.10–14; 2.90, CMG 8.1:181.15–22; 2.91, CMG 8.1:181.23–25; 2.92, CMG 8.1:181.26–183.2; 2.93,

CMG 8.1:183.3–22; 2.94, CMG 8.1.183.23–184.13; 2.95, CMG 8.1:184.14–185.10; 2.96, CMG 8.1:185.11–187.4; 2.97, CMG 8.1:187.5–15; Anthimus, *On the Observance of Foods* 76, ed. Liechtenhan, CML 8.1:27.12–28.11; Paul of Aegina, *Epitome of Medicine* 1.86, ed. Heiberg, CMG 9.1:62.9–19; 1.87, CMG 9.1:62.20–63.14; 1.88, CMG 9.1:63.15–23; 7.3, CMG 9.2:202.14–23; Symeon Seth, *Treatise on the Capacities of Foodstuffs* γ:31.7–32.11, ed. Langkavel. Cf. Galen, *On the Capacities of Foodstuffs* 3.14.1–16, ed. Helmreich, CMG 5.4.2:345.8–349.22 = Kühn, 6:681–89. On medical applications of milk, cf. M. Kokoszko, "Galen's Therapeutic Galactology (γαλακτολογία ἰατρική) in De Simplicium Medicamentorum Temperamentis ac Facultatibus," in *Latte e Latticini: Aspetti*, ed. I. Anagnostakis and A. Pellettieri, 33–48; M. Kokoszko and J. Dybała, "Medical Science of Milk Included in Celsus' Treatise De medicina," *Studia Ceranea* 6 (2016): 323–53; idem, "Milk in Medical Theory Extant in Celsus' De medicina," *Journal of Food Science and Engineering* 6.5 (2016): 267–79; Z. Rzeźnicka, "Milk and Dairy Products in Ancient Dietetics and Cuisine According to Galen's *De Alimentorum Facultatibus* and Selected Early Byzantine Medical Treatises," in *Latte e Latticini: Aspetti*, ed. I. Anagnostakis and A. Pellettieri, 49–55; Rzeźnicka and Kokoszko, *Milk*, 24–35, 47–50, 64–75.

129 On how to recognise high-quality milk, see Oribasios, *Synopsis for Eustathios* 5.3.1–2, ed. J. Raeder, *Oribasii Synopsis ad Eustathium, Libri ad Eunapium*, CMG 6.3:154.15–24 (Leipzig, 1926); Aetios of Amida, *Tetrabiblon* 4.5, ed. Olivieri, CMG 8.1:361.26–362.7; Paul of Aegina, *Epitome of Medicine* 1.3, ed. Heiberg, CMG 9.1:9.23–10.4.

130 Oribasios, *Medical Collections* 2.59.2, ed. Raeder, CMG 6.1.1:57.18–23; Aetios of Amida, *Tetrabiblon* 2.89.1–5, ed. Olivieri, CMG 8.1:181.10–14. Milk quality was also impacted by the quality and quantity of fodder, see Oribasios, *Medical Collections* 2.59.5, CMG 6.1.1:57.30–58.4; Aetios of Amida, *Tetrabiblon* 2.90, CMG 8.1:181.15–22. Also cf. Kroll, *Tiere*, 141; Yangaki, "On Clay Milking Vessels," 244.

131 That is why the physicians recommended drinking fresh milk, see Oribasios, *Medical Collections* 2.59.4, ed. Raeder, CMG 6.1.1:57.28–29; 3.15.1, CMG 6.1.1:76.15–17; Aetios of Amida, *Tetrabiblon* 2.252, ed. Olivieri, CMG 8.1:244.1–3; Anthimus, *On the Observance of Foods* 76, ed. Liechtenhan, CML 8.1:28.3–8; Paul of Aegina, *Epitome of Medicine* 1.87, ed. Heiberg, CMG 9.1:62.22; Symeon Seth, *Treatise on the Capacities of Foodstuffs* γ:31.21–24, ed. Langkavel; γ:32.6–7.

132 Oribasios, *Medical Collections* 3.30.4, ed. Raeder, CMG 6.1.1:88.26–31; Aetios of Amida, *Tetrabiblon* 2.95, ed. Olivieri, CMG 8.1:184.18–185.5; Anthimus, *On the Observance of Foods* 75, ed. Liechtenhan, CML 8.1:27.4–6; Paul of Aegina, *Epitome of Medicine* 1.88, ed. Heiberg, CMG 9.1:63.20–23; 7.3, CMG 9.2:202.18–19; Symeon Seth, *Treatise on the Capacities of Foodstuffs* γ:31.16–18, ed. Langkavel. Cf. Galen, *On the Capacities of Foodstuffs* 3.14.4, ed. Helmreich, CMG 5.4.2:346.2–7 = Kühn, 6:682–83. Also cf. M. Kokoszko, K. Jagusiak, Z. Rzeźnicka et al., "Pedanius Dioscorides' Remarks on Milk Properties, Quality and Processing Technology," *Journal of Archaeological Science: Reports* 19 (2018): 984–86.

133 Oribasios, *Medical Collections* 2.59.4, ed. Raeder, CMG 6.1.1:57.28–30; Aetios of Amida, *Tetrabiblon* 2.93, ed. Olivieri, CMG 8.1:183.17–22.

134 Cf. Anthimus, *On the Observance of Foods* 76, ed. Liechtenhan, CML 8.1:27.12–28.3. Anthimus recommended drinking unboiled milk with an addition of honey, wine, mead, or salt, arguing that they would prevent it form coagulating in the stomach. This means that these substances were perceived as preservatives preventing the natural coagulation process, which also takes place in milk going sour.

135 For instance, see Oribasios, *Medical Collections* 4.7.31–32, ed. Raeder, CMG 6.1.1:105.4–8; Aetios of Amida, *Tetrabiblon* 2.97, ed. Olivieri, CMG 8.1:187.12–15; Anthimus, *On the Observance of Foods* 71, ed. Liechtenhan, CML 8.1:26.7–11. On milk in the cuisine of antiquity and early Byzantium, see Z. Rzeźnicka, "Milk," 64–68; Rzeźnicka and Kokoszko, *Milk*, 76–83.

136 For instance, see Leontsini and Merianos, "From Culinary," 205, 215.

137 Aetios of Amida, *Tetrabiblon* 2.97, ed. Olivieri, CMG 8.1:187.13; Anthimus, *On the Observance of Foods* 70, ed. Liechtenhan, CML 8.1:26.1–6.

138 On the foodstuff, see Leontsini and Merianos, "From Culinary," 214–15. On rice in late antiquity and early Byzantium, see M. Kokoszko, K. Jagusiak, and Z. Rzeźnicka, "Rice as Food and Medication in Ancient and Byzantine Medical Literature," *BZ* 108.1 (2015): 129–56; Koder, *Die Byzantiner*, 224.

139 Oribasios, *Medical Collections* 2.60.1–3, ed. Raeder, CMG 6.1.1:58.30–59.5; Aetios of Amida, *Tetrabiblon* 2.98, ed. Olivieri, CMG 8.1:187.16–30. Cf. Galen, *On the Capacities of Foodstuffs* 3.15.1–14, ed. Helmreich, CMG 5.4.2:349.23–353.20 = Kühn, 6:689–96. On ὀξύγαλα, see Rzeźnicka and Kokoszko, *Milk*, 85–91.

140 Columella, *On Agriculture* 12.8.1–3, ed. and trans. H. B. Ash, E. S. Forster, and E. H. Heffner, Columella, *On Agriculture: A Recension of the Text and an English Translation*, 3 vols. (Cambridge, MA, 1941–55) 3:202, 204.

141 *Geoponika* 18.12.3, ed. Beckh, 492.11–12.

142 On the variety of cheeses available on the Constantinopolitan market, for instance, see D. Jacoby, "Mediterranean Food and Wine for Constantinople: The Long-Distance Trade, Eleventh to Mid-Fifteenth Century," in *Handelsgüter und Verkehrswege*, 128–29; Ch. Angelidi and I. Anagnostakis, "La concezione bizantina del ciclo del latte (X–XII secolo)," in *Latte e Latticini: Aspetti*, ed. I. Anagnostakis and A. Pellettieri, 152–57; M. Gerolymatou, "Τυρὶν κρητικόν, τυρὶν τούρκικον, τυρὶν ἀπὸ Βενετίας: Concerning the Cheese Trade in the 14th Century," in *Latte*, 179–84; Leontsini and Merianos, "From Culinary," 211–14.

143 For instance, see Bourbou, "Fasting," 100–101, 112, 114; B. Caseau, "Byzantium," in *A Companion to Food in the Ancient World*, ed. J. Wilkins and R. Nadeau (Chichester, 2015), 365; Leontsini and Merianos, "From Culinary," 205.

144 Constantine Porphyrogenitus, *Three Treatises, Text C*, 146, ed. Haldon, 102. For instance, see Leontsini and Merianos, "From Culinary," 209.

145 Constantine Porphyrogenitus, *Three Treatises, Text C*, 146–47, ed. Haldon, 102.

146 Psellos, *Letters* 13.35 (Michael Psellus, *Epistulae*, ed. S. Papaioannou, 2 vols. (Berlin, 2019) 1:81.1–3); 17.60.1:132.23–26; 29.108.1:233.15–17.

147 Psellos, *Letters* 105.492, ed. Papaioannou, 2:913.8–915.37.

148 Psellos, *Letters* 105.492, ed. Papaioannou, 2:915.38–916.58. Although we do not know which type of milk was used to produce the Paphlagonian cheese, it is possible that, just like the majority of Byzantine cheeses, it was made from goat's or sheep's milk. However, according to Ilias Anagnostakis ("Les trous dans le fromage: la description de Michel Psellos et la recherche contemporaine," in *Latte e Latticini: Aspetti*, ed. I. Anagnostakis and A. Pellettieri, 142), one cannot exclude that the cheese was made from cow's milk. Paphlagonia was known for its cheese as well as pork, cf. Leontsini, "Butter," 226; Ch. Messis, "Au pays des merveilles alimentaires: invitation à la table paphlagonienne," in *Latte*, 159–71.

149 Oribasios, *Medical Collections* 2.59.11–14, ed. Raeder, CMG 6.1.1:58.19–29; Aetios of Amida, *Tetrabiblon* 2.101, ed. Olivieri, CMG 8.1:188.12–189.5; Paul of Aegina, *Epitome of Medicine* 1.89, ed. Heiberg, CMG 9.1:63.24–28; Symeon Seth, *Treatise on the Capacities of Foodstuffs* τ:104.12–105.25, ed. Langkavel; Psellos, *Poem* 9.208–210, ed. Westerink, 197. Cf. Galen, *On the Capacities of Foodstuffs* 3.16.1–6, ed. Helmreich, CMG 5.4.2:353.21–355.19 = Kühn, 6:696–99.

150 Cf. Oribasios, *Medical Collections* 2.59.11–14, ed. Raeder, CMG 6.1.1:58.19–29; Aetios of Amida, *Tetrabiblon* 2.101, ed. Olivieri, CMG 8.1:188.29–189.4. On cheese in the cuisine of late antiquity and early Byzantium, see Rzeźnicka, "Milk," 69–70; Rzeźnicka and Kokoszko, *Milk*, 104–28.

151 Oribasios, *Medical Collections* 2.42.1.1, ed. Raeder, CMG 6.1.1:40.21–23; Aetios of Amida, *Tetrabiblon* 2.130, ed. Olivieri, CMG 8.1:200.6–8; Paul of Aegina, *Epitome of Medicine* 1.82, ed. Heiberg, CMG 9.1:60.2–3.

152 Lower consumption of fowl meat in relation to that obtained from quadrupeds is confirmed by the overwhelming majority of archaeological research conducted throughout the Byzantine empire, for instance, see Kroll, *Tiere*, 22 (fig. 6), 38 (fig. 15), 58 (fig. 20), 80 (fig. 29), 102 (fig. 37), 120 (fig. 50), 140 (fig. 59). So as to give the reader a better insight into the discussed question we will refer to the data from two archaeological sites, i.e. Novae and Yenikapı. In the former bird remains constituted 4.3% of all animal bones which were identified, ranking in the third place after those of domesticated mammals (87.4%) and fish (6.8%), see Makowiecki and Schramm, "Preliminary Results," 72 (fig. 2). As for the latter, the Constantinopolitan material shows that bird remains constitute merely 1.154% of all identified animal bones (versus 20.157% of cattle, 19.242% of sheep, 4.430% of pig, and 3.534% of goat remains), see Onar, Pazvan, Alpak et al., "Animal Skeletal Remains," 83 (fig. 2). As a relatively high percentage of bird bones were found

in Novae, in the area of Scamnum Tribunorum (ca. 10% of the total animal remains discovered in the area) and the Episcopal Basilica and Residence (4.3% of the total animal remains, see Makowiecki and Schramm, "Preliminary Results," 72 (fig. 2 – complex IV), 74 (fig. 3 – complex IV); Makowiecki, "Animal Economy," 133 (fig. 1); also cf. ibid., 132 (fig. 3)), where city elites were located (on Scamnum Tribunorum, see ibid., 133; on the Episcopal Basilica complex cf. our discussion on the value of young pork included in the present study (n. 70)), the data implies that in the Byzantine period fowl meat was mainly consumed by the better off there. The conclusion is supported by Kroll's (*Tiere*, 179) interpretation of the finds from the so-called "Byzantine Palace" in Ephesus, where a large number of chicken bones were excavated and said to have been on the menu of the local aristocracy. Additionally, our line of reasoning is supported by literary evidence made use of in the present study.

153 Oribasios, *Medical Collections* 2.42.5, ed. Raeder, CMG 6.1.1:40.30–41.5; Aetios of Amida, *Tetrabiblon* 2.130, ed. Olivieri, CMG 8.1:200.15–21. Cf. Paul of Aegina, *Epitome of Medicine* 1.82, ed. Heiberg, CMG 9.1:60.11–13.

154 Since the author of *Geoponika* (14.7.10, ed. Beckh, 412.11–12) argues that roosters should represent one sixth of the stock in the flock which was bred for egg production, we might presume that all roosters above that number were sent to slaughter. A similar production pattern is likely to have been followed in Stari Bar (present Montenegro) and Berenike (present Egypt), see Kroll, *Tiere*, 34, 121. When birds were kept primarily for meat, and eggs constituted a secondary benefit only, the difference between the number of male and female birds was usually slight. Cf. the ratio of roosters and hens (i.e., 39:47) found at Nicopolis ad Istrum, see Boev and Beech, "The Bird Bones," 249.

155 For instance, cf. Oribasios, *Medical Collections* 15.2.68, ed. Raeder, CMG 6.1.2:296.4–6; Paul of Aegina, *Epitome of Medicine* 7.3, ed. Heiberg, CMG 9.1:213.19–25 (7.3, CMG 9.1:213.23–25 – broth made of old roosters and hens). On the basis of *Geoponika* (14.7.16, ed. Beckh, 413.10–12) one may assume that in all likelihood such hens were older than two years.

156 *Timarion*, 1158–1159, ed., trans., and comm, R. Romano, Pseudo-Luciano, *Timarione* (Naples, 1974), 91.

157 *Timarion*, 1159–1162, ed. Romano, 91.

158 Oribasios, *Medical Collections* 2.42.1, ed. Raeder, CMG 6.1.1:40.23–24; Aetios of Amida, *Tetrabiblon* 2.130, ed. Olivieri, CMG 8.1:200.8–10; Paul of Aegina, *Epitome of Medicine* 1.82, ed. Heiberg, CMG 9.1:60.3–4.

159 Symeon Seth, *Treatise on the Capacities of Foodstuffs* o:80.4–23, ed. Langkavel.

160 Oribasios, *Medical Collections* 2.42.3, ed. Raeder, CMG 6.1.1:40.28–29; Aetios of Amida, *Tetrabiblon* 2.130, ed. Olivieri, CMG 8.1:200.13–14; Symeon Seth, *Treatise on the Capacities of Foodstuffs* φ:117.20–24, ed. Langkavel.

161 Oribasios, *Medical Collections* 2.42.4, ed. Raeder, CMG 6.1.1:40.30 – peacock meat; 2.43.1, CMG 6.1.1:41.7 – goose meat; Aetios of Amida, *Tetrabiblon* 2.130, ed. Olivieri, CMG 8.1:200.15 – peafowl meat; 2.131, CMG 8.1:200.28 – goose meat; Paul of Aegina, *Epitome of Medicine* 1.82, ed. Heiberg, CMG 9.1:60.7–9 – peafowl and goose meat; Symeon Seth, *Treatise on the Capacities of Foodstuffs* τ:106.2, ed. Langkavel – peacock meat; χ:121.17–18 – goose meat.

162 Oribasios, *Medical Collections* 2.42.4, ed. Raeder, CMG 6.1.1:40.29 – peafowl meat; 2.43.1, CMG 6.1.1:41.7–12 – goose meat; Aetios of Amida, *Tetrabiblon* 2.130, ed. Olivieri, CMG 8.1:200.14 – peafowl meat; Paul of Aegina, *Epitome of Medicine* 1.82, ed. Heiberg, CMG 9.1:60.8 – peafowl meat.

163 Oribasios, *Medical Collections* 2.42.4, ed. Raeder, CMG 6.1.1:40.30 – peafowl meat; 2.43.1, CMG 6.1.1:41.9 – goose meat; Aetios of Amida, *Tetrabiblon* 2.130, ed. Olivieri, CMG 8.1:200.15 – peafowl meat; Paul of Aegina, *Epitome of Medicine* 1.82, ed. Heiberg, CMG 9.1:60.8 – peafowl meat; Symeon Seth, *Treatise on the Capacities of Foodstuffs* τ:106.1, ed. Langkavel – peafowl meat.

164 Oribasios, *Medical Collections* 2.43.1, ed. Raeder, CMG 6.1.1:41.7 – goose meat; Aetios of Amida, *Tetrabiblon* 2.131, ed. Olivieri, CMG 8.1:200.27 – goose meat; Paul of Aegina, *Epitome of Medicine* 1.82, ed. Heiberg, CMG 9.1:60.9 – goose meat; Symeon Seth, *Treatise on the Capacities of Foodstuffs* τ:106.1, ed. Langkavel – peafowl meat; χ:121.17 – goose meat; Psellos, *Poem* 9.200–201, ed. Westerink, 197 – goose and peafowl meat.

165 Prokopios of Caesarea, *Anekdota*, 9.2, ed. J. Haury, "Historia que dicitur arcana," in *Procopii Caesariensis Opera omnia*, ed. idem, 3.1:60.5–9. The presence of geese in Constantinople has been proved by archaeological research, see Onar, "Animals," 340.

166 Constantine Porphyrogenitus, *Three Treatises, Text C*, 535, ed. Haldon, 128. From the same treatise we also learn that, in order to keep poultry during military campaigns, the imperial baggage train included nets (supposed to enclose the birds) and receptacles (holding water for them). Cf. Constantine Porphyrogenitus, *Three Treatises, Text C*, 155–57, ed. Haldon, 104.

167 The conclusion is based on the information from *Geoponika* (14.22.3, ed. Beckh, 425.14–15), where we read that geese lay eggs only three times a year, approximately twelve per each laying period. The information clearly indicates why goose eggs were in relatively short supply and why the population of geese was much more limited than that of chickens.

168 For instance, cf. *Geoponika* 14.22.1, ed. Beckh, 425.10–11.

169 They were eagerly hunted, in places where the abundance of water created a favourable habitat for the birds. Such conditions were characteristic, for instance, of Sagalassos (present Turkey), see Kroll, *Tiere*, 84.

170 The same line of reasoning, for instance, see Boev and Beech, "The Bird Bones," 249; Kroll, *Tiere*, 148; eadem, "Animals," 102.

171 This is confirmed, *inter alia*, by the fact that domesticating attempts regarding ducks are mentioned in a chapter from *Geoponika* that contains a quotation from Didymos of Alexandria, who lived in the fourth century. Also see Kroll, *Tiere*, 179; eadem, "Animals," 102.

172 *Geoponika* 14.23.4, ed. Beckh, 428.12–15.

173 Symeon Seth, *Treatise on the Capacities of Foodstuffs* v:71.17–18, ed. Langkavel.

174 Psellos, *Poem* 9.200–201, ed. Westerink, 197.

175 Since ducks were considered wild waterfowl, medical theory attributed their meat with properties typical for phlegmatic foods. For instance, cf. Aetios of Amida, *Tetrabiblon* 2.130, ed. Olivieri, CMG 8.1:200.25–26. For a similar suggestion, see *On Foods* 23, ed. Ermerins, 271.

176 Oribasios, *Medical Collections* 2.42.2, ed. Raeder, CMG 6.1.1:40.26–27; Aetios of Amida, *Tetrabiblon* 2.130, ed. Olivieri, CMG 8.1:200.11–12; Paul of Aegina, *Epitome of Medicine* 1.82, ed. Heiberg, CMG 9.1:60.6–7.

177 Anthimus, *On the Observance of Foods* 32, ed. Liechtenhan, CML 8.1:15.10–11. The physician uses the adjective 'tenerior' to describe this product. On the interpretation of this term, cf. C. Deroux, "Anthimus, De obs. cib. 32 (p. 15, 1.10–11 Liechtenhan): un texte correctement établi, mais en apparence seulement," *Latomus*, 65.4 (2006): 1011.

178 That they were in use is confirmed by Christopher of Mytilene (*Poem* 42.18–19, ed. and trans. M. De Groote, Christophoros Mitylenaios, *Versuum variorum Collectio Cryptensis* (Turnhout, 2012), 37), who writes about six duck eggs decorating a sophisticated dessert (most probably a pie).

179 Eustathios of Thessalonike, *Letter* 4. ed., intr. and indexes F. Kolovou, *Die Briefe des Eustathios von Thessalonike* (Munich/Leipzig, 2006), 14.109–15.122.

180 Eustathios of Thessalonike, *Letter* 5, ed. Kolovou, 17. 32–18.60. Descriptions of poultry dishes from both letters were recently analysed by Anagnostakis, see "What is Plate and Cooking Pot and Food and Bread and Table All at the Same Time?," in *Multidisciplinary Approaches*, ed. S. Y. Waksman, 211–27. When it comes to archaeological material it is less rich in details than literary sources. The excavations at Nicopolis ad Istrum, for instance, imply merely that birds (including chickens) tended to be boiled/stewed or baked (probably in the oven), whereas they were rarely prepared directly over a fire, see Boev and Beech, "The Bird Bones," 251.

181 On fattening geese, for instance, see *Geoponika* 14.22.7–8, ed. Beckh, 426.4–12; 14.22.11–15, 426.19–428.2. Also see Baron, "An Approach," 177.

182 Oribasios, *Medical Collections* 2.44.1–5, ed. Raeder, CMG 6.1.1:41.13–29 (2.44.1, CMG 6.1.1:41.14–15 – gizzards of fattened geese and hens).

183 Oribasios, *Medical Collections* 2.44.2, ed. Raeder, CMG 6.1.1:41.16–20.

184 Aetios of Amida, *Tetrabiblon* 2.254, ed. Olivieri, CMG 8.1:246.10–247.2 (2.254, CMG 8.1:246.16–17 – livers of fattened geese).

185 Cf. Anthimus, *On the Observance of Foods* 20, ed. Liechtenhan, CML 8.1:11.3.

186 Cf. the passage on pork.

187 Cf. nn. 182–183.

188 Anthimus, *On the Observance of Foods* 22, ed. Liechtenhan, CML 8.1:11.10–12.2. Despite such a negative opinion, it seems that the Byzantines enjoyed eating fatty meats, see Leontsini, "Butter," 285–317; eadem, "Διατροφικές συνήθειες," 41–68.
189 Anthimus, *On the Observance of Foods* 22, ed. Liechtenhan, CML 8.1:11.10–11.
190 Oribasios, *Medical Collections* 2.44.3, ed. Raeder, CMG 6.1.1.41:20–21; Aetios of Amida, *Tetrabiblon* 2.131, ed. Olivieri, CMG 8.1:200.27–29.
191 Cf. *Geoponika* 14.19.2–3, ed. Beckh, 423.16–424.8.
192 Philagrios is mentioned by Aetios of Amida as his source for the therapy of kidney calculi in Book 11 of *Tetrabiblon*. Cf. notes concerning Aetios' references to pheasant fat.
193 For instance, cf. Aetios of Amida, *Tetrabiblon* 5.122, ed. Olivieri, CMG 8.2:99.8–100.8 (5.122, CMG 8.2:100.4–5 – pheasant fat); 6.81, CMG 8.2:226.15–227.13 (6.81, CMG 8.2:226.21 – pheasant fat); Aetios of Amida, *Tetrabiblon* 11.5, ed. Ch. Daremberg and É. Ruelle, "Ἀετίου τοῦ Ἀμιδηνοῦ Βιβλίον ΙΑ," in *Oeuvres de Rufus d'Éphèse*, ed. Ch. Daremberg and É. Ruelle (Paris, 1879), 90.23–94.19 (11.5:91.16–17 – pheasant fat). The final extract was most probably derived by the author from Philagrios of Epirus.
194 Anthimus, *On the Observance of Foods* 22, ed. Liechtenhan, CML 8.1:11.10–12.2.
195 For instance, cf. Alexander of Tralles, *Therapeutics* 2:61.14; 2:219.8–9; 2:403.17; 2:509.14–15, ed., intr. and trans. T. Puschmann, *Alexander von Tralles: Original-Text und Übersetzung nebst einer einleitenden Abhandlung. Ein Beitrag zur Geschichte der Medicin*, 2 vols. (Vienna, 1878–79). A gradual increase in popularity of pheasants is proved by the archaeological data from Nicopolis ad Istrum, where only two bones were excavated in the layers dated between 250–450, whereas their number reached eleven in those dated between 450–600, see Boev and Beech "The Bird Bones," 244 (fig. 13.1). One should however note that for the time being it is the only archaeological site where a relatively high number of such remains were excavated, see Kroll, *Tiere*, 187; eadem, "Animals," 105. Such scanty archaeological evidence suggests that pheasant meat was rare and a luxurious foodstuff for the Byzantines. Our assumption is confirmed by written sources, in which pheasants were mentioned as such, for instance, by John of Damascus (*Sacra parallela*, 36–38 (PG 95:1509D)), who lived at the turn of seventh and eighth centuries, and by Michael Italikos (*Letter* 12, ed. Gautier, 138.15), active in the twelfth century. Pheasants as food of the rich, see Leontsini, "Οἰκόσιτα," 314.
196 Aetios of Amida, *Tetrabiblon* 3.88, ed. Olivieri, CMG 8.1:293.28–294.5 (3.88, CMG 8.1:294.2 – pheasant meat).
197 *Geoponika* 14.18.5, ed. Beckh, 422.21–423.1.
198 Rarity of the birds is confirmed by limited archaeological material, see Boev and Beech "The Bird Bones," 244 (fig. 13.1); Kroll, *Tiere*, 181; eadem, "Animals," 102.
199 However, as Symeon Seth (*Treatise on the Capacities of Foodstuffs* τ:106.4, ed. Langkavel) mentions peafowl fat in his entry on peafowl, we might presume that at least some birds of the species were fattened.
200 Eustathios of Thessalonike (*Commentary on the Iliad* 10.69, ed. M. van der Valk, *Eustathii archiepiscopi Thessalonicensis commentarii ad Homeri Iliadem pertinentes*, 4 vols. (Leiden, 1971–87), 4:573.10–11) is one of the authors whose information creates such an impression.
201 Anthimus, *On the Observance of Foods* 24, ed. Liechtenhan, CML 8.1:12.17–13.5. It is possible that Anthimus decided to include the said recipe in his treatise *On the Observance of Foods* for the sake of the target reader. We may presume that exquisite dishes were served at Theuderic's court on special occasions to showcase the king's wealth and prestige.
202 In the case of older animals, the process lasted between five and six days (cf. Anthimus, *On the Observance of Foods* 24, ed. Liechtenhan, CML 8.1:12.17–13.1), while the meat of younger and smaller birds was fit for consumption after one or two days (cf. Anthimus, *On the Observance of Foods* 24, CML 8.1:13.4–5). On tenderising peafowl meat, cf. Oribasios, *Medical Collections* 4.2.7, ed. Raeder, CMG 6.1.1:98.4–6.
203 The addition of pepper highlighted the exclusive nature of the dish.
204 *Geoponika* adds to the already presented medical data. Most of the information on hen eggs is contained in Book 14, especially in chapter 11 (*Geoponika* 14.11.6–7, ed. Beckh, 418.12–18), where, *inter alia*, various ways of preserving the product are described. The said techniques (and especially those requiring the use of salt) clearly indicate that the passage is not devoted to eggs as a source of chicks.

205 Eating goose eggs by the Byzantines is, *inter alia*, confirmed by Anthimus *On the Observance of Foods* 37, ed. Liechtenhan, CML 8.1:17.15–18.2 and *Geoponika* (see n. 167).

206 On egg consumption in the light of archaeological finds – hen eggs, cf. Kroll, *Tiere*, 178; eadem, "Animals," 102; goose eggs cf. eadem, *Tiere*, 179; eadem, "Animals," 102. Also cf. Leontsini, "Οἰκόσιτα," 310–11; Koder, *Die Byzantiner*, 219.

207 Athenaeus, *The Learned Banqueters* 2.58b, ed. and trans. S. D. Olson, Athenaeus, *The Learned Banqueters*, 8 vols. (Cambridge, MA, 2006–2012), 1:326.

208 Oribasios, *Medical Collections* 2.45.2, ed. Raeder, CMG 6.1.1:42.3–4; Paul of Aegina, *Epitome of Medicine* 1.83, Heiberg, CMG 9.1:60.18–19; Symeon Seth, *Treatise on the Capacities of Foodstuffs* ω:124.1–5, ed. Langkavel. Cf. Anthimus, *On the Observance of Foods* 35, ed. Liechtenhan, CML 8.1:17.8.

209 Most of the information was taken from Galen's *On the Capacities of Foodstuffs* 3.21.1–5, ed. Helmreich, CMG 5.4.2:359.17–360.25 = Kühn, 6:705–707.

210 Oribasios, *Medical Collections* 2.45.3, ed. Raeder, CMG 6.1.1:42.5–6; Aetios of Amida, *Tetrabiblon* 2.134, ed. Olivieri, CMG 8.1:201.20–22; Paul of Aegina, *Epitome of Medicine* 1.83, ed. Heiberg, CMG 9.1:60.19–20.

211 Oribasios, *Medical Collections* 2.45.3–4, ed. Raeder, CMG 6.1.1:42.6–9; Aetios of Amida, *Tetrabiblon* 2.134, ed. Olivieri, CMG 8.1:202.3–6. In the treatise *On Foods* (5, ed. Ermerins, 239) we read that hard-boiled eggs were called ὠὰ τῶν χυτρῶν, while Symeon Seth (*Treatise on the Capacities of Foodstuffs* ω:124.8, ed. Langkavel) termed them αὐτοκόλλητα.

212 Oribasios, *Medical Collections* 2.45.5, ed. Raeder, CMG 6.1.1:42.9–11; Aetios of Amida, *Tetrabiblon* 2.134, ed. Olivieri, CMG 8.1:202.6–8.

213 *Selection of Medicines* 53.12 (*Eclogae medicamentorum*, ed. Raeder, CMG 6.2.2:216.1–3). Cf. Paul of Aegina, *Epitome of Medicine* 3.42.2, ed. Heiberg, CMG 9.1:233.21–25.

214 Symeon Seth, *Treatise on the Capacities of Foodstuffs* ω:125.1–3, ed. Langkavel.

215 Oribasios, *Medical Collections* 2.45.6–7, ed. Raeder, CMG 6.1.1:42.11–19; Aetios of Amida, *Tetrabiblon* 2.134, ed. Olivieri, CMG 8.1:202.8–14; Paul of Aegina, *Epitome of Medicine* 1.83, ed. Heiberg, CMG 9.1:60.22–25.

216 Anthimus, *On the Observance of Foods* 35, ed. Liechtenhan, CML 8.1:16.13–17.8.

217 Cf. Oribasios, *Medical Collections* 4.11.1–14, ed. Raeder, CMG 6.1.1:107.29–109.20 (4.11.14, CMG 6.1.1:109.17–19 – turning eggs while boiling). The extract originates from Antyllos (ca. 100–260).

218 Anthimus, *On the Observance of Foods* 34, ed. Liechtenhan, CML 8.1:16.3–12. On the dish, see L. Plouvier, "L'alimentation carnée au Haut Moyen Âge d'après le De observatione ciborum d'Anthime et les Excerpta de Vinidarius," *Revue Belge de Philologie et d'Histoire* 80.4 (2002): 1367.

219 Anthimus, *On the Observance of Foods* 34, ed. Liechtenhan, CML 8.1:16.3.

220 Anthimus, *On the Observance of Foods* 34, ed. Liechtenhan, CML 8.1:16.10. For instance, Anthimus (*On the Observance of Foods*, 40, CML 8.1:18.8–10) mentions the meat of pike.

221 Anthimus, *On the Observance of Foods* 34, ed. Liechtenhan, CML 8.1:16.11–12.

222 Accordingly, our analysis has arrived at a similar conclusion to that produced by the existing research into stable isotope evidence, which reveals that there was a fairly unchanged diet in the Byzantine period, typical of different communities occupying the part of Mediterranean under Byzantine rule (however, different from the luxurious dietary habits of the upper classes). For animal protein it relied mainly on seafood (which is not covered in the present discussion), fresh and/or cured meat of domesticated quadrupeds (although the importance of each species in the diet, for the time being, has not been established in detail by means of the method), domesticated fowl (mainly chickens), eggs, but first and foremost on dairy products (i.e. mostly cheese), see Bourbou, *Health*, 152, 156, 159, 170; eadem, "Fasting," 106–107, 112–14; eadem, "The Biocultural Model Applied Synthesizing Research on Greek Byzantine Diet (7th–15th century AD)," in *Multidisciplinary Approaches*, ed. S. Y. Waksman, 356, 358.

223 On ham, see Rzeźnicka, Kokoszko, and Jagusiak, "Cured Meats," 245–53.

224 See Koder, *Die Byzantiner*, 220.

225 For instance, cf. *Geoponika* 18.19.7, ed. Beckh, 498.11–13.

226 For instance, see Rzeźnicka and Kokoszko, *Milk*, 111–12.

227 On the gradual extension of fasting days in Byzantium, see Caseau, "Quelques réflexions," 99–108.
228 For instance, see Bourbou, "Fasting," 102; Koder, "Everyday Food," 141–42; idem, "Natural Environment," 215; idem, *Die Byzantiner*, 209.
229 Cf. for instance, A.-M. Talbot, "Mealtime," 114–19; Koder, "Everyday Food," 142; B. Caseau, *Nourritures terrestres, nourritures célestes: La culture alimentaire à Byzance* (Paris, 2015), 258–80; Koder, *Die Byzantiner*, 209–11.

13

MORE THAN FOOD

Animal Bone Finds as a Source for Different Research Questions

Henriette Baron

The identified fish spectra generally indicate a local fishery in rivers such as the Danube and the Nile, or in other bodies of fresh water, as well as in the Mediterranean, the Black and the Red Sea. In the Red Sea desert in particular, fishing was of paramount importance for nutrition. A significant trade in fish can be recognised above all in the area between Egypt and the Levant. Wildlife utilisation is of minor importance and is characterised by the hunting of poultry in open landscapes and near bodies of water and the occasional killing of wild mammals such as hare, roe deer, deer or wild boar. Hunting, especially of poultry, is only more important in the Negev desert. Last but not least, I discuss a pest that constantly occurs in Byzantine animal bone materials: the black rat. It appeared not only as a storage pest, but also as an indirect carrier of the Justinianic Plague.

Introduction

Animal bone finds from archaeological excavations are an important source for research into Byzantine human-animal relations and the Byzantine fauna. Next to pottery, these are usually the most frequent type of find in terms of numbers. The skeletal materials largely comprise remains of domestic animals that were kept for meat, eggs, milk and wool or for labour purposes. By relating how these animals were kept, slaughtered and further processed as well as how products made from them were used, an insight is gained into the everyday lives of people.

A smaller portion of the finds usually comes from wild animals such as deer, gazelle, wild boar, poultry, molluscs and crustaceans as well as fish, which were also used economically and were integrated into food supply networks. Furthermore, these are suitable proxies for a reconstruction of landscapes and activity zones. Finally, a very small portion of the recovered finds comes from wild animals that were not used by humans but on the contrary beneficiaries of human infrastructures. A particularly noteworthy case is the black rat, which had a large impact on human life, as a storage pest as well as in spreading diseases such as the Justinianic Plague.

DOI: 10.4324/9781003055877-17

State of research and structuring of the material

The following remarks are based on the findings of a detailed survey of the zooarchaeological literature, which covered about 50 Byzantine sites.[1] Some of these were analysed intensively in several publications, others were published only in short reports.[2] Since then, the state of research has further improved and more recent studies are also included here. The number of Byzantine sites of different types that have been zooarchaeologically investigated is large enough to allow for some general statements on animal use, at least for the Early Byzantine period. For this phase, which begins with the division of the Roman Empire after the death of Theodosios I and ends with the final loss of Syria and Palestine in 642, the state of research is by far the best. On the one hand, this is due to the fact that the empire reached its greatest expansion for a short time with the Justinianic expansion. On the other hand, Early Byzantine archaeology benefits from the great excavation activity in the field of Roman archaeology. For the subsequent Middle Byzantine period until the conquest of Constantinople by Crusaders in 1204 and for the Late Byzantine Empire, which came to an end with the conquest of Constantinople in 1453, too little is known zooarchaeologically to make more general observations. For this reason, the following primarily presents findings on the Early Byzantine period.

The diversity of the natural areas of the Early Byzantine Empire had a considerable influence on both animal husbandry and the use of wild animals, as did regional economic traditions. For this reason, the results of individual sites for different regions of the empire are pooled according to larger geographical areas: these are Italy, the west coast of the Balkans with the Peloponnese and Crete, the Balkan Danube region as far as Thrace, Asia Minor, Syria and Palestine, Egypt, and North Africa.

Animal products and labour: domestic animals

By far the largest part of the meat consumed in Byzantine cities and settlements came from domestic mammals: sheep, goats, cattle and pigs. The proportion in which remains of these four species[3] occur provides a first insight into the livestock management and dietary strategies in different parts of the empire. In many areas the Byzantine period shows a very broad spectrum of different economies, which makes typification difficult. However, a cross-regional comparison carried out by Antony King for the Roman period[4] shows standard deviations in the percentages of sheep, goats, cattle and pigs similar to the Byzantine material. This testifies to an equally wide range of economic practices.

In the areas of the Early Byzantine Empire, sheep and goats were already predominant in Roman times.[5] An exception was the Danube region, which had already been characterised by intensive cattle husbandry in the Roman period (and beyond that back into the Iron Age). This picture did not change in the Early Byzantine period. Small ruminants and, in the Danube basin, cattle continue to be the best represented species in the bone assemblages, even if the proportions of the species change slightly over time, in comparison with the preceding period. However, these small changes do not need to be given great weight.

Between the fourth and the sixth centuries, the eating habits of the urban population changed: in the centuries before, the urban diet had been characterised by the consumption of pork, judging by the number of bones. For Late Antiquity, a predominance of bone remains originating from sheep and goats can be observed.[6] In all probability this change was largely completed in the sixth century. From then until the present day, the meat consumption pattern in the eastern Mediterranean region consists of a diet largely based on sheep and goats. Only in one area of the Byzantine Empire did pigs still have a consistently higher status than small ruminants (albeit it was still somewhat lower than that of cattle, which also kept providing meat): the Danube region, where the farming of sheep and goats had long played a very subordinate role.

A comparison of the proportions of these species shows a clear continuation of the livestock breeding traditions of Roman times with minor changes, this is especially so in the proportions of the economically less important species. However, such an examination only provides a rough insight into the economic practices of the Byzantine Empire. In order to discover more specific characteristics of Byzantine animal use, the most important species are examined in more detail below.

Sheep (Ovis ammon f. aries) *and goat* (Capra aegagrus f. hircus)

Sheep and goats were the main suppliers of meat in almost all regions of the Byzantine Empire.[7] Their high status in the context of Mediterranean livestock breeding is, among other factors, due to the fact that these animals are very undemanding in husbandry. Most of the assemblages found indicate a predominance of sheep over goats. Where goats predominate, this is often attributed to the assumption that there was only herbaceous or hard-leaved vegetation in the surrounding area, which does not provide a good food source for sheep. A study of sites in the Negev desert assumes a transition from nomadic sheep husbandry to sedentary goat husbandry during the Byzantine period for reasons of available pasture.[8]

The kill-off patterns usually indicate a culling of predominantly older small ruminants, which had been left alive long after reaching optimal meat yield. This points to the use of wool and milk during their lifetime. This is the second important reason for keeping these animals. Byzantine sources on dairy products testify to the high reputation of young cheese and also fresh milk.[9] In contrast to transhumant pastoralism, it was advantageous to lead the animals daily to green fallows and wooded areas close to the town or settlement, as recorded in the laws of the *Nomos Georgikos*, because this allowed farmers to access the milk continuously.[10] Since the lambs and kids suckled only to a small extent from the sixth week onwards and also grazed, milk production for humans could begin, without having to slaughter the offspring. The young animals, however, had a high market value: their meat was eaten at Easter and was considered a delicacy.[11] An upmarket, presumably costly diet, evident from high proportions of young animals, can be discerned for some individual complexes of the empire, e.g. at Carthage.[12]

In addition to milk, the use of sheep's wool and goat hair was certainly of great importance. In some cases, where an evaluation of the animal bones is well integrated into the archaeological context, an intensive wool economy can be evidenced. The absence of young animals in most areas of Amorium (Turkey) and the strong presence of old animals, accompanied by numerous finds of spindle whorls and weaving weights testify to wool production.[13] The *Geoponika* emphasises the aspect of wool production in sheep husbandry and particularly praises the smooth-haired sheep's hides of even colouring.[14] Goats were also ideally supposed to have long, white, thick fur, the particular advantage of which was the great resilience and tear resistance of products made from it. Thus, among other things, ropes and sacks were made, which were also used by seafarers, as they did not soak up water like those made of sheep's wool.[15]

After small ruminants had bought themselves a comparatively long life by providing wool and milk for a few years, they were then slaughtered. The appreciation of the meat of these older animals was probably limited. However, the upper social classes consumed the meat of sheep and goats, not only of delicious lambs and kids, as is attested by the bone finds from the bishop's palace in Novae, where they make up the largest part of the bone finds.[16]

As other livestock, sheep and goats were brought to the place of consumption on the hoof and were only slaughtered there, as attested by the regular occurrence of all skeletal elements, even those poor in meat. The animals were brought by shepherds over long distances to distant market places or driven by small farmers to nearby fairs. A supply of meat via a long-distance trade assumed for Constantinople cannot necessarily be transferred to other cities of the empire, since there was not such a large customer base as in the capital. Nevertheless, one can imagine the supply of other cities with livestock in a similar way, albeit on a smaller scale.[17] In the areas away from large urban centres and especially the military bases, other forms of organisation can also be assumed. In the provisioning of the Byzantine military, meat played a prominent role that certainly surpassed that in the provisioning of the civilian population.[18]

Finally, the importance of hides and skins should be briefly discussed. Even though their use is rarely discernible in the zooarchaeological evidence, which mostly consists of food remains, it can be assumed. A single small highlight is the find of a tannery from the 10th/11th century in Amorium, in which the hides or skins of unborn or newborn small ruminants were processed.[19]

Domestic cattle (Bos primigenius f. taurus)

Cattle are clearly less represented in the Byzantine faunal materials than the small ruminants.[20] This is probably due to the fact that keeping cattle in hot, dry areas with sparse vegetation and limited drinking water is costly and time-consuming. Somewhat better adapted to Mediterranean climates is the zebu, a humped cattle that is osteologically hardly distinguishable from domestic cattle of northern latitudes. Some vertebrae from Tell Hesban (Jordan) show indications that they may have come from the zebu.[21] The animal is frequently depicted on mosaics in the south-eastern Mediterranean region. Thus we may assume that it was widespread in Syria, Palestine, Egypt and North Africa. Higher numbers of cattle bones are found in areas where comparatively lush vegetation can be assumed. In addition to the lush pastures of the Danube region, which was the only region of the Empire whose climatic conditions virtually demanded cattle husbandry, one can observe a small intensive area of cattle husbandry in the Mount Carmel region.[22]

If there was no possibility of winter pasture, supplementary feeding with hay or leaves may have been necessary at this time of year, so that, especially at the beginning of the cold season, cattle were culled that were not intended to serve as offspring, milk cows or draught animals. In Berenike (Egypt), there were no sufficient natural conditions for keeping cattle. Breeders from the Nile Valley occasionally supplied the Red Sea port with cattle, as the size of the recorded animals suggests.[23]

The kill-off patterns throughout the empire indicate that cattle were kept both for meat and for labour. How great the liking for beef was is disputed.[24] However, the animal's role in the diet was moderate but constant, and pure beef cattle were also kept. Cattle slaughtered relatively young, i.e., primarily for meat production, are found with great regularity.[25] Among these, the most diverse age groups are usually represented, so that a herd economy adapted to the different purposes of use and economic conditions can be assumed in each case. Countless variants of herd management are conceivable and difficult to reconstruct in ignorance of the respective agricultural, natural and economic situation. However, it is striking that bones of really young calves, which would have already shown a better meat-feed ratio after a few more months of grazing, are mainly found in the cities. These prove a wealthy clientele that could afford veal.

Cattle were particularly indispensable as draught animals for the plough and the wagon. Where donkeys and mules were lacking, they were also used for milling and threshing. Cattle were cheaper to acquire and maintain for use in agriculture and transport than horses, and they could be loaded more heavily than horses. Hence, cattle were burdened with most of the draught work, as the horse kummet which allowed heavier loads was not yet invented.[26] In most of the find ensembles, skeletal remains of very old animals are found, whose joints, especially the ankles, show the typical strain arthropathies of hard-working draught cattle. Many of the cattle bones found at Iatrus-Krivina (Bulgaria), for example, stem from draught oxen and prove that they were used for agriculture.[27] The meat of working animals was also used, as evidenced by the high proportion of old animals in the food spectra of the Byzantine Empire, even if it was probably somewhat tough due to years of muscular exertion. For work as draught animals, oxen and cows are mainly used because of their more moderate temperament. Increased proportions of young animals or a noticeable predominance of females only rarely indicate that the animals were also used for milk.[28] However, the possibilities of sexing the skeletal remains in the small bovine bone material are usually limited.

Not only the edible parts of the body were utilised, but also everything else. The large, thick hides played a special role, especially in leather production, and in the case of calves, also in the production of parchment. Horn, sinews and the like were also used for various purposes. Zooarchaeologically, a processing of the long bones of cattle can be detected at many sites,[29] preferably of the metapodials with their long, straight and thick-walled shaft. These bones were used to make needles, polished plates and hinges.

Domestic pig (Sus scrofa f. domestica)

As explained above, a change in the meat consumption of the urban population took place in late Antiquity. The pig as the main meat supplier of urban centres is replaced by small ruminants. Nevertheless, cities continue to be the settlement types that show the highest proportion of pigs.[30] The pig shares in the Byzantine food spectra do not decline greatly. In the western half of the Early Byzantine Empire, i.e. in Italy, North Africa, Greece and the Dalmatian coast, where the pig was the second most important source of meat after small ruminants in Roman times, it continued to hold this position, even if its share in the bone finds decreases somewhat. In the Early Byzantine Danubian area, there is even evidence of an increase to some extent of pig breeding, which complements the traditional cattle breeding in this area.[31] In the eastern Mediterranean region, in Asia Minor, Syria and Palestine, the pig had already been less important than ruminants in Roman times. This did not change in the following centuries.

The special importance of the pig for the Byzantine economy is due to the suitability of pork for preservation and the fact that the pig does not provide secondary products. Its special value, also in monetary terms, can be attributed, apart from its culinary value, to the unsuitable vegetation conditions in the Mediterranean region for pig farming. In contrast to ruminants, pigs require a particularly energy-rich diet, so their ideal grazing land is forests, which were probably only available to a limited extent. Moreover, unlike ruminants, the animals could not be transported on the hoof over long distances to the place of consumption, as they are bad walkers. Therefore, they were usually only available in quantities in which they could be kept locally, or they were processed as dried or salted pork.

The taxation in kind *annona* which had been levied since the 3rd century as a result of inflation, was probably continued at least in part until the second half of the 8th century.[32] Where animal meat was taxed in kind, it was pigs, because they were valuable and producers were not deprived of any secondary products. There is nothing in the animal bone spectra to indicate this taxation upon pigs. There is no evidence of an over-representation of certain meat yielding parts that would indicate the import of preserved pork ham or the like.[33] Perhaps only completely boned meat lots were salted and distributed. Military posts on the Danube or in the Negev Desert do not show increased proportions of pork bones compared to civilian settlements in the same area, although meat, according to the Codex Theodosianus (438) and the Codex Justinianus (534), was part of the *annona expeditionalis* of Early Byzantine soldiers.[34] Possibly the *annona* was issued to the civilian population in coastal towns, which could be supplied with live animals by sea at little cost, while the inland areas were left to fend for themselves.

However, the pigs found in the cities do not necessarily have to be regarded as products of state care. There are also indications that the population tried to make itself independent of supply shortages by keeping pigs.[35] These could be caused by warlike conflicts in the hinterland, which forced a transfer of animal husbandry to the city. Since individual pigs are easy to feed with household waste, there is no more suitable domestic mammal for such urban animal husbandry. The pig, which had no significant use during its lifetime but could also be cost-intensive to keep, was killed sometimes younger, sometimes older, but at the latest when it reached its best feed conversion ratio. Although the animals represented a certain luxury anyway, at least where they could not be kept in masses, the consumption of especially young animals testifies to an even more elevated dining custom. At Tell Hesban, for example, foetuses and newborn piglets were eaten.[36]

Domestic fowl (Gallus gallus f. domesticus)

In some areas of the empire another domestic animal contributed significantly to the diet: the chicken.[37] Most of the ubiquitously found bird bones stem from this animal. As a rule, chicken remains take up more than 80 per cent of the bird bones; in Berenike and the Danube region alone, other species, especially wild birds, make up a somewhat larger proportion.[38] According to Byzantine authors, chicken was considered the best and lightest of all meats, healthy and easily digestible, and chicken eggs, along with those of wild fowl, were also very popular but also affordable.[39]

The chicken is such a frugal and adaptable animal that it could easily be kept from the fertile plains along the Danube to the desert. Nevertheless, the proportion of chickens in the bone spectra shows a range of variation that points to differences in the status of the animals in the respective regions. The state of research regarding bird bones, however, is much more heterogeneous than for large mammals. This is due to their small size: they are often overlooked during excavation work and only appear in large quantities when intensive sifting is applied.[40] Furthermore, bird bones are often not identified in detail due to the lack of a good reference collection.

The most intensive use of chickens can be found in two desert forts at the Dead Sea and in the city of Carthage.[41] The high proportion of chicken bones in the desert forts can presumably be explained by the fact that this animal, along with the pigeon, was the only domestic animal that could be easily kept in the immediate vicinity.[42] They are also quite numerous in Italian towns and settlements, and in coastal towns of the eastern and southern Mediterranean. It is likely that people also kept chickens in their backyards to provide themselves with eggs and meat. With a few exceptions, the Danube region, along with the western Balkan area between the Dalmatian coast and Greece, furthermore Asia Minor, the Euphrates region and Egypt (Berenike), are areas where chicken husbandry - at least according to the numbers of bones – played only a comparatively minor role.

Horse Equus equus f. caballus, *donkey* Equus africanus f. asinus *and their hybrids*

Equids, and camels, too, are very probably underrepresented in the animal bone spectra, as they were not usually eaten.[43] Furthermore, horses, donkeys and their hybrids are not easily distinguishable from each other osteologically, and so in many places the bones can only be identified to family level. Equids were indispensable for the transport of goods, locomotion and labour. Especially from Italy through Greece and the Danube region to Asia Minor, the entire workload rested on the shoulders of the equids (as well as cattle, which were used for heavier loads and the plough, see above), as camels were not common in this area of the Byzantine Empire.

Horses were not only kept for riding, but also as race and parade horses and as work animals. Occasionally, a further use emerges: horses were eaten. Traces of slaughter, which not only indicate the skinning of dead or killed animals but also testify to meat production, have been found on horse bones from Butrint (Albania), Iatrus-Krivina and Nicopolis ad Istrum (Bulgaria), Constantinople, Caesarea (Israel) and Limyra (Turkey).[44] With the exception of the Theodosian port of Constantinople, whose finds date to the Middle Byzantine period,[45] all these findings date to the Early Byzantine period. Horse meat consumption was uncommon in the Byzantine Empire and was probably due to emergency situations.[46]

However, most of the horse finds stem from animals that were used as riding and working animals. The high proportion of horse bones in the finds from the Danube region, which was marked by military conflicts, suggests that the animals were used for military purposes. In view of the soldier farmers living there, this area of activity may have overlapped in part with the agricultural one. In Nicopolis ad Istrum, for example, pathologies were discovered in the bones of the extremities, suggesting that the horses were used for draught purposes.[47] The animals could be harnessed for relatively light loads. In view of the very few records for the donkey in the Danube basin, the horse may also have been used in the donkey mill in this area.

Donkeys do, however, have advantages over horses: they are undemanding and, due to their small size, well suited for use in confined spaces, be it in the city as a pack donkey or in the donkey mill. One disadvantage of the donkey, however, is its slowness: when loaded, whether by luggage or a rider, donkeys never go faster than a walking pace, so that they are only suitable as mounts when there is no necessity for speed. The donkey has a fairly constant place among the pack animals throughout the empire.[48] It is of particular importance in Greece, where it can be found in comparably large numbers and was probably also the most suitable pack animal in the mountainous terrain of the Peloponnese and Crete. Quite high proportions of donkey bones are also recorded in Asia Minor, Syria and Palestine. In these areas of the south-eastern Mediterranean, the donkey is also a suitable animal for light draught work. On the light dry soils of this area, it can even be harnessed in front of the plough.

Hybrids were only very rarely found in the materials,[49] although the number of unidentified hybrids cannot be estimated. Compared to the horse, these animals have the advantage of a calmer temperament and, compared to the donkey, that of higher resilience. Mules in particular are considered to be very sure-footed, even in impassable landscapes such as mountains, and can carry heavy loads.

Dromedary – Camelus dromedarius, *Bactrian camel* – Camelus ferus f. bactriana *and their hybrids*

Two members of the camel family have been domesticated: on the one hand the one-humped dromedary, whose present range extends from the hot landscapes of North Africa across the Arabian Peninsula and Mesopotamia to Turkey, Iran and Pakistan, and on the other hand the Bactrian camel, which today is kept in an area stretching from the eastern Black Sea coast, through Kazakhstan and Mongolia to the Sea of Japan, thus encompassing rather cold-temperate, although also desert-like, stretches of land. As is the case with closely related species, these two camels are hardly distinguishable from each other osteologically, so that assignment to one of the two species is not always possible. ZooMS mass spectrography has proven suitable to determine the species of the finds.[50]

Camel finds occur in all regions of the Byzantine Empire except Greece and the west coast of the Balkans, mostly in single finds or small numbers.[51] A complete skeleton was found in Constantinople.[52] A notable case is Caričin Grad in Serbia, where over 40 camel bone remains were found, presumably used as a beast of burden and transport in the civil sector, and whose bones were processed after death.[53] Otherwise, only isolated finds from the tip of the extremities are known from sites in this area, which could also come from skins.

Camels had a certain status among the beasts of burden in the arid regions of Egypt and North Africa.[54] The skeletal elements of these animals appear there somewhat more regularly in the faunal materials than, for example, in the Danube region or Asia Minor, but for the most part do not appear in significantly large numbers. In Syria and Palestine, as well as in Carthage, camels competed with equids to a greater extent than in the arid regions of Libya and Egypt. However, a few sites, e.g. Upper Zohar in the Negev Desert west of the Dead Sea, indicate a particularly high status of the camel, as they were frequented by caravans.[55] The camels, however, not only had the role of beast of burden, but were also used as mounts. Whether they were eaten, milked or shorn cannot be proven.

Dog – Canis lupus f. familaris *and cat* – Felis silvestris f. catus

Dogs were kept throughout the Byzantine Empire and existed – as they had in Roman times – in all sizes and for various purposes.[56] They were used as shepherd dogs, hunting companions, watchdogs and lapdogs.[57] As a rule, dogs account for 1 to 2 per cent of the number of domestic animals found, i.e. they are regularly present, but mostly in small numbers. In the case of the dog (as well as the cat), the number of animals present may have been somewhat higher than the pure bone numbers suggest, since animals were not used for meat consumption throughout the empire and therefore do not appear in the food waste.

Cats are found in much smaller numbers than dogs. Their share of the domestic animal bone inventory averages only 0.3 per cent. This indicates that cats were not all that common. A special finding is the co-burial of a cat in a 6th/7th-century Byzantine tomb in the Balatlar Church in Sinop on the Turkish Black Sea coast. The animal lay next to the right thigh of a human and had apparently eaten a rat and a sparrow shortly before its death, as bone finds show.[58] In the case of the Danubian fort of Iatrus-Krivina, it is assumed that cats were skinned because the feet bones are missing and were probably left in the skin.[59] In the materials recorded here, this is the only reference to a use of cats beyond their purposes as pest exterminators and, possibly, pets.

Regionality and long-distance trade: fisheries

In early Byzantine times, the Mediterranean Sea was at the centre of the dominion and connected the far-flung parts of the empire. It also contained two important raw materials: fish and salt. The most important cities of the empire, the largest ports and the best infrastructures can be found on its coasts. The central location of the Mediterranean Sea and its importance as a communication route outshines the other bodies of water available to the empire, including two other inland seas in early Byzantine times: the Black Sea and the Red Sea. In addition, there were a multitude of freshwater routes that were lifelines of the empire. In early Byzantine times, the two most important rivers were probably the Danube and the Nile, but the resources of other rivers were probably also regionally significant. In addition, there were lakes and other bodies of standing water whose fish were apparently also valued.

Fish bone materials from all parts of the empire attest to the use of the respective local fish faunas, whether marine or limnic.[60] It is not uncommon to find both freshwater and marine fish near the coast. The various regional fisheries in the Byzantine Empire were diverse due to the respective local water conditions and fish stocks. Based on the animal bone finds, the importance of fish in the Byzantine diet can now be well assessed. This is due to the increasing use of sieves in excavations in the Mediterranean, sometimes on a large scale.

Black Sea and Bosphorus

The excavations at Cherson in the Crimea (6th to 12th/13th century) attest to a Black Sea fishery for brill *Scophthalmus rhombus* and anchovy *Engraulis encrasicolus*.[61] Furthermore, numerous thornback rays, *Raja clavate*, were also caught in Cherson. The special hydrological conditions in this inland sea result in a fauna that is somewhat different from that of the Mediterranean. Nevertheless, there was also an exchange between these two seas, from which Constantinople greatly profited: during their seasonal migration, mackerel and tuna were caught in the straits of the Bosphorus in set nets arranged in comb systems, the so-called *epochai*.[62] The annual fish migrations through the Bosphorus were a locational factor that may have contributed significantly to the capital's ability to survive for centuries. Fishermen were able to generate a surplus from these animals, which appeared in masses for short periods. This may have saved the local fish saltworks from losing its importance in post-Roman times. Large quantities of tuna and swordfish bones were recovered from excavations at the Theodosian harbour.[63]

Danube

Further north, on the always contested imperial border along the Danube, lies the area richest in freshwater fish species in Europe. This stream was intensively fished, as evidenced by numerous finds from such excavations, which involved the systematic use of sieves.[64] As a rule, primarily remains of the carp family (Cyprinidae[65]) occur. Among these, the wild carp *Cyprinus carpio*, now almost extinct, usually dominates, but many other species have also been detected. Besides cyprinids, pike *Esox lucius* and catfish *Silurus glanis* were frequently caught, perch of the family Percidae, i.e. Sander (*Sander lucioperca*) and European perch (*Perca fluviatilis*), were already much rarer and salmon (Salmonidae) as well as sturgeon (Acipenseridae) occur rather sporadically in the fish bone material. Targeted fishing of the latter probably began in the Middle Byzantine period, when caviar began to play a role.

Nile

At the opposite end of the Early Byzantine Empire, the Nile flowed into the Mediterranean. Few zooarchaeological inventories give indications of Nile fishing.[66] The materials show a fairly uniform spectrum of species, but with a clearly identifiable concentration on a particular species or family: sometimes it is catfish of the genus *Synodontis* (Mochokidae), sometimes cyprinids of the genera *Labeo* and *Barbus*, sometimes cichlids of the genus *Tilapia*. The state of research is still very fragmentary, but at present it looks as if the Coptic monasteries on the Nile had an organised fishery that targeted specific species. Since some of the finds come from storage contexts, and there are also isolated references to the import of Nile fish from other parts of the empire, the animals were probably dried or salted, i.e. processed into preserved fish. It seems that the fish salting trade was still practised to an extent that allowed the export of preserved fish, especially in the Nile region. Extensive trade as far as the capital of the Byzantine Empire and Asia Minor can only be detected for the family of the African predatory catfish Clariidae, while trade in the Nile perch *Lates Niloticus* (Latidae), which is native to Egypt, only extended as far as Palestine, according to current research. Since the eastern Mediterranean in the area of present-day Israel was particularly poor in species and the supply of freshwater fish was mainly ensured by catches from the Sea of Galilee, which could not cover the needs of the entire population, there was an increased demand for imported fish in this area.[67]

Red Sea

A little further east, on the Red Sea, the fringing reefs near the coast were intensively fished.[68] This seawater inlet of the Indian Ocean is home to many more fish species than the Mediterranean. In the Red Sea, a gigantic variety and abundance of edible animals can be found in the middle of the desert. Consequently, these were intensively exploited. The excavations in Berenike brought to light thousands of fish bones, which prove a very wide range of species: representatives of the families for sea bass and grouper (Serranidae), parrotfish (Scaridae), emperor (Lethrinidae), jack mackerel (Carangidae) and sea bream (Sparidae) are particularly common. Although the state of research does not yet allow for comparative analyses, it can be assumed that the Red Sea catch differed from place to place, depending on the nature of the substrate, whether sandy or rocky, the hiding places and marine vegetation conditions of the reefs, and the fishing techniques used.

For the emperor (Lethrinidae) and also triggerfish (Balistidae), which frequently occur in Berenike, an export to Palestine is recognisable. In view of the regular occurrence of Red Sea species in this area, it is conceivable that fish products from the Red Sea, which are more prevalent in Syria/Palestine compared to Nile fish from late Roman times onwards, reached the Levant via a link to a larger established trade network, possibly through the spice trade with Arabia, which had existed since Nabataean times and also flourished in Byzantine times.[69]

Mediterranean Sea

A certain heterogeneity can also be discerned for the Mediterranean Sea on the basis of the Byzantine fish bone material:[70] While sea bream, grouper and bass, for example, appear everywhere, mullet (Mugilidae) and drum (Sciaenidae) show a slight distribution focus in the Levantine Sea. Basically, a nearshore fishery for bottom-dwelling species is emerging. Species living in the open sea, especially tuna and mackerel (Scombridae), but also jack mackerel (Carangidae), are usually only found in small numbers.

Wildlife: hunting in different landscapes

In its early period, the Byzantine Empire encompassed peripheral areas of three continents. The biodiversity within the empire's borders is correspondingly immeasurable.[71] In addition to fish, economically exploited wild animals include hunting game such as deer (Cervidae), wild boar (Suidae), gazelle and wild goat (Bovidae), hare (Leporidae) and beaver (Castoridae), and less frequently also predators such as bear (Ursidae), wolf and fox (Canidae) or marten (Mustelidae). Birds were also hunted or caught with nets or glue rods. The most common families include duck, goose and swan (Anatidae), pheasant and partridge (Phasianidae), pigeon and dove (Columbidae), corvid (Corvidae) and birds of prey (mostly of the Accipitridae family). However, according to the animal bones recovered, hunting was probably not a significant part of Byzantine everyday life. The wildlife remains from archaeological excavations indicate a generally very small proportion of wild mammals and birds, which is likely to have been the result of occasional hunting excursions, lucky hits or purchases from the rural population: As a rule, wild mammal bones make up only a few percent of animal bone finds each. In the case of wild birds, the proportion is even lower, due to the small size of the bones.[72]

The composition of the archaeologically verifiable wild fauna reflects the diversity of the ecosystems in which the Byzantines moved: the high proportion of deer *Cervus elaphus* and roe deer *Capreolus capreolus* in the north of the Early Byzantine Empire testifies to a good tree population. However, it is important to bear in mind that many finds of these animal species, especially of the red deer, are antler fragments that reached the cities and settlements as craft materials. The hare *Lepus* sp., which is often similarly well represented, provide evidence of occasional hunting in the open cultural landscape. The high proportion of wild boar *Sus scrofa* among the hunting game in the sites along the lower Danube is impressive. Presumably, large parts of this fertile and climatically favoured region were covered by mixed forests. The animals probably also lived, like the beaver *Castor fiber*, which also appears in animal bone material from this area, in the softwood and hardwood floodplains along the river arms.[73] The animal bone ensembles from this region in particular also show a not inconsiderable proportion of waterfowl, indicating intensive use of these habitats close to water bodies. These are mainly ducks, but also large species such as pelican or heron, which certainly represented impressive prey.

In the south of the empire, from North Africa to the Levant, the proportion of hunting game is lower, but the variety of recorded species is much greater. Here we are obviously dealing (with the exception of the gazelle, which is quite well represented and presumably also killed to protect the fields) with animals that were not exposed to any purposeful hunting, but were rather killed by chance: antelope and barbary sheep (Bovidae), hare *Lepus* sp., deer (Cervidae), porcupine *Hystrix cristata*, rock hyrax *Procavia capensis*, weasel (Mustelidae), and along the Euphrates and Jordan rivers sometimes wild boar, etc. In these latitudes, on the other hand, birds were hunted more often. Catching birds was a way to get some meat on the plate despite unfavourable conditions for animal husbandry. Some waterfowl are always found near water bodies, but in the deserts and steppes it was mainly the Chukar Partridge *Alectoris chukar* or the Sand partridge *Ammoperdix heyi* that were hunted.[74] The dovecotes (*columbaria*), which are quite well documented at least for Palestine and Cappadocia, may also have attracted wild pigeons and doves.

While some forest species are found among wild mammals, they are rare among wild birds. Only few archaeological sites contain remains of species such as goshawk *Accipiter gentilis*, sparrowhawk *Accipiter nisus*, or wood pigeon *Columba palumbus*. They are also poorly represented in the more forested areas such as the lower Danube. This circumstance could be due to the fact that birds are more difficult to kill in forests, albeit light forests, than at water bodies and in the open country. In addition, wild boar or deer may have appeared to be more attractive prey in forests.

Unwanted cultural successor: the house rat *Rattus rattus*

Animal bone assemblages from archaeological excavations where no sifting was carried out often contain only the remains of large, economically significant animal species, while other animals that died in the cities remain invisible. For this reason, there are only a few clues from zooarchaeological materials about those animal co-inhabitants of Byzantine cities that died of natural causes and whose bones ended up in the ground without human intervention.[75]

A special role is played by commensal cultural predators, such as the black rat and the house mouse. The former in particular, which is easily identifiable, occurs regularly and sometimes in high numbers in Byzantine find material.[76] Originally immigrated from southwest Asia, as DNA analyses reveal, the black rat was at home throughout the Mediterranean in Byzantine times. The collapse of the Western Roman Empire led to a decline of black rat populations in its provinces, as the trade networks that had continuously brought new rats from afar into the trade hubs of the empire broke down.[77] In find assemblages from Byzantine sites rats appear for a little longer.[78] Numerous rat finds in the cargo area of the shipwreck from Ma'agan Mikhael B (648 to 750 AD) attest to its occurrence and its use of cargo ships.[79]

The role of the domestic rat in the spread of the Justinian Plague is controversially discussed.[80] The rats followed the grain on its way to the granaries located in the cities by ship across the Mediterranean, and thus added their share in spreading the disease directly to the most densely populated centres of the empire. In the warehouses, they destroyed supplies, thereby further aggravating the crises caused by the plague. The extent of the loss caused by rodents is difficult to estimate. In view of the very regular occurrence of subsistence crises in Late Antiquity and the early Byzantine period, on average every three years, the fact that rats and mice are mentioned only three times as triggering factors in the written sources indicates that the animals were relatively harmless.[81] However, a reduction in supplies due to pests was probably so commonplace and constant that it only led to large-scale crises when additional factors such as crop failures, natural disasters or sieges were added.

Notes

1 See in detail H. Kroll, *Tiere im Byzantinischen Reich: Archäozoologische Forschungen im Überblick*. Monographien des RGZM 87 (Mainz, 2010). For a short English summary of the results, see: H. Kroll, "Animals in the Byzantine Empire: An Overview of the Archaeozoological Evidence," *Archeologia medievale* 39 (2012): 93–121.

2 Since the following explanations are based on this summarising and comparative analysis, in order to take account of the character of the handbook, and since not all sites and related publications can be mentioned in the brief overview presented here, I mostly refer to this overview work from 2010, with a request to readers to further explore the primary literature listed there in detail.

3 Sheep and goats are grouped together because not all skeletal elements can be clearly assigned osteologically to one of the two species and they have similar uses.

4 A. C. King, "Diet in the Roman World. A regional inter-site comparison of the mammal bones," *Journal of Roman Archaeology* 12 (1999): 168–202.

5 Ibid.,183.

6 Kroll, "Tiere," 149–51.

7 Ibid., 157–61.

8 N. Marom, M. Meiri, Y. Tepper, T. Erickson-Gini, H. Reshef, L. Weissbrod and G. Bar-Oz, "Zooarchaeology of the Social and Economic Upheavals in the Late Antique-Early Islamic Sequence of the Negev Desert," *Nature Scientific Reports* (2019) 9:6702. https://doi.org/10.1038/s41598-019-43169-8.

9 A. Dalby, *Flavours of Byzantium* (Blackawton, 2003), 144, 147. See also the chapter by Kokoszko and Rzeźnicka in the present volume.

10 *Nomos Georgikos* 34, see J. Koder, *Nomos Georgikos, Das Byzantinische Landwirtschaftsgesetz: Überlegungen zur inhaltlichen und zeitlichen Einordnung, Deutsche Übersetzung*. Wiener Byzantinische Studien 32 (Vienna, 2020), 17–18; 56–57.

11 Dalby, *Flavours*, 148, see n. 9.

12 G. Nobis, "Die Tierreste von Karthago," in: *Die Deutschen Ausgrabungen in Karthago*, vol. 3, ed. F. Rakob (Mainz, 1999), 574–631; 578.

13 E. Ioannidou, "Animal Husbandry," in: *Amorium Reports, The Lower city enclosure*, vol. 3, *Finds, Reports and Technical Studies*, ed. C. S. Lightfoot and E. A. Ivison (Istanbul, 2012), 419–42. For further examples see Kroll, *Tiere*, 159.

14 *Geoponika* 8, 1, trans. T. Owen, ΓΕΩΠΟΝΙΚΑ: *Agricultural Pursuits* (London, 1805).

15 *Geoponika* 18, 9, trans. Owen.

16 D. Makowiecki and Z. Schramm, "Preliminary Results of Studies on Archaeozoological Material from Excavations in Novae (Season 1992)," in: *Novae: Studies and Materials*, vol. 1, ed. A. B. Biernacki (Poznan, 1995), 71–81; 74.

17 T. Kolias, "Die Versorgung des byzantinischen Marktes mit Tieren und Tierprodukten," in: *Handelsgüter und Verkehrswege. Aspekte der Warenversorgung im östlichen Mittelmeerraum (4. bis 15. Jahrhundert)*, ed. E. Kislinger, J. Koder and A. Külzer. Veröffentlichungen zur Byzanzforschung 18 (Vienna, 2010), 179–88; 183–84.

18 T. Kolias, "Eßgewohnheiten und Verpflegung im byzantinischen Heer," in: ΒΥΖΑΝΤΙΟΣ. *Festschrift für Herbert Hunger zum 70. Geburtstag*, ed. W. Hörandner, J. Koder, O. Kresten and E. Trapp (Vienna, 1984), 193–202.

19 E. Ioannidou, *Animal Husbandry*.

20 Kroll, *Tiere*, 161–65.

21 A. von den Driesch and J. Boessneck, "Final Report on the Zooarchaeological Investigation of Animal Bone Finds from Tell Hesban, Jordan," in: *Faunal Remains: Taphonomical and Zooarchaeological Studies of the Animal Remains from Tell Hesban and Vicinity*, ed. Ø. S. LaBianca and A. von den Driesch. Hesban 13 (Berrien Springs, 1995), 67–108; 72.

22 E.g. L. K. Horwitz, "Roman Through Ottoman Period Fauna from H. Shallale," in: *Shallale. Ancient City of the Carmel*, ed. S. Dar, BAR International Series 1897 (2009), 321–40.

23 W. Van Neer and A. Lentacker, "The Faunal Remains," in: *Berenike 1995: Preliminary Report of the 1995 Excavations at Berenike (Egyptian Red Sea Coast) and the Survey of the Eastern Desert*, ed. S. E. Sidebotham and W. Z. Wendrich (Leiden, 1996), 337–55; 348.

24 J. Koder, "Über die Liebe der Byzantiner zum Rindfleisch," *BZ* 102.1 (2009): 103–108.

25 Kroll, *Tiere*, 162–63.

26 A. Bryer, "The Means of Agricultural Production: Muscle and Tools," in: *The Economic History of Byzantium. From the Seventh through the Fifteenth Century*, vol. 1, ed. A. E. Laiou (Dumbarton Oaks, 2002), 101–14; 107–11.

27 N. Benecke, "Archäozoologische Untersuchungen an Tierresten aus dem Kastell Iatrus," in: *Iatrus-Krivina: Spätantike Befestigung und frühmittelalterliche Siedlung an der unteren Donau*, vol. 4: *Ergebnisse der Ausgrabungen 1992–2000*, ed. G. von Bülow, B. Böttger, S. Conrad, B. Döhle, G. Gomolka-Fuchs, E. Schönert-Geiss, D. Stančev and K. Wachtel (Mainz, 2007), 383–414.

28 E.g., Benecke, "Iatrus," and Horwitz, "Shallale."

29 Kroll, *Tiere*, 165.

30 Kroll, *Tiere*, 165–68.

31 E.g., Makowiecki and Schramm, *Novae*; Benecke, "Iatrus"; or M. J. Beech, "The Large Mammal and Reptile Bones," in: *Nicopolis ad Istrum, A late Roman and Early Byzantine City: The Finds and the Biological Remains*, ed. A. G. Poulter (Oxford, 2007), 154–97.

32 N. Oikonomides, "The Role of the Byzantine State in the Economy," in: *The Economic History of Byzantium. From the Seventh through the Fifteenth Century*, vol. 3, ed. A. E. Laiou (Dumbarton Oaks, 2002), 973–1058; 980–81.

33 In Nicopolis ad Istrum alone, a lack of meat-rich lots seems to indicate an export (see Beech, "Reptile Bones," 190).

34 Kolias, "Heer," 199.

35 For a discussion of Early Byzantine resilience strategies for urban food supply, see H. Baron, A. E. Reuter and N. Marković, "Rethinking Ruralization in Terms of Resilience: Subsistence Strategies in Sixth-century Caričin Grad in the Light of Plant and Animal Bone finds," *Quaternary International* 499 (2019): 112–28. https://doi.org/10.1016/j.quaint.2018.02.031.

36 Driesch and Boessneck, "Hesban." For other examples, see Kroll, *Tiere*, 168.

37 Ibid., 177–79.

38 Ibid., 177.

39 Dalby, *Flavours*, 143–44.

40 E.g., a particularly large number of bird finds were found in Upper Zohar, where intensive sifting was carried out: P. Croft, "Bird and Small Mammalian Remains from Upper Zohar," in: *Upper Zohar. An Early Byzantine Fort in Palaestina Tertia, Final Report of Excavations in 1985–1986*, ed. R. P. Harper (Oxford, 1995), 87–96.

41 Kroll, *Tiere*, 179.

42 Croft, "Upper Zohar." See also Marom et al., "Negev."

43 Kroll, *Tiere*, 168–72.

44 Ibid., 169–70.

45 V. Onar, H. Pazvant, G. Alpak, N. Gezer Ince, A. Armutak and Z. Kiziltan, "Animal Skeletal Remains of the Theodosian Harbour: General Overview," *Turkish Journal of Veterinary and Animal Sciences* 37 (2013): 81–85. https://doi.org/10.3906/vet-1111-7.

46 Compare the chapters by Schmidt, on the occasional discussion of a taboo on horse meat consumption in Byzantine textal sources.

47 Beech, "Reptile Bones," 175.

48 Kroll, *Tiere*, 171.

49 Ibid., 171–72.

50 N. Marković, V. Ivanišević, H. Baron, C. Lawless and M. Buckley, "The Last Caravans in Antiquity: Camel Remains from Caričin Grad (Justiniana Prima)," *Journal of Archaeological Science: Reports* (August 2021) 38:103038.

51 For the most recent overview, see Markovic et al., "Caravans."

52 Onar et al., "Theodosian Harbour."

53 Markovic et al., "Caravans."

54 Kroll, *Tiere*, 172–74.

55 G. Clark, "The Mammalian Remains from the Early Byzantine Fort of Upper Zohar," in: *Upper Zohar. An Early Byzantine Fort in Palaestina Tertia, Final Report of Excavations in 1985–1986*, ed. R. P. Harper (Oxford, 1995), 49–85; 63.

56 Kroll, *Tiere*, 174–75.

57 E.g. in Carthage, see Nobis, "Karthago," 582.

58 V. Onar, G. Köroğlu, A. Armutak, Ö. E. Öncü, A. B. Siddiq and A. Chrószcz, "A Cat Skeleton from the Balatlar Church Excavation, Sinop, Turkey," *Animals* 11 (2021): 288.

59 Benecke, "Iatrus," 395.

60 Summarised in Kroll, *Tiere*, 200–209.

61 W. Van Neer and A. Ervynck, "Fish Processing and Consumption at the Ancient City of Chersonesos (Crimean Peninsula, Ukraine)," in: *Archéologie du poisson: 30 Ans d'Archéoichtyologie au CNRS. Hommage aux travaux de Jean Desse et Nathalie Desse-Berset. XXViiie rencontres internationales d'archéologie et d'histoire d'Antibes, XiVth ICAZ Fish remains working group meeting*, ed. P. Béarez, S. Grouard and B. Clavel (Antibes, 2008), 207–17.

62 G. Dagron, "Poissons, pêcheurs et poissonniers de Constantinople," in: *Constantinople and its Hinterland. Papers from the Twenty-seventh Spring Symposium of Byzantine Studies, Oxford, April 1993*, ed. C. Mango and G. Dagron (Aldershot, 1993), 57–73.

63 G. N. Puncher, V. Onar, N. Y. Toker and F. Tinti, "A Multitude of Byzantine Era Bluefin Tuna and Swordfish Bones Uncovered in Istanbul, Turkey," *Collected Volumes of Scientific Papers ICCAT* 71.4 (2014): 1626–31.

64 Summarised in Kroll, *Tiere*, 212–16.

65 Because of their species diversity, fish and other wild animals are mainly presented here grouped by family. The family name can be recognised by the ending -idae.

66 Summarised in Kroll, *Tiere*, 216–19.

67 W. Van Neer, O. Lernau, R. Friedman, G. Mumford, J. Poblome and M. Waelkens, "Fish Remains from Archaeological Sites as Indicators of Former Trade Connections in the Eastern Mediterranean," *Paléorient* 30.1 (2004): 101–48.

68 For a summary of the Byzantine era fish remains mentioned in Van Neer and Lentacker, "The Faunal Remains," and other reports on the site, see Kroll, *Tiere*, 118–19 and 209–10.

69 Van Neer et al., "Trade," 138.

70 Summarised in Kroll, *Tiere*, 203–209.

71 For an overview, see H. Baron, "An Approach to Byzantine Environmental History: Human-Animal Interactions," in: *A Most Pleasant Scene and an Inexhaustible Resource: Steps Towards a Byzantine Environmental History*, ed. H. Baron and F. Daim. Byzanz zwischen Orient und Okzident 6 (Mainz, 2017), 171–98; 185–87. See also in detail Kroll, *Tiere*, 182–99.

72 For the, by contrast, prominent place of hunting in Byzantine literature, see the chapter by Nilsson and Messis in the present volume.

73 E.g. Benecke, "Iatrus" and Beech, "Reptile Bones."

74 Croft, "Upper Zohar."

75 E.g., Croft, "Upper Zohar."

76 For an overview, see Baron, "Interactions," 189, fig. 16.

77 H. Yu, A. Jamieson, A. Hulme-Beaman, C. J. Conroy, B. Knight, C. Speller, H. Al-Jarah, H. Eager, A. Trinks, G. Adikari, H. Baron, B. Böhlendorf-Arslan, W. Bohingamuwa, A. Crowther, T. Cucchi, K. Esser, J. Fleisher, L. Gidney, E. Gladilina, P. Gol'din, S. M. Goodman, S. Hamilton-Dyer, R. Helm, J. C. Hillman, N. Kallala, H. Kivikero, Z. E. Kovács, G. K. Kunst, R. Kyselý, A. Linderholm, B. Maraoui-Telmini, N. Marković, A. Morales-Muñiz, M. Nabais, T. O'Connor, T. Oueslati, E. M. Quintana Morales, K. Pasda, J. Perera, N. Perera, S. Radbauer, J. Ramon, E. Rannamäe, J. Sanmartí Grego, E. Treasure, S. Valenzuela-Lamas, I. van der Jagt, W. Van Neer,

J.-D. Vigne, T. Walker, S. Wynne-Jones, J. Zeiler, K. Dobney, N. Boivin, J. B. Searle, B. Krause-Kyora, J. Krause, G. Larson and D. Orton, "Palaeogenomic Analysis of Black Rat (Rattus rattus) Reveals Multiple European Introductions Associated with Human Economic History," *Nature Communications* (2022): 13:2399.

78 Yu et al, "Black Rat."

79 S. Harding, " 'Shipwreck' Scene Investigation and Bones: Subaquatic Diagenetic Taphonomy and Spatial Analysis of the Faunal Remains from the Ma'agan Mikhael B Shipwreck, Israel," *United Kingdom Archaeological Sciences Conference*. https://doi.org/10.5281/zenodo.6466919.

80 Yu et al, "Black Rat"; D. Stathakopoulos, *Famine and Pestilence in the Late Roman and Early Byzantine Empire. A Systematic Survey of Subsistence Crises and Epidemics*, Birmingham Byzantine and Ottoman Monographs 9 (Aldershot, 2004); L. K. Little (ed.), *Plague and the End of Antiquity: The Pandemic of 541–750* (Cambridge, 2007).

81 Stathakopoulos, "Famine," 45–55.

14

ANIMALS AS DIPLOMATIC GIFTS

From Species to Political Uses[1]

Nicolas Drocourt

Animals were not the most frequent of gifts throughout the Middle Ages, nor were they the most common, but neither were they uncommonly rare. This observation is true not only for the Western Christian World.[2] Indeed, the Byzantine Empire offers precise examples of this practice, along with information that leads historians to propose different interpretations of the reasons why sovereigns opted for animals instead of other goods, items, or objects. The following pages deal with this question by trying to provide an overview of a practice that occurred between Byzantium and its diplomatic partners from Late Antiquity to the last centuries of the Byzantine Empire.[3]

One should note how closely gift-giving was associated with diplomatic exchanges and peaceful encounters at that time. Many studies have demonstrated that gifts were a decisive component between diplomatic contacts. Not giving a gift or refusing to receive one was considered an insult in the practice of official negotiations. Since Marcel Mauss's survey on the function of gifts, much has been written about the social, anthropological, political, and even economic value of such exchanges.[4] In the context of Byzantine diplomacy, historians have noted that the gift itself could take different forms, depending in particular on the people and power the imperial authorities were dealing with.[5] The economic or commercial value of gifts has also been demonstrated, though this question is still open to discussion.[6] Furthermore, gifts and other goods or objects circulating with diplomatic delegations could have had a considerable cultural impact. All these aspects also concern animals when they were chosen as official gifts. Our purpose here is to present which kind of animals were selected and in which ways these gifts were interpreted or used. It concerns not only the animals sent by the Byzantine court abroad but also those received by the *basileis* and sent by different foreign sovereigns.

Diplomatic Animals on the Move

Horses were the most frequent animals chosen as diplomatic gifts, or at least the ones most frequently mentioned in our sources. Their presence covered the entire millennium of the empire. The famous reception of a Persian envoy in the mid-fifth century, detailed in the *De cerimoniis* of the tenth century, suggests some arrangements in the imperial palace if

DOI: 10.4324/9781003055877-18

the ambassador had horses among his gifts: in such cases, the three doors of the Great Consistorium had to be completely opened.[7] This reveals that such gifts were certainly frequent or regular already at that time.[8] In this context, and among numerous and luxurious gifts, the "thousand cross-breed pack horses of royal breed" sent by Chosroes II to Maurice are especially of note.[9] One should also underscore the presence of horses at the end of the period, especially with Mamluk sultans. As noted by scholars, they are regularly mentioned within the official treatises between Constantinople and Cairo.[10]

What seems remarkable is that horses appear to have been exchanged with almost all diplomatic partners of Byzantium. While Persian and Muslim partners provide many examples of such gifts, Western neighbours were not to be outdone.[11] The period of the Crusades may offer the best and most numerous examples.[12] Among various of these, when Duke Henry the Lion met Emperor Manuel I Komnenos in April 1172 during his pilgrimage to Jerusalem, he offered the latter "beautiful horses."[13] What is also remarkable is the fact that horses could be offered during a palatial encounter as well as outside the palace and outside Constantinople, even in a theatre of war. Just after being defeated in Myriokephalon by Sultan Kilij Arslan II, Manuel received from the sultan a horse with a silver-mounted bridle.[14] In 975, many horses were exchanged between John Tzimiskes and the Turkish lord of Damascus, while the former got the upper hand over the latter during one of his Syrian military campaigns.[15]

Furthermore, one should note that horses were frequently chosen among other equine animals. No less than 500 horses and 300 Egyptian asses were offered by Constantine IX to Tughril Beg, the Seljuk sovereign, in 1049. Bar Hebraeus adds that the Byzantine emperor also dispatched "a thousand goats, with black eyes and horns, which were nearly as large as asses." This gift is associated with peaceful relations at that time, when "the king of the Rhômâyê [the emperor] restored the great Mosque of the Arabs, which was in the royal city [Constantinople], and hung lamps therein."[16] A few years later, an Armenian text mentioned that the empress Theodora, ruling then as sole emperor, sent to the same Turkish sovereign some horses and "white mules," as well as a great number of valuable items and vestments in purple.[17] The Fatimid caliph was not forgotten in this regard. Already in 1046, Constantine IX sent him "a hundred and fifty beautiful she-mules," along with other selected horses, and 50 other mules carrying 50 pairs of boxes containing golden vessels as well as a great variety of fabrics.[18]

Finally, as further proof of a specifically equestrian culture of exchange, it is not rare to find in some sources mentions of objects associated with horses. The Persian king Chosroes II sent golden saddles studded with hyacinth and emeralds,[19] while a Byzantine emperor is known, thanks to an Arabic text, for his gift of four luxurious saddles to the Fatimid caliph al-Mu'izz. One of these saddles "was made of black brocade with cast-gold sides and stirrups, entirely studded with white jasper of fine quality." The emperor added straps made of black leather and a bridle, some parts of which were made of gold; and most importantly, the Arabic source testifies that this saddle and the bridle had belonged to Alexander the Great.[20] Other items produced in Byzantium and related to mechanical marvels may also be seen as part of equestrian culture such as one piece of art sent to Chosroes II by Emperor Maurice that represented a gold horseman riding on a silver charger galloping across a polo field.[21]

While horses played different roles in the Byzantine Empire, sometimes major roles, they were often associated with the practice of hunting, at least within elite circles.[22] Isaac II and Saladin exchanged not only horses and mules in 1192 but also other "hunting animals" (*zôa kynègetika*) – all these gifts were seized by a Genoese corsair near Rhodes.[23] These animals may have been dogs, like the ones sent by the *basileus* in 1052 to the Fatimid caliph.[24] In 806, the Abbasid sovereign offered four hunting dogs to Emperor Nikephoros I.[25] The choice of such animals is attested also within the relations between Constantinople and Western Christian courts. Thus, in 1178, Henry II Plantagenet sent similar animals to Manuel I.[26] We learn from the tenth-century Latin author and famous ambassador, Liudprand of Cremona, that his own father had been received as an envoy by Romanos I Lakapenos with "two dogs of a breed never seen before in that region."[27]

One should add here other animals closely associated with hunting: birds of prey. In 806, Nikephoros I was honoured by the caliph also with a gift of 12 falcons.[28] These animals were appreciated by the ruling elites of Byzantium and by their Muslim neighbours.[29] Hawks and saker falcons were sent by Imad ed-dîn Zengi to John II Komnenos.[30] These imposing birds appear in Arabic sources that describe at length the diplomatic relations between the Byzantines and Mamluks at the end of the Middle Ages. Emperor Andronikos II exchanged, among various other items, different species of falcons (notably *"burtasi"* falcons) with Egypt, while, in 1411, the Mamluk sultan received five falcons and one falconer from Emperor Manuel II.[31]

Rare or exotic animals were obviously selected as gifts by medieval sovereigns, be they Byzantine or not.[32] Rarity was one of the major aspects of official gifts since it could reinforce any sovereign's claim to superiority, or at least, nourish a kind of distinction. If we trust the eleventh-century Arabic *Book of Gifts and Rarities*, Emperor Michael VI sent Fatimid caliph al-Mustansir some animals (partridges, peacocks, cranes, herons, ravens, and starlings) that all had the particularity of being white – thus, certainly albino. The *basileus* added "huge bears that played musical instruments."[33] In 448, a tiger (*tigris*) was sent to Theodosius II from Axum.[34] This is not the only feline that we find in written sources; John II Komnenos was honoured with cheetahs by an emissary of the atabeg Zengi.[35]

Being the world's fastest land animal, the cheetah was used at that time for hunting. The truth of this assessment has recently been demonstrated not only for Byzantium's Muslim neighbours[36] but also for the Byzantine Empire.[37] A famous twelfth-century *ekphrasis*, composed by Constantine Pantechnes, celebrates an imperial hunt with cheetahs after describing a hunt with birds of prey.[38] Possessing cheetahs was a kind of privilege associated with aristocracy and the courtiers of particular sovereigns. Thus, it comes as no surprise to find these animals among diplomatic gifts; and diplomacy was certainly one of the main channels to procure them, even if the evidence remains scarce.[39] If one of the Latin testimonies can be trusted, five cheetahs were sent by the Ayyubid Sultan Saladin to Isaac II.[40] The same source indicates that on the same occasion, the Kurd sovereign also offered an ostrich to the *basileus* as well as "a little beast that makes musk" – in all likelihood, a musk deer.[41]

Much awaited in this category of rare animals, lions do not seem to have been so frequent as diplomatic gifts between Byzantium and its partners. One can find two lions, sent together with two antelopes, in the framework of contacts with China in 719.[42] At the very beginning of the twelfth century – if we can trust Albert of Aachen – the king of Jerusalem, Baldwin, sent "two lions which were tame" as well as "great pets of his" to Emperor Alexios I.[43]

Lastly, imposing animals such as elephants must also be included here. The relations with respect to these animals were asymmetric: as described or mentioned by various Greek and other sources, these animals were sent to Byzantium but never from Byzantium. It seems that there was one exception to this "rule." It concerns Late Antiquity, and recently Ekaterina Nechaeva and Mario Cristini have drawn our attention to it.[44] The khagan of the Avars received from Emperor Maurice "the most beautiful of the emperor's elephants." Theophylact Simocatta is our only source for this episode, which demonstrates that the emperor had an unknown number of elephants at his disposal. The chronicler further states that this elephant was sent at the request of the khagan himself, who then refused the gift, and the animal was sent back.

All the other elephants mentioned in our sources and sent as official gifts were received by the Empire as presents, even if they appear to have been more numerous at the beginning of the millennium. Thus, in 549, Emperor Justinian I received one through "an ambassador of the Indians," certainly to be understood as the Ethiopians.[45] Earlier, the Persian Shah Cavades I had dispatched an elephant to Emperor Zeno, as mentioned by the Syriac Pseudo-Joshua Stylites. Zeno died before the arrival of the embassy, and so it was his successor Anastasius who received this gift.[46] Obviously, this practice was more frequent than our sources let us know. Indeed, it is assumed that the "animals for triumphs" received from Persia by Theodosius I must have been elephants.[47] At the end of the twelfth century, during previously mentioned contacts, Saladin twice sent elephants to Isaac II. The second time, along with the ostrich and cheetahs, he chose a "baby elephant" to honour his correspondent.[48] Earlier, in 1053, the elephant sent by the Fatimid caliph to Constantine IX was not only recorded by Greek chroniclers but it also produced a great impression on contemporaries – as I will discuss later.[49]

This elephant did not travel alone with his mahout(s) and the Fatimid envoys: a giraffe was in its company. It is worth noting that both species were frequently offered together for a long period of time. In 496, Emperor Anastasius received an elephant and two giraffes dispatched by an unknown sovereign. The Latin author Marcellinus Comes argues that they came from India, but modern scholars suggest that they arrived from Axum or Himyar (South Arabia).[50] It is remarkable that a late fifth-century naturalist, Timotheus of Gaza, described what is probably these two giraffes and the elephant. In 1053, a giraffe's gait greatly impressed the courtier Michael Attaleiates, who described it in one of his works. In 1261, another giraffe apparently made a similar impression on its audience. Sent by Mamluk sultan Baybars to Emperor Michael VIII, this animal was considered "rare and odd" by George Pachymeres, who gave a long description of it.[51] Giraffes seem to have been one of the favourite gifts of this sultan to European courts.[52] Again in 1263, an elephant and a giraffe dispatched by the same sultan Baybars stayed in Constantinople – this time, however they were sent to the khan of the Golden Horde![53] In other cases, giraffe(s) could be offered alone, i.e., without an elephant, as in 1261. But this practice was not new. In the mid-fourth century, a giraffe had already been sent from Axum to Constantius II while he was sojourning in Antioch; others were dispatched to Constantinople in 457, also from Axum.[54] Around 573, Justin II received one from the Maccuritae, who also offered elephant tusks.[55]

Rarity logically led to more desired and valuable gifts. Such impressive and exotic animals come under this category. They were certainly all the more desired because the logistics of moving them led to many problems. Unfortunately, this aspect is not evoked in our sources, and modern historians must deal with it through deduction and hypothesis.[56] While the transport of horses or mules, by land or by sea, does not appear to have been problematic, the same cannot be said for other animals. Rare and fast animals first had to be caught before entering into this logic of diplomatic exchange. The transport of animals could lead to their death, especially when long distances had to be covered. If one trusts Philostorgius, who wrote at the beginning of the fifth century, an Indian king once sent an ape, "the so-called pan" – perhaps a baboon – to Emperor Constantius II. Due to the animal's fierceness, it was transported in a wicker cage. Unfortunately, it died before reaching the emperor. The ape was thus embalmed and offered as such to the emperor.[57] Imposing animals, such as elephants and giraffes, caused problems as well. The main issue was the need to transport them across thousands of kilometres. Mahouts and various specialised guides were necessary, and special logistics were required for their feeding and protection.[58]

For these animals, as well as many others, a sea voyage seems to have been a more practical route between Byzantium and most of its Muslim neighbours or partners, especially Egyptians, be they Fatimids, Ayyubids, or Mamluks. Nevertheless, while seaways were the quickest travel option, they were not the safest. The fate of the official travellers between Ayyubid Egypt and Constantinople in 1192 demonstrates this. Not only were horses, mules, and "hunting animals," as already mentioned, seized by pirates but so were wild and tamed animals en route from Africa and Egypt.[59] This case demonstrates that some logistical elements may be known only through accidents recorded in primary sources.[60] Many aspects of these animals' journeys are unknown, notably the percentage of losses due to death or disease along the way. It, therefore, remains impossible to answer the simple question: did all animals mentioned as official gifts in the sources effectively reach their recipients?[61]

The question of the cost of feeding animals during their journeys and once they had reached their destination also remains unanswered. High expenses may explain the preference for young animals, which would cost less and raise only minor difficulties when transported.[62] Once they had arrived, training and control were further issues to be solved, in particular, if the animals were meant to perform in a political or symbolic context (an aspect studied later in this chapter). The elephant received by Justinian in 549 was especially memorable because it broke out of its stable one night, killed many individuals, and maimed others.[63] Indeed, all of these logistical questions could appear to be less important than other aspects; nevertheless, as previously stated, mastering these elements of transport and care, as well as the costs they incurred, was certainly a major way for a sovereign to demonstrate his concrete ability to resolve major problems.[64]

Numerous Animals, Different Purposes

We should also consider the different purposes related to the offering of animals within the framework of diplomatic relations. An official gift is never insignificant. A careful reading of our sources together with a precise understanding of the context of an official contact usually leads modern historians to explain the reasons behind selections made by a sovereign and his administration. As noted by Nechaeva regarding Late Antiquity, gifts were strong indicators of the Byzantine Empire's attitude toward its partners, and items received also "marked the status of a barbarian ruler [...] at the international level."[65] This seems acceptable for later centuries as well and is true not only from a Byzantine point of view. Choosing an item as an official gift that would represent you as a sovereign was connected to other aspects, notably the importance and rank of the sovereign who would receive it, the context and precise geopolitical situation, the level of negotiations, and so on. All this appears particularly true when the gift was an animal.

Of primary importance was the gift giver's desire to please the recipient. This attitude was a way of thanking a diplomatic partner and/or of persuading him of the legitimacy of an agreement concluded, or to be concluded, with him. When Emperor Michael VI dispatched white animals to the Fatimid caliph, he would have certainly known that this colour received special attention from members of the dynasty: white was indeed what may be called its official colour.⁶⁶ From the same eleventh-century Arabic text comes an explanation that the Byzantine governor of Egypt, in response to Prophet Muhammad, who had invited him to become a Muslim, sent the Prophet "a gray she-mule equipped with saddle and bridle [...] a gray donkey [...]" and "a horse": the Prophet gave a name to each one, and the text specifies that the mule and donkey were Muhammad's "favourite mounts."⁶⁷ Although we may doubt the veracity of such details, the pleasure of transmitting and/or receiving certain animals as gifts is sometimes apparent in the official correspondence between two sovereigns. We learn this when reading a letter sent to the Mamluk sultan Barqûq by Manuel II Palaiologos in 1411: "As we know that your Majesty appreciates falcons, we dispatch five of these animals and one falconer."⁶⁸ This example reminds us that the choice of hunting animals was much appreciated by sovereigns who were known or renowned for their interest in, or even their passion for, hunting. One can interpret in this way the choice of dogs sent by Henry II to Manuel I.⁶⁹ This *basileus*'s passion for hunting with falcons is well known, and some gifts of this kind have already been mentioned.⁷⁰

While receiving an animal as a gift could be a pleasure, it could also be associated with the notion and reality of hunting. As the interest in this practice was largely shared by elites and sovereigns, whether Byzantine or not, it comes as no surprise to find that animals and a variety of objects associated with hunting were offered as diplomatic gifts. The imperial practice of hunting in Byzantium has already been mentioned. Clearly, it appears as a shared culture between political elites that went beyond many borders (i.e., geographic, cultural, or religious). Offering horses, notably rare breeds and thoroughbreds, either of "royal breed" or those qualified as *electi equii* in Latin sources (or *ekkritoi hippoi* in Greek sources), and dispatching precious items related to equestrian culture (saddles, bridles, etc.) seem to have been a perfect choice in order to satisfy a diplomatic partner. It is no coincidence that joint hunts were organized during official encounters between the *basileis* and foreign sovereigns.⁷¹

The sending and offering of horses must be considered customary, it was possibly more frequent between Byzantium and its oriental partners – e.g., Persians, Arabs, and Turks – than with others. One concrete example, complementary to those proposed in a previous study, indicates this to be true.⁷² According to the late eighth-century handbook on taxes written by Abû Yûsuf Ya'kûb, goods brought into Abbasid's territories for sale were tithed. However, there was an exception to this rule: any envoy "from the Christian prince" was not subject to this treatment, and his diplomatic status was recognizable from the fact that he was accompanied not only by a letter validating his mission but also by various goods, notably horses.⁷³

Concerning relations with the West, horses were less numerous as official gifts sent from or to Byzantium, at least until the twelfth century.⁷⁴ They could be offered in the context of military conflict. Since diplomacy has often been associated with war, notably during the middle-Byzantine period,⁷⁵ animals as official gifts should be seen as tools for war in some cases. This is particularly true with horses, as becomes clear upon a careful reading of the sources. In the case of Henry the Lion's gift to Manuel I in 1172, Arnold of Lübeck provides an insightful detail: the horses Henry the Lion offered were associated with "saddles and harness, hauberks [and] swords," among other gifts.⁷⁶ Therefore, they had a military purpose. Around 1088–1089, no fewer than 150 chargers were dispatched by the Count of Flanders to Alexios I Komnenos. They were added to the 500 knights also sent by the count to serve the military interests of the *basileus*.[77] The crusades may have increased the need for horses, and it is thus no surprise to see them mentioned both as gifts and for military use in some circumstances.[78] Not only did such animal gifts bring Latin and Byzantine interests closer together against Muslim neighbours but they could also serve Byzantine and Muslim interests, and even alliances, against Latin neighbours. During the period of the third crusade, if Magnus of Reichersberg is to be believed, no fewer than 1,050 "Turkish or Turcoman warhorses" were sent by Saladin to Isaac II.[79]

One should note that this attitude of choosing horses for diplomatic purposes only within a military campaign may have followed some normative recommendations in Byzantium. In the mid-tenth century, Emperor Constantine Porphyrogennetos wrote that 100 "complementary horses" should be included in the list of required items for an emperor who was going on campaign with his army. They were to be gelded and then used as gifts dispatched to foreigners.[80] But horses or other equine animals during military manoeuvres could also be put to non-military use, especially as pack animals. As already mentioned, 50 mules carrying "fifty pairs of boxes" and "covered with fifty pieces of silk thin brocade" were sent from the emperor to the Fatimid caliph in 1046.[81] In the *Book of Gifts and Rarities* from that same century, the heavy weight of diplomatic gifts is emphasized; and mules, she-mules, and/or asses are directly concerned as they carried these gifts.[82] The same document relates that in 1071, a Byzantine's gift included a piece of brocade "with compact thick embroidery," carried by "a single mule, which could not carry anything else besides."[83] The role of these mules as pack animals is clear, even though other animals could have played this same role and thus been chosen at the same time as diplomatic gifts. In 1172, Manuel I Komnenos offered Henry the Lion, who was back from Jerusalem, "fourteen mules, with gold and silver harness and silk saddle clothes," during an exchange which we will discuss later in this chapter.[84] Of course, other animals could have fulfilled this task during diplomatic moves as well: "fifty camel loads of costly gifts" were brought back from Baghdad to Constantinople by Byzantine emissaries as a gift from Abbasid caliph al Mûtasim to Theophilus.[85]

Beyond these initial reasons, other important purposes should also be mentioned. Various sources demonstrate that Byzantine emperors may have used animal gifts to serve their political interests and to demonstrate their superiority, especially by displaying the animals – and of course, the same logic is valid for other political circles outside of the Empire. The carrying and offering of animals could thus reveal the apparently good state of relations between two sovereigns or two courts. The gift of giraffes sent regularly to Constantinople from Axum has recently been interpreted as a sign of good relations between this kingdom and Constantinople following the Christianization decided by King Ezana at the beginning of the fourth century.[86] This observation is also valid for the elephant and giraffe sent by the Fatimid caliph to Constantine IX in 1053, just before Byzantine–Fatimid relations became strained. As underlined by Antony Cutler, the delivery of these two rare animals may be understood as a response to the shipment of grain dispatched to Egypt by the *basileus* in the time of famine.[87]

Of course, this kind of rare gift bestowed prestige on the receiver, but only after enhancing the apparent superiority of the sender. In this context, when the Byzantine court welcomed such exotic animals, it could potentially have put the court in a situation of inferiority or dependency, but imperial propaganda would then distort its interpretation into something more advantageous for the image of the emperor. In one of his famous *basilikoi logoi*, Michael Psellos alleged that an elephant recently received from Cairo knelt before Emperor Constantin IX, and even touched the ground with its forehead. This gesture was evidently a kind of *proskynesis*, i.e., the ritual prostration required by a subject in the presence of any *basileus*.[88] In another official speech, the same author and rhetorician recalled an old *topos*: foreign peoples were coming to the emperor from throughout the world, but instead of offering their usual items (silks, fabrics, and rare stones), they were giving animals that they possessed in great number, which they had refused to dispatch to previous emperors.[89] As such, this idea of different wild animals offered as gifts and/or as tribute to the emperor echoes the famous sixth-century ivory plaque now at the Louvre – the so-called Barberini ivory.[90]

For the case dating back to 1053, it should also be pointed out that these animals were not displayed solely to members of the court. Here, another contemporary witness should be mentioned: Michael Attaleiates. Along with other Greek chroniclers, he provided a rather long description of these two extraordinary animals.[91] While he acknowledged that the giraffe came from Egypt, he did not make this same observation for the elephant; on the contrary, he implied that this animal was exhibited at the initiative of only the emperor. "For the benefit of his subjects," explained Attaleiates, the emperor "brought strange animals from foreign lands, among them an elephant, the greatest of all four-legged creatures and an amazing sight to the Byzantines." This assertion was followed by a long anatomic description, this was also the case for the giraffe, which he called a "camel-leopard" and described as "a composite animal having the spots of the leopard but the size of the camel [...] and a thick, elongated neck."[92]

What seems important here is the way both of these animals were exhibited, so as to reinforce the *basileus*'s prestige and superiority among his subjects in Constantinople. The same logic remained true more than two centuries later in this same city, which had recently been reconquered by the "emperor of the Romans," Michael VIII Palaiologos. The giraffe he received from the Mamluk sultan, Baybars, was displayed daily at the agora, a spectacle that was "a delight for the audience."[93] This logic of exhibition can also be witnessed before 1261. The contemporaries and Greek authors who described it generally used the term θαῦμα (marvel).[94] Similar events even recurred across the centuries. The already mentioned Timotheus of Gaza spoke of giraffes and an elephant brought to Anastasios I in 496. Paraphrasing the naturalist's account, an anonymous eleventh-century author commented on the arrival of such animals by diplomatic means that "this was seen in our times too [...] these two animals [...] were at each opportunity shown to the people as a marvel, in the theatre of Constantinople."[95]

In October 549, the elephant received by Justinian arrived during the races, and then entered the hippodrome, as mentioned by Theophanes.[96] It is not hard to imagine how the emperor exploited this arrival to his advantage. In any case, there were already tight links between the exhibition of rare animals and the different places where mobs would gather in old Rome, and the emperors of the new Rome continued the practice.[97] At least until the twelfth century, the hippodrome of Constantinople remained a major public and political space, where official guests and prestigious foreigners were invited.[98] The memory of the display of elephants that were offered as gifts remained significant there throughout the mid-Byzantine period, and the various elephant statues in the city should also be noted.[99]

Exotic animals, whether or not received through diplomatic channels, may have taken part in the official ceremonies in Byzantium. Cheetahs, for example, played a role according to the fourteenth-century ceremonial treatise of Pseudo-Kodinos. When a princess-to-be-wed was led to the palace, cheetahs were part of the guard of honour near her, and they would ride pillion behind their handlers.[100] Cheetahs were also displayed at the hippodrome if one is to trust the testimony of al-Marwazî, an Arab naturalist at the turn of the twelfth century and an eyewitness hosted in Constantinople.[101] These animals performed different forms of hunting scenes, setting upon antelopes, like lions upon bulls and dogs upon foxes. Whether the animals were offered to the Byzantines as diplomatic gifts or not, their presence and the fact that they were displayed, also raises the question of menageries in or around Constantinople. This remains a point of discussion, at least for the period up to the mid-eleventh century when a giraffe and elephant were offered to Constantin IX and subsequently displayed.[102]

Recent archaeological excavations in the Theodosius harbour of Constantinople confirm the fact that exotic or more common animals were present there between the fifth and eleventh centuries, destined for transport or to remain in the city.[103] As with other exotic animals under the control of the Byzantine authorities, we assume that cheetahs were present in hunting parks such as the Philopation or the Aretai.[104] The latter may have housed various animals from all over the world, as depicted by the tenth-century Byzantine author John Geometres.[105] Of course, Byzantine envoys could have been confronted with the display of wild animals during a stay abroad, notably in Muslim territories. Such a display famously took place for the reception of an embassy in Baghdad in 917. The continuator of Tabarî, 'Arib, describes many goods and jewels as well as gold and silver objects laid out in the Abbasid's capital to amaze the Greeks, as did the "elephants, giraffes, lions, and panthers" that were paraded on the banks of the Tigris.[106]

With this last example, the question of competitiveness appears central. Diplomacy is often synonymous with rivalry and competition, and gifts play a major role. Exotic or rare animals enter into this logic. In his famous *Life of Constantine*, Eusebius of Caesarea mentions Indian embassies bearing gifts, and especially "animals of breeds different from those known among us."[107] If the Latin author and ambassador Liudprand is to be believed, his father reached Constantinople ca. 927 with gifts including "two dogs of a kind never seen before in that region."[108] But Liudprand says more: when the dogs were led before the emperor, they certainly would have "mangled him with their bites if the arms of many had not restrained them." The Latin envoy justifies such behaviour of the dogs by the fact that the emperor was dressed in an unusual costume, which led the dogs to consider that "he was not a man but some monster."[109] Of course, while modern readers must be careful with such remarks, by taking into account the degree of exaggeration and/or rewriting of past events, this game of competitiveness also occurred during official meetings in military campaigns.

The case is clear thanks to an Arabic account by Ibn al Qalanisi; when John Tzimiskes met Alptekin, master of Damascus, in 975, not only were the two sovereigns vying to be the most dexterous in equestrian exercises but they were also competitive in the field of gift-giving, in which horses played a central role.[110] Some of these horses were presented as animals that were only used in ceremonial circumstances; it should be noted that such a horse was offered by the Seljuk sultan to Manuel Komnenos just after the latter was defeated in 1176 by the sultan. Niketas Choniates called it a "Nisaean horse," a name referring to the work of Herodotus.[111] A similar horse appeared in an equestrian joust in the presence of the same emperor, Manuel I Komnenos, as described in an anonymous Vatican manuscript.[112]

Competitiveness with affirmation of superiority and prestige also occurred in other ways. Mastering the "recycling" of gifts is one of these: when the Fatimid caliph sent the emperor huge bears able to play musical instruments, he was also demonstrating this recycling, or reuse, at the heart of gift-giving.[113] Furthermore, this example underscores another aspect of diplomatic gifts exchanged with Muslim partners: the sending of rare items, be they animals or not. Exotic animals may be included *de facto* in the list of marvels, inherent in the well-known Islamic concept of *tuhaf*.[114] Another example of this aspect was *ars mechanica*: these "mechanical wonders" were figurative animals presented as *automata*, as previously mentioned for a sixth-century piece of art offered by Maurice Chosroes II.[115]

The question of iconographic representation of these animals offered as diplomatic gifts should also be mentioned here and is directly related to the impression they made on their contemporaries. In the sixth century, giraffes thus appeared in mosaics in two different places near Gaza; these representations are considered the direct consequence of the passing presence of these animals in the city and surrounding area a few years earlier.[116] The crude representation of an African elephant and its handler in a fragmentary papyrus of the same period has been connected to the elephant which travelled, with two giraffes, from "India" to Constantinople in 496.[117] Philostorgius saw a representation of a rhinoceros in Constantinople and gives a description of this animal *a priori* sent by diplomatic means.[118] Hunting imagery produced in Byzantium also reflected official interest in and curiosity for wild or exotic animals, and some such objects were circulated by diplomatic means.[119]

The presentation of animals as diplomatic gifts also implies or illustrates how strong the military or political rivalry which could have existed between Byzantine emperors and other sovereigns was. Three final examples will demonstrate this point before our conclusion. First, Arnold of Lübeck's description of the two official encounters between Duke Henry the Lion and Manuel I Komnenos remains very significant regarding this topic. As mentioned previously, during his first stay the duke offered Komnenos some horses along with a variety of military objects in April 1172. Back from Palestine, he met the emperor again a few months later. If we are to trust the Latin chronicler, Manuel "was overjoyed [...] and kept him [in the place where they met] with great honours for some days." Furthermore, "he gave him fourteen mules, with gold and silver harness and silk saddle clothes."[120] But the duke did not accept these gifts, even though the *basileus* "pressed him hard"; instead, Manuel eventually offered him "many precious relics of the saints, as he [the duke] has requested." These exchanges of gifts between the two sovereigns have led historians to various interpretations.[121] One can see it as a kind of emulation, where equine animals were central. As the mules would have been considered inferior to horses, this would explain the duke's refusal.[122] The scene also demonstrates – at least in the words of Arnold – that the duke, and not Manuel, had the last word in this matter.[123]

Moreover, in the same chronicle, and described between these two encounters, another passage appears noteworthy. The passage concerns the way the same duke, Henry the Lion, was received by the sultan of the Turks, Kilij Arslan II, on the duke's journey back from Palestine, but prior to his second meeting with Manuel I. Arnold of Lübeck describes a very cordial welcome by the sultan, who gave Henry many gifts of great value. Among them were "1,800 horses [...] for the duke to choose from them whoever he wanted." Thirty "most powerful horses were then brought forward, with silver reins and superb saddles made of precious cloths and ivory"; the sultan added to these, and others, gifts: "two leopards (*duos leopardos*) and horses and servants: for these leopards had been taught to sit on the horses."[124] In this case, the Latin chronicler also demonstrates that the sovereign of the Turks acted like a *basileus*, but that he was even more generous than Manuel I.[125]

When tensions arose, animals sometimes stood at the heart of opposition between two sovereigns, and they occasionally paid the price of this tension with their lives. Niketas Choniates depicted Alexios III Angelos's anger when he learned that the Seljuk sultan Kaykhusraw had seized two Arabian stallions previously sent to him by the Ayyubid sultan of Egypt.[126] Even though the Seljuk sovereign sent an embassy to Alexios asking for his forbearance, and explaining in particular that an injury to one of the two horses had prevented him from sending it along with the other one, the emperor decided then to imprison Roman and Turkish merchants from Ikonion, the capital of Kaykhusraw, as a hostile response to the seizure. This was a choice which eventually led to war between the Seljuk Turks and the Byzantines. This diplomatic case once again raises the issue of a third party's territory being crossed by embassies exchanged between two other states – i.e., exchanges that harmed the third party's state.

A similar case occurred in 1263, but this time it was the emperor who retained animals that had been exchanged as diplomatic gifts between two other sovereigns. From Egypt, Baybars sent an embassy to Berke, khan of the Golden Horde. Various Arabic accounts detailed the luxurious and numerous gifts associated with this embassy. These included zebras, monkeys, Arabian racehorses, rare camels, Egyptian donkeys, an elephant, and a giraffe.[127] This Mamluk embassy had to conclude an alliance against Hulagu and the Ilkhanid hegemony. Nevertheless, Emperor Michael VIII blocked the delegation when it reached Constantinople. Michael VIII did not wish to offend certain Ilkhanid envoys who were present at the time in the capital of the recently restored Empire, and he must certainly have been against the rapprochement of two distinct partners. Furthermore, the vessel that carried the Mamluk delegation remained on the Bosporus for more than a year and, during this time, most of the animals perished. We know then that Berke legitimately complained about this gruesome fate to another envoy of Baybars.

Conclusion

In the course of a millennium, numerous animals appeared as diplomatic gifts between Byzantium and its various neighbours. The sources, both Greek and non-Greek, that mention or describe these animals offer in some cases precise and valuable information concerning their origins, their species, and the way the Byzantine power used them.

Following the previous pages of this chapter, what seems remarkable at first sight is the continuity in the animals encountered in these diplomatic contacts throughout the centuries. Horses, and animals dedicated to hunting such as falcons and dogs, were regularly chosen in this context. This indicates the extent to which such practices were shared among medieval political elites, whether they were Christian, as in Byzantium, or not. Secondly, rare and/or exotic animals were also a choice of sovereigns to communicate with each other. Moving such animals raised many problems, especially for larger animals such as giraffes or elephants. But what seems most important is the political significance and the uses the Byzantine sovereigns made of them. Such meanings and uses were clearly related to the nature of these animals: the choice of a dog versus that of an elephant would certainly not have had the same meaning. Furthermore, in addition to the kind of continuity one may note in the choice of some of these animals, we can also find a certain degree of disparity and change. No more elephants are mentioned as official gifts after 1053 as far as I have found. This stands in contrast to the period of Late Antiquity, when elephants belonged to the culture, warfare, and iconography of the period, and were thus so numerous in diplomatic exchanges.[128]

The rarity of particular animals and their exhibition reinforce the idea that the emperor's own majesty was connected to these extravagant animals. Some of the animals, like the giraffe, may even be assimilated as the "property of kings."[129] Furthermore, such species do not offer any commercial nor economic value: how to estimate such value for an elephant, a giraffe, or a rhinoceros?[130] It is certainly no coincidence that these rare gifts are widely discussed in the sources we can still rely on so many centuries later – notably, the giraffe and the elephant sent by the Fatimid caliph to Constantinople in 1053. Also, the complex challenges of controlling and transporting these animals before they reached their destination were likely meant to demonstrate the superiority of those sending them. Lastly, as official gifts, these animals were sometimes directly affected by geopolitical rivalries and confrontations, as our last examples have demonstrated.

This reasoning transcends the Byzantine period as understood in a strict sense. Among the various gifts sent to the Mamluk sultan by Mehmet II to announce the latter's conquest of Constantinople in May 1453, is it any surprise to find two choice stallions, two elephants, and a zebra?[131]

Notes

1 The author would like to thank Petros Bouras-Vallianatos, Thierry Buquet, Stanley Burstein, Elisabeth Malamut, Tristan Schmidt and Nancy P. Ševčenko. He also expresses his warmest thanks to Don Stoudt for his careful reading of, and linguistic corrections to, this chapter.

2 See notably J.-M Moeglin and S. Péquignot, *Diplomatie et "relations internationales" au Moyen Âge, IXe-XVe siècle* (Paris, 2017), 213–17, and 247–49; M. Pastoureau, "Les ménageries princières: du pouvoir au savoir (XIIe-XVIe s.)," in *Les signes et les songes. Etudes sur la symbolique et la sensibilité médiévales*, ed. M. Pastoureau (Florence, 2013), 43–72.

3 A previous article has already tried to present and interpret sources, only focusing on the Middle Byzantine period: N. Drocourt, "Les animaux comme cadeaux d'ambassade entre Byzance et ses voisins (VIIe-XIIe siècle)," in *Byzance et ses périphéries. Hommage à Alain Ducellier*, ed. B. Doumerc and C. Picard (Toulouse, 2004), 67–93. The gift of exotic animals was a reality during diplomatic exchanges in Antiquity: L. Bodson (ed.), *Les animaux exotiques dans les relations internationales. Espèces, fonctions, significations* (Liège, 1998). See also M.-T. Mansouri, "Les animaux exotiques: des présents entre souverains musulmans et chrétiens au Moyen Âge," in *L'Homme et l'Animal au Maghreb, de la Préhistoire au Moyen Âge. Explorations d'une relation complexe*, ed. V. Blanc-Bijon, J.-P. Bracco, M.-B. Carre, S. Chaker, X. Lafon and M. Ouerfelli (Aix-en-Provence, 2021), 337–42.

4 See the general considerations of A. Cutler, "Les échanges de dons entre Byzance et l'Islam (IXe-XIe s.)," *JSav* (1996): 51–66; J. Nelson, "The Role of the Gift in Early Medieval Diplomatic Relations," *Le relazioni internazionali nell'alto medioevo, Settimane* 58 (2011): 225–48; D. Behrens-Abouseif, *Practising Diplomacy in the Mamluk Sultanate; Gifts and Material Culture in the Medieval Islamic world* (London and New York, 2014), 17–33; Drocourt, "Les animaux," 74.

5 On this specific theme, see the data on Late Antiquity analysed by E. Nechaeva, *Embassies-Negotiations-Gifts, Systems of East Roman Diplomacy in Late Antiquity* (Stuttgart, 2014), 174–95; for later periods: P. Schreiner, "Diplomatische Geschenke zwischen Byzanz und dem Westen ca. 800–1200: Eine Analyse der Texte mit Quellenanhang," *DOP* 58 (2004): 251–82; A. Cutler,

"Significant Gifts: Patterns of Exchange in Late Antique, Byzantine and Early Islamic Diplomacy," *Journal of Medieval and Modern Greek Studies* 38 (2008): 79–102.

6 A. Cutler, "Gifts and Gift Exchanges as Aspects of Byzantine, Arab and Related Economies," *DOP* 55 (2001): 245–278, *contra* A. E. Laiou, "Exchange and Trade, seventh-twelfth centuries," in *The Economic History of Byzantium from the Seventh through the Fifteenth Century*, ed. A. E. Laiou (Washington DC, 2002), 699–700.

7 Constantine Porphyrogennetos, *The Book of Ceremonies*, I. 89, ed. J. J. Reiske, *Constantinus Porphyrogenitus, De Ceremoniis aulae byzantinae libri duo*, (Bonn, 1829–1830), 405, and see now *Constantin VII Porphyrogénète, Le Livre des cérémonies*, sous la direction de Gilbert Dagron et Bernard Flusin (Paris, 2020), I. 98, vol. II. 398–99. The date of 551 for the arrival of this ambassador is proposed by Nechaeva, *Embassies*, 35 and 196.

8 For other examples of horses and mules as gifts between Persia and Byzantium, see Nechaeva, *Embassies*, 177–79, and the appendix on 234.

9 An eleventh-century Arabic compilation gives this information: *Book of Gifts and Rarities, Kitâb al-hadâyâ wa-l-tuhaf*, § 5, trans. G. al-Hijjâwî al-Qaddûmî (Cambridge, MA, 1996), 63 (and see philological note 10, 252; on this episode, see also the remarks of Cutler, "Gifts and Gift Exchanges," 251).

10 See the remarks of E. Malamut, "Les cadeaux entre souverains byzantins et étrangers aux XIIe-XVe siècles," in *De la guerre à la paix en Méditerranée médiévale. Acteurs, propagande, défense et diplomatie*, ed. E. Malamut and M. Ouerfelli (Aix-en-Provence, 2021), 239–65, at 251 and 263; more broadly: M. T. Mansouri, *Recherches sur les relations entre Byzance et l'Egypte (1259–1453) (d'après les sources arabes)* (Tunis, 1992), 126 ff.

11 Even if such Western neighbours seem less numerous: Drocourt, "Les animaux;" Schreiner, "Diplomatische Geschenke." Concerning northern neighbours, horses were sometimes chosen within the relations between Constantinople and the Huns, thanks to Priscus: see Nechaeva, *Embassies*, 198 and 249; some gold-gilded saddles were offered by Justinian to the Avars: Nechaeva, *Embassies*, 180 (after John of Ephesus).

12 See Drocourt, "Les animaux," 68 and 91.

13 *The Chronicle of Arnold of Lübeck*, 1. 3., trans. G. A. Loud (London and New York, 2019), 46. See notably H. A. Klein, "Eastern Objects and Western Desires: Relics and Reliquaries between Byzantium and the West," *DOP* 58 (2004): 283–314, at 285, n. 11 (and see later in this article).

14 Niketas Choniates, *History*, ed. J. L. van Dieten, *Nicetae Choniatae Historia* (Berlin and New York, 1975), 189. This horse is called "Nisaean" by Choniates, who gives additional information about it that will be treated later. The Sultan also offered a long, two-edged sword to the *basileus*.

15 M. Canard, "Les sources arabes de l'histoire byzantine aux confins des IXe et Xe siècles," *REB* 19 (1961): 284–314, at 295 (after Ibn al-Qalānisī); M. M. Vučetić, *Zusammenkünfte byzantinischer Kaiser mit fremdem Herrschern (395–1204). Vorbereitung, Gestaltung, Funktionen* (Berlin, 2021), 266–67 (and its appendix #44, 53–56).

16 *The Chronography of Gregory Abū'l Faraj*, trans. E. A. W. Budge, vol. 1 (Oxford, 1932), 206; *The Annals of the Saljuk Turks: selections from al-Kāmil fī'l-Ta'rīkh of 'Izz al-Dīn Ibn al-Athīr*, trans. D. S. Richards (Richmond, 2002), 144, who adds to the animals already mentioned "300 Shihri mules." On this episode and its context see: A. Beihammer, *Byzantium and the Emergence of Muslim-Turkish Anatolia, ca. 1040–1130* (Farnham, 2017), 92–93.

17 Aristakès de Lastivert, *Récits des malheurs de la nation arménienne*, trans. M. Canard and H. Berberian, (Bruxelles, 1973), 88. This contact occurred before the seizure of Bagdad (December 1055) by Tugril Beg according to Aristakès.

18 *Book of Gifts*, § 82, trans. al-Hijjâwî al-Qaddûmî, 108–9; on which, see Cutler, "Gifts and Gift Exchanges," 271.

19 Nechaeva, *Embassies*, 197.

20 *Book of Gifts*, § 99, trans. al-Hijjâwî al-Qaddûmî, 114.

21 Nechaeva, *Embassies*, 179–80 and the references to an Arabic text, the *Book of Beauties and Antitheses (Kitab al Mahasin wal Azdad)* by Pseudo-Jāhiz.

22 See T. G. Kolias, "The Horse in the Byzantine World," in *Le cheval dans les sociétés antiques et médiévales*, ed. S. Lazaris (Turnhout, 2012), 87–97; on the practice of hunting, the bibliography is large, from E. Patlagean, "De la chasse et du souverain," *DOP* 46 (1992): 257–63 to T. Schmidt, *Politische Tierbildlichkeit in Byzanz Spätes 11. bis frühes 13. Jahrhundert* (Wiesbaden, 2020),

196–238, with further bibliography. See also the chapter by Ch. Messis and I. Nilsson in this volume.

23 *Acta graeca Medii aevii sacra et profana*, ed. F. Miklosich and J. Müller, 6 vols. (Vienna, 1860–1890), vol. 3, 38; F. Dölger, *Regesten der Kaiserurkunden des oströmischen Reiches von 565–1453, 2. Teil, Regesten von 1025–1204* (Munich, 1995), n° 1612. The ambassadors and merchants who were on the boat were killed. Twenty-seven saddles studded with precious stones and pearls are also mentioned, cf. Cutler, "Gifts and Gift Exchanges," 267, n. 117.

24 *Book of Gifts*, § 85, trans. al-Hijjâwî al-Qaddûmî, 110: "Salukis [hunting dogs] and guard dogs (*dhabībiyyah*)," and see 298, n. 4.

25 M. Canard, "La prise d'Héraclée et les relations entre Hârûn al-Rashîd et l'empereur Nicéphore Ier," *Byzantion* 32 (1962): 345–79, at 360.

26 See the references in Schreiner, "Diplomatische Geschenke," 282; A. A. Vasiliev, "Manuel Comnenus and Henry Plantagenet," *BZ* 29 (1929–1930): 233–44, at 243; Malamut, "Les cadeaux," 261.

27 *The Complete Works of Liudprand of Cremona*, trans. P. Squatriti (Washington DC, 2007), 119 (*Antapodosis*, III, 23).

28 M. Canard, "La prise d'Héraclée," 360.

29 Cf. A. Külzer, "Some Notes on Falconry in Byzantium," in *Raptor and Human – Falconry and Bird Symbolism Throughout the Millenia on a Global Scale*, ed. K.-H. Gersman et O. Grimm (Hamburg, 2018), 699–706.

30 Kemal ed dîn (Kamāl al-dīn), *Extraits de la chronique d'Alep (Zubdat al-halab fî ta'rîkh Halab)*, in *Recueil des historiens des Croisades, Historiens orientaux*, 5 vols. (Paris, 1872–1906), vol. 3, 674.

31 Mansouri, *Recherches*, 124, 134, and 238; Behrens-Abouseif, *Practising Diplomacy*, 129; Malamut, "Les cadeaux," 262.

32 On the topics of "exotic" and "exotic animals," see T. Buquet, "Les animaux exotiques dans les ménageries médiévales," in *Fabuleuses histoire des bêtes et des hommes*, ed. J. Toussaint (Namur, 2013) 97–121, at 98, and N. P. Ševčenko, "The Giraffe that Came to Constantinople," in *The Eloquence of Art. Essays in Honour of Henry Maguire*, ed. A. Olsen Lam and R. Schroeder (Abingdon/New York, 2020), 336–49, at 348, n. 37.

33 *Book of Gifts*, § 85, trans. al-Hijjâwî al-Qaddûmî, 110.

34 Nechaeva, *Embassies*, 199 (after Marcellinus Comes), and 250.

35 Kamāl al-dīn, 674; Drocourt, "Les animaux," 69.

36 This is especially well-known for Syria and Iraq: T. Buquet, "La belle captive. La girafe dans les ménageries princières au Moyen Âge," in *La bête captive au Moyen Âge et à l'époque moderne*, ed. C. Beck and F. Guizard (Amiens, 2012), 63–90 (and see the illustrations); Buquet emphasizes that "selon la littérature cynégétique arabe médiévale [...] on dressait principalement les guépards à chasser le lièvre et la gazelle, car, plus rapide que le chien, le guépard peut rattraper ces deux animaux en pleine course sur terrain découvert."

37 N. Nicholas, "A Conundrum of Cats. Pards and their Relatives in Byzantium," GRBS 40 (1999): 253–98, at 259–62; A. Papagiannaki, "Experiencing the Exotic: Cheetahs in Medieval Byzantium," in *Discipuli dona ferentes. Glimpses of Byzantium in Honour of Marlia Mundell Mango*, ed. T. Papacostas and M. Parani (Turnhout, 2017), 223–57, at 231–39.

38 See M. E. Miller, "Description d'une chasse à l'once par un écrivain byzantin du XIIe siècle de notre ère," *Annuaire de l'association pour l'encouragement des études grecques en France* 6 (1872): 28–52, and more recently Nicholas, "A Conundrum of Cats," 260, and 290–96; on this text, see the commentaries of Papagiannaki, "Experiencing the Exotic," 230, 238 and 240 (with comparison to Arabic and Indian sources); comparison to other Arabic sources: T. Buquet, "Le guépard medieval, ou comment reconnaître un animal sans nom," *Reinardus* 23 (2011): 12–47, at, 16–17; Buquet, "La belle captive."

39 Papagiannaki, "Experiencing the Exotic," 239, and the reference to the *Book of Gifts and Rarities*, in which cheetahs not only appear as animals for hunting but also as gifts between members of the Arab ruling elite; see also 241–42.

40 *Magni Presbiteri Chronicon*, ed. W. Wattenbach. MGH SS, vol. XVII (Hannover, 1861), 512, *sub anno* 1189; trans. G. A. Loud, *The Crusade of Frederick Barbarossa: the History of the Expedition of the Emperor Frederick and Related Texts* (Farnham/Burlington, 2010), 155. These

cheetahs were sent together with other exotic animals. They are mentioned as *"leopardos"* and may indeed be cheetahs, as interpreted by Papagiannaki, "Experiencing the Exotic," 241, but the date of 1192 given for this diplomatic gift is not correct, and must be backdated, certainly to around 1189–1190: see A.-M. Eddé, *Saladin* (Paris, 2008), 285; the date of 1188 is proposed by T. Buquet, "Aspects matériels du don d'animaux exotiques dans les échanges diplomatiques," in *Culture matérielle et contacts diplomatiques entre l'Occident latin, Byzance et l'Orient islamique (XIᵉ-XVIᵉ s.)* ed. F. Bauden, (Leiden and Boston, 2021), 177–202, at 181.

41 Coming from central Asia, as suggested by Loud, *The Crusade of Frederick Barbarossa*, 155, n. 10.

42 See Drocourt, "Les animaux," 70, and further bibliography; M. S. Kordosis, *T'ang China, the Chinese Nestorian Church and "Heretical" Byzantium (AD. 618–845)* (Ioannina, 2008), 90, suggests that these lions came from Central Asia, and not from Byzantium.

43 Albert of Aachen, *Historia Ierosolomitana*, 8.47, ed. S. B. Edgington (Oxford, 2007), 636–37. On the other hand, Byzantine lions appear in Latin sources within the context of relations between Byzantines and Latins during the same period: N. P. Ševčenko, "Wild Animals in Byzantine Parks," in *Byzantine Garden Culture*, ed. A. Littlewood, H. Maguire and J. Wolschke-Buhlman (Washington DC, 2002), 69–86, at 79–80. Lions seemed much more numerous in Western Christian menageries, at least at the end of the period: see Buquet, "Aspects matériels du don," 180.

44 Theophylact Simocatta, *History*, 1, 3, 9–10, ed. C. de Boor and P. Wirth, *Theophylacti Simocattae Historiae* (Stuttgart 1972), 45–46; Nechaeva, *Embassies*, 183; Cutler, "Gifts and Gift Exchanges," 255–56. See more recently the enlightening interpretations of M. Cristini, "Elephant Diplomacy: A disturbing Gift to the Khagan of the Avars," *Viator*, 53 (2022): 49–59

45 Theophanes, *Chronicle*, ed. C. de Boor, *Theophanis Chronographia* (Leipzig, 1883–1885), 226–27, For the interpretation of "Indians" as Ethiopians, see: *The Chronicle of Theophanes Confessor. Byzantine and Near Eastern History A.D. 284–813*, trans. C. Mango and R. Scott (Oxford, 1997), 331, n. 2; same information in John Malalas, *Chronography*, ed. I. Thurn, *Ioannis Malalae Chronographia*, CFHB 35 (Berlin and New York, 2000), 411.

46 See Nechaeva, *Embassies*, 196 and 249 (after Josh. St., 19).

47 Nechaeva, *Embassies*, 196 (with the reference to a previous study by A. Cutler).

48 *Magni Presbiteri Chronicon*, ed. Wattenbach, 511 and 512; trans. Loud, 154 and 155.

49 Michael Attaleiates, *The History*, transl. A. Kaldellis and D. Krallis (Cambridge, MA and London, 2012), 87–89; John Skylitzes, *Synopsis of Histories*, ed. H. Thurn, *Ioannis Scylitzae Synopsis historiarum*, CFHB 5 (Berlin and New York, 1973), 475; Michael Glykas, *Annals*, ed. I. Bekker, *Michaelis Glycae Annales* (Bonn 1836), 597.

50 See Nechaeva, *Embassies*, 199–201 who indicates references and discusses that point, and for what follows on Timotheus of Gaza; see also Ševčenko, "The Giraffe," 336; P.-L. Gatier, "Des girafes pour l'empereur," *Topoi* 6.2 (1996): 903–41, at 919.

51 Georges Pachymérès, III.4, ed. A. Failler, trans. V. Laurent (Paris, 1984), vol. I, 238–39. On the context, see Malamut, "Les cadeaux," 261; Ševčenko, "The Giraffe," 343.

52 Behrens-Abouseif, *Practising Diplomacy*, 95–6, and see for other Mamluk sovereigns: 140–142; see the map proposed by Buquet, "Aspects matériels du don," fig. 8.1, 179. For the giraffe as an official gift in other contexts, see the examples and remarks of Mansouri, "Les animaux exotiques," 338–40.

53 This case will be developed and analysed below; see Behrens-Abouseif, *Practising Diplomacy*, 62–63.

54 On these two cases: Gatier, "Des girafes pour l'empereur," 921–23 with further bibliography.

55 Gatier, "Des girafes pour l'empereur," 920; Nechaeva, *Embassies*, 201 and 251. The origins of the Maccuritae remain obscure; they may have come from the middle of Nubia. Tusks were certainly welcomed by the Byzantines: it is the source of ivory, a very valuable and desired material and gift.

56 See the results of the overview proposed by Buquet, "Aspects matériels du don," who has a broader view than the sole Byzantine Empire.

57 Philostorgius, *Church History*, trans. P. R. Amidon (Atlanta, 2007), 3, 11, 48 (whose comments on n. 37 cannot be accepted); see also Gatier, "Des girafes pour l'empereur," 918, who suggests that this ape was a baboon, and who considers that it came from Axum and was indeed destined for Constantius II. Horses would often die during transport on diplomatic missions; see the example developed by Nelson, "The Role of the Gift," 245.

58 This was also true for other centuries outside of the Middle Ages, see Ševčenko, "The Giraffe," 337–38.

59 The specific detail is given by Isaac II in one of his letters to Genova: *Acta graeca*, 3: 41; Dölger, *Regesten*, no. 1616.

60 Perhaps, by the thirteenth century, Latin sources had become more precise and complete, as the history of the elephant offered by the Mamluk sultan to Louis IX of France and then by the latter to Henry III of England would suggest: see Pastoureau, "Ménageries impériales," 52–54.

61 See the remarks of Nechaeva, *Embassies*, 201; Buquet, "Aspects matériels du don," 185. It is also clear that for the last centuries of Byzantium, we lack the necessary accounting documents to resolve these questions of Western diplomacy: ibid., 178.

62 As underlined by Buquet, "Aspects matériels du don," 181, who explains by this consideration the presence of an *elephantum parvulum* among the gifts of Saladin to Isaac II in 1188.

63 Theophanes, *Chronicle*, ed. de Boor, 227; John Malalas, *Chronography*, ed. Thurn, 411.

64 See the conclusion of Buquet, "Aspects matériels du don," 196–97, after Behrens-Abouseif, *Practising Diplomacy*, 144–45.

65 Nechaeva, *Embassies*, 204, who adds that it also marked him among his people: receiving an emperor's gifts bestowed prestige on the receiver – as long as the emperor was not considered weak.

66 As Cutler stated, "Gifts and Gift Exchanges," 267, n. 115, after the reference to the *Book of Gifts*, § 85, 110. On this very aspect, see also J. Hathaway, *A Tale of Two Factions: Myth, Memory, and Identity in Ottoman Egypt and Yemen* (Albany, 2003), 97; the colour white created a visual contrast to the black of their Abbasid enemies (I owe this reference to Christine Mazzoli-Guintard, whom I thank).

67 *Book of Gifts*, § 7, trans. G. al-Hijjâwî al-Qaddûmî, 64 – the three animals were respectively named "Duldul," "Ya'fûr" and "Lizâz." Other expensive gifts were among the objects sent to Muhammad.

68 Mansouri, *Recherches*, 263 (French translation of the letter), but see also on 124 and 134.

69 As has Vasiliev, "Manuel Comnenus," 243. On the dogs dispatched from Constantinople to Fatimid caliph al-Mustansir, the so-called *salūqīyya* and *zabībiya*, see the remarks of Drocourt, "Les animaux," 80.

70 See the references given by Külzer, "Some Notes on Falconry," 704; see also Schmidt, *Politische Tierbildlichkeit*, 217–28.

71 Vučetić, *Zusammenkünfte byzantinischer Kaiser mit fremden Herrschern*, I, 269–70, with further bibliography. Even for a foreign delegate who was not a sovereign, such hunts in the presence of the emperor could be organised, as the case of Liudprand of Cremona demonstrates, in 968 (*Legatio*, § 37–38); see Ševčenko, "Wild Animals in Byzantine Parks," 72–73. On the ambivalence of the two wild goats Liudprand received as gifts, see the remarks and references of F. Bougard, *Liudprand de Crémone, Œuvres*, transl. F. Bougard (Paris, 2015), 541, n. 136; Schreiner, "Diplomatisch Geschenke," 267.

72 See the examples and remarks of Drocourt, "Les animaux," 89–91.

73 Abū Yūsuf Ya'kūb, *Kitāb al-kharādj*, trans. E. Fagnan, *Le livre de l'impôt foncier*, (Paris, 1921), 291, mentioned by Cutler, "Gifts and Gift Exchanges," 266.

74 See the overview of Schreiner, "Diplomatische Geschenke," 2004, and the recent commentaries of Malamut, "Les cadeaux," 262–63, for the last centuries of Byzantium. Of course, in Western Christianity horses were often associated with power, sovereigns, and hunting (see Voisenet, 2000, 40–43), and horses played a significant role during some official encounters, displaying the power and superiority of sovereigns – directly or by the means of their envoys: N. Drocourt, "Le cheval, animal diplomatique entre Byzance et l'Occident (IXe-XIIIe s.)," in *Byzance et l'Occident VI. Vestigia philologica*, ed. E. Egedi-Kovács (Budapest, 2021), 113–130.

75 On this very point see N. Drocourt, "La diplomatie byzantine (IXe-XIIe siècle): instrument pour la paix ou arme de guerre?" in *De la guerre à la paix en Méditerranée médiévale. Acteurs, propagande, défense et diplomatie*, ed. E. Malamut and M. Ouerfelli (Aix-en-Provence, 2021), 183–206.

76 *The Chronicle of Arnold of Lübeck*, 1.3, trans. Loud, 46. On this military dimension, see Malamut, "Les cadeaux," 263.

77 Anna Komnene, *The Alexiad*, 7.7, transl. E. R. A. Sewter, revised with an introduction and notes by P. Frankopan (London, 2009), 202); see Malamut, "Les cadeaux," 263.

78 See the examples and references given by Drocourt, "Les animaux," 75–76.

79 *Magni Presbiteri Chronicon*, ed. Wattenbach, 511; transl Loud, 154.

80 *Constantine Porphyrogenitus. Three Treatises on Military Expeditions*, ed. J. F. Haldon (Vienna, 1990), 100–101; one may remark that these kinds of gifts are here associated with some types of clothes: ibid. 108–109.

81 See n. 18. The content of the boxes is also described (gold vessels, brocade, girdles bordered with gold; high turbans and so on.)

82 See for instance *Book of Gifts*, § 85, 91 and 105, trans. G. al-Hijjâwî al-Qaddûmî, 110, 112 and 116–117.

83 *Book of Gifts*, § 105, trans. G. al-Hijjâwî al-Qaddûmî, 117.

84 *The Chronicle of Arnold of Lübeck*, 1.12, trans. Loud, 57. For mules (and other equine animals, such as donkeys) in the role of carrying objects and items, either associated with diplomatic purposes or not, see the recent study by T. G. Dawson, "Baggage Animals – the Neglected Equines. An introductory Survey of their Varieties, Uses and Equipping," in *The Horse in Premodern European Culture*, eds. A. Ropa and T. Dawson (Berlin and Boston, 2019), 45–54.

85 *The Chronography of Gregory Abū'l Faraj*, trans. Budge, 139 (ca. 841).

86 See Gatier, "Des girafes pour l'empereur," 920–21; A. Becker, "La girafe et la clepsydre. Offrir des cadeaux diplomatiques dans l'Antiquité tardive," *Mondes(s). Histoire, espaces, relations* 5 (2014): 27–42, at 35.

87 Cutler, "Gifts and Gift Exchanges," 253; Dölger, *Regesten*, no. 912.

88 Michael Psellos, *Panegyric Orations*, no. 4, ed. G. T. Dennis, *Pselli Orationes panegyricae* (Stuttgart and Leipzig, 1994), 62; see Drocourt, "Les animaux," 86. One should perhaps compare this reference to the one in Aristotle's *History of Animals*, 630b, which asserts that elephants were the gentlest of all the wild animals, that they could both learn and understand, and that they were even taught to kneel before the king. The same observation in John of Ephesus, *Lives of the Eastern Saint, Syriac text*, II, 48, trans. E. W. Brooks (Paris, 1923) 83–84, quoted by G. Dagron, *Constantinople imaginaire. Etudes sur le recueil des "Patria"* (Paris, 1984), 166, n. 27: John describes that some inhabitants were very impressed by certain elephants sent by the Persians and imagined that the animals were making some kind of prayer with their trunks in front of the churches of Constantinople, as they prostrated themselves in front of the emperor at the Hippodrome.

89 Michael Psellos, *Panegyric Orations*, no. 1, ed. Dennis, 13; following this assertion, Psellos describes the elephant and giraffe from an anatomical perspective, especially the elephant's trunk and the way the giraffe walks. For more than simply the prostration alone, this description has also been interpreted as an illustration of the pretensions of the *basileus* to rule the entire *oikoumēnē*; see N. Radošević, "L'Oecumène byzantine dans les discours impériaux du XIe et XIIe siècle," *ByzSlav* 54.1, (1993), 158–59. See also the chapter by Tristan Schmidt in this volume.

90 On which see A. Cutler, "Barberiniana: Notes on the Making, content and Provenance of Louvre, OA 9063," in *Tesserae: Festschrift für Josef Engemann. Jahrbuch für Antike und Christentum* 18 (1991): 329–39. On the lower side of this plaque, a lion, a tiger, and an elephant are kowtowing to their superior, the emperor, usually interpreted by modern historians as Justinian I.

91 *Michael Attaleiates*, transl. Kaldellis-Krallis, 86–9 (and for what follows), see above in n. 49 for the other two Greek chroniclers whose information remains scarce.

92 See the remarks established by Gatier, "Des girafes pour l'empereur," 926–27; Ševčenko, "The Giraffe," 339–40.

93 Georges Pachymérès, III, 4, ed. A. Failler, trans. V. Laurent, vol. I, 238–39; Malamut, "Les cadeaux," 261; Gatier, "Des girafes pour l'empereur," 927, suggests that this kind of exhibition is far from any Roman triumph, examples of which are provided by previous Antique and Late Antique sources.

94 See the references and remarks by Ševčenko, "The Giraffe," 340 and 343

95 *Timothy of Gaza on Animals, Περὶ Ζῴων: Fragments of a Byzantine paraphrase of an Animal-Book of the 5th century a.d.*, eds. F. S. Bodenheimer and A. Rabinowitz (Paris and Leiden, 1949), 31; Ševčenko, "Wild Animals in Byzantine Parks," 77, n. 39; Ševčenko, "The Giraffe," 340.

96 Theophanes, ed. de Boor, 226–27.

97 See G. Dagron, *L'hippodrome de Constantinople. Jeux, peuple et politique* (Paris, 2011), 20 with further bibliography; Gatier, "Des girafes pour l'empereur," 916–17, who asserts that the giraffe "sert, à Rome, à manifester la domination de l'Empire sur les pays lointains."

98 N. Drocourt, *Diplomatie sur le Bosphore. Les ambassadeurs étrangers dans l'Empire byzantin des années 640 à 1204* (Louvain, 2015), 586–89.

99 Dagron, *L'hippodrome de Constantinople*, 102; Dagron, *Constantinople imaginaire*, 43 and 166. Some diplomatic animals may have been sent for their participation during triumphal ceremonies, see Nechaeva, *Embassies*, 249. Constantinopolitans and guests also meet animals caught during military campaigns, like the example given by Heraclius, who came back from a military attack against Persia with some elephants and chose to display four of them at the hippodrome: Nikephoros, § 19, 66–67.

100 *Pseudo-Kodinos and the Constantinopolitan Court: Offices and Ceremonies*, ed. R. J. Macrides, J. A. Munitiz and D. Angelov (Farnham and Burlington, 2013), 268; cf. Papagiannaki, "Experiencing the Exotic," 242.

101 Papagiannaki, "Experiencing the Exotic," 231; see Figure 3 for an ivory representation of a cheetah riding pillion.

102 A. Littlewood, "Gardens of the Palace," in *Byzantine Court Culture from 829 to 1204*, ed. H. Maguire (Washington, DC, 1997), 13–38, at 35 (but see below, in n. 103); Ševčenko, "Wild Animals in Byzantine Parks," 75–81. On the concept of a "menagerie" in the Middle Ages, with some reservations: Buquet, "Aspects matériels du don," 189.

103 V. Onar et al., "Animal skeletal remains of the Theodosian harbor: general overview," *Turkish Journal of Veterinary and Animal Science*, 37 (2013): 81–85, who recorded nine elephants, one gazelle, two monkeys, nine bears, 32 ostriches, 246 camels, 503 mules, 4,018 sheep, 4,209 cattle and 6,816 horses, among other species (I thank Johannes Preiser-Kapeller for this reference).

104 Papagiannaki, "Experiencing the Exotic," 248–49, who emphasizes that both parks were "enclosed landscapes, not heavily forested, stocked with game and visible from the city."

105 Following the interpretation of H. Maguire, "A description of the Aretai Palace and its Garden," *Journal of Garden History* 10 (1990): 209–13, at 210–11; Ševčenko, "Wild Animals in Byzantine Parks," 81.

106 Full quotation of Cutler, "Gifts and Gift Exchanges," 269, and references n. 122; A. A. Vasiliev, *Byzance et les Arabes*, vol. II, part 2, *La dynastie macédonienne, extraits des sources traduites* (Brussels, 1950), 61–62; on this reception see also *Book of Gifts*, § 162, trans. G. al-Hijjâwî al-Qaddûmî, 152.

107 Eusebius, *Vita Constantini* 4. 50 quoted by Nechaeva, *Embassies*, 198 and 250.

108 Liudprand, *Antapodosis*, III, 23, transl. Squatriti, 119; F. Bougard, in Liudprand, transl. Bougard, 475, n. 67, assumes that these animals certainly came from northern Europe and notes that Anglo-Saxons had to deliver two of them, every three years, to Pavia, at the end of the tenth century.

109 On this episode, see also the interpretations of P. Buc, *The Dangers of Ritual: Between Early Medieval Texts and Social Scientific Theory* (Princeton, 2001), 22–24: here, the animals needed to be impressed by the sovereign, thus creating a sort of symbolic inversion; the attitude of the dogs revealed the emperor's false identity (see Liudprand, trans., Bougard, 2015, 475, n. 69).

110 Canard, "Les sources arabes," 295–96; Drocourt, "Les animaux," 90.

111 Niketas Choniates, *History*, ed. van Dieten, 189; Nisa was a plain of Media. See also A. Cutler, "Significant Gifts: Patterns of Exchange in Late Antique, Byzantine and Early Islamic Diplomacy," *Journal of Medieval and Modern Greek Studies* 38 (2008): 79–102, at 93.

112 L. Jones and H. Maguire, "A Description of the Jousts of Manuel I Komnenos," *BMGS* 26 (1992): 105.

113 Since bears do not come directly from Egypt; see *Book of Gifts*, § 85, trans. G. al-Hijjâwî al-Qaddûmî, 110; on different forms of "recycling" in this context: Cutler, "Gifts and Gift Exchanges," 264–65; see also Buquet, "Aspects matériels du don," 181–82.

114 See the references in Behrens-Abouseif, *Practising Diplomacy*, 17–18, and 129 in the case of 1312, when the emperor sent falcons, eagles, *tuhaf* and 42 loads of wool.

115 Nechaeva, *Embassies*, 179, and see n. 21 above.

116 See the deduction and proposition of Gatier, "Des girafes pour l'empereur," 933–41.

117 S. M. Burstein, "An Elephant for Anastasius. A Note on P. Mich. Inv. 4290," *The Ancient history bulletin*, 6-2 (1992): 55–57; see the remarks of Gatier, "Des girafes pour l'empereur," 920, n. 101.

118 Philostorgius, 3. 11, 47; Gatier, "Des girafes pour l'empereur," 929.

119 Ševčenko, "Wild Animals in Byzantine Parks," 70, n. 7 and 79, n. 47.

120 *The Chronicle of Arnold of Lübeck*, 1. 12, trans. Loud, 57.

121 See notably: Klein, "Eastern Objects and Western Desires," 284–87 (with previous references), who emphasizes to what degree a part of historical fiction is associated with this passage; Schreiner, "Diplomatische Geschenke," 260 and 281–82; Malamut, "Les cadeaux," 252–53.

122 On this interpretation, see Drocourt, "Le cheval," 122.

123 Various accounts in medieval sources – whether in Greek, Latin, Arabic or other languages – tend to depict the superiority of one camp over the other through various attitudes, gestures, or the gifts exchanged; see Drocourt, *Diplomatie sur le Bosphore*, 37 ff.

124 *The Chronicle of Arnold of Lübeck*, 1.9, trans. Loud, 52.

125 To the best of my knowledge, this passage has not been integrated into the reflection and interpretation of previous historians.

126 Niketas Choniates, *History*, ed. van Dieten, 493–94.

127 See the references in Behrens-Abouseif, *Practising Diplomacy*, 62–63, and the commentaries of Buquet, "Nommer les animaux exotiques de Baybars, d'Orient en Occident," in *Les non-dits du nom: onomastiques et documents en terre d'Islam. Mélanges offerts à Jacqueline Sublet*, ed. C. Müller and C. Roiland-Rouabah (Beyrouth, 2013), 375–402, at 389, and more recently Buquet, "Aspects matériels du don," 188.

128 As underlined by Cristini, "Elephant Diplomacy," 54–55.

129 As formulated by Ševčenko, "The Giraffe," 346, n. 14; same idea ("pas de girafes sans empereur") in Gatier, "Des girafes pour l'empereur," 922; see also the consideration of Buquet, "Aspects matériels du don," 177. The same idea concerning elephants: their use and ownership were considered an imperial prerogative, as recently noted by Cristini, "Elephant Diplomacy," 53.

130 See N. Drocourt, "Quelques *exotica* dans les cadeaux diplomatiques entre Byzance et ses voisins (VIIIᵉ-XIIIᵉ s.)," *Topoi* 26 (2023): 79–100.

131 Behrens-Abouseif, *Practising Diplomacy*, 88, n. 141; on the menagerie of Constantinople after the Ottoman's conquest of the city, Buquet, "Aspects matériels du don," 190.

15

ANIMALS WE LOVE

Pets and Companion Animals in Byzantium?

Przemysław Marciniak

Whilst writing a letter to a judge from Thrakesion, Michael Psellos stated "Believe me, if I saw your little dog (κυνάριόν σου) being abused by someone, I would help it, in whatever manner I could, and I would chase away the abuser."[1] This passage has several interesting implications. First, it acknowledges the obvious, the Byzantines kept dogs as pets; second, these pets could be of value and importance to their owners. And finally, cruelty towards animals could meet with a reaction. Of course, it is important to acknowledge uncertainties within these conclusions: was it a guard dog, or perhaps a hunting hound? Is Psellos using this example because of its inconceivability (I would help *even* your dog)? Be that as it may, it is beyond any doubt that the Byzantines, similarly to other humans throughout the centuries, enjoyed the company of animals. The question remains: what was exactly the nature of such relationships, especially when it comes to animals which we today describe as 'pets'?[2]

The question mark in the title of this chapter is meant to express the uncertainty surrounding the idea of 'pets' in the medieval (and more widely premodern) period.[3] While some scholars take the term for granted,[4] others debate whether such a relationship between an animal and a human was conceivable in the premodern period. Yi-Fu Tuan has argued that that "a particularly sentimentalized" view of pets was developed in nineteenth-century western civilization as a response to the challenges of a changing society.[5] In this sense our modern view of pets is very much different from the notion of pets in the premodern period. Sian Lewis and Lloyd Llewellyn-Jones claim that ancient sources "demonstrate that pet relationships, in the sense that we understand them, can be found in the literary and archaeological records of Egyptian, Greek and Roman society (...)".[6]

In Byzantium, similar to Antiquity, there was no word which could be rendered as 'pet' (but this holds true for many modern languages as well). The closest one would be perhaps *a domestic animal*. Such a phrase can be found in the Byzantine sources, although it would be more properly translated as 'an animal living in/around the house' (τὸ κατοικίδιον ζῷον). Yet the range of animals, which can be described with the help of the adjective *katoikidion* is very wide. It can include a rooster (ἀλέκτωρ),[7] a hen (ὄρνις) or a cat (ὁ κατοικίδιος αἴλουρος).[8] This adjective does not describe the relationship between an animal

DOI: 10.4324/9781003055877-19

and a human but rather underscores the fact that a given animal was to be found *in* or *around the house* (and not even necessarily as a domesticated one, as the Suda refers to a mouse).[9] In other words, there did not exist a specific terminology reserved for animals whose aim was not strictly utilitarian (hunting, pest control).

Moreover, many of the available sources underscore the utility of an animal. Demetrios Pepagomenos opens his veterinary treatise *Kynosophion* with the praise of dogs by saying that canines serve humans as guardians, during the hunt or for both purposes. Pepagomenos says specifically πρὸς ὑπηρεσίαν (for service).[10] Similarly, in the fourteenth-century vernacular satire *Synaxarion of the honourable donkey* an old lady is said to love her children (ἔχει ἀγάπην), the cat Parditzis and the hen Kavaka because they were useful. The cat caught mice and the hen laid eggs (τὸν κάτον διὰ τοὺς ποντικούς, τὴν ὄρνιθαν διὰ τὰ αὐγά της).[11] Social companionship is not mentioned in these texts as it was obviously not something practical or useful. However, the idea of keeping animals for purposes other than hunting or as guardians arises in Byzantine sources, which mention the Maltese dogs kept for pleasure (οἱ δὲ ἐπὶ τέρψει, ὡς τὰ Μελιταῖα κυνίδια).[12] A similar function of an animal is implied in John Italikos' monody on his partridge (see below).

The aim of this chapter is not to partake in this (probably mostly inconclusive) debate but rather to demonstrate the peculiar Byzantine way of constructing the relationship between animals and humans. The second part of this chapter offers a short survey of dogs and cats as the most usual pets of the Byzantine period, mostly focusing on information not discussed in earlier studies.

Animals We Love[13]

Understandably, regardless of the existence of the proper terminology, much depends probably on the definition of a pet. Christine Overall notes: "To start, we could define a pet, perhaps perversely, as an animal whom meat-eating human beings would, for moral reasons (and other things being equal, that is, absent extreme conditions of famine), regard as unthinkable to eat."[14] In Byzantium, the issue of edible and non-edible animals had religious underpinning and is therefore a less helpful indicator for treating an animal as a pet.[15]

The Apocalypse of Ps.-Methodios (seventh century) describes cats, snakes, and frogs as impure and, therefore, uneatable.[16] Yet, some instances of recycling animal parts can be found. The cat's skin (fur) seems to have been used in Byzantium, although there are only a handful of direct mentions of this: one Bartholomew of Edessa (possibly from the thirteenth century) mentions some sort of clothing (or blanket) made of cat skin.[17] A magical treatise specifies a special belt made from the same material.[18] An anonymous Byzantine dream book states, "Finding a cat's skin signifies profit from the property of every thief."[19]

Importantly, a pet is an animal with which a human can form a unique, emotional bond. Unfortunately, there are no extant Byzantine testimonies, with perhaps one exception to which I will return, of relationships between humans and animals, which we consider as typical pets (that is cats and dogs). Nonetheless, there are other examples worth mentioning. In the anonymous oration addressed to John II Komnenos, its author describes the gloomy fate of animals, which drown in the mud of a street in Constantinople.[20] This oration includes the description (more a general example than a report as it seems) of an unsuccessful attempt at rescuing an animal, some kind of beast of burden, which in the end dies and is eaten by the wild dogs. At some point the author says that the owner of this animal laments over it as if it were a dead child (ἴσα καὶ ἐπὶ τέκνῳ νεκρῷ ὀλοφυρόμενος ἐπὶ ζώῳ).[21] This is a very powerful comparison, as the death of a child differs from the loss of other family members.[22] Yet again, we must perhaps account for the fact that such an animal was crucial in securing the family's economic livelihood. This does not exclude, however, an emotional attachment and treating an animal as part of the extended family. Humans can form bonds with animals not necessarily associated with pets, such as working animals or animals bred for meat, but as noted "those attachments tend to be more distant and may be actively avoided".[23]

Some of the descriptions, which imply an attachment to an animal, lack the emotional aspect. For instance, John Skylitzes notes that a talking parrot of Basil I (r. 867–886) was kept "for view and pleasure" of those who saw and heard it (πρὸς θέαν καὶ τέρψιν τῶν θεωμένων καὶ ἀκουόντων).[24] The animal is here merely a pleasant commodity, not an object of any kind of affection or emotional attachment. John Tzetzes mentions a partridge he received as a gift from Alexios Pantechnes. Tzetzes portrays this partridge as tame (τὸν πέρδικα τὸν χειροήθη),[25] implying that it was more likely domesticated rather than wild.[26] Tzetzes finds its presence pleasing but nothing in the letter points to any sort of emotional attachment. He describes the bird as "ὡραῖον" (graceful) underscoring its external appearance.

A different partridge was the topic of a monody penned by Michael Italikos.[27] Here the emotional bond with the bird is presented most convincingly when Italikos describes its death caused by some illness. At the end of this passage, he notes "If I didn't want to be seen as weak, I would have cried, and I would have shed many tears because of the bird" ("εἰ μή γε ὑφεωρώμην τὸ μαλθακίζεσθαι, καὶ ἐδάκρυσα ἂν καὶ πολλὰ τῷ ὄρνιθι κατέσπεισα δάκρυα").[28] It is interesting to note that the grief over the animal – and thus a tangible expression of the emotional bond – is apparently an unwelcome reaction (although Italikos does not clarify if it was crying alone or crying over the animal that would have made him look weak). In fact many arguments Italikos made in his monody sound as if he tried to justify the importance of his animal companion.[29] Italikos' unwillingness to cry stands also in direct opposition to the earlier portrayal of a man lamenting and crying over the demise of his animal. This might be simply a rhetorical difference or perhaps a difference mirroring the different social strata approaches to human–animal relationships.

There is one more defining feature of a pet: a name. Giving an animal a personal name is meaningful: "having a name symbolically and literally incorporates that animal in the human domestic sphere".[30] What unifies the stories mentioned above is the fact that their animal protagonists remain nameless. There seem to be very few instances where sources preserve the name of the individual animal. This stands in sharp contrast to the situation in Antiquity and in Western Europe. One such exception is the cat of the empress Zoe (978–1050), described in a passage from the *Histories* by John Tzetzes, named Mehlebe.[31] Yet another example comes from a letter of Emperor Theodore II Laskaris. He sends his courtier George Mouzalon a small dog Hyellovourtis (Ὑελλοβούρτης, no. 159).[32] Apostolos Karpozilos notes in his article on the Byzantine realia " [...] we should mention a 'Scythian' dog (κυνίδιον) called Kleopas which Laskares gave for temporary upkeep to Mouzalon [...]". This would be an exciting piece of information as Kleopas is a human name. However, this seems to be an incorrect interpretation of the source: Kleopas is mentioned in the *Synopsis Chronicle* of Theodore Skutariotes as a person, not an animal.[33] Theodore of Gaza (c. 1398–c. 1475) in his encomium of a dog does not mention any name. Although this particular case could be reasonably explained by the fact that Theodore reserved the right to name the dog for its future owner (the recipient of the letter).[34]

Laskaris is possibly exceptional when it comes to revealing the name of his dog, but not unique in naming an animal close to him. John Skylitzes tells the story of the Constantinopolitan patriarch Theophylact's unfitting passion for horses:

> He had this absolute passion for acquiring horses (he is said to have procured more than two thousand of them) and their care was his constant concern. He was not satisfied with feeding them hay and oats but would serve them pine-seeds, almonds and pistachios or even dates and figs and choicest raisins, mixed with the most fragrant wine. [...] It is said of him that once when he was celebrating the great supper of God on the Thursday of Holy Week and was already reading the prayer of consecration, the deacon charged with the task of caring for the horses appeared and gave him the glad tidings that his favourite mare – he mentioned its name – had just foaled. [Theophylact] was so delighted that he got through the rest of the liturgy as quickly as possible and came running to Kosmidion where he saw the newly born foal [...].[35]

While this entire passage is meant to discredit Theophylact, it also provides the readers with an interesting piece of information. It is explicitly said that the deacon mentioned the name of the mare (προσθεὶς καὶ τὸ ὄνομα). In other words, it confirms what is a reasonable guess – that at least the animals with which humans felt a special connection had names (this may also include horses of soldiers, etc.). Skylitzes handed down a story, which although primarily serving as an illustration of the righteousness of emperor Theophilos, also includes a brave and trustworthy horse unjustly seized from his owner. But the name of the animal is never revealed throughout the story. Interestingly enough, a few centuries earlier, the names of the horses of a famous charioteer, Porphyrios, had been inscribed for posterity on his monument: Aristides, Palaistiniarches, Purros, and Euthynikos.[36]

However, in many cases, even when explicitly referred to, as in Italikos' monody, the animals remain nameless. The names that were handed down to us seem to be more an exception than a rule. The animals in Byzantine sources are portrayed as what is today referred to as 'absent referents'. While this term, coined by Carol Adams, originally referred to animal meat, it seems to aptly describe the animal world as portrayed in Byzantine literature.[37] Animals have no names; they serve specific functions as carriers, hunt helpers, or embody certain features that define their species, such as loyalty in the case of dogs. The recurrent argument in texts featuring animals is their usefulness for humankind.[38] They existed *en masse*, as meat, as a means of transport, as pests, or entertainers, but rarely as individuals. And yet, this seems only a half-truth. It is clear that people had emotional bonds with animals and gave them individual names.

The question remains: Why are Byzantine sources so dismissive when it comes to the names of animals? Should it be explained simply by the image of a strict, God-given hierarchy, which the chronicles and historians projected in their works? But the equally Christian Western literature includes a plethora of examples of animals with names.[39] I would be inclined to think the main reason was to create the impression of the absolute human control over nature. To some extent this is the pattern set by the biblical story where

Adam gave animals general and not individual names. In narratives where animals are protagonists, animal names are more frequent (but still not as frequent as in the Western sources). Mice in the *Katomyomachia* are named, and a cat in the fourteenth-century satirical work mentioned earlier about an honourable donkey is called Parditzis.[40] On the other hand, texts such as those written by Italikos and Manasses are more about the human pet-owners than pets themselves. Be that as it may, as a result, most Byzantine animals remain nameless for us.[41]

Of Dogs and Cats

Although humans choose to bond with various animals from birds to lizards to various mammals, dogs and cats remain the most popular amongst our pets. Apart from horses (and ironically mice), these two species – cats and dogs – received most attention from scholars of the Byzantine period.[42]

Recent archeological excavations confirm that both cats and dogs co-existed with humans within society.[43] It is wise to assume that at least some of them might have been what today we call 'pets'. John Tzetzes describes the luxurious life of Mehlebe, Zoe's cat, then adds that today "some little dog" is similarly well fed and taken care of while there are people who do not have bread. While Tzetzes is obviously annoyed about this social injustice, his testimony confirms that pets were part of the Byzantine world.[44]

Even more elusive testimony, this time regarding a cat, might be found in the twelfth-century *Ptochoprodromika*. In the text, the protagonist mentions "our cat" (τὸ κατούδιν μας στήσας εἰς τὸ τραπέζιν).[45] While this gives us very little to work with, this description might signal a special bond with the animal. A possible – non-literary – confirmation regarding a cat as a pet comes from the excavation at the Balatlar Church complex in Sinop (6th/7th c.), where a female house cat was buried alongside her possible human owner.[46] Recent archaeological excavations from the Yenikapı, Istanbul (the ancient harbour of Theodosios in Constantinople) brought to light remains of 859 dogs and 78 cats.[47] This makes it the biggest assemblage of such remains from the Byzantine period. However, its value for the study of pets might be limited. As the authors of the study on canine remains from the Yenikapı demonstrate, most of the remains come probably from stray dogs and tell us little about dogs (or cats for that matter), which were kept as pets.[48] Therefore, the literary evidence still remains the most important source for studying Byzantine dogs and cats as pets.

One dog breed consistently appears in our sources throughout the centuries: the Maltese dog, Μελιταῖον κυνίδιον. Considering how much the physical appearance of dog breeds changes over time, it would be unwise to assume that a dog called the Maltese dog remained the same throughout centuries and across geographic locations. J. Busuttil noted that people "commonly called it νανούδιον – a little dwarf".[49] However, there are only two instances where sources confirm this information, and both come from the Byzantine period, possibly even from the same author – Arethas of Caesarea.[50] While the ancient origins of this name cannot be excluded, they also cannot be undeniably confirmed. On the other hand, Byzantine sources repeated the Aristotelian information (*Hist. Anim.* 612b.10) that the Maltese dog was as small as an *iktis* (a small bird).[51] The dog's small appearance seemed to be its defining feature as described.

Information about this breed's nature is much scarcer. As mentioned earlier, the *Suda* offers that some breeds were kept for pleasure, like the Maltese dogs (οἱ δὲ ἐπὶ τέρψει, ὡς τὰ

Μελιταῖα κυνίδια).[52] Their pet-like status appears to emerge in etymological explanations provided by lexica. The name is supposed to come from the word μελῶ (meaning 'I care').[53] Moreover, in an oration authored by Plethon, Maltese dogs are contrasted with brave dogs (ἀντὶ κυνῶν ἀλκίμων Μελιταῖά τε κυνίδια) thus suggesting their docile nature.[54] Interestingly, in the entry on Maltese dogs the *Suda* mentions one more breed kept for pleasure, Μελιτηροὶ κύνες (repeated later only once by Apostolios in his collection of proverbs). This seems, however, to be a variant of the more widespread term Μελιταῖον κυνίδιον.[55]

While this is a mere speculation, it would be tempting to think that at least in some cases *the Maltese dog* does not denote a specific breed but rather serves as a general umbrella term describing various small dogs.[56]

Our knowledge about Byzantine cats is even more limited than our knowledge of dogs.[57] What we have are fragments of information scattered here and there, which, in the end, fail to coalesce into a reliable depiction of the feline world in the Byzantine period.

The Byzantines might not have spilled much ink on describing animals (and pets in particular) but that does not mean that they did not observe or understand their nature. Michael of Ephesos, while discussing animals using their legs as hands, remarks that domestic cats strike and seize mice with their paws, thus using them as hands (οἱ κατοικίδιοί τε αἴλουροι τοὺς μῦς μετὰ τούτων τύπτουσί τε καὶ κατέχουσιν).[58] Gregory of Sinai, a late Byzantine theologian, describes various forms assumed by demons. Demons of desire turn into pigs or asses, and demons of *akedia* (which may mean apathy, melancholy, and boredom) into cats.[59] Gregory's description must have been based on observation or a stereotypical thinking about animals; cats spending most of the day sleeping were excellently suited to incarnate *akedia*.

Similarly, literary evidence both confirms and reinforces the notion of cats as gluttonous animals (on par with mice), see for instance the saying preserved in Apostolios' collection "Ἔνθα περιττὴ τροφή, πολλοὶ μύες καὶ γαλαῖ" (where there is abundant food [there are] many mice and cats).[60]

Cats were perceived as effective pest control – the fable included in the *History* by Nikephoros Gregoras opens with the following sentence: "ἦν τις ἀνὴρ τὴν τέχνην σκυτεὺς γαλῆν ἔχων τὸ χρῶμα λευκὴν, ἢ τῶν κατ' οἶκον μυῶν καθ' ἡμέραν ἐθήρευεν ἕνα" (There was a man, a shoemaker by trade, who had a white cat. Every day she caught one of the mice in the house).[61] This is also indirectly confirmed by literary works such as the *Schede tou myos* and the *Katomyomachia* where cats hunt and play with mice as they do in the real world.[62]

However, unlike dogs, cats are more independent and therefore their history as pets is perhaps more elusive. Moreover, there is a terminology issue. The most commonly used term for a cat in the earlier times was 'αἴλουρος'. However, in the later period, the borrowed term 'κά(τ)τα', originating from the Latin 'cattus', gained popularity.[63] To add to the complexity, another term, 'γαλῆ', could probably refer not only to (domestic) weasels but also to cats, making the accurate identification of the animal challenging, if not impossible.[64] The already mentioned Mechlebe is described as 'gale'. However, if the animal remains uncovered in the Theodosius harbour could give us any pointers, no bones of weasels or ferrets have been discovered. This absence could suggest that these species were not particularly widespread, at least in the capital, and 'gale' could denote 'a cat' in the sources.

As a wide range of sources confirm, cats were kept by humans and enjoyed special privileges. In *An Entertaining Tale of the Quadrupeds* a rat throws various accusations against the cat and says (vv. 159–161):

Σὺ δέ αἰσχρέ, παμμίαρε, ἀλευροκαταχέστη,
ἐκεῖ ὁποὺ σέ ταγίζουσιν, ἐκεῖ ὁποὺ σὲ ποτίζουν
καὶ ἀγαποῦν καὶ ἔχουν σε καὶ ὁμαλίζουσίν σε [...].

"But you, you shameful, filthy flour-shitter,
even as people feed and water you,
and love you, keep you by their side, and pet you (...)"[65]

What the rat describes: love (ἀγαποῦν), closeness (ἔχουν σε), and caressing (ὁμαλίζουσίν σε) along with the willingness to forgive various misdeeds are the usual signs of the emotional bond with a pet.

Both literary and archaeological materials provide us with an incomplete view of reality. Put differently, some of the available sources tend to offer a rather distorted perspective, or, perhaps more accurately, a version of reality altered according to the particular beliefs of a writer. Authors, purposefully or not, projected religion-based views of the world where humans were considered more important than non-human beings. However, when Tzetzes notes that some dogs were treated better than humans, it clearly shows that such religious views were not necessarily shared unanimously.

However, even sources that, at first glance, do not seem concerned with animals can provide valuable insights. Even if the above-mentioned Kleopas was not a dog, it can still be argued that equating a human servant/soldier with a dog shows that dogs might have been highly valued and respected. Information included in a wide range of non-zoological sources was based either on empirical observations or general knowledge of animals such as cats and dogs, which, in turn, proves that such animals were reasonably widespread and well-known.

The Byzantines obviously kept animals as companions, they named and cared for them. Understandably, this was partly due to their usefulness (providing company to a human may also be viewed as something useful) but true emotional bonds were not excluded.

Funding

This chapter was written as part of the project UMO-2019/35/B/HS2/02779 funded by the National Science Center Poland.

Notes

1 Michael Psellos, *Letters* no. 311, ed. S. Papaioannou, *Michael Psellus, Epistulae*, 2 vols. (Berlin and Boston, 2019), 721.

2 Throughout the text I tend to use the term 'pet' and 'companion animal' interchangeably, following the example of other authors; see for instance L. Calder, "Pet and Image in the Greek World: The Use of Domesticated Animals in Human Interaction," in *Interactions between Animals and Humans in Graeco-Roman Antiquity*, ed. Th. Fögen and E. Thomas (Berlin–Boston, 2017), 61–88. However, as suggested by Clare Palmer 'companion animal' can also be used in the narrower

sense and it "refers to a sub-category of pets: animals kept to provide companionship, understood to mean a mutual, positive, and interactive engagement, for instance for play and for reassurance. While 'companion animal' in the broader sense refers to all the animals kept voluntarily in homes – mammals, birds, fish, reptiles, amphibians, insects – in the narrower sense, companion animals are primarily dogs and cats, and perhaps small mammals (such as rabbits) and birds." C. Palmer, "Companion Animals," in *The International Encyclopedia of Ethics*, ed. H. Lafollette (London, 2015), s.v. However, none of the available definitions seems to be fully convincing and accordingly a certain degree of ambiguity remains.

3 On a variety of animals kept in the Antiquity, see F. Lazenby, "Greek and Roman Household Pets," *Classical Journal* 44 (1949): 245–52 and 299–307.

4 See for instance L. Bodson, "Motivations for Pet-Keeping in Ancient Greece and Rome: A Preliminary Study," in *Companion Animals and Us: Exploring the Relationships Between People and Pets*, ed. A. L. Podberscek, E. S. Paul, and J. A. Serpell (Cambridge, 2000), 27–41.

5 Y. F. Tuan, *Dominance and Affection: The Making of Pets* (New Haven, CT, 1984), 112.

6 S. Lewis and L. Llewellyn-Jones, *The Culture of Animals in Antiquity. A Sourcebook with Commentaries* (Abingdon and New York, 2018), 713.

7 See for instance *Etymologicum Genuinum*, eds F. Lasserre and N. Livadaras, *Etymologicum magnum genuinum. Symeonis etymologicum una cum magna grammatica. Etymologicum magnum auctum*, vol. 1 (Rome, 1976): Ἀλέκτωρ: Τὸ ζῷον τὸ κατοικίδιον. This explanation is excerpted from the work of the grammarian Philoxenus (first century B.C.).

8 Aelius Promotus (second century) in his work *On poisonous animals* 27.20 mentions ἡ κατοικίδιος γαλλία (a house weasel), S. Ihm, *Der Traktat περὶ τῶν ἰοβόλων θηρίων καὶ δηλητηρίων φαρμάκων des sog. Aelius Promotus. Serta Graeca* 4 (Wiesbaden, 1995).

9 *Suda*, μ 1475 (οἱ κατοικίδιοι μύες), ed. A. Adler, *Suidae lexicon*, 4 vols. (Leipzig, 1928–1935). On Byzantine mice see Katarzyna Piotrowska's chapter in this volume.

10 Demetrios Pepagomenos, *On Healing Dogs*, ed. R. Hercher, *Claudii Aeliani de natura animalium libri xvii, varia historia, epistolae, fragmenta*, vol.2 (Leipzig 1866), 588.

11 *Synaxarion of the Honourable Donkey* 171–72, in U. Moenning, "Das Συναξάριον τοῦ τιμημένου γαδάρου: Analyse, Ausgabe, Wörterverzeichnis," *BZ* 102 (2009): 138–48.

12 Suda μ 519, ed. Adler.

13 There seems to be at least one example of a lion (mentioned by Albert von Aachen in the twelfth century), which might have been treated as a pet, see Nancy P. Ševčenko, "Wild Animals in the Byzantine Park," *Byzantine Garden Culture*, ed. A. Littlewood, H. Maguire, and J. Wolschke-Bulmahn (Washington, D.C., 2002), 79–80 and T. Schmidt, *Politische Tierbildlichkeit in Byzanz. Spätes 11. bis frühes 13. Jahrhundert* (Wiesbaden, 2020), 81–82.

14 Ch. Overall, *Introduction*, in *Pets and People. The Ethics of Our Relationships with Companion Animals*, ed. Ch. Overall (Oxford, 2017), 18*.

15 See B. Caseau, *Nourritures terrestres, nourritures célestes – La culture alimentaire à Byzance* (Paris, 2015), 28, 70.

16 E. Kislinger, "Byzantine Cats," in *Animals and Environment in Byzantium (7th–12th c.)*, ed. I. Anagnostakis, T. Kolias, and E. Papadopoulou (Athens, 2011), 174. Kislinger mentions, however, that the starving defenders of Monemvasia in 1252 consumed both cats and mice.

17 Bartholomew of Edessa, *Refutation of Agarenian* 32.13, ed. K.-P. Todt, *Bartholomaios von Edessa. Confutatio Agarenian*. Corpus Islamo-Christianum. Series Graeca 2 (Würzburg, 1988). However, this mention is made in the Muslim context so it might well be an accusation.

18 R. P. H. Greenfield, *Traditions of Belief in Late Byzantine Demonology* (Amsterdam, 1988), 287. See also Baron's chapter in this volume regarding the archaeological evidence of using cats' skin.

19 S. M. Oberhelmann, *Dreambooks in Byzantium: Six Oneirocritica in Translation with Commentary and Introduction* (Aldershot and Burlington, 2008), 177.

20 M. Loukaki, "Une rue centrale de Constantinople, piège funeste pour les marchands voyageurs et leurs animaux: les doléances d'un Constantinopolitain à Jean II Comnène," *Anekdota Byzantina. Studien zur byzantinischen Geschichte und Kultur. Festschrift für Albrecht Berger anlässlich seines 65. Geburtstags*, ed. I. Grimm-Stadelmann, A. Riehle, R. Tocci, and M. M. Vučetič (Berlin and Boston, 2023), 413–26.

21 Ibidem, 419. The model for this statement seems to be Euripides, *Rhesus* 896: τέκνον σ' ὀλοφύρομαι.

22 J. Fischhoff and N. O'Brien, "After the Child Dies," *The Journal of Pediatrics* 88.1 (1976): 140–46. For the discussion of mourning of the deceased child in Byzantium, see D. Ariantzi, *Kindheit in Byzanz. Emotionale, geistige und materielle Entwicklung im familiären Umfeld vom 6. bis zum 11. Jahrhundert* (Berlin and Boston 2012), 331–34.

23 J. A. Serpell, "The Human-Animal Bond," in *The Oxford Handbook of Animal Studies*, ed. L. Kalof (Oxford, 2017), 82.

24 John Skylitzes, *Synopsis of History*, ed. J. Thurn, *Ioannis Scylitzae synopsis historiarum*. CFHB, 5 (Berlin, 1973), 46.40–44.

25 John Tzetzes, *Letters* no. 93, ed. P. L. M. Leone, *Ioannis Tzetzae epistulae* (Leipzig, 1972), 135.19

26 Similarly, Demetrios Pepagomenos speaks about a 'tame duck' (νῆσσαν χειροήθη), Demetrios Pepagomenos, *On the Healing of Falcons* 303.8, ed. R. Hercher, *Claudii Aeliani de natura animalium libri xvii, varia historia, epistolae, fragmenta*, vol. 2 (Leipzig, 1866).

27 Michael Italikos, *Monody on the Partridge*, ed. P. Gautier, *Michel Italikos. Lettres et Discours* (Paris, 1972), 102–104. Horst Schneider notes that Italikos means more precisely a rock partridge and notes that this might be the only detailed description of this bird dating back to either Antiquity or the Middle Ages, see H. Schneider, "Michael Italikos' *'Monodie auf ein totes Steinhuhn'*. Ein byzantinischer Text im Fokus moderner, Human–Animal Studies," *Das Mittelalter* 2023, 28(2): 429–47. On this text see also P. Agapitos, "Michael Italikos. Klage auf den Tod seines Rebhuhns," *BZ* 82 (1987): 59–67.

There is yet another bird monody from the Komnenian period authored by Constantine Manasses who mourns the loss of his goldfinch. However, Manasses' text is probably a metaphor of his fate as an author, or to use the words of Ingela Nilsson as "an elaborate literary play on the creation of *personae*", see I. Nilsson, *Writer and Occasion in Twelfth-Century Byzantium* (Cambridge, 2021), 81–82. For this reason, I've decided not to include it in this chapter.

28 M. Italikos, *Monody on the Partridge*, ed. Gautier, 104.9

29 For a detailed analysis see Schneider, "Michael Italikos," 429–47.

30 M. DeMello, *Animals and Society. An Introduction to Human-Animal Studies* (New York, 2002), 156.

31 John Tzetzes, *History* 5.12.525–539, ed. P. L. M. Leone, *Ioannis Tzetzae historiae* (Naples, 1968). This was an Arabic name, which could mean "talon" or "claw", see A. Kaldellis, *A Cabinet of Byzantine Curiosities: Strange Tales and Surprising Facts from History's Most Orthodox Empire* (Oxford, 2017), 31–32.

32 Theodore Laskaris, *Letters*, ed. N. Festa, *Theodori Ducae Lascaris Epistulae CCXVII* (Florence, 1898). A. Karpozilos, "Realia in Byzantine Epistolography XIII – XV C.," *BZ* 88.1 (1995): 68–84, at 71. I owe this information to Kelly Mavrommati.

33 See Μεσαιωνικὴ Βιβλιοθήκη, vol VII, ed. K. Sathas (Venice, 1894), 524.7.

34 Theodore of Gaza in his encomium never mentions the small dog's name, Theodore of Gaza, *Praise of the dog*, ed. and trans. K. Staikos and I. Tzifa, Κύνος ἐγκώμιον (Athens, 2017). On the text see C. A. Gibson, "In Praise of Dogs: An Encomium Theme from Classical Greece to Renaissance Italy," in *Our Dogs, Ourselves: Dogs in Medieval and Ealy Modern Art, Literature, and Society*, ed. L. D. Gelfand (Leiden, 2016), 31–35.

35 John Skylitzes, *A Synopsis of Byzantine History, 811–1057*, trans. J. Wortley (Cambridge, 2011), 234–35.

36 A. A. Vasiliev, "The Monument of Porphyrius in the Hippodrome at Constantinople," *DOP* 4 (1948): 27–49, at 34.

37 M. R. Calarco, *Animal Studies. The Key Concepts* (London and New York, 2021), 2–4. This term appears in C. Adams, *The Sexual Politics of Meat. A Feminist-Vegetarian Critical Theory* (New York, 1990).

38 See for instance Theodore of Gaza, *Praise of the Dog*, ed. Staikos and Tzifa, 50.

39 K. Walker-Meikle, *Medieval Pets* (Woodbridge, 2012), 16–18.

40 *Synaxarion of the Honourable Donkey* 171, ed. Moenning.

41 Koukoules provides some information concerning dog names, which – upon further investigation – turn out to be misleading and the texts he lists are either not Byzantine or do not include the

information he promises, see Ph. Koukoules, *Βυζαντινών βίος καὶ πολιτισμός*, 6 vols (Athens, 1948–1955), 5. 390.

42 See for instance A. Rhoby, "Hunde in Byzanz," in *Lebenswelten zwischen Archäologie und Geschichte. Festschrift für Falko Daim zu seinem 65. Geburtstag*, ed. J. Drauschke, E. Kislinger, K. Kühtreiber et al. (Mainz, 2018), 807–20.

43 H. Kroll, "Animals in the Byzantine Empire: An Overview of the Archeozoological Evidence," *Archeologia Medievale* 39 (2012): 98.

44 John Tzetzes, *History* 5.12.525–39, ed. Leone.

45 *Ptochoprodromika III*, 255–27, ed. H. Eideneier, *Ptochoprodromos, Einführung, kritische Ausgabe, deutsche Übersetzung, Glossar* (Cologne, 1991).

46 V. Onar et al., "A Cat Skeleton from the Balatlar Church Excavation, Sinop, Turkey," *Animals* 11.288 (2021).

47 V. Onar, "Animal Skeletal Remains of the Theodosius Harbor: General Overview," *Turkish Journal of Veterinary and Animal Sciences* 37 (2013): 81–85, at 83.

48 V. Onar, C. Cakırlar, M. Janeczek et al., "Skull Typology of Byzantine Dogs from the Theodosius Harbour at Yenikapı, Istanbul," *Anatomia Histologia Embryologia* 41.5 (2012): 350.

49 J. Busuttil, "The Maltese Dog," *Greece & Rome* 16.2 (1969): 205–208, at 205.

50 *Scholia in Protrepticum et Pedagogum*, ed. O. Stählin and U. Treu, *Clemens Alexandrinus*, vol. 1. Die griechischen christlichen Schriftsteller 12 (Berlin, 1972), 337.24 and *Scholia in Lucianum* 17.19.2, ed. H. Rabe (Leipzig, 1906).

51 See for instance Eustathios of Thessalonike, Comments on Homer's Iliad, ed. M. van der Valk, *Eustathii archiepiscopi Thessalonicensis commentarii ad Homeri Iliadem pertinentes*, vols. 1–4 (Leiden, 1971–1987), 3.84.8.

52 Suda μ 519, ed. Adler.

53 Ibid.

54 Plethon, *Counsel for Despot Theodore of Peloponese*, ed. S. P. Lampros, *Παλαιολόγεια καὶ Πελοποννησιακά, Δ* (Athens, 1930), 134. 10.

55 Michael Apostolios, *Collection of Proverbs* 11.24, ed. E. L. von Leutsch *Corpus paroemiographorum Graecorum*, vol. 2 (Göttingen, 1851 repr. Hildesheim, 1958).

56 As the dog remains from the Yenikapı excavations seem to suggest, smaller and medium-sized dogs prevailed in Constantinople, V. Onar, M. Janeczek, G. Pazvant et. al., "Estimating the Body Weight of Byzantine Dogs from the Theodosius Harbour at Yenikapı, Istanbul," *Kafkas Üniversitesi Veteriner Fakültesi Dergis* 21.1 (2015): 55–59.

57 See Henriette Baron's chapter in this volume where she claims that animal remains suggest that cats were not common in Byzantium. And while they seem to be less common than dogs, the textual evidence seems, at least partly, to contradict this statement.

58 Michael of Ephesos, *Commentaries on Aristotle's Parts of Animals*, ed. M. Hayduck, *Michaelis Ephesii in libros de partibus animalium, de animalium motione, de animalium incessu commentaria*. Commentaria in Aristotelem Graeca 22.2 (Berlin, 1904), 49. 6–7.

59 PG 150, col. 1258

60 Michael Apostolios, *Proverbs* 7.41, ed. E. L. Leutsch, *Corpus Pareomiographorum Graecorum* (Hildesheim, 1965), 2, 405. This saying is much older since, according to Stobaios, it should be ascribed to Diogenes the Cynic (though Stobaios' version slightly different: "Διογένης ἔλεγεν τῶν οἴκων ἔνθα πλείστη τροφὴ πολλοὺς μῦς εἶναι καὶ γαλᾶς", Stobaios, *Anthology* 3.6.37, eds. O. Hense and C. Wachsmuth, *Ioannis Stobaei anthologium*, 5 vols. (Berlin, 1884–1912).

61 Nikephoros Gregoras, *Roman History*, ed. I. Bekker and L. Schopen, *Nicephori Gregorae historiae Byzantinae*, 3 vols., Corpus scriptorum historiae Byzantinae (Bonn 1829–1855), 1. 216. Although fairly Aesopic in tone, this fable has no clear antecedents, see B. E. Perry, *Aesopica. A Series of Texts Relating to Aesop or Ascribed to Him or Closely connected with the Literary Tradition that Bears His Name* (Urbana and Chicago, 1952), 493.

62 For *Schede tou myos* see P. Marciniak, "A Pious Mouse and a Deadly Cat: The *Schede tou Myos*, attributed to Theodore Prodromos." *GRBS* 57 (2017): 507–27; for the *Katomyomachia* as the

source of the zoological observations see P. Marciniak and K. Piotrowska, "How to Lament a Fallen Mouse? A Parody of Ancient Lament in the *Katomyomachia* by Theodore Prodromos," *BZ* 117.1 (2024): 157–68, at 168.

63 Kislinger, "Byzantine Cats," 170–72.

64 See for instance *Scholia in Batrachomyomachia* 9.21 (γαλέης), ed. A. Ludwich, *Die Homerische Batrachomachia des Karers Pigres: nebst Scholien und Paraphrase* (Leipzig, 1896).

65 *An Entertaining Tale of Quadrupeds*, trans. N. Nicholas and G. Baloglou (New York, 2003), 169.

16

IT WAS NOT EASY BEING A MOUSE IN CONSTANTINOPLE

Some Notes on the Role of Mice in Byzantine Life and Literature

Katarzyna Piotrowska

"I caught sight of two mice, big and fat and smooth, just like the pigs men rear at home and feed on flour and bran. Gasping for breath at the strangeness of this sight, I turned to my companion and said, 'My dear, good, friend, everything in Hades seems well and truly hateful and abominable, the sorts of things that in life particularly lend themselves to be cursed. And these mice that you have here I find the most unbearable thing of all. To someone like me who loathes these creatures more than every other loathsome thing, it seemed some sort of consolation to come down here in that I would at least be rid of their troublesomeness. But if I have to fight against them here as well, then I demand another death and another descent to a different Hades' ".[1]

In Byzantium, mice were probably omnipresent. Apparently, one might have expected them even in the afterlife, as we learn from the twelfth-century satire quoted above. The fictional character, Timarion, who reports his journey to Thessalonica and subsequent – quite unwilled – journey to Hades, sums up what we would call a human-mice relationship in Byzantium. But the disgust expressed by Timarion is not the only Byzantine perspective concerning mice. Since antiquity these "popular" rodents feature in literary texts produced in the Greek-speaking world. It is small wonder that Byzantine literature is also full of mice.

There is no coherent study of the role of the mouse in Byzantine literary texts. The only work partly covering this topic is an article by Caterina Carpinato, "Topi nella letteratura greca medievale".[2] The author declares that her study presents mice in demotic literature, but she also analyses some texts written in Hochsprache.[3] Przemysław Marciniak describes one individual mouse in his paper entitled "A Pious Mouse and a Deadly Cat: The *Schede tou Myos*, attributed to Theodore Prodromos."[4] This article offers an in-depth analysis of a witty and funny text written by one of the most famous Byzantine poets. The same poet penned some more texts where mice are protagonists, the most obvious being his *Katomyomachia*, a short poem analyzed by many scholars.[5] But apart from a substantial number of studies on *Katomyomachia*, there are not many other surveys on mice in

DOI: 10.4324/9781003055877-20

Byzantium. This chapter does not attempt to fill this gap completely, I only intend to show the role of the mouse from several perspectives: presenting its (perceived) nature, its relationship with humans and other animals, and finally its symbolism.

Naming and taxonomy

Modern taxonomy differs significantly from ancient and medieval zoological concepts. Therefore, it is necessary to define what the term 'mouse' (μῦς) meant and which animal or groups of animals were so described in ancient and, subsequently, Byzantine times. Firstly, it is important to notice that μῦς is an umbrella term covering a considerable part of the order of rodents. Moreover, there are also other names for the mouse. According to the definition in the "Liddell-Scott-Jones Greek English Lexicon" (LSJ), ὁ μῦς signifies a *mouse* or a *rat*, and this combination reflects the knowledge of naturalists before the nineteenth century (even Linnaeus described these both animals collectively), hence a false conviction existed that rats did not appear in Europe before the Middle Ages.[6] In Modern Greek, there is still one common term for both rodents, be it μῦς or ποντικός (ποντίκι).

The rat was not the only animal described by the same term as the mouse. In the *History of Animals*, Aristotle states that "the mice in Egypt have bristles like hedgehogs. There is another kind of mouse which walks on its hind legs: the front legs are short, and the hind ones long: these exist in great numbers. There are also many other kinds of mice".[7] The former seems to be a spiny mouse (Acomys), and the latter is quite easily identified as a jerboa. The ancient philosopher also describes another breed of mice that is said to have lived "in a certain district of Persia",[8] characterized by its unnatural way of reproduction – "when a female mouse is cut open, the female embryos are seen to be pregnant". The same breed is mentioned at a later date by Aelian.[9] In the Aelianic description we read that "the traders there [Teredon in Babylonia] bring their [mouse] skins to the Persians, for they are soft and when sewn together make tunics that keep men warm. And these garments they call candytanes or 'clothes-presses' according to custom".[10] The Aelianic story inspired the Byzantine poet Manuel Philes (14th c.) to include these mice, which he calls "Kanatanian mice", in his collection of iambs on the properties of animals.[11] However, this particular breed of mouse is difficult to identify because the main characteristic given by all the authors – the way it reproduces – seems to be an impossible exaggeration rather than a real feature of an animal.

Be this as it may, the term μῦς still defines some other rodents, especially when it is combined with an adjective, as for example, ἀρουραῖος. The μῦς ἀρουραῖος is a field mouse well known from the Aesopic Fables,[12] where it is juxtaposed with its urban counterpart, μῦς ἀστικός. A similar pair appears in the Byzantine satire quoted above, though there are different reasons for this juxtaposition. In the ancient fable the field mouse symbolized a modest but safe life, whereas the "city" mouse stood for an attractive and easy life which brings danger of premature death. In the *Timarion*, on the other hand, a field mouse (here: ἡ μυγαλῆ) represents the plainness of the real world while the mouse (ὁ μῦς) inhabits the underworld and is "unconcerned about the field mouse and her problems".[13] The word used by the author of *Timarion* – μυγαλῆ – may signify a field mouse or a shrewmouse. The latter is not a rodent but rather a close relative of a mole. This term forms an interesting combination of the words μῦς and γαλέη (γαλῆ after contraction), which signifies some small animal of the weasel variety (be it marten, polecat or foumart). Therefore, the word μυγαλῆ includes two mortal enemies in one name, but the name-keeper is neither of them.

In terms of appearance, though, a shrew fits the description. Furthermore, Clemens of Alexandria reports (after Nicander) that μυγαλῆ is blind and links it with a mole.[14] Both shrews and moles are known for their poor vision, so this information may confirm the identification. It is worth noting, though, that blindness was generally ascribed to mice since the word μυωπίας signifies a *shortsighted person*.[15] Nevertheless, in the Byzantine satire μυγαλῆ is simply another kind of mouse, and it represents a less *asteios* (urban and refined) class of mouse than the proper one (μῦς) which lives in the underworld.[16]

Μῦς ἀρουραῖος (a field mouse) and μῦς κατοικίδιος (a house mouse) form yet another pair of mice. This pair can be found in the *Geoponica*.[17] Here again, adjectives describing their usual habitat indicate the difference between the mice. Μῦς κατοικίδιος πᾶσι γνωστός ἐστι, as we read in the *Cyranides*, the fourth-century compilation of works on magic and medicine still popular in twelfth century Byzantium.[18] While the *Geoponica* advises how to get rid of both kinds of mice, the *Cyranides* proposes how to use house mice to prepare ointments curing such sicknesses as earache or alopecia. The reader does not receive any further information about the characteristics of the mice, but it does not seem necessary since *a house mouse is known to everyone*. Hesychius (6th c.) uses the term μῦς κατοικίδιος as an explanation for a less obvious appellation, σμίνθα.[19] Interestingly enough, we find a very similar word 'σμίνθος' a few lines later in his Lexicon, and this one is explained simply as μῦς.[20] While σμίνθος is a very ancient name for a mouse of Mysian or Aeolian origin,[21] the version σμίνθα does not appear in any other text.

Byzantine authors were not very precise when it comes to taxonomy, nor did they use any systematic description of the physical characteristics of the rodents. Some information was excerpted from ancient scholars, such as the statement of Michael Glykas (12th c.) about the size of a mouse's heart which is based on Aristotle's treatise.[22] In *De partibus animalium (On the Parts of Animals)* the Greek philosopher argues that some animals, including mice, are more cowardly because they have large hearts.[23] Glykas simply repeats Aristotle's argumentation and does not even mention his source (a few lines earlier, he alludes to Aelian while speaking about a hare being a hermaphrodite[24]). Information given by the *Suda* concerning a mouse's good hearing is based on Aristophanic scholia.[25] As an explanation of some Aristotelian statement, Michael of Ephesus gives an unsurprising comment that the mouse's tail is narrow.[26] There was also a conviction that mice mate very often, especially white mice, which were considered female.[27] The idea was of ancient origin but still very popular in Byzantium, as can be observed, for instance, in proverbs (see below). However, a mouse's most prominent and commonly recognized characteristics were their cowardice and greediness. These are the main themes of every appearance of mice described in literature, be it entirely fictitious or based on the authors' personal experiences with rodents.

Human-mice relationship: between literature and reality

While in ancient Greek literature poets take a fairly moderate approach to the little thieves of their supplies, Byzantine authors are much less forgiving. Preserved in *Anthologia Graeca* (AG VI 303) a poem by Ariston (3rd c. B.C.) tells of a poet who reproaches mice at his humble abode, and advises them gently to find another, better-equipped one. The poem belongs to the so-called beggar poetry, so it says more about the poet's persona than his real relationship with mice. But the rodents in this text seem to be rather companions in his poverty than voracious destroyers. The same theme is explored by Christopher of Mitylene, an eleventh-century Byzantine poet, in his poem *On the mice in his house* Εἰς τοὺς ἐν τῇ <οἰκίᾳ αὐτοῦ μῦς>.[28] Compared to his ancient predecessor, Christopher is much more aggressive in chasing mice away. Aggressive, frustrated and unsuccessful – he tries all possible ways to get rid of these horrific, big, fat (as pigs) and indomitable creatures, and he fails. How he describes these little and relatively harmless animals reflects the general approach to mice in Byzantium. The representatives of the human world (at least the urban writers whose testimony we possess) consider mice loathsome and unbearable. They all μυσάττονται τοὺς μῦς, (*feel disgust at the mice*). The above-mentioned poem by Christopher of Mitylene serves as an excellent example (even if it is not very well preserved).

Another, longer and even more dramatic text that similarly expresses this sentiment is a letter by Archbishop Eustathios of Thessalonica (12th c.), addressed to John Komnenos. The letter is a report of constant disaster due to the presence of mice in Eustathios' house.[29] The archbishop describes his miserable situation caused by these unbearable little creatures by using all possible rhetorical tools and making a real mountain out of a molehill, or a mouse hole in this case. Eustathios depicts this human-mice war as if it were a real tragedy, while in fact the unbeatable foe is the most trifling thing on earth.[30] The way he hyperbolizes his enemy results in a tragi-comic impression he intends to make on the addressee. He presents the world of mice as a parallel reality, which, to his huge disgust, intersects with his own: these mice are his κατάσιτοι – they eat everything that falls from the table, and Eustathios himself feels as if he were only a παράσιτος of the mice because it is him who eats the leftovers. The highest drama, though, is that he attempts to fight the unseen – he knows that the mice are there, but he never sees the rodents ("how happy was Diogenes who saw them!" – he exclaims at one point[31]). He tries (using mainly words) to defeat those which he can only hear, but they will not let him live nor rest because οὐ μόνον ἐκδαπανῶσι τὰ τοῦ ζῆν ἐπιτήδεια, ἀλλὰ καὶ ἀπ' αὐτοῦ τοῦ ὕπνου μοι παρατρώγουσιν[32] – *they not only exhaust his supplies but also nibble at his sleep*. It happens so because they are nocturnal creatures. They never show up during the day when someone might chase them away, but they lurk in the darkness as they are "a color of the night".

The letter is undoubtedly the product of Eustathios' rhetorical skills, being full of literary and cultural allusions. However, it also reveals the most distinctive aspects of the human-mice relationship: the rodents are gluttonous, eat human supplies, destroy equipment – including books, as Christopher of Mytilene complains (v. 48) – and they act at night.

Yet another common opinion expressed by Eustathios in this letter is an ancient belief, apparently still present in Byzantium, that mice generate underground.[33] Eustathios writes about the loathsome mice that they are τοὺς βριαροὺς γηγενεῖς[34] – *earthborn strong ones*. A natural connection between mice and the ground where they build their holes resulted in a folk belief that rodents come from underground. Timarion learns this when he comes to the underworld on his way back from the same Thessalonica where Eustathios despaired so badly because of dreadsome mice. After expressing his strong disgust and prejudice against the rodents, Timarion is rebuked on the matter. His guide in Hades, quite surprised with his words, explains:

My dear friend, I am amazed at your ingenuousness and manifest ignorance of what is what. Don't you realise that all mice are earthborn and that they spring up from small cracks in the earth that occur in times of drought? Hence it is more logical that they should live underground and multiply in Hades rather than up on earth in the land of the living. They don't come down to us from up there, they go up to the surface of the earth from our world down here. So you mustn't be astonished to find mice living here with us but should realise that they are our regular companions.[35]

However, the mice from Hades behave quite unlike their counterparts from the real world, or at least those described by Eustathios. They are civil and polite, respect the old man's lineage and his sleep, and eat only what he had not eaten.[36] But even in this upside-down world, one thing remains unchanged – a man can never see the mice which live at his expense.

The world of mice

In all the texts discussed so far, mice are considered a part of the human world (be it fictive or real) and are presented from the human perspective. However, in Byzantine literature, a different (story)world exists. It is a world in which humans are rarely mentioned: the world of mice. The rodents in this world interact mainly amongst themselves, the only other animal that makes an appearance is a cat. In ancient Greek literature, mouse protagonists interact with other animals[37] (and humans).[38] This also happens in the medieval works of non-Greek origin (e.g., *Stephanites and Ichnelates*).[39] Interestingly enough, in Byzantine literature even the cat is not always truly present, in *Katomyomachia* (the *Cat and Mice War*) we only hear about it and never see it acting. What is more, there is no certainty if the mice tormentor is a cat and not some other weasel-like animal. Similarly, as with the mouse, there is taxonomical chaos in describing cats.[40] The ancient name, also attested in Byzantine literature, is αἴλουρος (in both genders, with the variation αἰλουρίς for feminine), and it signifies a cat. However, there is a possibility that it also refers to a weasel. The more Byzantine name (attested from the sixth century onwards) is κάτ(τ)α (with the male version κάτ(τ)ος or κάττης[41]), and this term features in the works of Theodore Prodromos, *Katomyomachia*.[42] This is perhaps the most complex Byzantine storyworld populated by mice. And it is entirely fictive – the protagonists are fully anthropomorphized, which leads to some absurdity (a mouse riding horses and practicing with chivalric weapons or praying to the gods).[43] Their real animal properties are used only as tools to create the comic effect (e.g., mice speaking names are based on their favorite menu).

Katomyomachia is a well-composed mock-epic poem, it contains a long list of elements specific to drama and epic, for example, there is a scene of a dream in which a multi-layer anthropo- or rather myo-morphism is observed. The poem's main hero, a mouse named Kreillos (*Ham-lover*), dreams that he sees Zeus. The god is presented in a personification of a mouse because this is the dream of a mouse.[44] Kreillos threatens mouse Zeus (who looks like the wise elder, Tyroleichos – *Cheese-licker*), saying that if he does not make the mice victorious in the battle, Kreillos himself will go to the temple and eat the sacrifices prepared for the god. After hearing this story, Tyrokleptes (*Cheese-thief*), a mouse who has just expressed great fear and did not want to leave the mouse hole under any circumstances, now declares that he will go and join the eating party together with his wife and children. The entire scene is full of literary allusions decipherable to anyone who reads ancient literature – the basis of Byzantine education. The mouse Zeus is an analogy to Nestor-Zeus from the *Iliad*, and the threat of being cut off from the sacrifices has its prototype in Aristophanes' *Birds* (1230–1233, 1515–1552). Also, the idea of mice eating a god's stuffs in the temples is already found in Hellenistic *Batrachomyomachia* 174–186. However, it is not exactly the same situation with Prodromic mice.

In pagan antiquity mice lived in the temple of Apollo; being as sacred as his priests, they ate the sacrifices prepared for the god. The Christian reality, however, looks completely different. Basil the Great warned: Ὅρα οὖν μὴ μῦς ἤ τι ἅψηται τῶν θείων μυστηρίων – *watch that a mouse does not touch the holy mystery*.[45] Saint Augustine wondered why God had created such things as mice, he concluded that they were (together with frogs, flies and worms) "all beautiful in their specific kind".[46] But they were only animals, and as such they could not take part in the Holy Sacrifice. Western theologians raised a question (introduced by Berengar in the 11th century)[47] *quid mus sumit* – What does a mouse touch? – based on an imagined (or genuine in that time) situation when a mouse creeps into a church and eats some of the bread that has been consecrated for the Eucharist. The question eventually became part of the conflict between the proponents of transubstantiation and consubstantiation, but this issue goes far beyond the framework of this text. Still, it shows how the Christian concept of holiness and animality differed from what happened in the ancient temple of Apollo – full of mice fed by the pilgrims.[48]

However, as already said, Prodromic mice are not rodents anymore, they resemble humans in all respects, and their gluttonous behavior understood as a characteristic of a mouse, is only a part of the game, the pretext to introduce intertextual allusions. Apart from those already mentioned, there is also a biblical allusion: the narrative of Bel from the 14th chapter of the Book of Daniel. The prophet demonstrates to the king and priests that the sacred meal of Bel, worshiped as a living god, is in fact furtively consumed at night by the priests together with their wives and children. The Prodromic mouse Tyrokleptes declares precisely the same – "I will come [to eat the sacrifices] with my wife and children". Thus, the motive known from Aristophanes' comedy is upgraded here to fit a new reality.

Nevertheless, it is a mouse character which makes this allusion possible, and the entire text is built on what the Byzantines thought of mice. Not only greediness but also darkness, living underground and fear are the factors that characterize the mice heroes of *Katomyomachia*. Mice were already associated with darkness in ancient times.[49] Still, in the Prodromic text, this association is used to identify night with enslavement – the mice are forced to sit in their holes and fight their fear to break free and see the light of the sun (vv 1–13). The darkness is a symbol of fear and separation from the world. Kreillos compares himself and his mouse companions to the Cimmerians of Pontus, who live in the dark for half a year and have very poor vision on account of this (ὡς οἱ ζοφώδεις Κιμμέριοι τοῦ λόγου, οἱ Ποντικῶν ἔχοντες ἀμβλυωπίαν – *as the dark Cimmerians from the story, those of Pontus who have weakening sight*). The wording here brings to mind the correspondence of the Cappadocian Fathers, namely a letter from Gregory of Nazianzus to Basil the Great.[50] Gregory writes that he "will admire your Pontus and your Pontic darkness, and your dwelling place so worthy of exile" (Ἐγὼ δέ σου τὸν Πόντον θαυμάσομαι καὶ τὴν ποντικὴν ξουφηρίαν καὶ τὴν φυγῆς ἀξίαν μονήν).[51]

The passage from *Katomyomachia* can be read on at least three levels: there is a witty word game with the name Ποντικοί which means the inhabitants of Pontus, but also ποντικός/ποντικί, which means 'a mouse' in Greek demotic.[52] Secondly, by using the exact wording as Gregory of Nazianzus, Prodromos places his mice in the position of φυγή – exile. We can see them as creatures of the night, animals naturally characterized by the darkness, but on the other hand, they are exiles, the prisoners of this darkness. And even if this parallel to the Church Father is not so obvious, there is another literary allusion, readable by anyone – the Κιμμέριοι τοῦ λόγου are the Cimmerians from the Odyssey to whom the hero travels to learn his destiny by means of *katabasis*, going to the underworld.

Prodromos neatly weaves this net of references by using all the ideas Byzantines associated with a mouse: the (vernacular) name, poor vision, darkness, the underworld and fear. Fear is also the main factor influencing the mice's behavior in the poem's first part: they are cowards, so they keep sitting in their holes. Or at least this is how Kreillos thinks of them. He addresses the mice, expressing his dissatisfaction:

Τί τὸν τοσοῦτον, ἀνδρικώτατοι, χρόνον
μένοντες εἴσω τῶν ὀπῶν ἀεννάως
δείμῳ σύνεσμεν καὶ φρίκῃ καὶ δειλίᾳ
καὶ δυσμόρως δίιμεν οἰκεῖον βίον,
μηδὲ προκύψαι τῆς ὀπῆς ᾑρημένοι,
ἀλλ᾽ οἰκτρότατοι καὶ φόβου πεπλησμένοι
βίον σκοτεινὸν ἀθλίως μυωξίαις
ζῶμεν, καθώσπερ οἱ πεφυλακισμένοι.

Why do we, oh bravest men,
constantly sit in our holes for such a long time
living in fear, shaking like cowards
and passing our life in private so ill-fortunate?
Why did we choose not to take a glance from the hole,
but pitiable and filled with terror
live our dark wretched life in the mouse-holes
as if we were thrown into prison.[53]

A century later, Manuel Philes reuses this idea in a short poem on a sea mouse, which despises non-water mice (as he called them) because of the same vices:

and he blames the non-water mice,
who live bitterly in the dark, even if they deprive themselves from the crumbs;
those whom have destroyed the crying cat
and a tablet and noose and a trap and a piece of fat.[54]

The sea mouse is not a rodent, not even a mammal, it is a kind of a marine worm called a mouse because of its resemblance to a house mouse. It is an active predator, and the description we find in Philes' poem is reasonably correct – the poet writes about pairs of appendages near its mouth (comparing them to arrows) and about its ability to hunt other sea creatures much bigger than itself. However, the comparison of these two creatures, the courageous and the cowardly one, is purely rhetorical because they share nothing but the name.[55] In Philes' poem, mice live their bitter life in darkness because they are afraid of a cat (or a weasel) or a mouse trap,[56] and in this, they are closer to the mice from *Batrachomyomachia* (112–117) who have the same fears. Prodromos' mice fear only a cat, though one never appears in the poem.

We meet both cat and mouse and learn about their characters in another text ascribed to Theodore Prodromos. This text is known as *Schede tou Myos* because there are two *schedos* (didactic exercises) telling the same story. Here the author uses the idea of mice's greediness to comment on human behavior. A gluttonous mouse caught by a cat amid the banquet leftovers introduces himself as an ascetic monk. Again, the feature of the mouse is only a pretext to construct a satire based on ancient literature and the Bible, in this case, the Book of Psalms.[57] The mouse here is a glutton and a coward, and the cat acts according to its nature – it plays with its victim sadistically, having the intention to kill and eat the mouse in the end. While in the real world it is a physical game, in the Byzantine *schedos*, the cat bullies its victim using quotations from the Bible.

Two centuries later, the same animals exchange much less subtle and sophisticated arguments, which actually reflects changes in the everyday life of Constantinople,[58] rather than any change in the approach to felines or rodents. The conversation in question is to be found in *An Entertaining Tale of Quadrupeds* (Παιδιόγραστος διήγησις τῶν ζῴων τῶν τετραπόδων) which was written in vernacular Greek and is described as an animal epic or epic bestiary.[59] The *Tale*'s plot involves the presentation of a sequence of animals disputing in pairs and one of these is a cat-mouse pair. They exchange scurrilous insults and would be happy to attack each other physically, but they cannot as they are restrained by the oath they made at the beginning of the assembly. The text differs from *Schede* in all respects, except for its protagonists. However, although they use entirely different language – both in register and style - the insults are the same. The cat accuses the mouse (here translated as a rat):

> You long-tailed, long-nosed, and long-whiskered fool!
> What's all this shaking to-and-fro your whiskers?
> You seek a hole to sneak in, here and there,
> Pot-licking cheese-munching bread-squanderer!
> Foul rat, there's nothing that you don't defile:
> Figs, raisins, buttermilk, milk, meat, and fish,
> Mollusks and foods like that, and grain and pulses,
> And countless other goods consumed by people,
> You either eat, you filth, or piss and shit on–
> What you don't spill and scatter with your feet!
> Or should you find a pitcher of oil unsealed,
> You dip your tail and draw the oil back out,
> And licking at your tail, you fill your belly.[60]

And the mouse responds similarly, though beginning with St. Paul's words (2 Cor, 12).[61] This tiny nuance would not be enough to make a literary comparison between the texts under discussion, but they have in common something else, namely the list of goods the cat presents. In the *Schede tou Myos* the cat craftily asks the mouse, a self-proclaimed hermit: "How come you don't sing this – 'I want olive oil and not sacrifice, butter from cows, milk from sheep [with young lamb fat?]'[62] and this 'These are more dear to me than honey'[63] and this 'I have anointed and rubbed my head with the fatty olive oil'[64] and the like".[65] Most of these products are components of the names of mice used in the Hellenistic Pseudo-Homeric *Batrachomyomachia*[66] and the Byzantine *Katomyomachia*[67] and *Schede tou Myos*.[68] The menu always consists of oil, dairy, meat, fish, and all dry goods such as flour, groats, dried fruits, and sometimes even wine and honey – in brief, all human food accessible to the rodents.

As stated above, Byzantine mice do not interact with many other animals apart from their natural enemy. They are not predators themselves; therefore, they usually take on the role of a victim. Because of their small size, they are also considered a symbol of something low and unimportant, and as such, they appear among others in proverbs together with a turtle[69] or an elephant who – depending on the author – cares[70] or cares not[71] for the mouse. In both cases, the mouse symbolizes the littleness and plainness of a thing, though not always considered a weakness anymore – in the Christian world, one should admire all of creation, even the smallest of things. Basil the Great states: Οὐ μόνον δὲ ἐν τοῖς μεγάλοις τῶν ζώων τὴν ἀνεξιχνίαστον σοφίαν ἔξεστι κατιδεῖν, ἀλλὰ καὶ ἐν τοῖς μικροτάτοις οὐδὲν ἔλαττον συναγεῖραι τὸ θαῦμα[72] – *It is not in great animals only that we see unapproachable wisdom; no less wonders are seen in the smallest,*[73] and as an example, he gives a mouse, admired despite its size, because it φοβερός ἐστι τῷ ἐλέφαντι – *is feared by the elephant.*[74]

The mouse as a symbol

Different times and cultures have their own patterns of symbolism, based either on plain associations or more complex allegories. The Byzantines inherited many symbols from their ancient ancestors. I do not aim to extract some purely 'Byzantine' aspects of the symbolism of a mouse, I intend to present what was still current in the Greek-speaking world in the Middle Ages. The primary sources of this section are the collections of proverbs and dream books composed in Greek during the Byzantine era.

As mentioned above, the mouse symbolizes something small and unimportant. This allegory often appears in the proverbs, for example, in the collections of Michael Apostolios and Macarius Chrysocephalus. Apart from the already mentioned elephant-mouse proverb, we read that:

- One was pregnant with a mountain and gave birth to a mouse: when one expects huge results and only very little happens to him.[75]
- A mouse hidden in dried figs bit Brasidas: it means that one should beware even of the small.[76]
- A mouse escaped after having bitten a boy: it means that even the little ones can hurt the big ones.[77]
- Mouse-dung: worth nothing.[78]

This allegory is based on a simple observation of the size of the rodent.[79] There is, however, another group of proverbs that reveals some knowledge about their mode of procreation. Mice were considered to be very prolific,[80] one of the proverbs offers the explanation according to which mice are extremely fertile, they are very often pregnant, and they bear many offspring. What is more, if they taste salt, they immediately bear very many offspring.[81] In modern languages there is the idiom *to breed like rabbits*, apparently for Byzantines it would have been *to breed like mice*. Already in ancient Greece, it was noticed that mice, and especially white mice, copulate eagerly.[82] This observation leads to an association between white mice and those who lack self-control regarding erotic passions. This is how the white mouse became proverbial (Apostolios 11, 87: μῦς λευκός: ἐπὶ τῶν ἀκρατῶν περὶ τὰ ἀφροδίσια ἡ παροιμία εἴρηται – white mouse: *The proverb is applied to those who lack self-control when it comes to sex*).

This is interesting for at least two reasons – firstly, in modern symbolism, we do not associate white mice with fertility and sex drive, they appear as a sign of alcohol abuse – in some modern languages, *to see white mice* means simply *to be drunk*.[83] Seeing white mice is reported as one of the symptoms of delirium tremens, which explains the idiom. Why the ancient Greeks assumed that white mice are more compelled to mate than other rodents remains unknown; however, it is easier to understand why they automatically decided that these mice were females. The white mouse as a symbol of sex drive first appears in a fragment of a comedy by Philemon (4th–3rd centuries B.C.),[84] and Greek comedy generally associates females with increased sex drive as we read in Aristophanes' *Lysistrata* or *Assemblywomen*. Interestingly enough, dreaming of mice is never explained by lust or any other allusions to love life.

In Byzantine dream books, we read about mice as symbols of profit made by business with foreigners or people outside one's family.[85] The appearance of the mouse is also linked to a character trait of a dreamer, which could have been "previously hidden", something that Maria Mavroudi explains "on the basis of the analogy to mice coming out and destroying things in a household while people are sleeping: nobody sees the mice's destruction, only the result of their presence".[86] This explanation corresponds very well with what was present in Eustathios' letter or Christopher's poem. However, there are also some interpretations made by Byzantines themselves, as in the case of the *Oneirocriticon* of Manuel II Palaeologus, where we read: *If mice appear in someone's house roaming up and down, all the secrets [μυστήρια] of the homeowner will be revealed. For the mouse [μῦς] derives its name from the [verb] μύω – that is, to shut one's eyes [καμμύω].*[87] This interpretation based on etymology (be it true or false) requires a learned audience and is not necessarily rooted in the shared imagination of society, so I would consider this symbol as more incidental than widespread. If we search for the latter, the *Oneirocriticon* of Germanus sums it up with a simple statement: *A mouse, when dreamt of, is an indicator of evils.*[88]

This may all give the quite solid impression that it was not so easy being a mouse in Byzantium – mice were considered useless pests and it was difficult to find a good reason for their existence since they caused so many troubles. Nevertheless, their greatest opponent, Eustathios of Thessalonica, on the edge of despair, finds one, if not a justification, a consolation at least:

> *They comfort me by one thing only – I regard them as the omen that my chamber is stable, because, as the saying goes, as much as they run out of the falling house, no doubt they remain in that one which is to stand.*[89]

Funding

This chapter was written as part of the project UMO-2019/35/B/HS2/02779 funded by the National Science Center Poland.

Notes

1 *Timarion* 18, trans. B. Baldwin, *Timarion* (Detroit, 1984), 54–55.
2 C. Carpinato, "Topi nella letteratura greca medievale," in: *Animali tra zoologia, mito e letteratura nella cultura classica e orientale*, ed. E. Cingano, A. Ghersetti, and L. Milano (Padua, 2005), 175–92.

3 Carpinato, "Topi nelle letteratura," 177 (*Timarion*), 182–84; *Schede tou Myos*, 185–86 (*Katomyomachia*).

4 P. Marciniak, "A Pious Mouse and a Deadly Cat: The *Schede tou Myos*, attributed to Theodore Prodromos." *GRBS* 57 (2017): 507–27.

5 M. Lauxtermann, "Introduzione," in: *Teodoro Prodromo La battaglia della gatta e dei topi (Katomyomachia)*, ed. Ch. Faraggiana di Sarzana, M. P. Funaioli (Roma, 2021), 9–35; P. Marciniak and K. Warcaba, "Theodore Prodromos' Katomyomachia as a Byzantine Version of Mock Epic," in: *Middle and Late Byzantine Poetry: Text and Context*, ed. A. Rhoby and N. Zagklas (Turnhout, 2018), 97–110; F. Meunier, *Théodore Prodrome. Crime et châtiment chez les souris* (Paris, 2016), to mention only the most recent studies. The basic survey on *Katomyomachia* remains H. Hunger: *Der byzantinische Katz-Mäuse-Krieg* (Graz, 1968).

6 P. Chantraine, *Dictionnaire étymologique de la langue grecque. Historie des mots*, vols. 1–4 (Paris, 1968–1977), s.v. μῦς: "mais ne désigne pas le rat proprement dit qui n'est entré en Europe qu'au Moyen Age". On the history of rats, see for instance P. L. Armitage, "Unwelcome companions: ancient rats reviewed," *Antiquity* 68 (June 1994): 231–40 or J. Burt, *Rat* (London, 2005).

7 Aristotle, *History of Animals* 6.581a.2–5, ed. and trans. A. L. Peck, *Aristotle, Historia Animalium*, vols. 1–3 (Cambridge, MA, 1965–1991); later on, he describes mice living in Arabia, probably jerboa again: "there are many mice bigger than the field mice, having forelegs a span long and hind legs as long as up to the first joint of the fingers" (Aristotle, *History of Animals* 7.606b.7–9, ed. Peck).

8 Aristotle, *History of Animals* 6.580b.29–31, ed. Peck: τῆς δὲ Περσικῆς ἔν τινι τόπῳ ἀνασχιζομένης τῆς θηλείας τῶν ἐμβρύων τὰ θήλεα κύοντα φαίνεται – "There is a place in Persia where when a female mouse is cut open the female embryos are seen to be pregnant".

9 Aelian, *On the Nature of Animals* 17.17, ed. and trans. A. F. Scholfield, *Aelian, On Animals* (Cambridge, MA, 1958): ἐὰν ἁλῷ μῦς κύουσα, κᾆτα ἐξαιρεθῇ τὸ ἔμβρυον, αὐτῆς δὲ διατμηθείσης ἐκείνης εἶτα μέντοι καὶ αὐτὸ διανοιχθῇ, καὶ ἐκεῖνο ἔχει βρέφος. - *If a pregnant Rat is caught and the foetus is removed, and after the dissection of the female the foetus in turn is opened, it too is found to contain a young Rat.*

10 Ibid.

11 Manuel Philes, *On the Characteristics of Animals* 1.1488–1493, ed. F. Dübner and F. S. Lehrs, "Manuelis Philae versus iambici de proprietate animalium," in: *Poetae bucolici et didactici* (Paris, 1862), 3–68.

12 Aesop, *Fabulae Dosithei* 16, ed. A. Hausrath and H. Hunger, *Corpus fabularum Aesopicarum* (Leipzig, 1959), 129: μῦς ἀρουραῖος καὶ μῦς ἀστικός; possibly confused with a bank vole or a hamster but *modus vivendi* of all these rodents fits the fable equally well.

13 *Timarion* 19, ed. R. Romano, *Pseudo-Luciano, Timarione* (Naples, 1974), 49–92, translation after B. Baldwin, *Timarion. Translated with Introduction and Commentary* (Detroit, 1984), 55.

14 Clement of Alexandria, *Protrepticus* 4.51, ed. C. Mondésert, *Clément d'Alexandrie. Le protreptique* (Paris, 1949²), 52–193.

15 Cf. also George of Pisidia, *Hexaemeron* 1127, ed. L. Tartaglia, *Carmi di Giorgio di Pisidia*, Torino, 1998): τὸν μῦν [...] τὸν τυφλὸν τὸν παμφάγον.

16 They are well born and polite – they patiently wait till their feeder, the old man, falls asleep and only then they approach him. See *Timarion* 19, ed. Romano.

17 *Geoponica* 13.4–5, ed. H. Beckh, *Geoponica*, (Leipzig, 1895).

18 Cyranides 2.26, ed. D. V. Kaimakes, *Die Kyraniden* (Meisenheim am Glan, 1976). The text was translated into Latin by Pascalis Romanus, the Latin interpreter for Manuel I Komnenos.

19 Hesychius, *Lexicon*, μ 1258, ed. P. A. Hansen, *Hesychii Alexandrini lexicon*, vol. 3, (Berlin, 2005).

20 Hesychius, *Lexicon*, μ 1268. For the etymology of the name μῦς (σμῦς) derived from σμίσ- (σμίνθος) see P. Chantraine, *Dictionnaire étymologique de la langue grecque*, 1028 (s.v. σμίνθος).

21 Aelian, *On the Nature of Animals* 12.5, ed. M. García Valdés, L. A. Llera Fueyo and L. Rodríguez-Noriega Guillén, *Claudius Aelianus de natura animalium* (Berlin, 2009); *Scholia D in Iliadem* A 39/Z⁵, 1–18, ed. H. van Thiel, *Scholia D in Iliadem. Proecdosis aucta et correctior secundum codices manu scriptos edita* (Cologne, 2014).

22 Michael Glykas, *Annals*, ed. I. Bekker, *Michaelis Glycae Annales* (Bonn, 1836), 103.

23 Aristotle, *On the Parts of Animals*, ed. P. Louis, *Aristote, Les parties des animaux* (Paris, 1956), 667a.

24 Aelian, *On the Nature of Animals* 13.11, ed. García Valdés et al.

25 *Suda*, μ 1036 (s.v. Μίδας), ed. A. Adler, *Suidae lexicon* (Leipzig 1928–1935).

26 Michael of Ephesos, *Commentaries on Aristotle*, ed. M. Hayduck, *Michaelis Ephesii in libros de partibus animalium, de animalium motione, de animalium incessu commentaria* [*Commentaria in Aristotelem Graeca* 22.2.] (Berlin, 1904), 52.

27 *Suda*, μ 1475 (s.v. μῦς λευκός), ed. Adler; Michael Apostolios, *Collection of Proverbs* 11.87, ed. E. L. von Leutsch, *Corpus paroemiographorum Graecorum*, vol. 2, (Göttingen, 1851 repr. Hildesheim, 1958), 538; Photius, *Lexicon*, μ 619, ed. C. Theodoridis, *Photii patriarchae lexicon (E—M)*, vol. 2 (Berlin, 1998). All after Pausanias, Ἀττικῶν ὀνομάτων συναγωγή μ 28, ed. H. Erbse, *Untersuchungen zu den attizistischen Lexika* [*Abhandlungen der deutschen Akademie der Wissenschaften zu Berlin, Philosoph.-hist. Kl.*] (Berlin, 1950), 152–221.

28 Christopher of Mitylene, *Poems* no. 103, ed. M. De Groote, *Christophori Mitylenaii Versuum Variorum Collectio Cryptensis* (Turnhout, 2012).

29 Eustathios of Thessalonike, *Letters* no. 6, F. Kolovou, *Die Briefe des Eustathios von Thessalonike. Einleitung, Regesten, Text, Indizes* (Munich, 2006).

30 Cf. later where the symbolism of mice is discussed. Similarly, since a mouse is considered to be a model coward, what we read in Christopher's poem makes the same impression of contrast (Christopher of Mitylene, *Poems* no. 103.65–66, trans. F. Bernard, Ch. Livanos, *The Poems of Christopher of Mitylene and John Mauropous.* Dumbarton Oaks Medieval Library 50 (Cambridge, MA, and London, 2018): "they shed their innate cowardice, and threw away all their dread as far as they could").

31 In this allusion Eustathios combines two stories about philosopher Diogenes confronting a mouse, both transmitted by Diogenes Laertius: "When mice crept on to the table he addressed them thus, 'See now even Diogenes keeps parasites'" and "Through watching a mouse running about, says Theophrastus in the Megarian dialogue, not looking for a place to lie down in, not afraid of the dark, not seeking any of the things which are considered to be dainties, he discovered the means of adapting himself to circumstances." Diogenes Laertios, *Lives of Eminent Philosophers* 6.2.40 and 6.2.22, trans. R. D. Hicks, *Diogenes Laertios. Lives of Eminent Philosophers*, vol. 2 (Cambridge, MA, 1925). There is no need to say that Eustathios' (and generally Byzantine) views on mice differ profoundly from that of Diogenes.

32 Eustathios of Thessalonike, *Letters* no. 6. 13–14, ed. Kolovou.

33 Basil the Great, *Homilies in Genesis* 9.2.21–31, ed. S. Giet, *Basile de Césarée. Homélies sur l'hexaéméron*, 2nd edn. *Sources chrétiennes* 26 bis. (Paris, 1968): Οὐ γὰρ μόνον τέττιγας ἐν ἐπομβρίαις ἀνίησιν, οὐδὲ ἄλλα μυρία γένη τῶν ἐμφερομένων τῶ ἀέρι πτηνῶν, ὧν ἀκατωνόμαστά ἐστι τὰ πλεῖστα διὰ λεπτότητα, ἀλλ᾽ ἤδη καὶ μῦς καὶ βατράχους ἐξ αὐτῆς ἀναδίδωσιν. Ὅπου γε περὶ Θήβας τὰς Αἰγυπτίας ἐπειδὰν ὕση λάβρως ἐν καύμασιν, εὐθὺς ἀρουραίων μυῶν ἡ χώρα καταπληροῦται. Τὰς δὲ ἐγχέλεις οὐδὲ ἄλλως ὁρῶμεν ἢ ἐκ τῆς ἰλύος συνισταμένας· ὧν οὔτε ὠὸν οὔτε τις ἄλλος τρόπος τὴν διαδοχὴν συνίστησιν, ἀλλ᾽ ἐκ τῆς γῆς ἐστιν αὐτοῖς ἡ γένεσις. The same information is repeated by M. Glycas, *Annales*, ed. Bekker, 84, 15–23.

34 Eustathios of Thessalonica, *Letters* no. 6.3–4, ed. Kolovou.

35 *Timarion* 19 (transl. Baldwin, p. 55).

36 Ibid. The old man who (unconsciously) feeds the mice is not known by name because he was condemned to *damnatio memoriae*.

37 The most obvious example is pseudo-Homeric *Batrachomyomachia* where mice befriend and later fight frogs and at the end meet crabs; also in Aesopic fables mice meet other animals such as a lion, a fox, a weasel and a frog.

38 For example, *Greek Anthology* XI, 391 where a mouse talks to and even laughs at Asklepiades Philargyros.

39 L.-O. Sjöberg, ed., *Stephanites und Ichnelates* (Stockholm, 1962), 151–244, bk. 3.

40 On cats in Byzantine literature see E. Kislinger, "Byzantine Cats," in *Animals and Environment in Byzantium (7th–12th c.)*, ed. I. Anagnostakis, T. G. Kolias, and E. Papadopoulou (Athens, 2011), 165–78.

41 For other versions of this name see E. Trapp et al., *Lexikon zur byzantinischen Gräzität, besonders des 9.-12. Jahrhunderts*, vol. 4 (Vienna, 2001), 812–13.

42 In more recent manuscripts we find a name γαλῆ instead of κάτα in verse 27 of the poem and on this basis the text was called *Gelaomyomachia*. The interesting thing is that in the eighteenth-century MS (Agen, Bibliotèque municipal, fonds principal, 20) we find scholia explaining γαλῆ as κάτα. There exists another text entitled *Galeomyomachia* preserved in papyrus (Michigan inv. 6946, Fajum) and edited by H. Schilbi, "Fragments of a Weasel and Mouse War," *ZPE* 53 (1983): 1–25. It is doubtful though if it was known in Byzantium.

43 On the anthropomorphism of the mice in *Katomyomachia* see L. R. Cresci, "Parodia e metafora nella Catomiomachia di Teodoro Prodromo," *Eikasmos* 12 (2001): 197–204.

44 The theory by Xenophanes of Colophon (Xenophanes F 15, D-K) about animals imagining their gods in the shape of themselves was known in Byzantium through Clement of Alexandria.

45 Basil the Great, *Sermo ob sacerdotum instructionem* 190, ed. P.-P. Joannou, *Fonti. Fasciolo ix. Discipline générale antique (iv-ix s.). Les canons des pères grecs*, vol. 2 (Rome, 1963), 187–91. Similarly, we read in the *Life of St. Irene of Chrysobalanton*, 80, that when after some ceremonies a mouse was seen running around the altar, it was necessary to clean the holy place because of a disgrace brought by the presence of the mouse, see J. O. Rosenqvist, ed., *The Life of Saint Irene Abbess of Chrysobalanton: A Critical Edition with Introduction, Translation, Notes and Indices* (Uppsala, 1986), 1–112.

46 Augustine, *On Genesis, Against the Manichees* 1.26.1–3: "Ego vero fateor me nescire mures et ranae quare creatae sint, aut muscae aut vermiculi; video tamen omnia in suo genere pulchra esse." Translation after P. C. Miller, *In the Eye of the Animal. Zoological Imagination in Ancient Christianity* (Philadelphia, 2018), 158.

47 G. Macy, "Of Mice and Manna: Quid Mus Sumit as a Pastoral Question," *Recherches de théologie ancienne et médiévale* 58 (Janvier-Décembre 1991): 157–66.

48 See for example Aelian, *On the Nature of Animals* 12.5, ed. García Valdés et al.

49 *Greek Anthology* 6, 302 σκότιοι μύες - *mice that love the dark*.

50 Gregory of Nazianzus was the literary model for Prodromos, see N. Zagklas, "Theodore Prodromos and the use of the poetic work of Gregory of Nazianzus: Appropriation in the service of self-representation," *BMGS* 40.2 (2016): 223–42.

51 PG 37, 25a; trans. by C. G. Browne and J. E. Swallow in: *A Select Library of Nicene and Post-Nicene Fathers*, ed. P. Schaff and H. Wace, vol. 7 (Buffalo, NY, 1894), ep. IV, 446.

52 Pseudo-Zonaras, *Lexicon*, μ 1375, ed. J. A. H. Tittmann, *Iohannis Zonarae lexicon ex tribus codicibus manuscriptis* (Leipzig, 1808; repr. Amsterdam, 1967): μῦς. ὁ ποντικος. G. Babiniotis *Λεξικό της Νέας Ελληνικής Γλώσσας*, s.v. ποντίκι.

53 *Katomyomachia* 1–7, ed. Hunger.

54 Manuel Philes, *On the Characteristics of Animals* 1.1909–1912, ed. Dübner, Lehrs:

καὶ μέμφεται μῦς τοὺς ἀνύδρους, ἐν σκότει
οἳ ζῶσι πικρῶς, κἂν συλῶσι τὰς ψίχας·
οὓς καὶ γαλῆ τρίζουσα, καὶ πλὰξ, καὶ βρόχος,
καὶ παγὶς ἐξέτριψε, καὶ βραχὺ στέαρ

55 Philes based his description on Aelian, *On the Nature of Animals* 11.4, ed. García Valdés et al. The same comparison is made by Michael Apostolios, *Collection of Proverbs*, 11.82, ed. Leutsch, 537.

56 These are exactly the same fears which Christopher's mice do not have (Christopher of Mitylene, *Poems* no.103. 61–64, trans. F. Bernard, Ch. Livanos: "And as far as the cats and the traps are concerned, they saw the cats as companions and the traps as a kind of buffet. So far were they from having fear.")

57 On *Schede tou Myos*, see P. Marciniak, "A Pious Mouse and a Deadly Cat." For a discussion on the similarity of this text to a fragment of *Description of the Earth* by Manasses, see 510–12.

58 N. Nicholas and G. Baloglou: *An Entertaining Tale of Quadrupeds. Translation and Commentary*, (New York, 2003), IX-X*.

59 The latter is a genre proposed by Nicholas and Baloglou, *An Entertaining Tale of Quadrupeds*, 13–14. There exists one more cat-mouse text written in vernacular Greek (Ὁ κάτης καὶ οἱ ποντικοί) but it has to be considered post-Byzantine, since it was composed in sixteenth-century Crete.

60 Ibid., 167 (Greek text at 166).

61 Ibid., 134.

62 *Book of Hosea* 6.7

63 *Psalms* 18.11

64 Various Psalms: 88.21; 91.11; 22.6.

65 Translation after Marciniak, "A Pious Mouse and a Deadly Cat," 526.

66 Ψιχάρπαξ – Filchcrumbe; Τρωξάρτης – Champbread; Λειχομύλη – Lickmill; Πτερνοτρώκτης – Hamchamper; Λειχοπίναξ – Lickplatter; Τυρογλύφος – Cheesegraver; Ἐμβασίχυτρος – Paddlepot; Τυροφάγος – Cheese-gobble; Πτερνογλύφος – Hamgraver (translations of the names after *Homeric Hymns. Homeric Apocrypha. Lives of Homer*, ed. and trans. M. L. West (Cambridge, MA, 2003), 262–93).

67 Ὁ Κρεΐλλος: τὸ κρέας + ὁ ἰλλός (squinting at ham), ὁ Χαρτοδάπτης: ἡ χαρτης + ὁ δάπτης (paper eater), ὁ Τυροκλέπτης: ὁ τυρός + ὁ κλέπτης (cheese thief), ἡ Λυχνογλύφος: ὁ λύχνος + γλύφειν (lamp carving), ἡ Λαρδοκόπος: ὁ λάρδος + κόπτειν (lard cutter), ὁ Σιτοδάπτης: ὁ σῖτος + ὁ δάπτης (grain eater), ὁ Ψιχάρπαξ: τὸ ψιχίον + ἁρπάζειν (crumb robber), ὁ Ψιχολείχης: τὸ ψιχίον + λείχειν (crumb licker), ὁ Κολλικοκλόπος: ὁ κόλλιξ + ὁ κλόπος (roll thief).

68 Ὁ Ἐλαιοπότης: ὁ ἔλαιος + ὁ πότης (olive oil drinker), ὁ Λαρδοφάγος: ὁ λάρδος + φαγεῖν (lard eater), ἡ Παστόλειχος: παστος + λείχειν (salt licker).

69 Arsenios, *Apophthegmata* 13.61d, ed. Leutsch, 592: Οὐ μέλει τῇ χελώνη μυῶν – *the turtle does not care for mice.*

70 Basil the Great, *Homilies in Genesis* 9.5.71–73, ed. Giet: οὐ μᾶλλον ἄγαμαι τὸν ἐλέφαντα τοῦ μεγέθους, ἢ τὸν μῦν, ὅτι φοβερός ἐστι τῷ ἐλέφαντι; repeated also by Michael Glycas, *Annals*, ed. Bekker, 100.1–4.

71 Michael Apostolios, *Collection of Proverbs* 7.8, ed. Leutsch, 396: ἐλέφας μῦν οὐκ ἀλεγίζει: ἐπὶ τῶν τὰ μικρὰ καὶ φαῦλα ὑπερορώντων – *an elephant does not trouble himself about a mouse: on those who take no notice of small and easy things.* In the ancient version of this proverb (second century) we read ἐλέφας μῦν οὐκ ἁλίσκει – *an elephant does not conquer a mouse,* but the explanation is exactly the same, see Diogenianus Grammaticus 2.66, ed. Leutsch, 29. Similarly in Macarius Chysocephalus (fourteenth century) 3.75, ed. Leutsch, p. 163: ἐλέφας μῦν οὐκ διώκει – *an elephant does not chase a mouse.*

72 Basil the Great, *Homilies in Genesis* 9.5.65–66, ed. Giet.

73 Translated by B. Jackson in: *A Select Library of the Nicene and Post-Nicene Fathers,* ed. P. Schaff and H. Wace, vol. 8. (Buffalo, NY, 1895), 105.

74 On an elephant's fear of mice see C. A. Zafiropoulos, "What Did Elephants Fear in Antiquity," *Les Études Classiques* 77 (2009), 241–66.

75 Michael Apostolios, *Collection of Proverbs* 18.57, ed. Leutsch, 733; Macarius Chrysocephalus, *Paroemiae* 8.94, ed. Leutsch, 227.

76 Michael Apostolios, *Collection of Proverbs* 11.82, ed. Leutsch, 537.

77 Ibid., 11. 88. ed. Leutsch, 538

78 Photius, *Lexicon*, μ 622, ed. Theodoridis; *Suda* μ 1424, ed. Adler.

79 Similarly, there are some other simple allegories: ἔνθα περιττὴ τροφή, πολλοὶ μύες καὶ γαλαῖ – *where there is abundance of food, there are many mice and cats* (Apostolios, *Collection of Proverbs* 7.41, ed. Leutsch, p. 405) or εἴληφεν ἡ παγὶς τὸν μῦν: ἐπὶ τῶν ἀξίως ἑαλωκότων – *a trap has to catch a mouse: on those who are justly caught* (Apostolios, *Collection of Proverbs* 6. 86, ed. Leutsch, p. 388).

80 Cf. the poem by Manuel Philes on Kanatanian mice discussed above.

81 Michael Apostolios, *Collection of Proverbs* 11.82.2–5, ed. Leutsch p. 537: Οἱ μύες εἰσὶ μὲν καὶ ἄλλως πολύγονα ζῷα, καὶ ἀθρόα τῇ ὠδίνι πολλὰ τίκτουσιν · εἰ δέ πως καὶ ἁλὸς γευσάμενοι τύχοιεν, ἐνταυθα δήπυ καὶ πάμπολλα ἀποκυΐσκουσιν.

82 Pausanias the Atticist, *Attic Dictionary,* ed. H. Erbse, *Untersuchungen zu den attizistischen Lexika* (Berlin, 1950), 152–221 s.v. μῦς λευκός: οἱ κατοικίδιοι μύες ἄγαν πρὸς τὴν ὀχείαν κεκίνηνται καὶ μάλιστα οἱ λευκοί. Οὗτοι δέ εἰσιν θήλεις. ἐπὶ τῶν ἀκρατῶν περὶ τὰ ἀφροδίσια ἡ παροιμία εἴρηται – *white mouse: domestic mice are very much impelled to mate, and especially the white ones; these are females. The proverb is applied to those who lack self-control when it comes to sex;* repeated by Photius (μ 819 ed. Theodoritis) and the *Suda* (μ 1475, ed. Adler). A completely different symbolism of a white mouse is to be found in *Barlaam and Josaphat*: we meet two mice, one black and one white, both nibbling tree roots, the explanation of this scene is that they symbolize night (black mouse) and day (white mouse) constantly consuming human life (R. Volk, *Die Schriften*

   des Johannes von Damaskos Historia animae utilis de Barlaam et Joasaph (spuria): Text und zehn Appendices, VI/2 (Berlin, 2006), chapter 12).

83 As it is in German and Polish.
84 T. Kock, *Comicorum Atticorum fragmenta*, vol. 2 (Leipzig, 1884), fr. 126: μῦς λευκός, ὅταν αὐτήν τις ἀλλ᾽ αἰσχύνομαι λέγειν, κέκραγε τηλικοῦτον εὐθὺς ἡ κατάρατος, ὥστ᾽ οὐκ ἔστι πολλάκις λαθεῖν.
85 *Somnia Danielis* 16.25, in F. Drexl, "Das Traumbuch des Propheten Daniel nach dem cod. Vatic. Palat. gr. 319," *BZ* 26 (1926): 292–314: Ποντικοὺς ἰδεῖν ἢ γαλᾶς ἐν τῷ οἴκῳ σου ξένα πράγματα μετὰ κέρδους ὑποδέξῃ – *If you dream that mice or cats are in your house, you will be involved in business with foreigners and make a profit*, transl. M. Oberhelman, *Dreambooks in Byzantium. Six Oneirocritica in Translation with Commentary and Introduction* (Farnham, 2008), 109.
86 M. Oberhelman, *Dreambooks in Byzantium*, 76–77.
87 *Oneirocriticon of Manuel II Paleologus* 4, numeration and translation after S. M. Oberhelman, *Dreambooks in Byzantium*, 196.
88 *Oneirocriticon of Germanus, Patriarch of Constantinople* 135, numeration and translation after S. M. Oberhelman, *Dreambooks in Byzantium*, 159.
89 Eustathios of Thessalonike, *Letters* 6.32–35, ed. Kolovou: ἐκεῖνο με καὶ μόνον παραμυθούμενοι, ὅτι μοι τὸν οἰκίσκον ἐν μονίμῳ ἵσασθαι οἰωνίζονται. Εἰ γὰρ καμνούσης οἰκίας, ὡς λόγος, ἐκφεύγουσιν, ἐστηξομένης πάντως προσμένουσιν.

INDEX

products 272–73; *On the Observance of Foods* (Anthimus), and references to 265, 271–72, 275, 277; partridge, and symbolism of 21; pheasants and pheasant fat, consumption of 275–76, 294n195; pigs, and pork consumption 269, 271, 283n54, 283n45–n46; poultry and wildfowl consumption 267, 273–77, 291n152; public festivals and feasting 270; recipe books, and food-related treatises 265–66; seasonality and availability, impact on 278; sheep and goat, consumption of and evidence for 299–300; society and culture, and status of 21; sources for 265–66; 'unclean' animals, and risks of consumption 238, 250, 255, 280n14; wealth and status, and impact on 277–78; wine and vinegar, and use of 288n108

foxes: *De Animalibus* (Timotheus of Gaza), description of 45; lion skin metaphors, and contrasts with fox pelts 135; satirical texts, and literary references to 175–76

Galen of Pergamon 266; *On the Capacities of Foodstuffs* 266, 279n5, 279n7; *On the Capacities of Simple Drugs* 266

Genesios, Joseph 136; *Reign of the Emperors* 191

Geoponica 52, 53, 266, 275; cheesemaking, and references to 272; mice, and references to 345; peafowl, and references to 276; sheep wool production and animal husbandry, references to 299–300

George of Pisidia 58, 64, 68–69

George the Monk 136, 137

giraffes: diplomatic gifts of 24, 147, 251, 316, 319–20; manuscripts, depictions of and narratives relating to 253–56; *Topographia Christiana* (Indicopleustes, Cosmas), description of 67

Glykas, Michael 345; *Annales sive Βίβλος χρονική* 70–71

Gregoras, Nikephoros 135; *Roman History* 136

Gregory of Nazianzus 27, 254; animal illustrations and imagery of 241; *Orationes* 74n3; Turin Homilies 250, 254

Gregory of Nyssa 73

hagiographical narratives *see also* literary tradition and ancient texts; *Passions*; animal dreams and visions, and depictions of 28–29; dragons, and representations of 26; hunting, and literary references to 188–89, 190; Life of Antony (Athanasios of Alexandria) 107; Life of Mary of Egypt 107; overview of 107; *Passions*, or martyrdom narratives 108–10; and zoological studies 5

hedgehogs: *De Animalibus* (Timotheus of Gaza), reference to 48; *Physiologus*, reference to 48

Henry the Lion (Duke of Saxony and Bavaria) 318–19, 322–23

Herodotus: *Histories* 1; lions, and depictions of 133

Hesychius 345

Hexaemera 4, 59, 62–63; *Annales sive Βίβλος χρονική* (Glykas, Michael) 70–71; *De opificio mundi* (Philo of Alexandria) 63; *De opificio mundi* (Philoponus, John) 65–66; *Hexaemeron* (Ambrose of Milan) 64–65; *Hexaemeron* (Anastasius of Sinai) 69–70; *Hexaemeron* (Cosmas of Jerusalem) 70; *Hexaemeron* (George of Pisidia) 58, 64, 68–69; *Hexaemeron* (Jacob of Edessa) 67–68; *Hexaemeron* (Neophytos of Cyprus) 72; *Hexaemeron* (Pseudo-Eustathius) 65; *Hexaemeron* (Theophilus of Antioch) 63–64; *Synopsis Chronike* (Manasses, Constantine) 71–72; *Topographia Christiana* (Indicopleustes, Cosmas) 66–67

Hippiatrica 3, 52, 53

horses and horse studies: Arabian horse breeds 18–19; bone finds from 303–04; diplomatic gifts and exchange 19, 313–14, 318–19, 321; elite societies, and representations of 18–19; *Hippiatrica* 3; horsemeat, and human consumption taboos 17, 34n21, 303; literary and visual representations of 17–19, 139–42; manuscripts, depictions of and narratives relating to 242; neighing, and interpretations of 150; pet keeping and animal domestication 146, 335; sports and entertainments, and use of horses in 146; Thessalian horse breeds 18; warhorses and military roles 140

human-animal studies: concepts of 2; development of 2–3

human-animal intercourse, literary references to, 17, 18

hunting 6; animal bone finds, and evidence for 307–08; bird hunting, and accounts of 195–96, 197–98; cheetahs, and use of in 197–98, 315; crane hunting, and literary accounts of 193, 197; *Digenis Akritas*, and depictions of 201–03; diplomacy and diplomatic gifts relating to 318; dogs, and roles of 199–200, 204; as an elite sport 192–95; falcons, and use of in 192–93, 194, 196–97, 198, 203; hagiographical narratives, and depictions of 189, 192; and human-male domination of wild animals 19–20; hunting *ekphraseis* (12th century) 195–201; 'imperial' hunting, and accounts of 190–92; legal sources for, and references to 163; literary references, and 'warrior-hunter' imagery